Man and the Environment

An Introduction to
Human Ecology and Evolution

MAN AND THE

Second Edition

ENVIRONMENT

Arthur S. Boughey

Department of Population and Environmental Biology
University of California, Irvine

MACMILLAN PUBLISHING CO., INC.
New York

COLLIER MACMILLAN PUBLISHERS
London

Macmillan Publishing Co., Inc.
866 Third Avenue, New York, New York 10022

Collier-Macmillan Canada, Ltd.

Library of Congress Cataloging in Publication Data

Boughey, Arthur S
  Man and the environment.
  Includes bibliographies and indexes.
    1. Human ecology.   2. Human evolution.
I. Title.
GF41.B67 1975   301.31   73–22623
ISBN 0–02–312770–8

Printing: 1 2 3 4 5 6 7 8   5 6 7 8 9 0

Acknowledgments for copyrighted material begin on page 547.

PREFACE

Human ecology is a very young and yet an old science. There are indeed a few who question whether it is really a science at all, but rather an agglomerate of three, perhaps four, or even five disciplines. As with all other pedagogies that spring from the union of several older traditions, the earliest publications on human ecology were mostly anthologies. Increasing numbers of such volumes are constantly appearing. They contain appropriately edited collections of relevant readings garnered from diverse sources and well depict the diffuse nature of present studies in human ecology. Thus some will consider the preparation of a comprehensive text on human ecology by a single author at this time premature, even presumptuous. Nevertheless, many recent developments in the social sciences, in engineering, and in biology demand that an effort be made toward closer integration of their overlapping areas.

These several and until now separate developments can readily be perceived. First, social scientists are increasingly turning from the introspection and qualitative theorizing that long dominated their thought to an emphasis on quantitative studies of populations and communities. Second, engineers, who generally regard themselves as a "service group," have understood that although many of the present environmental crises can be provided with engineering *solutions, decisions* on economic policies and political issues must be sought elsewhere. Third, there is a change in biology. Medical biologists have tended previously to emphasize the morphology, biochemistry, physiology, and behavior of the *individual,* while nonmedical biologists pursued experimental investigations in which the use of *Homo sapiens* as an experimental animal was usually proscribed. All biologists, whatever their particular specialty, have now realized the need to set aside one recognized area of studies in which there is a concentration on human populations. Finally, in the new multidisciplinary study of paleoanthropology, immense and exciting advances are being made toward a better understanding of human evolution and the natural processes that have fashioned us into what we presently are and continue to help shape us into whatever we will elect to become.

Although these converging interests of the social, biological, and medical sciences, together with engineering, represent largely independent incursions into human ecology, it is apparent that they must meet on some common philosophical ground. The rational focus for these varied approaches is on the fundamental ecological concept of the *ecosystem.* Societies of *Homo sapiens,* however affluent and sophisticated, are tied inescapably to the micro- and macroecosystems of this planet. They are subject to the same ecological laws that govern the performance and behavior of the other populations included within the various environments of these ecosystems. The urgent need to consider the now perilous situation of all contemporary ecosystems, and the patent degradation of their environments, alone provides sufficient justification for the preparation of a unified text on human ecology, however imperfect the initial attempts.

The dilemma confronting the author of such a work has been very eloquently expressed by Gordon Orians in rationalizing his introductory survey of the discipline of biology. He presents the philosophy of what Konrad Lorenz has termed the "aha" experience, which epitomizes what is basically the only valid *raison d'être* for the university professor. The "aha" situation arises when a previously disordered sequence of circumstances is suddenly perceived as falling into a logical relationship. It has to be invoked in each one of us separately. The experienced pedagogue can endeavor to set the scene that will trigger the response, but cannot impose it. John Platt several years ago pleaded for the encouragement of just this kind of synergistic interaction by the formation of task groups organized to solve the many critical environmental problems that now confront us. His expressive words bear repetition here: "In the past we have had science for intellectual pleasure, and science for the control of nature. We have had science for war. But today, the whole human experiment may hang on the question of how fast we press the development of science for survival."

An introductory text such as this must contain material extracted—with due permission and acknowledgment—from many sources. It should be designed to set the scene for Lorenz's "aha" response and to promote synergistic reactions in the minds of its readers. An appropriate metaphor is the "closed" and "open" call systems described in these pages. A compilation of facts by a specialist, however accurate and up to date, is forever a "closed" system. Further stimulation is required before these externally uncoordinated facts come to be internally arranged into integrated sets of relevant *new* facts and thereby convert this previously closed system to an "open" one. Such an "open" system will go far beyond these pages if this text succeeds in triggering a response to relationships beyond those to which attention has actually been drawn.

This work is designed for use as a basic text in human ecology or as a supplement for related courses. Preferably, the reader should have some introductory understanding of elementary biology from either a high school or a college course. Key references are provided for further reading in the bibliographical material at the end of each chapter, and a glossary explains the usage of such expressions as are readily definable. The appendixes contain information too specialized to incorporate into the main text.

The bibliographies are not an exhaustive coverage of particular aspects of human ecology. Often only the most recent of a series of publications can be cited. Basic accounts comprehensible to the nonspecialist have been specifically selected for inclusion, rather than more advanced and detailed studies however competent. More complete bibliographical lists are to be found in recent collections of readings, including *Readings in Man, the Environment, and Human Ecology* prepared for use with this book.

An introductory text must have some coherent and continuously developing theme. To me, and I hope to the reader, this ecological account of the origins and final emergence of our contemporary urban technocracy is engrossingly fascinating, and it is also very relevant. It is high time that we look critically and dispassionately at ourselves, our nature, and our behavior as this has come to us from the past, at our present relations with our environment and with one another, and at our future prospects for survival in the universe about us, which we are now with a few first hesitant steps finally beginning to explore.

A. S. B.

CONTENTS

14 *The Future* 507
Expectations and obligations · Resource
modeling

Introduction

The term *human ecology* is not yet so well established in contemporary idiom that it requires no explanation. Various alternative terms, usually involving the words "population" and "environment," are still current, and their use tends to cause some confusion. The parent discipline of ecology—the study of populations and communities relative to one another and to their environment—has itself only recently emerged as a distinct science. It is sometimes even more simply described as the study of ecosystems. Whatever the basic definition of the parental science, it is possible to interpolate and define human ecology as the study of the development of human populations and societies and their interactions with one another and with their total environment.

The Purpose of Human Ecology

Traditionally, human ecology has been defined and approached in a slightly different manner. Human societies have amassed, particularly over the last few millennia, a unique, complex, and immense store of rituals. Two groups within our culture, the humanists and the social scientists, have over the years emerged as specialists, the one to record, the other to study these rituals. These two groups have gradually become even further specialized, and others have appeared in such areas as the health sciences, engineering, agriculture, forestry, and the earth sciences. The concerns of these late-comers are with the technical applications of cultural rituals, or, in terms that will be explained later, with the production of artifacts and the institution of sociofacts based on mentifact developments.

Insofar as the ecologist relates to these various groups in human ecology, it is with particular reference to the evolution of human form and behavior, and to the use and abuse of the resources of the ecosystems that our cultures have learned to exploit.

To this holistic concept of the scope of human ecology has been added an urgent demand for more specific examination of this inexorable process of resource exploitation. Within the past 20 years it has become apparent that we have produced too many people, too many pollutants, too much waste, too many poisons, too much stress. At the same time we have too little food, energy, shelter, education, health, and understanding. We are squandering our global resources of fossil fuels, mineral ores, productive lands, wildlife, air, water, landscape, wilderness, and biotic diversity. Disaster looms on every horizon, both for our own population and for the ecosystems we occupy.

Ecological studies on the early history of our ancestral societies show that we were not always so wantonly destructive of natural systems, or so prodigal with their resources. As will be discussed later, the evidence available to us from fossil forms, such as it is, and from comparative studies of contemporary but presumably ancient rituals suggests the previous existence of a number of regulatory feedback mechanisms that prevented our over-exploitation of irreplaceable resources.

The study of human ecology, as it is most commonly understood by ecologists, therefore includes both a backward view of these early ecosystems in which human forms played an integral but less destructive role and a forward look necessitating an urgent consideration of the disastrous effects of this latest human intrusion into contemporary environments.

We have much remaining to be learned about ourselves. We are beginning to trace our primate origins and relate these to our biochemical reactions and behavioral patterns. Yet we have hardly begun to understand how far we are still controlled by innate primate behavior, or to what extent our species has been subjected to group selection favoring social as opposed to individual survival. We must also explore further the degree to which we still need social identity, social hierarchies, competition, and continuing selective adaptation, especially in terms of the urban environments that now constitute our primary habitat.

Paul Shepard, in a critical review of current progress in human ecology, quotes a passage from a publication by Richard Chase: "The difference between primitive and civilized mentality is not absolute, there is no chasm between them. . . . we live in the same world as the savages. Our deepest experience, needs, and aspirations are the same, as surely as the crucial biological and psychic transitions occur in the life of every human being and force culture to take account of them in aesthetic forms." Shepard nevertheless takes a somewhat restricted view of human ecology, which he considers will never be more than a troika discipline comprised of individual physiological studies, general landscape and ecosystem ecology, and exploratory investigations of human nature as a feedback system. There is indeed the distinct possibility that after the sudden great surge of interest in human ecology of a few years ago, momentum could now be lost, and concern could fall away, leaving what little attention remains focused on these three areas. The approach followed in this text aims at a more intimate integration and a more permanent relationship than this. It attempts to achieve both a wider and a deeper role for human ecology, recognizing that Chase's dictum concerning the primitive and the savage provides an axiomatic base from which to start.

Our power is now absolute. We are the only animal species possessed of a cultural strategy that enables us to escape from the confines of one unique ecological niche, and indeed even from the bounds of all the niches established on this planet. Cultural control of ourselves and our microenvironment permits us to adjust readily to the special requirements of any ecological niche. There is thus no living heterotroph with which we could not compete. We have already annihilated or drastically reduced in number and dispersal area many carnivores, both large and small; we have similarly eliminated many autotrophs. Our cultural strategy permits us such rapid adaptation that we can escape the ultimate entrapment that follows commitment to a niche of limited variablity, a limitation that has already driven many of our ancestrally related species into ignominious rarity or final extinction.

The latest environmental crises are not always the result of direct or indirect interaction of human populations. The pathways of interference are sometimes more tenuous, not always completely clarified. Human society has become so complex and so competitive that the short-term gains of one group may be allowed to override the long-term necessity for all, which is not even perceived, let alone taken into consideration. Human ecology cannot *resolve* such problems. A text such as this can only attempt to outline them, explore their possible origins, and indicate the causal relationships that may provide a more complete understanding of the technical problems for which solutions must rapidly be found, if we and our descendants are to continue to survive on this planet.

Only with the development of the ecosystem concept in ecology was it realized that populations, communities, and ecosystems had to be studied together, the so-called *holistic approach* first presented by Jan Smuts a half century ago. Holism supposes that a system is characterized by *interactions* between its component parts as much as by these parts themselves. Fortunately this realization of the holistic nature of ecology was soon followed by an appreciation that the techniques of systems analysis could be applied to such studies. Previously the experimental ecologist carefully isolated and manipulated one population, one factor, or one interaction while he held everything else steady. (Figure 1-4). It is now clear that by employing the systems approach, unmanipulated populations, communities, and ecosystems can be investigated, provided their operation is broken down into very simple steps, reproducible by iterative logical procedures in computer simula-

FIGURE 1-1. *The old and the new approaches in ecology.* A. Schematic representation of the procedures of experimental ecology on the one hand, contrasted with the systems approach on the other. B. A very simple matrix representation of an ecosystem expressed in pseudoalgebraic language for the purpose of constructing a computer simulation model. Straight arrows represent transfer of matter, wavy arrows transfer of energy. Energy from the sun maintains the system. It will be difficult for the uninitiated to comprehend this model, but an examination of the next four diagrams may help.

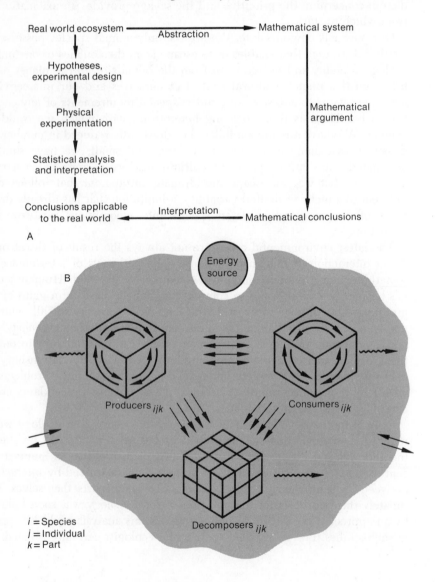

tion models. A scheme based on such a systems approach is shown in Figure 1-1.

The more ecosystems come to be studied with this holistic approach, the more ecologists realize that the effect of human occupation is all-pervasive. There is no population, community, or ecosystem left on earth completely independent of the effects of human cultural behavior. Now this influence has begun to spread beyond the globe to the rest of our planetary system and even to the universe itself.

This text is therefore a generalized and basic account that attempts to explain the complementary development of human populations and the earth's ecosystems, the interaction between them, the current state of imbalance that has become suddenly manifest through many disturbing environmental occurrences, and the courses of action remaining to us if ultimate disaster is to be avoided.

Many human societies are again populated by a majority of young people, as all have been in their initial growth stages following each major cultural advance. In a world that boasts of democratic process, the part that youth is allowed to play, and the extent of total resources to which young people are presently permitted access, as Margaret Mead has emphasized, can only be described as limited. If human society has any meaning at all, it is to insure the persistence under more favorable conditions of each new generation. Despite this, it is always contemporary youth who have to bear the brunt of society's mistakes and failures; it is they who are expected to take actions we ourselves were not prepared to take. Judith Blake, in a presentation to the First Science and Man Conference for the Americas held in Mexico City in 1973, put this matter very succinctly: "Instead of having to spend their maturity adjusting to a world they never made, [young people] . . . increasingly have transformed this world into one requiring the elders to adjust." This is the transfer from what Mead and other anthropologists call postfigurative learning to prefigurative learning, which it is not possible to discuss further here but which is part of the societal succession story briefly presented in Chapter 5.

The rallying cry of "law and order" sometimes has been disparaged as the resort of the "haves" when dealing with the "have nots," the old when confronting the young. We must have peace if our society is to remedy in time its many defects; but it has to be an acceptable peace, the result of an agreed allocation of the earth's power and resources, not the adjudication of a past generation as to what may be secured from life by a younger one. A move to restore this ecological imbalance has recently been taken in several countries such as the United States and Britain, where the legal age for obtaining one's majority has been reduced to 18 years.

Many human ecologists today find themselves trapped by their own reflections into passing moral judgments such as these that trespass into the confines of other disciplines, in this instance philosophy and political science. This will also be apparent from many of the references provided at the end of every chapter in this text. For the time being at least, human ecologists have had to become both moralists and soothsayers—if not

PLATE 1. *An undisturbed ecosystem* in the Wasatch Mountains of Utah. Occasional fires interacting with other environmental factors produce a mosaic of communities dominated by coniferous trees, deciduous trees, and sagebrush. Light grazing on this unfenced range by cattle does not radically change the steady-state relationship of these communities.

jeremiahs and prophets of doom at least forecasters of dire disaster unless specific political decisions are made to legislate against particular eventualities. These ecological prophets tend to become exasperated by the unwillingness to listen on the part of many members of our society. The glossy magazines are full of the good life and advertisements for swimming pools, recreation homes, dune buggies, camping trailers, motor homes, sports gear, and world cruises. The press features articles on the 4-day week or even the 3-day week, and the uses of greater leisure. A college education for everyone is forecast before many decades have passed. "Leisure Worlds" and similar retirement communities set eligibility for senior citizenship at 52 years of age or even less, little more than 20 years after graduate school.

The activist human ecologist striving to establish some prerequisites for the achievement of these utopias tends to be brushed aside by vested interests and lobbyist concerns of one faction or another. When an individual or organization seeks government support or action for the realization of some social goal, it is a common experience to be confronted by a formidable and

PLATE 2. *Insertion of a homestead* into the type of ecosystem illustrated in Plate 1 causes only local disturbance to its communities. The meadows along the valley bottoms by the stream are mown for hay, the winter feed of the beef cattle. Electricity or oil is not used for heating because of the relative cheapness and availability of timber.

PLATE 3. *Main Street, U.S.A.* The urban ecosystem is usually completely alien to communities of the natural ecosystems of the region. Although this small town in Wyoming lies close to the site of the homestead illustrated in Plate 2, its communities and its microenvironments are almost entirely different.

very discouraging bureaucratic array resulting from diversification in administrative responsibilities, the apparent price of democracy in government.

The Ecosystem Concept

It has been noted in the preface that a basic knowledge of biology is assumed for the purposes of this text, and this would include some ecology. However, in view of the central importance of the *ecosystem concept* in human ecology, it appears advisable to review briefly its salient features, and some of the fundamental processes that relate to it.

Living organisms on this planet do not exist under natural conditions in physical and biological isolation. There is an interplay both among the various populations themselves and with the physical and chemical components of the environment. The functional system that results from this interplay is known as an *ecosystem*. In terms of energetics, i.e., energy transfer, the ecosystem is a living complex of interlocking processes characterized by many cause-effect pathways. This complex of interactions provides feedback mechanisms that modify the rate of growth and development of the populations of living organisms involved. The feedback mechanisms tend always to produce a steady state or *homeostasis* in the system as a whole.

There are therefore two basic elements in a given ecosystem, the living or *biotic* one and a nonliving or *abiotic* one. Our whole planet may be regarded as a single ecosystem; its biotic components are the living populations of plants, animals, and microbes. These are sometimes described as forming the *biosphere*. By contrast, the abiotic components are the atmosphere—the oxygen, carbon dioxide, water vapor, and other gases and suspended particles of the air—together with the various geological, chemical, and physical features of sea and land that comprise the totality of habitats on this globe —the *ecosphere*.

Although we can regard the ecosphere of our planet as a single ecosystem, and for some purposes it is most convenient to do so—as when considering atmospheric or water pollution—for many other purposes it is preferable to imagine a considerably less all-embracing unit. At the opposite extreme, a single rain puddle can be regarded as an ecosystem, one that could indeed be vital in particular studies on human ecology. For example, among the populations of such a microecosystem could develop larval stages of the vector of a disease with a high incidence in adjoining human populations.

Whatever size is taken for an ecosystem, the decision is always based on convenience. The ecosystem is a *concept* applied to a somewhat arbitrary series of functions of the world about us. Our earth is not in fact made up of an interlocking mosaic of ecosystems with clearly defined borders, awaiting only a scientific determination of their precise limits. We may sometimes be able to recognize certain topographical boundaries such as rivers, mountain ranges, or oceans, or some physical barrier such as a thermocline, cultivation clearing, or fire burn, providing a discrete outline to an ecosystem that we are describing. However, we must not permit the occasional occurrence of such discrete boundaries to obscure the fact that there is a continuous ecological gradation throughout the world, rather than a jigsaw pattern of naturally defined units.

Ecosystem Organization

An ecosystem, as we define it, must contain the following elements:

1. Abiotic environment
2. Population of autotrophic and/or heterotrophic organisms
3. Energy source and utilization
4. Nutrient input and/or cycling

In Figure 1-2 these elements are related to one another schematically. A majority of naturally occurring ecosystems utilize solar radiation as their energy source; these are autotrophic or autotroph-based ecosystems. An energy input in the form of solar radiation is not, however, an essential feature of a self-perpetuating ecosystem. More especially in complex ecosystems there are various subsidiary ecosystems that utilize energy obtained from organic matter, for example, the system that develops on cow dung in a pasture. The deep layers of the sea provide another example, for they receive no radiant energy. A pond in an evergreen forest may receive very little sunlight, deriving its energy from forest detritus that blows or falls into it. A heavily fertilized pool in a tropical savanna may be supplied with energy derived from the excreta of its hippopotamus population, which forages out every night over the surrounding land. Such decomposer, heterotrophic, or detritus-based ecosystems are supplied with chemical energy obtained from the "rain" of organic matter that continuously falls, settles, or is otherwise introduced from the outside. For some purposes it is useful to conceptualize a city as a detritus ecosystem, dependent on external sources of energy and other resources.

In the case of autotrophic ecosystems, which utilize radiant solar energy, this energy is absorbed into the ecosystem and transformed by the group of organisms collectively known as *producers*. In natural ecosystems these are represented mainly by green plants. When the ecosystem is self-perpetuating, the autotrophs or producers must be associated with another element, the heterotrophic *reducers* or *decomposers*. These break down the organic production or *biomass* of the consumer populations and *recycle* back into the system the nutrients necessary for the growth of the producers. The decomposers are normally various *saprobes,* that is, microbial populations generally specifically related to some particular reaction in the total process of organic decomposition.

Another heterotrophic element is also present in most ecosystems, namely, the *consumers,* which remove a portion of the biomass built up by the activities of the producers. These are usually animal populations. In a mature or climax situation it is believed that ten times as much energy is circulating

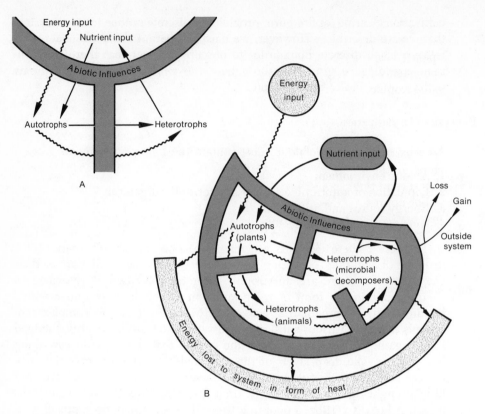

FIGURE 1-2. *Representations of ecosystem fundamentals. A.* The flow of energy and nutrients between the biotic and abiotic components of a schematic elemental self-sustaining ecosystem. This oversimplifies any real system by ignoring *biotic* influences on living populations and by disregarding energy and nutrient exchanges across the external boundaries of the system. *B.* A slightly more complicated schematic ecosystem allows for external energy and nutrient exchange, but still takes no account of biotic influences. These latter are especially vital when considering, first, regulation of population size and, second, evolutionary relationships.

through the subsidiary decomposer ecosystems as through the producer-consumer pathways.

The consumers themselves are generally a complex system of *primary consumers,* or herbivores, and *secondary consumers,* or carnivores. The latter are commonly divided into primary carnivores and secondary carnivores, feeding on the primary carnivores, and sometimes third- and even fourth-level carnivores. The last consumer level of an ecosystem is comprised of the *top carnivores.* Each of these consumer levels, together with those represented by producer and decomposer populations, is known as a *trophic level* (Figure 1-3).

The whole biological history of the earth, from the origin of life, which occurred perhaps 3½ billion years ago, to the contemporary world, is the story of the evolution, diversification, and elaboration of ecosystems that have varied continuously in time and space. Evolutionary processes have operated especially on two particular types of interfaces in these developing ecosystems. One kind of interface is formed by competing populations at the same trophic level, a competition that leads either to extinction or to evolutionary diversification. The second kind is produced by interactions between the organisms in different trophic levels. This concerns populations at a particular trophic level that are used as food and those at the next higher trophic level consuming the food. In particular it involves such relationships as the predator-prey interaction. Although evolution also occurs at this second type of interface, extinction is much less frequent.

For approximately one thousandth of all biological time, the presence of our human populations has exerted additional selection pressures on ecological relationships without our consciously understanding or reflecting on the ultimate results of such pressures.

Information Accumulation

Evolutionary development in response to selection pressures from several sources has continuously fed information into the world's ecosystems; this information has been stored in the ecosystems and has directed their further

FIGURE 1-3. *Diagram of an Eltonian pyramid,* named after Charles Elton, a British ecologist who first conceived this way of expressing the relationship between the several *trophic levels* of an ecosystem. Conventionally this pyramid scheme is used to model specific aspects of an ecosystem such as energy relations (as in this case), numbers of organisms, size of organisms, or biomass of organisms, thus contrasting with a more general model of the type shown in Figure 1-2. This pyramid illustrates energy acquisition by an ecosystem and energy loss, which must ultimately be equal in amount. At each trophic level energy is lost, so that a reduced amount can be passed on. Eventually insufficient energy remains to support further trophic level.

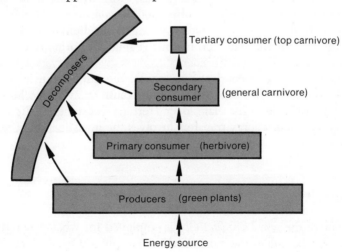

elaboration. Indeed, instead of looking at evolution as a process in which adaptations are continuously occurring in response to changing features of the living and the physical environment, we can regard this process as a continuing accumulation of biological information within ecosystems.

As evolution proceeds, this information is sometimes inadequate to allow particular populations to adapt successfully to environmental change, whether natural or man-made. Segments of this accumulated information are therefore lost with the ensuing biological extinction. As we shall see later, one feature of the modern world that greatly disturbs ecologists is the increasing loss of information resulting from species extinction as we lose stability in ecosystems because of human interference with their feedback mechanisms. The direct or indirect consequence of inserting ourselves at particular trophic levels is to destroy much of the diversity that has resulted from millions of years of biological variation and adaptation.

Energy Flow

The energy flow between the various trophic levels of an ecosystem is controlled by a number of factors, some physical, others biological. The laws of thermodynamics determine that there are some losses during energy transfer owing to its conversion to heat energy and subsequent dispersal into the environment. Considerable losses of energy are involved in the metabolic process of respiration, which the consumer levels usually further increase because of the locomotor activity of the animals involved. Various calculations and experiments suggest that as a result only some 10 per cent of the energy absorbed by one trophic level can be transmitted to the next higher trophic level in an ecosystem.

The energy pyramid in an ecosystem therefore soon reaches an apex. This severely limits the number of successive trophic levels that can be accommodated in a given ecosystem, and explains the conventional representation of the Eltonian pyramid illustrated in Figure 1-3. As a consequence of this inevitable energy transfer loss, terrestrial ecosystems rarely have more than two to four levels of secondary consumers. Moreover, the higher up the trophic level at which we insert our human population, the less energy is available. The herbivorous vegetarian (primary consumer) is able to extract ten times more energy from a given ecosystem than the carnivorous beef eater (secondary consumer), the salmon fancier (fourth or fifth level consumer) has access only to one tenth or even one hundredth of the energy available to the clam lover (tertiary consumer) (Figure 1-4). Study of these various energy transfers involves the investigation of what is termed *productivity*.

Productivity

Although the word *productivity* is employed for what generally is acknowledged to be one of the most fundamental properties of all ecosystems, there

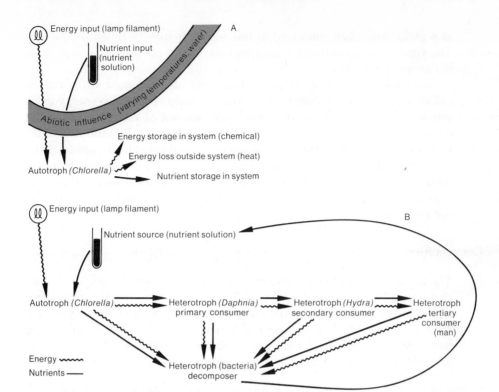

FIGURE 1-4. *Simulated ecosystems.* In *A* an alga (*Chlorella*) is maintained in a nonsustaining ecosystem to determine the effect of one abiotic factor (temperature) on population size, holding certain other abiotic factors (nutrient level, energy input) constant, while ignoring other abiotic and biotic factors (accumulation of autotoxic substances). *B* is a model of a self-sustaining experimental ecosystem that could be used to study two variables, the effect of varying the population density of the producer organisms and the effect of varying the rate of predation by a tertiary consumer. As in *A*, some relationships are still ignored, for example, the possible effects of autotoxic substances and energy loss or gain outside the system.

is still no precise definition of this term. It is also difficult to measure ac-curately experimentally.

Developing from a consideration of the energetics of an ecosystem, pro-ductivity can be regarded as a *rate*, expressed as the heat energy (calories) produced per unit area (1 square meter) per unit time (1 year). This is the *gross productivity,* and it results entirely from the photosynthetic ac-tivities of the producer organisms, which are finite, thus setting an absolute limit on the total biotic population of the world ecosystem. Also, the pro-ducers utilize some of the gross productivity in respiratory processes. When losses due to respiration by producers are deducted from the gross produc-tivity, *net productivity* is the result.

Another way of measuring productivity is to regard it not as a rate, but as a *yield*, and when conceived in this way it is usually called *production*. The yield of an agricultural field in bushels of corn per acre is an expression of *biomass production*.

These ways of looking at productivity are all concerned with what is called *primary productivity*, that which results from the energy-fixing activity of the producer trophic level. The amount of energy contained in other trophic levels is known as *secondary productivity*. Because it involves only about one tenth of the total energy content of net primary productivity, in terms of gross primary productivity it can be largely ignored. It is not, however, insignificant in respect to the consumer trophic levels in an ecosystem. It varies with another aspect of ecosystem structure—community composition.

Communities

The biotic element of an ecosystem is divisible into a number of populations that normally show some interaction. These populations can be at the same trophic level, in which case the populations tend to be competing for the same resources. Or they may be at different trophic levels, with one group using the other as a food resource. In either case the assemblage of inter-related populations is known as a *community*.

Ecological Succession

When a given ecosystem is analyzed in terms of its community representation, it rarely is found to be completely stationary. More usually it apparently is composed of a mosaic of communities, each of which replaces another in an orderly and predictable sequence. These sequential community changes comprise *ecological succession,* a process associated with an increase in species diversity and in the total standing crop. There is, as a consequence of the last circumstance, an increase in respiration that results in a decrease in net primary productivity.

The earliest states in an ecological succession, the *pioneer communities,* are therefore characterized by a high level of net primary productivity. When the situation has become stabilized and there is no further succession, the *climax* has been reached. At this time respiration and other energy losses at all trophic levels balance the caloric value of the energy fixed by the producer trophic level. There is no further succession unless some change occurs in the biotic or abiotic elements of the ecosystem. *Such climax communities are clearly of little use, as they are, to human societies, because there is no net productivity to be exploited.* It is primarily into the pioneer communities with their high net productivities that human societies have entered. Or else they have broken down the climax communities so as to replace them with exploitable pioneer communities.

Ecologists at one time were greatly concerned with successional relationships of communities and with the recognition and definition of climaxes. A greater emphasis on the energetics of ecosystems and an emphasis on the evolutionary processes affecting populations have now tended to diminish the significance of such studies.

In the world of today we must go out of our way to find examples of *primary succession,* an uninterrupted succession of seral communities beginning with the invasion of a virgin habitat like open water or bare alluvium by pioneer communities and culminating in a stable climax community. Nor are we certain of the real stability of many of the communities once thought to be climaxes. Quite often it is clear that some abiotic factor, such as the occurrence of wildfires, or a biotic one, such as animal grazing, has halted succession temporarily and holds the seral communities in a delicately balanced steady state.

Moreover, succession is usually of a secondary, not primary, type, starting from abandoned farmland, a clear-felled forest, or some other such partially colonized substrate rather than from a virgin habitat. In addition, although there is an undeniable increase in species diversity as one community succeeds another in the process of ecological succession, some ecosystems appear to contain feedback mechanisms that insure a maximum diversity by producing a steady-state pattern of seral stages. It has been shown, for example, that a maximum range of game animals in certain African preserves is obtained from such mosaics, rather than from uniform climax communities. In very much the same way, in the Pacific Southwest of the United States, the chaparral ecosystems become less diversified if no periodic burns are allowed. Under "natural" conditions chaparral is a mosaic of successional stages of recovery from lightning-set wildfires.

Such considerations of ecosystem relationships will be discussed further in other sections of this text. One of the important concepts introduced will be the interpretation of the various levels of human societies as a successional series.

Succession in Geological Time

Ecological succession can thus be regarded as a concept arising from a consideration of the energetics of contemporary ecosystems and of their environments, but there are other ways of looking at this phenomenon. One is to regard succession as an increase in *information content* of the ecosystem. Regarded from this standpoint, there can be no such community stage as a "climax," because information content is open-ended; potentially at least it continuously increases with the passage of time. The populations within a community maintain a store of variation, which is interacting continuously with biotic and abiotic elements of the ecosystem to effect further *adaptation,* and therefore further *evolution.* Each evolutionary step increases the information store of the community and represents further *niche diversification* within it. Our own populations have uniquely acquired the ability to adapt to niche diversification by cultural evolution. This permits us to utilize the information store of the ecosystem and of our own culture and so modify our techniques that we can adjust culturally, or behaviorally, while retaining the same biological form of structure.

We have changed comparatively little in either form or physiology over 3 or 4 million years, despite the passing of 100,000 or more human generations. Culturally this time interval has taken us from a unique omnivorous niche to one where we can at will either insert ourselves into or control any ecological niche. This ability has already been noted, and we will refer to it again a number of times. Meanwhile, another aspect of the ecosystem concept must be reviewed briefly, *ecosystem structure*.

Ecosystem Structure

Biologists often refer to "plant communities" or "animal communities." These are well-established terms that appeared long before the ecosystem concept was developed. The interactions between the two types of communities were until recently largely ignored. Starting afresh from the basic concept of a biotic component in an ecosystem, and ignoring these community concepts as previously conceived, the several types of populations—producers, consumers, decomposers—can best be related in a structural network of interdependencies. We can recognize that within each ecosystem there is usually a definable system of *food webs* in which particular populations of organisms at each trophic level are bound to one another in a consumer-consumed relationship.

Food Webs

Food webs are elaborations of simple food chains that embody consumer-consumed relationships. Food chains, resulting from a simple association between two or more populations at successive trophic levels (Figure 1-3), represent situations most frequently encountered under experimental laboratory conditions (Figure 1-4). They may also fit the case of some agricultural, horticultural, and forestry circumstances where the cultivator endeavors to create a synthetic ecosystem in which competition from noneconomic populations at each trophic level is avoided by restricting the food webs to those species that are being exploited economically. Examples would include an alfalfa-cow-man ecosystem, or a wheat-man ecosystem (Figure 1-5). Such synthetic ecosystems represent the ultimate in the destruction of community diversity by human populations.

In destroying this diversity and disrupting the energy pathways that regulated the relationships between organisms at various trophic levels, and dispersing much of the information content of the original ecosystem, we have unfortunately eliminated also many other features, including sometimes the mechanisms that governed our own population size. This is the most serious but by no means the only consequence of our increasing dominance of all major ecosystems; nor is it the only disastrous consequence. There are unhappily many others.

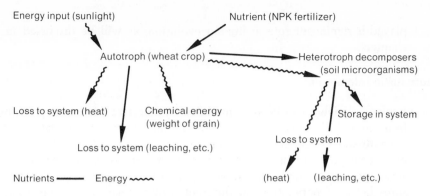

FIGURE 1-5. *Model of a simple agricultural ecosystem,* where customarily experimental information is restricted to such items as the nutrient input (artificial nitrogen-phosphorus-potassium fertilizers) in relation to yield of grain harvested. Although this model includes other interactions, for example, the storage of nutrients in the soil, these are often largely disregarded. Failure to take into account what nutrients are lost to the system as shown in this model has led to one of our present environmental crises, as will be seen later when water pollution is discussed. This model entirely ignores the abiotic factors that will influence energy transfer in this system.

Ecological Niches

Food webs and food chains are comprised of communities of interdependent populations. As we have already noted, the interface between populations at the same trophic level in a given food web or in other food webs of the same ecosystem is a site of interaction. In this instance the interaction is of a competitive nature, for each population competes for one or more common resources such as light, water, food, or space. Ecologists have presented a concept, known as *competitive exclusion,* which supposes that as a result of this interface interaction, either one population eliminates all others or one or more of the competing populations undergo evolutionary adaptation so as to achieve *resource partitioning.* In other words, the competing populations minimize the competitive interaction at the common interface by specialization on a range of resources different from those of their competitors. Each population then has a unique set of interactions with its total environment, physical and biological. Ecologists refer to this total set of interactions as the *ecological niche.* Although every population is considered to have its unique ecological niche, most commonly it is the *species* population category that is so characterized. It must be emphasized that quite apart from biological factors such as competition, niches have been determined also by selection pressure emanating from the physical features of the environment. No two species can have the same ecological niche, because by the competitive exclusion principle either one of them would be driven to extinction or one or both species would evolve into a form with a different ecological niche. This process of trait modification under competitive pressure is known as *character displacement.* It appears to have played a significant role in human evolution, as will be discussed in later chapters.

Mutualistic Interactions

Ecological niches can become modified by selective pressures quite distinct from competitive ones. The second interface between populations, as we have already considered, is one in which consumer-consumed relationships, or as ecologists are beginning universally to call them, *predation* relationships, potentially develop. However, increasing numbers of interrelationships between populations at different trophic levels are being discovered that are not predator-prey interactions. Sometimes also, less frequently, there are relations between populations at the *same* trophic level that are not competitive ones. In these instances the two populations are interacting in such a way that they are favored more by natural selection when they are interacting then when they are not. Such interactive behavior is termed *mutualistic.* Classical examples are the combination of an algal species and a fungal one to form a lichen, the development of nitrogen-fixing bacteria in nodules on the roots of legumes such as peas and beans. More recently explored mutualisms are, for example, those in the wide field of pollination biology or seed dispersal, or more specific arrangements, as that between yucca moths and yucca plants and between egrets and Cape buffalos. Human societies, as we shall see, provide innumerable examples of multualistic interactions between species. Some we have sought, as in the case of crop plants and domestic animals. Others originally developed without our instigation, as was the case with that between cattle and dung beetles in the Old World tropics.

Ecosystem Stability

These several types of ecological interactions reinforce the selective pressures that tend to prevent change in a given ecosystem. Not only is it difficult in a complex ecosystem to interrupt the flow of energy between trophic levels, but resource partitioning tends also to minimize any minor perturbations of the physical environment. Likewise, the existence of complex food webs will permit *prey switching* when there are limited fluctuations in the biological environment. Moreover, the existence of many mutualistic interactions tends to dampen biological variations, as will direct predator-prey dependencies.

These several buffering effects that confer stability on an ecosystem are not *progressively* overridden by a successive increase in the level of factors tending to produce instability. Disruption of ecosystem function comes *suddenly;* there is a threshold level for each distributing factor, or for the synergistic combinations of several factors. Ecologists sometimes term this circumstance a *nonlinear dose-response* relation. This kind of response is encountered especially when poisons such as toxic heavy metals or pesticides are introduced into ecosystems. The classical case of such disturbance has been described under the code name Operation Cat Drop.

This operation took place in the Dayak areas of Borneo. In this region, malaria had been endemic until the World Health Organization (WHO) decided this disease should be controlled by destroying its mosquito vectors. The Dayaks form what in Chapter 5 will be described as an Early Agricultural society; they live in "longhouses," large communal huts each housing up to 500 people. WHO sprayed every longhouse with DDT, malarial vectors were vastly reduced, and the incidence of malaria dropped dramatically. Unfortunately, there were other nonlinear dose responses to the introduction of DDT into the Dayak ecosystem. Cockroaches in the longhouses absorbed DDT, which then became concentrated in the lizards that ate them. The domestic cats ate the lizards in the way domestic cats always do, and still further concentrated the DDT. By this time the concentration was so high it proved lethal to the cats. With the death of most of the Dayak cats, rats, which had been a portion of their diet, began to increase. There was a parallel increase in the number of rat parasites such as fleas, lice, and other forms. Some of these parasites were vectors for sylvatic plague, also endemic among human populations of the Dayak territory. Whereas the Dayaks were no longer suffering greatly from malaria, they were increasingly exposed to sylvatic plague. Operation Cat Drop was an attempt to parachute in a new population of domestic cats.

Unfortunately disturbance to the Dayak ecosystem was not so easily stopped. The thatched roofs of the longhouses were subject to damage by a caterpillar normally held in check by natural parasites and predators. The DDT killed both caterpillars and their parasites and predators, but the caterpillar, as always in these cases, was the first population to recover. A caterpillar explosion resulted, and when the heavy rains began the roofs of the Dayak huts had been so ravaged they collapsed.

As a generalized rule, ecologists thus consider that stability in an ecosystem is positively correlated with its diversity. This rule has to be modified in certain circumstances, but by and large it holds true. Arctic ecosystems, which have poor diversity, are generally unstable. Many tropical ones have a high diversity and are remarkably stable. Man-made ecosystems, especially agricultural ones, usually have a very low diversity and are readily disturbed. A cornfield, for example, could be wiped out by one of several plant diseases, a cattle feedlot could be destroyed by a disease such as hoof and mouth.

Natural Selection

Ecologists are therefore very concerned about loss of diversity in ecosystems on two counts. First there is the loss of information content as each species is eliminated, second there is the resultant decreased stability, which may lead to the loss of the whole ecosystem. A very considerable amount of attention has recently been paid by theoretical ecologists to the effects of selection on the constituent populations of particular ecoystems, and the issue of whether ecosystems can also be subjected to selection as whole systems. Traditionally it has been assumed that, as Darwin was the first to establish, natural selection operates to insure a greater survival of offspring from the best-adapted individuals in given populations. Because of genetic variation, no two individuals in a population that reproduces sexually can be exactly the same. One will be better suited to a particular environment than another; in Darwinian jargon it is said to be more *fit*. *Fitness* is expressed through the higher rate of offspring production, and this is how a population gradually comes to show adaptation to change or, by contrast, maintains stability in a stable environment.

In a rapidly changing environment, say toward the close of an ice age, or when a new area is invaded, as is possible, for instance, when the Bering bridge between Asia and North America is exposed by falling sea levels, many more variants than usual will be *fitter* for one or the other segments of the new environment. Natural selection then operates through differential reproduction of these various *fitter* forms to produce a swarm of new populations. This process is called *adaptive radiation* and many examples of it are known; it provides by niche diversification for the *diversity* that confers stability on an ecosystem. The most often quoted example of adaptive radiation is provided by the pouched mammals (marsupials) of Australia, which because of its early isolation was not invaded by the usual range of placental (eutherian) mammals. So ecological niches in Australia are occupied by *ecological equivalents*. That is, a marsupial such as the Tasmanian devil has much the same form and function as a placental carnivore like, say, the ocelot.

In human ecology it has to be assumed that natural selection has operated on *fitter* individuals. It is tacitly assumed that it has also operated on *fitter* societies, and that these societies may contain populations that by analogy could be termed ecological equivalents when comparisons are being made between various human societies. The concepts of *fitness, adaptive radiation, natural selection,* and *ecological equivalents,* together with *competition* and *predation* discussed previously, are basic ecological concepts of unavoidable application to human ecology. Although theoretical and experimental work is constantly requiring their modification at least in part, a treatment of human ecology such as is provided here must assume the general validity of these principles. A recent review paper by the author discusses these theoretical implications at greater length.

In all these ecological processes, the decisive factor ultimately controlling ecosystem development is the relative number of individuals in each of the component populations. If ecosystem stability is to be achieved, a steady state in the size of each population has to be reached. Each population must be *stationary*, that is, its numbers must be controlled within definable limits by regulatory feedback mechanisms. Latterly this stationary phase of a population has been known as its "K" mode. Its genetic strategy in this mode, as well as its population dynamics, may be quite different from that in its "r" mode. The latter is the phase of exponential growth, or explosive natural increase. When one or more populations of an ecosystem are observed to be in an "r" phase, instability is indicated. Some perturbation that has neutralized particular population regulatory mechanisms must have occurred. It is obvious that naturally occurring populations have been selected for both dynamic and genetic strategies that react with the environment to optimize their operation during either the "r" or "K" modes and to optimize the timing of the switch from one to the other phase. It is equally obvious that our human population has been recently exposed to environmental perturbations and is currently responding to environmental pressures which are now directing an "r" mode to "K" mode switch.

During the last few years there has been an extensive theoretical exploration by some ecology schools of several of these various aspects of ecosystem stability, including competitive interactions within one trophic level, predator-prey relationships between different trophic levels, the effect of seasonality and of heterogeneity in resource availability, and the extent of occurrence of mutualistic interrelationships. These theoretical mathematical modeling analyses are only now being generally applied to field situations, and it will be a number of years before they can be incorporated into human ecology theory. Beyond initially supplying the ecosystem concept itself, ecologists have not in fact yet made very substantial contributions to human ecology. It seems quite likely that community matrix modeling, when its fundamental mathematical basis has been fully explored, and its various dynamic and genetic implications have been properly investigated, will provide an indispensable tool for the investigation of environmental problems. An excellent but rather advanced review of progress in this area has recently been published by R. M. May. This work details some of the modifications that, as mentioned here, will have to be made in the traditional and somewhat simplistic views of the relationship between ecosystem diversity and stability.

Waste Products

A further consequence of the disruption of natural cycling processes is the production of wastes at a far greater rate than the microbial populations of the decomposer trophic level can contain. Partly this is the result of the huge biomass of wastes that we now form, but often it results from presentation of wastes in the form of synthetic substances new to populations of decomposer organisms.

Given sufficient evolutionary time, decomposer organisms will usually adapt to the utilization of virtually any kind of chemical substance, the principle known as *microbial infallibility*. We are now depositing many new materials that simply accumulate, because no decomposer organisms have yet adapted to break them down and circulate the nutrient element back into our ecosystems. Quite often we also insert actually toxic substances into this recycling process at one trophic level or another. This interferes with or even stops the process of nutrient cycling, and actually kills individual organisms in populations at higher trophic levels, where *selective concentration* causes accumulation of these poisons beyond the tolerance levels of the particular organisms, as in the case of Operation Cat Drop just described. Successive trophic levels utilize without change certain of the inorganic and organic substances that they obtain from the populations of the trophic levels on which they feed. So green plants obtain substances such as nitrates, which have been synthesized from molecular nitrogen by bacteria. From these nitrates green plants synthesize amino acids, which they build into proteins. Animals that consume the green plants may initially break these proteins down again into their constituent amino acids, but they cannot themselves synthesize these compounds. So any compound not excreted or broken down as it passes through each successive trophic level of an ecosystem will be present in the same total quantity in the last level as it was at the level where it first entered. Because the pyramid of biomass, as we have already noted, like all ecological pyramids, decreases progressively with each successive trophic level, a given persisting compound will be dispersed over a much smaller quantity of biomass in the last trophic level than it was in the first. Thus for example lead, which is not excreted at any trophic level, may not be toxic to the algae that first absorb it but can attain fatal concentrations by the time it reaches our own trophic level. Various radionuclide concentrations in the water of a river may be at acceptable levels in terms of potential damage to a human swimmer, even if some water is swallowed, but seafood from its estuary might far exceed permissible radioactivity levels in human food.

Unfortunately also, the capacity for biogeochemical cycling diminishes as we pass backward through an ecosystem to earlier successional ecological stages. The heaviest nutrient losses to the outside through failure of the recycling processes in an ecosystem occur in just those early successional stages that are the favored sites for human occupation. This may lower still further the nutrient status of particular ecosystems.

As for ourselves, we have become exceedingly complex animals, and the biological information that has been incorporated into our genetic and cultural systems is correspondingly vast. There sometimes appears to be only a very tenuous connection between the basic principles of ecosystem structure just considered and the function and behavior exhibited by our populations and their individual members. Nevertheless, there is no reason to doubt that it is from these basic principles that all human behavior stems, as indicated earlier by the quotation from Chase. In succeeding chapters we shall attempt to follow the origin and development of some of these behavioral patterns, and the manner in which accumulating cultural information has been applied to the adaptation of techniques permitting an ever-intensifying exploitation of the energy sources available in natural ecosystems. All too frequently either we have remained ignorant of the consequences of this exploitation, or we have deliberately ignored them. In either case, our actions have brought us close to the brink of disaster as far as the continuing survival of our species is concerned.

BIBLIOGRAPHY

References

Alexander, M. "Biochemical ecology of soil microorganisms," *Ann. Rev. Microbiol.*, 18:217–52, 1964.

Bormann, F. H., and Likens, G. E. "The nutrient cycles of an ecosystem," *Scientific American,* **223**(4): 92–101, 1970.

Boughey, A. S. "Interaction between animals, vegetation, and fire in Southern Rhodesia," *Ohio J. Sci,* **63** (5): 193–209, 1963.

Boughey, A. S. "Human Ecology," *Proc. EDRA conf. Milwaukee,* 117–29, 1974.

Chase, R. *Quest for Myth,* Baton Rouge: Louisiana State University Press, 1949.

Cole, L. C. "The ecosphere," *Scientific American,* **198**(4): 83–92, 1958.

Harrison, T. "Operation Cat Drop," *Animals,* **5**: 512–13, 1965.

Holling, C. S. "Stability in ecological and social systems," in G. M. Woodwell and H. H. Smith (eds.), *Diversity and Stability in Ecological Systems,* Brookhaven Symposia in Biology No. 22, 1969, pp. 128–41.

May, R. M. *Stability and Complexity in Model Ecosystems,* Princeton, N.J.: Princeton University Press, 1973.

Mead, M. "The generation gap," *Science,* **164** (3876): leader, April 11, 1969.

Odum, E. P. "The strategy of ecosystem development," *Science,* **164**: 262–70, 1969.

Shepard, P. "Whatever happened to human ecology?" *Bioscience,* **17**: 901–11, 1967.

Smuts, J. C. *Holism and Evolution,* New York: Macmillan, 1926.

Van Dyne, G. M. "Ecosystems, systems ecology, and systems ecologists," *Oak Ridge National Lab Rep.,* **3957**: 1–31, 1966.

Further Readings

Boughey, A. S. *The Ecology of Populations,* 2nd ed., New York: Macmillan, 1973.

Ehrlich, P. R., Ehrlich, A. H., and Holdren, J. P. *Human Ecology: Problems and Solutions,* San Francisco: Freeman, 1973.

Helfrich, H. W. (ed.) *The Environmental Crisis,* New Haven: Yale University Press, 1970.

Hutchison, G. E. "The biosphere," *Scientific American,* **223**(3): 44–53, 1970.

Odum, E. P. *Fundamentals of Ecology,* 3rd ed., Philadelphia: Saunders, 1971.

Turner, F. B. (ed.) "Energy flow and ecological systems," *American Zoologist,* **8**: 10–69, 1968.

Watt, K. E. F. *Ecology and Resource Management,* New York: McGraw-Hill, 1968.

Whittaker, R. H. *Communities and Ecosystems,* New York: Macmillan, 1970.

Young, J. Z. *An Introduction to the Study of Man,* Oxford, England: Clarendon Press, 1971.

The Earliest Stages of Hominid Evolution

2

Any study of human ecology first must relate the single surviving species of our genus in time and space to other biological populations and the environments of the ecosystems in which they have evolved. From researches of this kind it is already possible to identify many operative factors. These provided the selection pressures that have led—through response and adaptation—to the emergence of *Homo sapiens* as a distinct species population. The later evolutionary stages of our species have been characterized more especially by *cultural speciation*, adaptations in *behavioral* traits as contrasted with physical ones. One particular behavioral feature, an enormous capacity for social learning, has led to the assembly of vast information stores in the form of *ritual*. The accumulation of ritual, in the guise of expressive form in the fine arts and humanities, has become so great that we tend sometimes to ignore its behavioral origin and imagine that it has an independent existence. It is little more than a hundred years since the true relationship between ritual and human behavior first was perceived, less than a decade that we generally have been able to examine our own evolution dispassionately without being castigated as materialists. This change of attitude has resulted partly because of increasing evidence demonstrating our evolutionary relationships, but also because physical concepts of the universe are coming increasingly to be based on stochastic rather than deterministic models. This has added an element of chance to materialistic processes and removed one of the main sources of previous philosophic objections to the evolutionary approach.

Origin of Life on Earth

Our universe is estimated to be 14 billion years old. Our planetary system, including the earth, is now believed to be approximately 4–5 billion years old, and living matter is thought to have originated somewhere between 4 and 3½ billion years ago. There are indications that photosynthetic autotrophs, green plants in the form of blue-green algae, have existed for more than 3 billion years; heterotrophs (animals and decomposers) may not have appeared until the beginning of the Cambrian period, 600 million years ago. It is thus possible to consider that for about a half billion years there have been self-sustaining ecosystems composed of producer (plant), consumer (animal), and reducer (microbe) trophic level populations and a nutrient recycling system. These ecosystems have been characterized by an ever-extending diversity, as adaptation in response to selection pressures has led to further niche diversification of their component populations. Moreover, there has been a steady increase in net primary productivity, making possible a greater structural complexity. Thus the fossil record shows a general evolutionary progression of animal dominants in these increasingly complex ecosystems, as fishes, amphibians, reptiles, mammals, and finally man succeeded one another in geological time.

Cyclic Orogenies

This process of ecosystem adaptation and diversification has not evolved at a continuous and steady pace, but appears to have proceeded by a series of quantum advances, associated usually with massive extinctions of species populations. It seems also to have been associated with geological cataclysms, first extensively discussed by C. E. P. Brooks, which occur cyclically at intervals of approximately 250 million years (Figure 2-1). These cataclysms are preceded by a period of mountain-building (orogeny) and volcanic activity (volcanism), followed by a series of ice ages. For a minimum period of 4 or 5 million years during such cataclysms, the climates and topography at the earth's surface become considerably modified, as do sea levels. At such times temperatures are generally lower, rainfall is heavier, and topography of the land varies in altitude from sea level to 10,000 meters or more. Following these cyclic orogenies, erosion slowly wears down the mountain masses and warmer climates melt the polar ice caps. The sea level, which has lowered some 100 meters or so because of the freezing of vast amounts of water at the poles, returns to its previous height. Land again

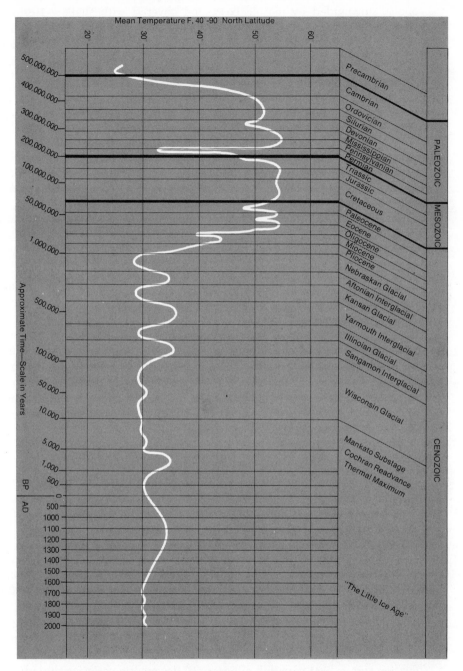

FIGURE 2-1. *Cyclic cataclysms in the history of the earth* showing the temperature fluctuations estimated to have occurred during the last three cataclysms at about 500, 250, and 5 million years BP. The time scales are successively lengthened, so that the geological intervals they represent increase from the Paleozoic to the Cenozoic to Recent time. The climatic regime the Earth presently enjoys is one of the temporary cool periods during otherwise long uninterrupted hot intervals.

becomes, for much of the remaining 250-million-year intervals, flat, monotonous, surrounded by marine bays, and covered with shallow freshwater lakes.

Although we still are only just beginning to relate these geological and climatic events to ecosystem evolution, it appears possible that quantum advances in biological development may have been associated with them. It is a comparatively simple matter to correlate the biological changes and the geology of the Pleistocene ice ages, which began in general about 1 or 2 million years ago and ended, if indeed they have yet done so, perhaps 10,000 years ago (the first traces of the Pleistocene glaciations date to between 2.5 and 3.0 million years ago). When we look at the penultimate cataclysm, which occurred approximately 250 million years ago, in the Permocarboniferous periods toward the close of the Mesozoic, there is much less surviving biological and geological evidence, although some still remains. For example, in southern Africa it is possible to find rocks that have been uncovered only recently from overlying morainic material by erosion, which still bear the characteristic scratches made when a glacier moves boulders across the face of a smoothed rock surface. Such scratches from the Pleistocene ice age are a very familiar and characteristic feature of any glaciated region of North America and the north temperate zone as a whole.

Ice-age episodes such as those that occurred in the Permocarboniferous and the Pleistocene must have profound effects on the contemporaneous ecosystems. Many of the animal consumers would become extinct, unable to adapt sufficiently rapidly either genetically or phenotypically to the changed conditions. Plant (producers) species would be removed from vast areas where their genetic plasticity was insufficient to permit the evolution of forms suitable for the new conditions. During periods of glacial recession, morainic soils would afford a rich habitat for any surviving pioneer plant species that could invade them. New plant communities with dominants of quite different life forms would have the opportunity to develop at first in these virgin habitats without the limits imposed by the continuous cropping of herbivore populations.

Evolution of Tropical Rain Forest

In speculating as to what may have occurred at such times of climatic amelioration, it may be supposed that following the close of the Permocarboniferous glaciation, about 230 million years ago, one of the new forms of plants with the potential to exploit pioneering situations was the group we now call *flowering plants*. The unique possession of the propagule known as a *seed*, a highly efficient water-conducting tissue, the *vessels*, and a reproduction system that gave full opportunity for genetic *variation* provided this newly evolved group of plants with an ability to adapt to changing circumstances far greater than had been possessed by producers in any earlier ecosystems. By the onset of the period that geologists term the Cretaceous, 120 million years ago, the success of these new forms was demonstrated in their ubiquitous spread throughout the warmer regions of the world. Beginning from this time flowering plants formed tropical rain forests not entirely dissimilar in structure and composition to the moist forests of the humid tropics of today.

By a rather similar evolutionary progression the reptilian forms of the Late Paleozoic (Permian) came to be associated in the Early Mesozoic (Jurassic) with mammal-like reptiles, and especially a group known as therapsids. These were carnivorous forms, but their teeth appear to have been modified to permit cutting food into small pieces before swallowing, rather than gulping it down in large chunks and then having a long digestive period. Other changes in the vertebrae, limbs, and feet departed from the reptilian plan and approached that of mammalian forms.

During the Jurassic many reptilian groups became extinct, and by the Cretaceous, mammals of a somewhat opossum-like appearance had begun to assume their place as dominants of the tropical forest ecosystems. These were fur-covered homeotherms (warm blooded) contrasting with the naked-skinned poikilothermous (no constant temperature regulation) reptiles.

Diversity of Form and Structure

The evolution of multistoried tree communities presented a range of aerial habitats that had never previously been available on the earth's surface (Figure 2-2). This unique diversity was exploited fully by one particular group of homeotherms, which had been simultaneously exhibiting niche diversification in the flowering plant rain forest ecosystem. These were the *primates*.

Just as flowering plants had shown specialized features that enabled

FIGURE 2-2. *Tropical rain forest structure.* This schematic profile shows the three-layer stratification that typifies African tropical forest but is not so characteristic of that of the other equatorial regions.

them to form multistoried forest communities over vast areas of the earth's surface, so the primates also possessed unique features that allowed them to adapt more readily to differentiating habitats and to evolve with them by the process evolutionists call *adaptive radiation*. This parallel development has been graphically described by E. J. H. Corner, whose central thesis is the evolution of mutualistic adaptations between primates and tropical rain forest trees based on attractive and edible fruit that provided food for the consumer populations and effective dispersal for the producers.

Evolution of Primates

By the close of the Mesozoic era some 60 million years ago, reptiles had ceased to be the dominant animals of most ecosystems, and their place was being taken by a placental mammalian group, the Insectivora, whose sur-

FIGURE 2-3. *Prosimian primates.* The map below shows the present world distribution of nonhuman primates, indicating the modern range of (A) treeshrews, (B) lemurs, (C) lorises, (D) tarsiers.

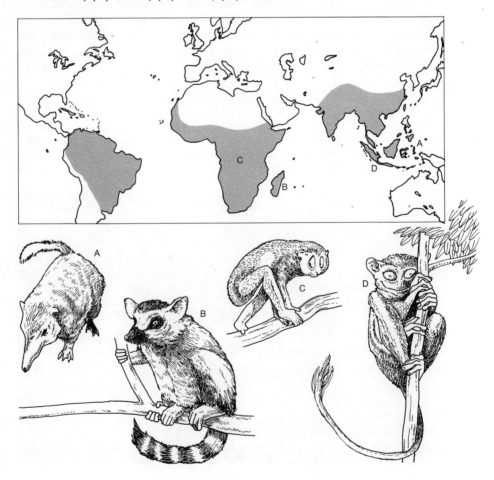

FIGURE 2-4. *Monkey grade and nonhominid anthropoid primates.* *A.* Old World monkey. *B.* New World monkey. *C.* Gibbon. *D.* Orangutan. *E.* Chimpanzee. *F.* Gorilla.

viving representatives are hedgehogs, moles, and shrews, and by a second and initially more successful group, the rodents. Primates evolved from the insectivore group during the Paleocene and Eocene epochs of 60–35 million years ago (Figure 2-3 and 2-4).

Primate Characteristics

Some eight groups of features characterize primates but are not constant features of every primate. All of them are, however, relevant to a survey of human evolution because, to a greater or lesser degree, they are characteristic of our own biological organization.

1. Manual Grasping Capabilities

Manual grasping results partly from the substitution of flat nails for claws on the digits of the four limbs, which retain the original pentadactylism of early mammals (five digits, articulating on two bones, relating to one bone attached to the skeletal frame). Associated with this grasping capability is a greater tactile sensitivity of the lower surfaces of the digits and an ability to oppose these, especially the two largest (finger and thumb, big and first toe, Figure 2-5).

The retention of the clavicle (collar bone) further enhances the maneuverability of the manual grasping equipment.

FIGURE 2-5. *Evolution of primate grasping abilities in hand of forelimb.* A. Tree shrew (*Tupaia*), little more than a multiple grappling hook. B. Loris (*Loris*), a very strong grip. C. Tarsier (*Tarsius*), a jumper with disclike swellings at end of "fingers" to assist grasping. D. Marmoset (*Callithrix*), fingers all operate in the same plane, "hand" a clasping structure only. E. Macaque (*Macaca*), opposable "thumb," clutching capacity very extensive. F. Chimpanzee (*Pan troglodytes*), also an extensive clutching capacity, but reduced compared with macaque because of shorter "thumb."

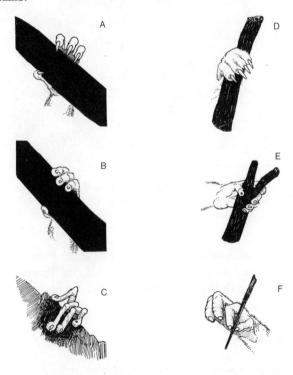

2. Emphasis of Visual Sense

The development of visual sense was apparently a corollary of the above-mentioned manual dexterity, for small objects could be grasped and held up for visual examination instead of being smelled or touched *in situ*. There was also a need to see less movable objects, which could be climbed on or over. All primates have binocular vision; possibly all diurnal primate species have some form of color vision, but this has not yet been fully determined.

3. Elaboration of the Cerebral Cortex

The elaboration of the cerebral cortex was required by the increased need for rapid and extensive eye-limb coordination centers. Together with the reduction in the size of the snout, made possible by reduction of the olfactory in favor of the visual sense, and the frontal position of the eyes required for binocular vision, this gives a characteristic shape to the cranium (brain case) not encountered in any nonprimate group.

4. Lengthening of Gestation Period

A longer gestation period is associated with a further elaboration of the uterus and the placental structures, developing presumably as a response to the increased complexity of the cerebral cortex and optical equipment, which necessitates a longer period of prenatal development (Table 2-1).

5. Prolonged Infancy

A prolonged infancy provides for a lengthened interval before sexual maturity during which development of mental and physical equipment can be completed. More important, perhaps, the long period of infancy affords the opportunity during play periods to perfect under parental supervision many of the brain-limb instruction maneuvers essential for adult survival. Moreover, the prolonged intimate relationship between parent and offspring provides the opportunity for an increased transmission of behavior learned by imitation. This is an infinitely less hazardous way of acquiring behavioral patterns than individual trial and error, and more flexible than depending on complete genetic control of every individual trait.

6. Comparative Longevity

Comparative longevity is associated possibly with the long infancy, which would be wasted without a reasonably extended potential breeding life. Longevity is perhaps also linked to the marked tendency toward single

TABLE 2-1 *Gestation Periods in Selected Mammals*

The length of the gestation period is correlated with the size of the embryo at birth, but in all the examples cited below the body size of the primate is much smaller compared to that of other mammals with similar gestation periods. In other words, the gestation period for primates is far longer than would be expected on the basis of their size alone.

Species	Length of gestation (Days)	Species	Length of gestation (Days)
Virginia opossum (*Didelphis viginiana*	12	Chacma baboon (*Papio ursinus*)	180–190
Golden hamster (*Mesocricetus auratus*)	16	Brown bear (*Ursus americanus*)	210
Mouse (*Mus musculus*)	20	Gibbon (*Hylobates lar*)	210
Rat (*Rattus norvegicus*)	22	Hippo (*Hippotamus amphibius*)	240
Rabbit (*Oryctolagus cuniculus*)	31	Chimpanzee (*Pan troglodytes*)	216–260
Weasel (*Mustela nivalis*)	35	Orangutan (*Pongo pygmaeus*)	220–270
Gray squirrel (*Sciurus vulgaris*)	40	Porpoise (*Phocaena phocaena*)	220–270
Tree shrew (*Tupaia* sp.)	46–50		
Cat (*Felis maniculata*)	63	Cow (*Bos taurus*)	278
Guinea pig (*Cavia cobaya*)	67	Man (*Homo sapiens*)	280
Lion (*Panthera leo*)	110	Gorilla (*Gorilla gorilla*)	250–290
Lemur (*Lemur catta*)	120–140	Horse (*Equus caballus*)	340
Spider monkey (*Ateles paniscus*)	139	Sperm whale (*Physeter catadon*)	365
Marmoset (*Callimico jacchus*)	140–150	Giraffe (*Giraffa camelopardalis*)	450
Sheep (*Ovis aries*)	150	Elephant (*Elephas maximus*)	660
Langur (*Presbytis entellus*)	170–190		

births. This could be a response to the considerable hazards of arboreal life, which are increased considerably if attention must be divided between more than one offspring.

7. Sociality

Sociality obviously is observable only in *living* primates but by inference it was general also in extinct forms and more developed and complex than in any other animal group. Perhaps this resulted from the vastly increased ability to observe and react to visual stimuli; it also provides a further opportunity for juveniles to learn.

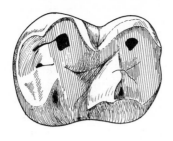
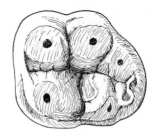

A   B

FIGURE 2-6. *Molar cusp patterns* distinguish the Old World monkeys (Cercopithecoidea) from the hominoids (Hominoidea). The crowns of Old World monkey molars have *four* cusps in two pairs, a ridge joining each of the paired cusps (*A*). In hominoids (*B*) there are *five* cusps, and a Y-shaped depression in the crown separates one cusp from the two pairs either side of it. This last pattern is not found in any fossil form occurring before the Miocene about 25 million years ago.

8. Cusp Pattern

The molar teeth of primates retain a simple cusp pattern characteristic of the more primitive mammals (Figure 2-6).

Adaptive Radiation in Primates

The adaptive radiation that primates exhibit appears to have resulted from their locomotor adaptations for climbing. Such adaptations to an arboreal life are the only distinguishing features that all primates have in common. These adaptations, as indicated above, include a lengthening of the digits to permit clasping, substitution of nails for claws, emphasis on vision with corresponding reduction in the sense of smell and size of the snout, and consumption of soft food that permits a reduction in the size of the teeth. The insectivore digestive systems were capable of coping with an omnivorous diet, which could range from any form of vegetable matter, exception cellulose and lignin, to animal proteins.

The unspecialized limb structure enables primates to use their extremities as climbing, clasping, and feeling organs. From a lateral position, the eyes have become frontal, thus providing for the development of stereoscopic vision. This would be of very high selective value in a multistoried forest community where leaping from one branch to another requires precise judgment as to distance and the penalty of any misjudgment is immediate death or severe damage. Rather similarly, the hazards to the young inherent in an arboreal life generally have reduced to one the number of offspring produced at a birth. This has become associated with a period of parental care after birth that also can be utilized for instruction in acquired behavioral characteristics (Figure 2-7). Perhaps frequently associated with this behavioral instruction would be a social structure of both a genetic and socially learned form. V. C. Wynne-Edwards has discussed how such a social system might evolve.

It is only in the past few years that we have begun to understand, as a result of experimentation in the laboratory and observation in the field, the social life of primates, and particularly of monkeys and apes. In the United States, for example, there are now eleven primate centers where primate behavioral studies are being carried out. This research is revealing with increasing authority the significance of early experience in determining the adult primate responses, and field observations are reinforcing the conclusions from such experimental observations. Understanding the behavioral patterns and social life of primates is crucial to human ecology, where direct experimental evidence is scanty, because experimentation on ourselves is necessarily limited.

FIGURE 2-7. *Primate evolution* essentially features cultural evolution. Selection has favored variants with a larger brain, whose development necessitates a longer gestation period and for an extension of the time for infant dependency and for the play years when the juvenile is perfecting adult movements and coordinating central nervous system–muscular activity. The lemur is lowest in the evolutionary scales shown here, man the highest. Although *H. sapiens* takes twice as long to reach sexual maturity as the gorilla, the period each *H. sapiens* female is potentially fertile is also twice as long.

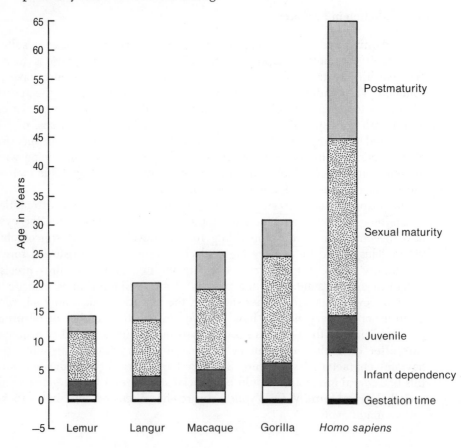

Early Primates

Early primates, commonly known as the *prosimians,* and including tarsiers, lorises, and lemurs, were extensively distributed over both the Old and New Worlds in the Paleocene, apparently attaining a maximum dispersal and radiation in the Eocene, after which the majority passed to extinction. No primate, indeed no small mammal, fossils are known from the African Eocene, but from fossil remains elsewhere it is apparent that during this period three further groups evolved, two of them forming respectively the Old World and New World "monkeys." These two groups were always independent, populations of each remaining in distinct continents. The third group that evolved included the apes. Their evolution was associated with the perfection of binocular color vision and a consequent major advance in arboreal locomotor activity that permitted the transfer of exploratory and feeding functions from the snout to the forelimbs. One of the more common general classifications of these various primates is shown in Table 2-2.

TABLE 2-2 *Classification and Relationships Within the Primate Order*

The figure after the common name of a group indicates the number of living genera it is usually considered to contain.

Suborder	Infraorder	Superfamily	Family
Prosimii (prosimians)			
	Tupaiiformes (tree shrews, 4)		
	Tarsiiformes (tarsiers, 1)		
	Lorisiformes (lorises, 5)		
	Lemuriformes (lemurs, 9)		
Anthropoidea (monkeys, apes, and men)			
		Ceboidea (New World or platyrrhine monkeys, 13)	
		Cercopithecoidea (Old World or catarrhine monkeys, 13)	
		Hominoidea (hominoids)	
			Pongidae (pongids or apes, 5)
			Hominidae (hominids, 1)

Brachiation

Associated with this transfer of function was the evolution in the apes of a new locomotor procedure known as *brachiation*. This is a swinging arm-over-arm motion that has reached its greatest development in modern gibbons. Brachiation required a change from a *pronograde* (horizontal) body position to an *orthograde* (upright) posture, which in turn resulted in changes in the muscular attachment of the internal organs, the breathing muscles, rib case, and so forth. Although several groups of monkeys can occasionally brachiate, this mode of locomotion associated with an orthograde posture is characteristic of apes. It is typified in the surviving groups of these forms, the gibbons and simiangs (southeast Asia), orangutan (Borneo), chimpanzee (Africa), and gorilla (Africa). The last three, sometimes known collectively as the *great apes,* are also capable of extensive bipedal progression; this is the preferred locomotor procedure for the last modern anthropoid group, man.

Dentition

Old World anthropoids, that is, monkeys, apes, and man, are characterized by an identical dental formula. Proceeding on each side from front to rear each jaw has two incisors, a canine, two premolars, and three molars, conventionally written as

$$\frac{2:1:2:3}{2:1:2:3} = 32$$

This permanent dentition is preceded by a juvenile deciduous dentition expressed by the formula

$$\frac{2:1:2}{2:1:2} = 20$$

A summary of the variation in the permanent dental formula found in primates is provided in Table 2-3. It can be seen that the number and form of permanent primate teeth vary little between one group and another, but that there is a general tendency toward a small reduction in number, usually by the loss of a pair of incisors or premolars. Once a particular pair of teeth has been lost (in an evolutionary sense), it cannot be replaced by the same teeth. Therefore, in examining any evolutionary relationship, the number of teeth in a postulated ancestral form *must* be the same as or larger than that in the form whose ancestry is being traced.

Cusp Patterns

The morphology of mammalian teeth has been investigated in great detail. Many aspects of the general shape, roots, and crown are constant within a group, and evolutionary relationships can be determined reliably from them. One such pattern relates to the *cusps,* the portions of the crowns

TABLE 2-3 *Primate Dentition*

Traditionally, the form of the permanent mammalian dentition is expressed in a *dental formula*. This enumerates for one side the teeth in the upper (above) and lower (below) jaws, starting with the number of incisors, and continuing successively with the number of canines, premolars, and molars. As this selection of dental formulas shows, primates exhibit remarkably little variation in the general plan of their dentition. Only the aye-aye, a highly specialized lemur, shows a major departure from the basic pattern.

For detailed comparative studies it would be necessary to take into account also the deciduous dentition or "milk teeth."

Primate group	Formula	Total number of teeth
PROSIMIANS		
Tupaiiformes (tree shrews)	2.1.3.3. 3.1.3.3.	38
Tarsiiformes (tarsiers)	2.1.3.3. 1.1.3.3.	34
Lorisiformes (lorises)	2.1.3.3. 2.1.3.3.	36
Lemuriformes (lemurs)		
Lemuridae (diurnal lemurs)	2.1.3.3. 2.1.3.3.	36
Lepilemur (sportive lemur)	0.1.3.3. 2.1.3.3.	32
Indridae (indris)	2.1.2.3. 1.1.2.3.	30
Daubentonioidea (aye-aye)	1.0.1.3. 1.0.0.3.	18
ANTHROPOIDS		
Ceboidea	2.1.3.3. 2.1.3.3. or 2.1.3.2. 2.1.3.2.	36 32
Cercopithecoidea	2.1.2.3. 2.1.2.3.	32
Pongids (gorilla)	2.1.2.3. 2.1.2.3.	32
Hominids (*Homo sapiens*)	2.1.2.3. 2.1.2.3.	32

that project beyond the general surface. As illustrated in Figure 2-6, this cusp pattern can be used to distinguish hominoids from other primates.

Coevolution

In the early Cenozoic we therefore find increasing evidence of a parallel evolution between the developing plant structures of tropical ecosystems and the primate groups that are adaptively radiating within them. The arboreal life to which all these primate forms exerted strong selection pressures for those adaptations conferring a greater awareness of this essentially three-dimensional, complex, and physically hazardous environment. The most significant evolutionary trend was a progressive development of the primate brain, associated with nervous and muscular mechanisms that operated rapidly and precisely, yet with versatility. This involved enlargement of that portion known as the *cerebral cortex,* which is associated with sensory representation, i.e., with visual interpretation and association.

In a recent paper Matt Cartmill has suggested that arboreal adaptations in early primates were more coincidental than direct. Nonprimate animals, such as chameleons and squirrels, can show similar climbing dexterity. The primary significance both of binocular vision and of grasping forelimbs, Cartmill proposes, is that they enhance insect-capture abilities. Stealthy and controlled, rather than rapid, movements are required when stalking insects; binocular vision provides the distance assessment necessary in the final capture movement. This visual predation hypothesis which Cartmill presents does seem to further refine the theory of primate evolution; it supplements rather than contradicts the account as presented here.

Appearance of Grasslands

As the Cenozoic proceeded, the general onset of environmental aridity that appeared progressively after each series of ice-age episodes began to affect the evolutionary trends exhibited by the plants and animals of the rain forest. On the margins of the multistoried forest communities, arid periods would favor, instead of trees and shrubs, smaller plants that could survive the dry winter season in a dormant condition either as some form of underground food-reserve storage organ or as persisting seeds. These *geophytic perennials* and *annual herbs* when growing together in large stretches constituted an entirely different type of community from forest, one that we now usually place under the general ecological term of *grassland.*

As the Tertiary period advanced, grassland began to appear not only on forest margins but also in the drier interior sections of the tropical rain forests throughout the area that they covered. It is this second "ecological" type of grassland that was probably of far greater significance in hominid evolution than the "geographically" marginal type. Regions in which these drier climates prevailed would have a considerable rainfall during 6 or 7 months of the year, whereas the other months would be dry and relatively rainless. River valleys, which under the previous more continuously well-watered regimes would have supported trees able to survive waterlogging, would be colonized by new types of plants. These would have to be capable of withstanding the waterlogged condition of the valley soil during the rainy season, but equally capable of sustaining the drought conditions of the dry season. Even in modern times, few tree groups have evolved that are capable of this, species of the genus *Syzygium* being a notable exception. For the most part it was annual herbs and perennial geophytes that evolved into this new niche and formed the seasonally flooded valley grasslands that many workers believe are the main natural grasslands of undisturbed tropical African habitats (Figure 2-8).

Grassland Pioneers

These evolving tropical grassland ecosystems provided an entirely new habitat for animal species. It may be surmised that the first to exploit the food resources of this new habitat were the more mobile animal forms—

FIGURE 2-8. *Evolution of savanna grassland in the African tropics during the Mesozoic.* In *A*, the Late Paleozoic, a well-distributed rainfall of some 1500 millimeters per annum maintained a water table in the flood plain bordering the rivers at about or above soil level. This section of the *catena* supported *periodically inundated* forest, with tree dominants adapted (by stilt roots, etc.) to saturated soils. There are no grasslands. In *B*, total rainfall may still be as high as 1500 millimeters per annum but there is a prolonged *dry season* during which the water table is lowered considerably in the flood plain. The tree species able to survive the saturated soils of the rainy season cannot withstand the lowering of the water table in the dry season. New life forms, grasses and other herbaceous plants, evolve to occupy this new habitat.

A

Ground-water table

B

Ground-water table

insects, for example. Among the more successful of such colonizing animal groups appears to have been the terrestrial stock already mentioned, the rodents. Their potent capacity for adaptive radiation into these early grasslands is attributed to their immense reproductive abilities. It is significant that many modern rodent species are either entirely insectivorous, entirely seed-eating, or, if they are mainly vegetarian, supplement their diet with insects. The genetic flexibility of the rapidly evolving primate stock would permit them also to colonize these new grassland habitats, and it is not surprising that in the geological record soon after the appearance of grasslands we begin to find primate fossils in such areas. There is substantial evidence to support the conjecture that seeds provided a major element of their diet, as will be discussed.

Early Apes

The most extensive series of fossil primates associated with these late Tertiary grasslands has been found in equatorial regions of Africa, and especially in Uganda. These fossils date to the Miocene period, which lasted from 25 million to 12 million years ago, and have been placed in the aggregate genus *Proconsul*. This name is said to have been chosen because of a resemblance between the teeth and jaws of these fossil creatures and those of a chimpanzee at one time living in the London zoological gardens who was named "Consul."

Proconsul

Proconsul is not classifiable as an Old World monkey because of its brain size, dentition, and skeletal features (Figure 2-9). Nor is *Proconsul* clearly

FIGURE 2-9. *Various features of* Proconsul. *A.* Composite drawing of the skull. *B.* Reconstruction of the head showing the receding forehead, small nose, and prognathous face. *C.* Upper molar showing characteristic monkey cusp pattern (see Figure 2-6). The mixture of monkey and ape characters suggests that *Proconsul* resembled the common ancestor of both. Probably also it is not separable taxonomically from forms placed in the genus *Dryopithecus*.

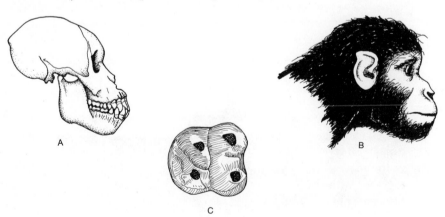

A

C

B

identifiable as an ape or hominid form. It is therefore often considered to have represented a stage of primate evolution before the ape stock (pongids) differentiated from the human one (hominids).

Definition of Terms

The use of some classificatory terms at this stage is unavoidable. The term *primates* is capitalized when it refers to the zoological *order* Primates; when referring to most of this group in general, but not in a taxonomic sense, it is uncapitalized.

Pongidae is the *family*, including living and fossil great apes, whose most obvious but not most significant characteristic is that they are tailless. Such forms are often termed *pongids* here; *pongid* may also be used as an adjective.

All human species in the genus *Homo* and all "subhuman" forms are placed in the family Hominidae, commonly termed the *hominids*. This term also may be used as an adjective. Hominidae and Pongidae can be combined in a superfamily Hominoidea, the hominoids, and this noun and its derived adjective are occasionally used here. The terms *subhominoid* and *protohominoid* are sometimes applied to forms like *Proconsul* but are not used in this text. Some authorities place *Proconsul* in pongid subfamily Dryopithecinae, which includes a number of Miocene-Pliocene fossil apes of which about 550 fossil specimens are known. They may also, as already noted, place all *Proconsul* material in the genus *Dryopithecus*. Table 2-2 shows a generalized version of the taxonomic relationship and classification of these various primate forms. For technical reasons relating to the skills associated with particular disciplinary training, hominids, which are primarily Pleistocene forms, have been investigated by anthropologists and anatomists, whereas prehominids such as Dryopithecinae, which are essentially Tertiary forms, have been studied by vertebrate paleontologists. This has not been an inflexible dichotomy, but it has dispersed papers over several professional journals, and produced significant differences in approach. Unfortunately, the Pleistocene specialists have suffered from taxonomic oversplitting, so that some thirty genera and many species of hominids have been named and described. This oversubdivision has resulted especially from the fragmentary nature of most ostelogical material, the difficulty of dating it, and the relative absence of knowledge concerning breeding barriers in populations known only from fossils.

Tool-Using Capacity

The anatomical and mental attributes exhibited by all living apes, especially the relative freedom of movement of their forelimbs, indicate that all members of this group, extinct and living, probably shared the limited capacity for *tool using* that behaviorists have been able to demonstrate in the modern great apes. There is every reason to suppose that *Proconsul*, together with the other ape forms whose diversity attained its zenith in the Miocene, possessed this limited tool-using ability.

Behavioral Patterns

It is difficult enough to relate the behavior of living primates to human behavior and to draw conclusions about possible analogous and homologous relationships. When we are dealing with an exclusively fossil group such as the genus *Proconsul* represents, comparative evidence is circumstantial indeed. Nevertheless, in such groups of fossil primates behavioral patterns must have been of extreme importance in evolutionary development; speculation has to be made on whatever circumstantial evidence is available. Among living forms there is always a strong resemblance discernible between human behavior and that of the other primates whether observations have been obtained in the field or in the laboratory (Figure 2-10). Phenomena such as mother-baby recognition, play periods, and weaning trauma appear to be constant features of all the primate groups, and it is very easy for these patterns to become part of anthropocentric arguments if care is not taken to avoid doing so. It is very easy also to emphasize the particular animal behavioral patterns that parallel human behavior, while ignoring those which do not conform to expectation.

There appears to be universal agreement that major features of vertebrate social behavior—including gregariousness, social communication, territori-

FIGURE 2-10. *Social communication in langur monkeys* and its development. The social communication system in langurs, explored by P. Jay, includes *calls,* such as the alert bark, that would provide information for animals out of sight. The grimace, by contrast, is an effective signal only to animals in direct visual range. A terrestrial social primate form might tend to extend the latter type of communication, an arboreal troop whose members' field of view soon becomes obscured by trees might elaborate the call system.

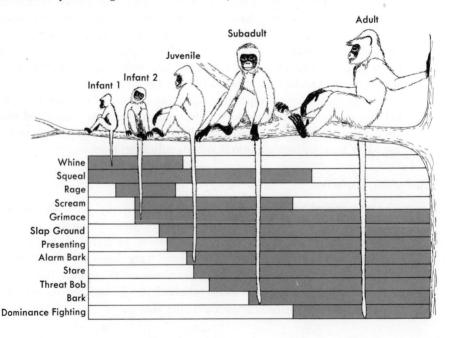

ality, dominance hierarchy, sexuality, parental care, and social play—also characterize primate behavior. Extensive studies on rhesus monkeys [*Macaca mulalta*] and other primates demonstrate that as a result of affinities among members of a group and aggression toward individuals outside the group, primates live in discrete and highly stable social units. The implications of this will be referred to later, when hominid evolution and human diversity are considered.

Social Life of Arboreal Primates

Working on langur monkeys in India, Phyllis Jay has described the social life of an Old World monkey troop. She lived in the territories of four different langur troops and was able, after 2 years of familiarization with their members, to predict much of their behavior. She observed a dominance hierarchy among males in each troop and a somewhat less pronounced hierarchy among females, to which individual females engaged in raising a young infant did not conform. A newborn infant was an object of curiosity to the females in the troop, who gathered round the mother to inspect it and take turns holding it.

The growing infant rode about by clutching its mother's front. It stayed close to her at all times, performed the same actions she did, and learned by imitation. As it became more competent at walking, the infant tended to wander away from its mother and play with others of its age group.

This social mixing apparently is necessary in order for the young langur to develop a well-adjusted personality. H. F. Harlow was able to demonstrate experimentally the need for such social interaction in young animals. He discovered that if rhesus monkeys were reared in isolation and not allowed to play, even if they were within sight of other members, they became neurotic and then psychotic. They could not subsequently make any kind of social contact, nor were the males ever able to copulate. It therefore appeared, first, that maternal interest and care were required during raising if young females were to become in their turn competent mothers, and, second, that periods of play with other infants were necessary for the proper development of social behavioral patterns, even of sexual behavior. Recently further experiments by H. F. Harlow and observations by other workers suggest that deprivation phenomena may be more complex than originally thought.

After weaning at about 15 months, langur juveniles were found to associate in one of two groups according to sex. The female juveniles got to hold and sometimes look after infants of other females for a time. The males indulged in increasingly vigorous competitive play, during which their adult hierarchical position became established.

A langur troop is essentially an *arboreal* population. Although individuals are completely accustomed to movement on the ground, this never takes them more than a few paces from a tree. When danger threatens it is a case of *chacun pour soi*. Mothers snatch up their babies, and every member of the troop leaps for the nearest tree. Males have very little to do with babies and juveniles in the troop, and there has been no adaptation of social behavior in response to group selection for such interests. The situation is very different in a terrestrial primate species population.

Social Life of Terrestrial Primates

The work on the langur monkey illustrates some of the behavioral patterns of an essentially *arboreal primate* species. The behavior of *terrestrial primates* provides a marked contrast and may therefore indicate more appropriately the nature of behavior in *Proconsul*. Studies have been made of several baboon species, in particular the Asiatic macaques and African species such as the chacma baboon (Figure 2-11). The characteristic and common feature of all such terrestrial social primate species is an individual *aggressive* temperament. This is not something exhibited as a behavioral response invoked by particular environmental stimuli but is an integral part of the individual animal's personality. It is exhibited not only in the manner in which the troop is constantly ready to defend itself and its individual members against a predator, but also in the fierceness and the insistence with which the dominance hierarchy is maintained (Figure 2-12).

Dominance

The normal mode of behavior within a macaque or baboon troop is peaceful, but peace is maintained only because it is enforced by the dominant members. These alpha animals have not only a specific function in defending the group against external aggression, but also a parallel function in policing

FIGURE 2-11. *An African chacma baboon* (Papio ursinus) *troop moving out across its feeding ground.* The dominant clique of dogs (A) at the center protects the females with nursing young (B); younger males take point (C). Individual dog baboons (D, E) high up in the dominance hierarchy squire individual females in the final stages of estrus (F, G) or bring up the rear (H); juveniles cavort on the flanks (K, L). If threatened, the dominant pair of dogs (A) will come right out of the center and attack, with the primary intention of driving off rather than killing the predator, which is usually a leopard. See text for further explanation.

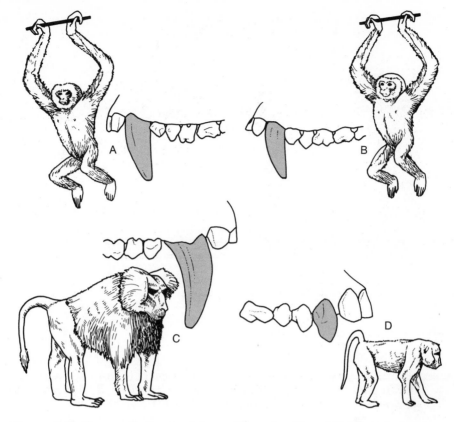

FIGURE 2-12. *Group selection for defense.* The male gibbon (*Hylobates*), shown in *A*, differs little in body size and dentition from the female, shown in *B;* defense against predators in such social *arboreal* primates usually is on an individual basis. In a terrestrial social primate such as the baboon (*Papio*) the male (*C*) is much larger and stronger than the female (*D*) with the canines especially developed as very effective tearing and piercing weapons. This adaptation of the dog baboon favors the survival of the troop as well as the individual male, and therefore provides an example of group selection.

the troop to prevent internal strife. The source of strife would be individual competition for essential resources such as food and water. Any gregarious animal species is faced with this problem and often responds the same way by instituting rank orders in a social hierarchy. Dominance can thus be regarded as a behavioral mechanism which has been favorably selected in gregarious species because it facilitates resource partitioning.

Internally, however, it is a *threat* of force rather than the use of it that is normally employed to maintain order. In macaques, for example, a "threat thermometer" has been formulated (Figure 2-13). Starting with the mildest form of threat, a stare, it rises through an open-mouthed glare to "bobbing," which is the most extreme form of threat short of actual physical contact. Supplementing these threats are subsidiary behavioral performances, the drawing back of the ears, grunts, forward steps, slapping the ground, and a number of other ritual procedures and actions.

FIGURE 2-13. *The "threat thermometer" in macaques* (Macaca). The lowest-level threat is *staring* (A), followed by an open-mouthed *glare* (B). The most extreme threat not involving physical contact is *bobbing* (C).

If dominance is accepted by the individual being threatened, there are various forms of submissive gestures that signal acceptance. Among them are looking away, crouching close to the ground, actual flight, abandoning food, offers of grooming, and permitting symbolic mounting.

Dominance in baboons and macaques is further complicated by the fact that, unlike dominance in other social animals such as birds and certain other mammals, there appears to be in some species a system of dominance by a *clique* rather than by individual members (Figure 2-14). This may be because aggression is so strongly developed in these terrestrial primates that it would be impossible for one individual to maintain dominance in the face of such strong rivalries. Moreover, individual dominance would pose a survival problem for a troop subject to predation and attack; it is likely that the troop would be disorganized at least long enough for serious damage to be done to it if the one alpha animal were suddenly incapacitated. If, on the other hand, dominance is shared by a small establishment of alpha animals, the troop is not exposed to the possibility of this catastrophe.

The female hierarchy is not so clearly established in baboons as is the male one. This is partly related to the cyclic hierarchical situation imposed by the female reproductive cycle. As long as the female has an infant with her in close attendance, she remains, as it were, outside the female dominance hierarchy. Likewise, a female in estrus is also in a special hierarchical category.

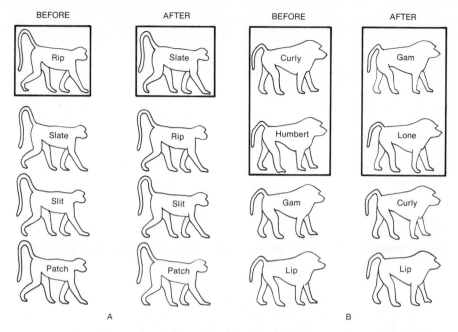

	BEFORE		AFTER		BEFORE		AFTER
	Rip		Slate		Curly		Gam
	Slate		Rip		Humbert		Lone
	Slit		Slit		Gam		Curly
	Patch		Patch		Lip		Lip

A B

FIGURE 2-14. *Dominance cliques.* A langur troop observed by P. Jay had the simple individual dominance hierarchy illustrated in *A*. The animals, to which she gave the names indicated, exchanged positions when Slate browbeat Rip into surrendering dominance to him. In a Japanese macaque troop, as reported by Imanishi, two males, Curly and Humbert, combined to form a dominance clique (*B*). When Humbert disappeared, Gam formed a clique with a male, Lone, from outside the troop, and the two displaced Curly at the head of the dominance hierarchy.

Hereditary Dominance Cliques

Nevertheless, one effect of even this loose hierarchical structure in the female population of a troop is a hereditary position for offspring in the troop hierarchy. A male offspring born to a dominant female is less continuously harassed by the bickering and squabbling of the females, and will be more serene and arrogant, likely to occupy a high place in the order when adult. The male offspring of a female in the lowest hierarchical cluster will be pushed around as an infant and in adulthood is more likely to retain a similarly low place in the male hierarchical order. This hereditary acquisition of hierarchical status may result in a clique of siblings, associated with nephews and nieces coming to monopolize the uppermost hierarchical positions—a primate example of nepotism. Similar but more flexible hereditary dominance cliques have been reported in social nonprimate species, more particularly carnivorous ones such as Cape hunting dogs and hyenas. Indeed, some behavior is so widespread among social animals that a British ethologist, J. H. Crook, has proposed a rule that can be expressed as follows: *in similar environments all social species will tend to develop similar social structures.*

It is tempting to read anthropomorphic analogies into this behavior of baboon and other terrestrial primate troops. It is equally dangerous to argue from these behavioral patterns to the probable behavior of extinct apes such as *Proconsul*. It is possible that any similarities in either case are entirely coincidental or a common response to selection pressures that evoke rather similar adaptations, in the same manner that ecological equivalents can be developed with apparently similar morphological adaptations. This possibly is emphasized by the general occurrence of social organizations closely resembling one another even in nonprimate species, as what may be termed "Crook's law" emphasizes. Recent studies by G. B. Schaller on African nonprimate social predators including lions, cheetahs, hunting dogs, and hyenas reinforce the close interaction between social organization and the environment.

The study of primate behavior was to some extent inspired by a belief that it would shed light on human behavior. In the event, studies of non-human primate behavior have been most revealing of the ecology of non-human primates. Within the last 2 years an excellent summary statement by J. F. Eisenberg and his collaborators at the Smithsonian Institution, Washington, D.C., has reviewed present knowledge of this behavior. We have to regard primates as a related group of consumer populations exploiting a series of ecosystems and subjected to a range of ecological pressures. In response to these pressures they develop various behavioral patterns. These run the whole gamut from solitariness to various social groups including unimale, age-graded, and multimale structures. Primate physiology is correspondingly varied, and like the larger herbivorous consumers such as ruminants, primates can show modifications to cope with the digestion of cellulose. So there are primates that are specialized vegetarians, or fruit-eaters, or a mixture of both. There are others requiring a high protein diet, or omnivores that regularly include a fair proportion of animal protein in their diet. The territorial systems characteristic of each primate species relate to the communications pattern, which partly reflects the need to maintain the spatial patterns.

Troop size appears to relate to the productivity of the ecosystem occupied. Terrestrial troops in regions of high but seasonally distributed rainfall appear to have the largest numbers. Perhaps because of this they usually have multimale social structures. The dominant males provide leadership and protection against predators, maintain spacing between troops, and reduce territorial competition between other troop members. Although they probably impregnate most of the females, they have few or no parental responsibilities.

The life of the females is dominated by infant care. The advantage of social life in a troop for a male is a potentially greater opportunity for the propagation of his own genes. For a female the assistance in food resource location, protection against predators, and perhaps help with juvenile care must be the main advantages. If we are seeking analogies between non-human primate behavior and human behavior, it is tempting to look most closely at large terrestrial troop-forming species of the African savannas such as the chacma baboon.

The same kinds of argument apply in even greater force to the behavior of infants—in particular to the significance of single as opposed to multiple births—and the unique feature of the grip that the infant, using all four limbs, maintains on its mother's hair, either on her back or on her front. This behavior could be a matter of a common response to a similar need in any arboreal animal population.

Clasping

The question of clasping has also been the subject of experimental investigation, and H. F. Harlow has reported a number of experiments on surrogate "mothers." In one of these half the surrogates had bare wire bodies, whereas the other half had similar structures covered with terrycloth. Variation was further developed by providing a milk supply in one half of the terrycloth-covered surrogates. To these various surrogates Harlow introduced newborn macaque babies. By far the strongest attraction was found to lie in the terrycloth, rather than in the milk; the infants spent considerable periods huddled up against the surrogates provided with terrycloth, whether or not there was also a milk supply. In this and other experiments it was possible to demonstrate that the infant macaque's primary need was for the security provided by contact with its mother's body. Without this the infant would be frightened by any strange objects, but if he could return to his cloth-covered surrogate, he would eventually overcome his fear of anything new and proceed to investigate it.

Evolution of Social Activities

There is one further major difference between an arboreal and a terrestrial primate in the significance of the clinging action of infants. Although both groups tend to seek refuge in trees, the time the arboreal species will spend in them is usually a longer portion of the day. The effect of a single moment's slackening of the grip of the infant on its mother will usually be fatal, and the clinging relationship develops uniquely between the mother and her offspring. In terrestrial groups, which will resort only momentarily to trees to escape danger, and for longer but more relaxed periods to pass the night, a continuous and persistent clinging is not such a vital necessity, and this protective mother-child relationship is extended to any member of the troop, including the males. If an infant is threatened on the ground by some predator, any animal in the troop will remove it to safety.

There is a reasonable supposition that the essentially terrestrial *Proconsul* forms exhibited behavioral patterns not very dissimilar to these reported for baboons and other primates. To what extent the patterns were transmitted to hominids will be further discussed in another section.

Social Communication and Other Behavior

As to the question of communication, Hockett and Ascher speculate that at the level of *Proconsul* this was probably by means of a *call system* quite similar to that of modern gibbons. It is argued that the gibbon call system has evolved little beyond the *Proconsul* stage, whereas in other anthropoids, where the call system appears less complex, there are more subtle means of variance. It seems possible, however, that the gibbon call system evolved to meet the needs of an arboreal social group whose members were normally unable to *see* one another. In a terrestrial social group, individual members would not usually be out of sight, and visual communication could have lessened the value of an advanced call system.

Of the several social activities in terrestrial primates already discussed, two aspects must be examined further, sexuality and food sharing. Sexuality in baboon troops, like social hierarchy, was initially described as it pertains to the chacma baboon. This species lives in large multimale troops in African savanna ecosystems. Under these circumstances, females coming into estrus will present to and copulate with any nearby male in the troop. Toward the close of the estrus cycle, the female generally associates with and copulates with one particular male of the dominant clique. As fertility is highest toward the end of estrus, it is this male that probably impregnates her. Other than this there is no proprietary interest exhibited by the males, no pairing of sexually active mates, no possibility of males being able to recognize and so to react to their own biological progeny.

Sexuality is not as rigid a feature of a species, primate or otherwise, as was once imagined. In fact, it tends to respond to the environment as Crook's law maintains. For example, the mating behavior of wild turkeys in the United States varies according to the relative availability of food resources between one ecosystem and another. Gelada baboon troops living on mountain slopes in Ethiopia fragment during the dry season into small units of several females and their young associated with an adult male. Supposedly this represents a more effective way of partitioning a seasonally limited food resource, but it has the effect of introducing proprietary rights of one male over a harem of females and his own offspring. Survival of this group and natural selection processes will be directly related to the genes of this one male. Hamadryas baboons inhabit an even more rigorous habitat in Ethiopia and the drier neighboring territory of Somalia. The harem structure has become permanent and an adult hamadryas male jealously shepherds his females at all times.

Despite this apparent association of breeding patterns with resource availability in various species of baboon, food-sharing behavior such as is observable in hunting dog packs is never observed. Nor is it reported among chimpanzee troops with respect to the *vegetarian* elements of their diet. However, chimpanzees are occasional meat eaters, mostly through scavenging but also as a consequence of sporadic hunting activity. A successful chimpanzee hunter will usually share out some of the prey, in response to begging. This chimpanzee food sharing appears to be quite unrelated to any loose dominance hierarchies that exist, or to kinship ties.

Evolution of *Proconsul*

The orgins of *Proconsul* are not clear, but some fragmentary remains have been found of interesting primates in the Fayoum oasis in upper Egypt which date from the Oligocene epoch. From the anatomical structure of the forelimbs of some of these fossils, it is concluded that they may have been early brachiating forms. Like all apes they were tailless.

The location of such fossil forms so far north in Africa may appear strange. However, somewhat scanty remains in the form of macrofossils suggest that in mid-Tertiary times a tropical rain forest–type vegetation extended this far beyond the Tropic of Cancer, reaching to the shore of the Mediterranean, which was then, because of marine transgression, located much farther south than now. It might be expected that the ancestors of ground apes would be the rapidly evolving group of brachiating apes, and that terrestrial adaptations would occur in precisely such marginal forest habitats where interplay between the increasingly arid climate and the adaptation of more fit plant types would take place, and where grassland ecosystems were developing for the first time.

Characteristics of *Proconsul*

From the skeletal structure of the foot and leg and the form of the pelvic girdle, it is surmised that *Proconsul* could stand approximately upright and was able to shamble along at least for a time in this position. The assumption of an upright habit was not yet fully facilitated by these skeletal adaptations, and probably for an appreciable amount of the time the various *Proconsul* forms ran on all fours in a manner similar to that of a modern chimpanzee.

The dentition of *Proconsul* in all its forms includes massive canine teeth (Figure 2-9), which implies that this genus of ground apes probably had not yet assumed the extensive use of artifacts as defensive tools, but like its brachiating arboreal ancestors defended itself by biting. Studies on the origin of tool use employ the decreasing size of the canines as evidence for the substitution of artifacts in defense instead of teeth, and therefore the development of tool-using and later tool-making species. The significance of the bipedal habit also lies in its indication of this tool-using ability. An upright posture is a disadvantage in terms of running speed, but it frees the arms to use and carry tools. The structure of the hand, in particular the arrangements for the opposition of the thumb and fingers, also has some bearing on tool use and tool making.

An entirely new hypothesis regarding the function of primate canine teeth has been argued by Clifford Jolly. To restate the essential features of his theory, he first notes that canine teeth appreciably longer than the rest of the dentition need to be accommodated in slots in the opposing jaw. This is the reason for the gap in the teeth that can be seen, for example, in Figure 2-15B. When the jaws are brought together as the mouth is closed, the canines lock in their slots so that no *chewing* motion is possible. That is, no extensive sideways grinding action between opposing molars can be achieved, as for example a cow manages when chewing its cud. Jolly emphasizes that the most nutritious plant food resources plentifully available in the new grassland habitats appearing in the Late Tertiary would be *small* seeds, abundantly produced by grasses and other annual and perennial herbs. For birds to utilize these small seeds involves no anatomical change; seed-eating birds have a powerful crop in which seeds are ground. For mammals, by contrast, it is necessary to develop a dentition with a grinding action to work the lower against the upper molars.

On this theory presented and extensively documented by Jolly, it is possible to imagine Late Tertiary apes partitioning their food resource. A gorilla-type group would continue to be foliage eaters, a chimpanzee-type group would retain its largely fruit-eating habit. Neither group would require any modification of the canine dentition, nor did this occur in the *Proconsul*-type forms, which became largely terrestrial. A third type of ape group has to be postulated. This group would have reduced canines, so that it would be able to *chew* small seeds. An ape form with a dentition basically similar to the modern human form would accordingly have access to this new and bountiful food supply of the grassland. Once this selective advantage of a chewing dentition was exploited, the other selective features of bipedalism, release of forelimbs for nonlocomotor activity such as tool use and so forth, would continue the canine reduction trend.

This would leave us about 8 or 9 million years ago with some human ancestral group roaming the warmer African grasslands. It would look very much like *Proconsul* fitted with a human dentition, and its diet would mainly be grass and forb seeds; this ecological niche would remove it from direct resource competition with any other primate. We have perhaps never seriously pondered before why several million years later human populations are found to be harvesting and grinding up wild grass seeds, and why some neolithic populations developed cultivated grasses as their main food resources. Our human dentition may well reflect an early ancestral commitment not only to a bipedal terrestrial life but also to a basically granivorous diet. As will be seen in the next few pages, a Miocene fossil candidate for this granivorous niche has been forthcoming.

Proconsul was not a single species of ground ape. At least two forms are known to have occurred, which we can demonstrate by lumping the larger skeletons found into one species and the smaller ones into another. It is probable that these early ground apes, even if they originated from a single common ancestor, diversified rapidly in the new valley grassland habitat. On such slender evidence as the skeletal remains, which is all we have today, it is difficult to make any conclusions about the behavioral patterns of these first ground apes. It can, however, be conjectured that they were social animals associated in troops with a nucleated territoriality. That is, the territory was not defended at its boundaries, as is the case with many kinds, but each group was centered in a particular area, with commonly overlapping use of marginal portions.

Pigmentation in *Proconsul*

Because of the dense hairy coat that *Proconsul* species supposedly possessed, they would have had to forage in the grasslands in the early morning and late afternoon, passing the night and the heat of the day in the forest margins, probably roosting in trees at night as do the modern chimpanzees, gorillas, and baboons. They would have been diurnal animals with little or no night vision. Because of the comparatively low light intensity beneath the canopy of the forests that their ancestors had occupied, they would not at first have been heavily pigmented. Probably one of the earliest adaptations to a terrestrial life in forest openings was to develop a heavy pigmentation in the naked parts of the skin, especially the face, to prevent overproduction of vitamin D when they ventured into grassland habitats. The same kind of pigmentation is seen in modern tropical apes, which similarly are exposed to full sunlight in forest clearings. This issue will be returned to later.

Predation

With the possible exception of occasional attacks by leopard-like animals that may have evolved by this time, and by large snakes, the arboreal ancestors of *Proconsul* would have suffered little predation. Some genera of carnivores, including the big cats (*Panthera*), are known from Late Miocene–Early Pliocene deposits (10–15 million years BP). On the grasslands there would likewise have been no immediate evolution of a predator sufficiently strong and courageous to brave the bites from the large canines of the dominant male members of a *Proconsul* troop. The many rodent species there would be far easier prey for any predator. The island of Madagascar may support this contention, for the prosimian populations there have no predators apart from snakes. Grasslands do not appear to have been part of the original landscape of Madagascar either, and neither are antelope, bovine, or equine species found there. Perhaps the big cat predators of the African mainland did not evolve until the larger herbivorous herd animals became established on grassland and savanna.

Diet of *Proconsul*

The general nature of the diet in the grassland ecosystem would probably have been little different in form for terrestrial primates from that in the forest. It might have consisted of insects and underground storage organs of plants, varied with edible fruits at particular times of year. Obviously some forest foods would continue to be used. As herbivorous animals, large and small, adapted to the grassland habitat, *Proconsul* may also have scavenged dead animals, or even chased and caught smaller ones, particularly rodents. This chasing activity could have had considerable evolutionary significance, as will be discussed later. In one respect the available food of the grasslands differed from that of forest ecosystems—there was an abundance of small seeds. Rodents, insects, and birds could exploit this immediately. Ostriches, for example, can be observed working over a grassland nibbling off the grains in the grass inflorescences; their crops will cope adequately with this food source. Terrestrial primates, as already noted, would have to grind these small seeds with their molars. They could not do so if peglike canines such as those of baboons were present, preventing extensive rotary movement of the lower jaw against the upper.

Distribution of *Proconsul*

Although the composite group of ground apes placed in the genus *Proconsul* must have been common and extensively distributed in some parts of tropical Africa throughout the Miocene, no representatives of it are known from any other part of the world, nor have any related forms yet been found in the succeeding epochs of the Tertiary or the Pliocene. There was therefore a gap of some 20 million years before further and more recent forms of ground apes were known to have occurred; only in the past few years have exciting new discoveries and fresh interpretations begun to fill this gap.

Ramapithecus

There are now grounds for supposing that this restriction of *Proconsul* to Africa may have been the consequence of taxonomic proliferation of dry-

FIGURE 2-15. *Comparison of dentition of* Ramapithecus *with that of other anthropoids.* A. Capuchin (*Cebus*). B. Orangutan (*Pongo*). C. *Ramapithecus.* D. Man. All are scaled so as to reduce to the same length. The U-shaped arc in the capuchin and orangutan contrasts distinctly with the curved arc of *Ramapithecus* and man. Simons considers this as supporting the contention that *Ramapithecus* is the oldest ancestral hominid known and a distinct genus from *Proconsul.* The lack of conspicuous canines suggests one of three possibilities: *Ramapithecus* was essentially an arboreal form, it lacked troop dominance hierarchies, or it was able to defend itself with artifacts.

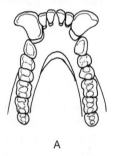

A

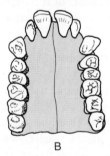

B

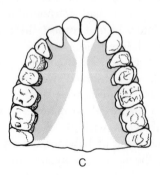

C

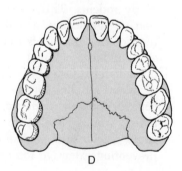

D

opithecine genera, and it may be more apparent than real. From sites in Europe, Africa, and Asia that are considered to have had a relatively warm climate in the Miocene, many fossil "ape" genera have been described. E. L. Simons believes that all these could be reduced to at most four genera, *Dryopithecus, Sivapithecus, Proconsul,* and *Ramapithecus.* A species of the last genus, *R. brevirostris* from the early Pliocene of the Siwalik Hills of India, has dental features remarkably similar to those of the Pleistocene australopithecines to be discussed in Chapter 3 (Figure 2-15). L. S. B. Leakey described a contemporaneous East African specimen, *Kenyapithecus wickeri,* which in Simons' opinion is not significantly different from forms described under *Ramapithecus,* a deduction followed also in this text.

An elegant method of resolving the taxonomic problems associated with these various dryopithecine forms has recently been described by R. B. Eckhardt. From analysis of the causes of morphological and metric variation in primate skeletal structures, it is apparent that the main variations arise from differences in age, sex, geographical occurrence, and chronological time. Eckhardt was able to secure a measure of variation accruing from the first three sources by reference to modern chimpanzee populations. Then by assuming that his fossil forms belonged to only one phyletic line, he calculated the amount of morphological change necessary per generation to secure the range of variation shown over 6 million years by dryopithecines. The amount of change necessary proved to be very slight, and well within the rate of change per generation that has been produced experimentally in laboratory populations of fruit flies.

Eckhardt concludes from his calculations on fossil teeth crown dimensions that among the various dryopithecines extant in Asia and Africa from Miocene to Middle Pliocene times there was only one hominoid line. He also concludes that no compelling evidence exists for supposing that the hominoid *Ramapithecus* is ancestral to any later hominid forms.

Nevertheless, in the absence of further evidence it seems reasonable to suppose for the present that *Ramapithecus,* perhaps segregated into Asiatic and African species, is the hominoid the closest yet known to ancestral hominoid forms. Moreover, *Ramapithecus* seems both the most acceptable and most applicable of the various names available for this dryopithecine group of Late Tertiary fossils. When skeletal remains other than fragmentary jaw material and teeth become available, it may be possible to state the phyletic relationships of *Ramapithecus* with a greater degree of confidence. Certainly this Middle Miocene to Pliocene dryopithecine hominoid looks a more likely candidate than *Proconsul* to have been the originator of the hominid line. It is still not impossible to consider *Proconsul* as a Miocene ape that occupied open terrestrial habitats before the actual divergence of pongid and hominid lines.

Separation of Pongid and Hominid Lines

Although fossil remains from the Pliocene are rare, an elegant method of approaching such a time gap in a phylogenetic progression has been described by V. M. Sarich and A. C. Wilson using immunological cross-

TABLE 2-4 *Chronology of Advanced Primates*

The absolute age determinations shown below represent generally acceptable dates, most of which have been obtained by the carbon-14 or potassium/argon method. Dispersal areas of various forms will probably eventually be revealed as much wider than those indicated. Because of this, and the appearance or disappearance of particular forms at considerably different times in different areas, some forms that are generally believed to be derived from another particular form may be shown as appearing in the geological record before it.

Taxon name	Dispersal area	Approximate age in millions of years BP
Proconsul	East Africa	?14
Ramapithecus wickeri	Kenya	14
Ramapithecus punjabicus	Northern India	10
Homo africanus	Sub-Saharan Africa	2.5–3.0
Paranthropus robustus	Southern Africa	1.7–2.0
Paranthropus boisei	East Africa, Ethiopia	1.1–3.7
Homo habilis	East Africa, Ethiopia	1.5–5.5
Homo erectus	Old World	0.3–2.0
Homo sapiens neanderthalis	Old World	0.04–0.3
Homo sapiens sapiens	Cosmopolitan	present–0.025

reactions. This method is based on the clear structural relationship between proteins and genes and the supposition that quantitative comparative studies of protein structure will aid in clarifying the genetic relationship.

Sarich and Wilson have demonstrated that albumin evolution in primates is a very regular process and that lineages of equal time depth show similar degrees of change in their albumins. Therefore, the degrees of change exhibited would appear to be a function of time. Through a mathematical relationship between an index of dissimilarity determined by immunological methods, they fix the time of divergence of two species of primates.

The results of such an examination are shown in Figure 2-16, in which it can be seen that Old World monkeys and an undifferentiated hominid-pongid stock shared a common ancestor in the Oligocene. This is consistent with the supposition that populations referred to the genus *Proconsul* represent terrestrial forms of a group of arboreal apes that became highly successful in the Oligocene coincident with their development of the locomotorfeeding adaptation of *brachiation*.

The albumins of the gorilla, chimpanzee, and man are considered to be equidistantly related. The three genera into which these forms are presently placed, *Gorilla, Pan,* and *Homo,* are thought to have separated in the Late Pliocene, whereas the orangutan (*Pongo*) diverged from this anthropoid group in the Early Pliocene; the gibbons and siamangs did so at the very beginning of the Pliocene, only finally branching into individual genera (*Hylobates* and *Symphalangus*) at the close of the Pleistocene.

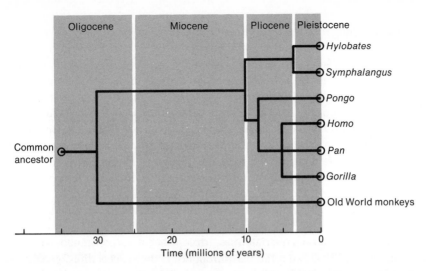

Figure 2-16. *The time of divergence of pongid and hominid stocks* as estimated using immunological data from contemporary anthropoid groups. Sarich and Wilson suggest a date of 5 million years BP for this separation, but, as seen from their diagram, they place this event in the early part of the Pliocene. Revised calculations of Pliocene dates could therefore modify this time estimate.

These biochemical investigations, which can be further substantiated by an examination of other amino acid sequences of proteins, clearly have interesting applications in confirming deductions from fossil remains, and in providing hypotheses of phylogenetic relationships even if fossils are lacking. In the particular instance of the origin of the genus *Homo*, they appear to substantiate the general time scale for evolution of the earliest members of this genus from a prosimian primate stock during the early Tertiary.

The next chapter describes the appearance of the genus *Homo* and the relationship and evolution of its presently known forms. *Proconsul* is the last of the identified fossil forms to show a mixture of hominid and pongid characteristics. Later anthropoid ape fossils are considered to represent groups that have clearly evolved into populations referable either to the genus *Homo* or some closely related form, or to one of the various great ape genera.

Although there is this general agreement between the *sequence* of evolutionary events as between paleontological and biochemical evidence, there remains a confusing discrepancy as to *absolute timing*. Wilson and Sarich have now shown that quantitative comparisons can be made of hemoglobins and DNA in addition to serum albumins. They insist that there is a concordance between their hemoglobin and albumin results, confirming that the most probable date for separation of the human and African ape lineage was 4 to 5 million years ago. As will be discussed in the next chapter, forms assignable by some workers to the genus *Homo*, and agreed by all to be *hominid* as opposed to *pongid*, were already in existence 5 million years ago. This would suggest a Miocene rather than a Late Pliocene date for the separation of the hominid and pongid lineages, a discrepancy of at least 10 million years, as Uzzell and Pilbeam have described.

The current state of our knowledge of Late Tertiary hominoid forms leaves us with the distinct impression of three major evolutionary lines. One, ancestral to New World monkeys, by development of a prehensile tail was evolving into an arboreal group with extreme agility. A second, the progenitors of Old World monkeys, generally retained the tail, but as a balancing organ. Although most of this second group remained preponderantly arboreal, a few, as will be explained, became terrestrial. A third group, the apes, evolved brachiation and lost the tail, which could serve no function with this particular locomotory habit. A number of this third group became terrestrial, a few of them simultaneously evolving into bipedal forms (Figure 2-17).

The bipedal terrestrial apes, dryopithecines, continued to evolve and display radical adaptations throughout the succeeding geological time. Members of the first two groups, that is, the monkeys, and all but the bipedal terrestrial forms of the third group, apes, either became extinct or underwent no further major evolutionary modification after the Miocene. This is a very familiar problem in macroevolution, why one form continues to evolve while other related ones persist virtually unchanged. Assuming that DNA changes, i.e., gene mutations, continue at more or less the same rate in all allied forms, why should virtually all the consequent changes be rejected by natural selection in the one instance and strongly favored in the other. As far as our human species is concerned, features that the first terrestrial anthropoid apes, as represented by *Proconsul,* have contributed to our own morphology, physiology, and behavior, and that have therefore persisted for some 25 million years, may be summarized as follows:

1. A pentadactyl limb structure, with hind limbs modified for bipedal movement, together with an orthograde body position
2. A clavicle, permitting extensive movement of the forelimbs, and opposable thumbs, both facilitating tool using
3. A diurnal habit, associated with stereoscopic color vision and a drastic reduction in olfactory capabilities, associated with increased conceptual brain development
4. Single births, with long gestation and infancy periods, the latter relating to learning through play and imitation
5. A social grouping with dominance hierarchies, inherited dominance cliques, and a communication system
6. Selection for social characters such as group identity and defense, infant protection, and group occupation of a nucleated territory
7. Development of pigmentation on exposed body surfaces to prevent overproduction of vitamin D
8. An omnivorous diet, associated with a dentition of identical dental formula, but with canines in *Homo* not projecting beyond the level of the other teeth

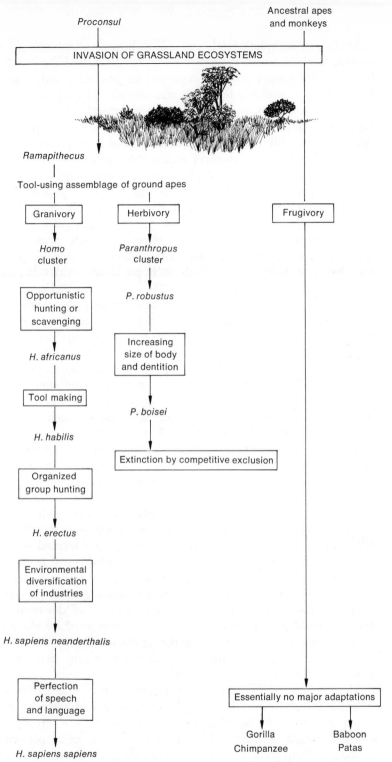

FIGURE 2-17. *Tentative scheme of human evolution.* Many such schemes have been presented, and all must be subject to modification as more information becomes available. This particular scheme relates sequentially the final major evolutionary stages of terrestrial primates so far afforded general recognition. It also identifies what was probably the strongest selective influence of the many that controlled primate evolution from one anthropoid grade to another. For chronological dates see Table 2-4.

As far as morphological and physiological adaptation is concerned, ecological evolution in *Proconsul* had almost reached the point where any further development must be regarded as producing only minor changes. The dominant emphasis hereafter was on cultural adaptation. In a tropical rain forest, resources are scattered evenly throughout the environment and distributed on a structural pattern. Animals occupying particular micro-environments within the general forest structure are presented with a fairly uniform distribution pattern of exploitable resources. In a savanna, most often the resources are unevenly patterned, providing what ecologists describe as a *patchy environment*. The immediate or proximate needs of a primate troop— food, water, and shelter—are not only unevenly distributed, but some may be scarcer than others. At the social level we are considering in this chapter, the primate group copes only with proximate needs; its members live from day to day with immediate survival problems. The knowledge of how to cope with such problems resides in the collective memory of the troop. Individual roles have to be learned and transmitted in turn by each member. Although primate troops are therefore problem-solving units in this proximate sense, they have also been subjected to natural selection for their group reaction to ultimate problems. Thus there is some emphasis on social behavior, such as the establishment of rank orders, learning practices, and the care of the young which, while dealing with a range of proximate problems, can also ensure that the group is "fitter" than groups that do not have a distinct way of responding to proximate problems. Natural selection would operate more favorably on such a group, and this "fitter" behavior would be more likely to survive.

An observed example of natural selection is provided by the painstaking and long-term studies the Japanese have made on an island population of macaques. Individuals of this population initially did not swim in the sea. Then one youngster discovered how to swim; other youngsters, both male and female, learned to swim by imitation of this one. Most older animals did not learn. As it was to turn out, swimming was an advantage to the troop, increasing its food resources. Young females learned to swim, and in turn their young copied them. Young males, who travel from one troop to another as they change hierarchial status, could introduce the new technique to new troops. Had the acquisition of a swimming habit proved a disaster, all would not have been lost. The older females of the troop, who never learned, could have produced a new generation sired by older males, who never learned to swim either, and the whole swimming practice could have been abruptly stopped with the death without issue of those individuals who had adopted the disastrous habit.

The intimate relationship between primate groups and their savanna environment with regard both to the proximate and ultimate problems its patchiness poses for survival can be expressed by a simple model. The chacma, gelada, and hamydryas baboons, whose social patterns were described earlier, form a series corresponding to increasing dispersal and scarcity of resources in the savanna environment. Within these baboon troops, and in other terrestrial monkey and the terrestrial ape societies of gorillas and chimpanzees, this is the essential pattern of behavior. Each member exists in the present, and essentially for itself. That is, its traits are basically egotistical; altruistic features are few. Those that do exist are essentially features favorable to the ultimate survival of the individual *group*. Altruistic features are nevertheless real, as was conceded recently by E. O. Wilson.

When we examine a phylogenetical scheme, such as that in Figure 2-17, we can find no surviving savanna species which has any altruistic features over and beyond those displayed at this early primate level, except our own species. Thus we can begin to ponder whether altruism, like an open communication system, arose uniquely in our direct ancestors, or whether, like tool making and tool using, it was at one time more liberally distributed among terrestrial savanna primates, but was lost when our direct ancestors drove all other forms with this feature to extinction by competitive exclusion.

However this may be, even at the baboon troop level we can see the rudimentary behavioral structure necessary for the survival of a primate terrestrial group. To the preceding list of ancestral characteristics that have come down to us from the past, we must add the critical behavioral feature of altruism, derived from group social behavior and patterns. Whether we share these altruistic behavioral patterns with other surviving terrestrial primate species by inheritance from a common ancestor or whether the observable similarities result from similar responses to similar environmental pressures, we cannot for the moment determine. We are left with a picture of our earliest terrestrial ancestors occupying African savanna areas and forming troops adequately equipped to locate the necessary daily food, water, and shelter. The long-term survival of these savanna troops was ensured by the selection of altruistic individual and group behavior. These troop forms were fully capable of coping successfully with their environment. Even at the monkey grade level, as baboon troops, they survived the vicissitudes of the Pleistocene Ice Ages and did not become faced with extinction by competitive exclusion until their savanna habitat itself was threatened with destruction by a highly evolved modern species of their terrestrial ape relatives which had acquired unique cultural patterns of behavior. If our species had not developed this propensity toward habitat degradation, the other terrestrial primates at the monkey and ape grade levels, with their several group behavioral patterns, so often basically similar to our own, would have remained in occupation of their particular ecological niches for many millennia to come.

By the Miocene, numerous of our own morphological and physiological characteristics had thus been determined. They were subsequently subjected to little more modification than was the *totality* of characteristics displayed by Old and New World monkeys. Our ancestral stock being primarily modified by *cultural* adaptation, any further *biological* adaptation that occurred was essentially associated with mental capacities and with improving manual dexterity and speech, which interacted with and facilitated further cultural adaptation.

Over the last few years experimental investigations have been conducted on the DNA mutation rate in various primate species. The technique applied in these studies involves the formation of hybrid DNA molecules between

TABLE 2-5 *Date of Divergence of Various Pairs of Higher Primates*

The dates have been estimated by determining the extent of nucleotide substitution sequences, that is, the number of gene mutations that have occurred since the separation. As always in biochemical comparison the man-chimpanzee relationship appears the closest.

Species whose DNA molecules are compared	Millions of years since divergence
Man–chimpanzee	30
Man–gibbon	60
Man–rhesus monkey	90
Man–capuchin monkey	130
Man–galago (a prosimian)	160

the species under consideration and a comparison of the differences in nucleotide sequences in the two strands. The observed differences provide an indication of the time that has elapsed since the two species shared a common ancestor. Table 2-5, extracted from a recent publication prepared by Kohne and co-workers, illustrates some of the data obtained in this way.

Although some adjustments may eventually have to be made in this time series, it well illustrates that gene differences distinguishing apes from prosimians must have taken about 100 million years to establish, whereas a mere 30 million years sufficed to separate human and chimpanzee populations. This stresses the point, already made several times in this text, that human societies as problem-solving units owe most of their effectiveness to cultural rather than biological evolution. Had it not been so, there might even now have been no more difference between our societies and chimpanzee groups than can be observed between monkeys and arboreal apes.

BIBLIOGRAPHY

References

Altman, S. A., and Altman, J. *Baboon Ecology: African Field Research,* Chicago: University of Chicago Press, 1970.
Bishop, A. "Use of the hand in lower primates," in J. Buettner-Janusch (ed.), *Evolutionary and Genetic Biology of Primates,* Vol. 2, New York: Academic Press, 1964, p. 133.
Bond, G. *Past Climates of Central Africa,* London: Oxford University Press, 1962.
Boughey, A. S. *The Origin of the African Flora,* London: Oxford University Press, 1957.
Brooks, C. E. P. "Geological and historical aspects of climate change," in *Compendium of Meterology,* Boston: Malone, 1951, pp. 1004–18.

Carpenter, C. R. "A field study in Siam of the behavior and social relations of the gibbon," Comparative Psychology Monographs No. 16, 1940 (republished in C. R. Carpenter, *Naturalistic Behavior of Nonhuman Primates,* University Park, Pa.: Pennsylvania State University Press, 1944).

Cartmill, M. "Rethinking Primate Origins," *Science,* **184:** 436–43, 1974.

Chance, M., and Jolly, C. *Social Groups of Monkeys, Apes and Men,* London: Cape, 1970.

Corner, E. J. H. "The durian theory or the origin of the modern tree," *Ann. Bot.,* **13:** 367–417, 1949.

Crook, J. H. "The socio-ecology of primates," in S. L. Washburn, and P. Dolhinow, *Perspectives on Human Evolution Two,* New York: Holt, Rinehart and Winston, 1972, pp. 281–347.

Echlin, P. "The origins of plants," *New Scientist,* **42:** 286–89, 1969.

Eckhardt, R. B. "Population genetics and human origins," *Scientific American,* **226**(1): 94–103, 1972.

Eisenberg, J. F., Muckenhirn, N. A., and Rudran, R. "The relation between ecology and social structure in primates," *Science,* **176:** 863–74, 1972.

Haldane, J. B. S. "The argument from animals to men: an examination of its value in anthropology," Huxley Memorial Lecture, *J. Roy. Anthrop. Inst.,* **86:** 1–14, 1956.

Hall, K. R. L. "The sexual, agonistic and derived social behavior patterns of the wild chacma baboon *Papio ursinus,*" *Proc. Zool. Soc. London,* **139:** 131, 1962.

Harlow, H. F., and Harlow, M. K. "Social deprivation in monkeys," *Scientific American,* **207**(5): 136–46, 1962.

Hockett, C. F., and Ascher, R. "The human revolution," *Curr. Anthrop.,* **5**(3): 135–68, 1964.

Jay, P. "The common langur of North India," in I. DeVore (ed.), *Primate Behavior,* New York: Holt, Rinehart and Winston, 1965, pp. 197–249.

Jolly, C. J. "The seed-eaters: a new model of hominid differentiation based on a baboon analogy," *Man,* **5**(1): 5–26, 1970.

Kohne, D. E., Chiscon, J. A., and Hoyer, B. H. "Evolution of primate DNA: a summary," in S. L. Washburn, and P. Dolhinow, (eds.), *Perspectives on Human Evolution Two,* New York: Holt, Rinehart and Winston, 1972, pp. 166–68.

Kummer, H. *Social Organization of Hamadryas Baboons,* Chicago: University of Chicago Press, 1968.

Leakey, L. S. B. "A Miocene anthropoid mandible from Rusinga, Kenya," *Nature,* **152:** 319, 1943.

Leakey, L. S. B. "A new lower Pliocene fossil primate from Kenya," *Ann. Mag. Nat. Hist.,* **4:** 689, 1962.

Macinnes, D. G. "Notes on the East African Miocene primates," *J. E. African and Uganda Nat. Hist. Soc.,* **17:** 141–81, 1943.

Moore, W. J., Adams, L. M., and Lavelle, C. L. B. "Head posture in the hominoidea," *J. Zool.,* **169:** 409–16, 1973.

Petter, J. J. "L'ecologie et l'ethologie des lemuriens Malgaches," *Memories du Museum d'Histoire Naturelle,* t27, fasc. 1, Paris, 1962.

Petter, J. J., and Hladik, C. M. "Observations sur le domaine vital et la densité de population de *Loris Tardigradus* dans les forêts de ceylan," *Mammalia,* **34**: 394–409, 1970.

Sarich, V. M., and Wilson, A. C. "Immunological time scale for hominid evolution," *Science,* **158**: 1200–03, 1967.

Schaller, G. B. "Predators of the Serengeti," Part I: "The social carnivore," *Nat. Hist.,* **81**(2): 38–49, 1972; Part 2: "Are you running with me, hominid?" *Nat. Hist.,* **81**(3): 61–68, 1972; Part 3: "The endless race of life," *Nat. Hist.,* **81**(4): 38–43, 1972.

Seward, A. C. *Plant Life Through the Ages,* Cambridge, England: Cambridge University Press, 1931.

Simons, E. L. "Some fallacies in the study of hominid phylogeny," *Science,* **141**: 879–89, 1963.

Simons E. L. *Primate Evolution,* New York: Macmillan, 1972.

Tavener-Smith, R. "Glacial phenomena in the lower Karroo rocks of the Kandabure coal area, Zambesi Valley," *Rec. Geol. Surv. Northern Rhodesia,* 19–22, 1955.

Tobias P. V. "Olduvai Gorge and hominid evolution—a review of *Olduvai Gorge,* vol. 3, by M. D. Leakey," *So. Afri. J. Sci.,* **68**(8): 225–27, 1972.

Tuttle, R. (ed.) *The Functional and Evolutionary Biology of Primates,* Chicago: Aldine-Atherton, 1972.

Uzzell, T., and Pilbeam, D. "Phyletic divergence dates of hominoid primates: a comparison of fossil and molecular data," *Evolution* **25**(4): 615–35, 1971.

Washburn, S. L., and DeVore, I. "The social life of baboons," *Scientific American,* **204**(6): 62–71, 1961.

Wilson, A. C., and Sarich, V. M. "A molecular time scale for human evolution," *Proc. Nat. Acad. Sci. U.S.A.,* **63**: 1088–93, 1969.

Wilson, E. O. "Group Selection and Its Significance for Ecology," *Bioscience* **23**: 631–38, 1973.

Wynne-Edwards, V. C. "Intergroup selection in the evolution of social systems," *Nature,* **200**: 623, 1963.

Further Readings

Altmann, S. A. (ed.) *Social Communication Among Primates,* Chicago: Chicago University Press, 1967.

Campbell, B. G. *Human Evolution,* Chicago: Aldine, 1966.

Colbert, E. H. "The ancestors of mammals," *Scientific American,* **180**(3): 40–43, 1949.

Crook, J. H. (ed.) *Social Behavior in Birds and Mammals,* New York and London: Academic Press, 1970.

Davis, P. R. "Hominid fossils from Bed I Olduvai: a tibia and fibula," *Nature,* **201**: 967–68, 1964.

Davis, P. R., and Napier, J. "A reconstruction of the skull of *Proconsul africanus* (R.S. 51)," *Folia Primat.,* **1**: 20–28, 1963.

Day, M. *Guide to Fossil Man,* London: Cassel, 1965.

Dolhinow, P., and Sarich, V. M. (eds.) *Background for Man,* Boston: Little, Brown, 1971.

Eimerl, S., and DeVore, I. *The Primates,* New York: Time-Life, 1965.

Emiliani, C. "Ancient temperatures," *Scientific American,* 198(2): 54–62, 1958.

Flint, R. F. *Glacial and Quaternary Geology,* New York: Wiley, 1971.

Goodman, M. "Immunochemistry of the primates and primate evolution," *Ann. N.Y. Acad. Sci.* 102: 219, 1962.

Le Gros Clark, W. E. *The Fossil Evidence for Human Evolution,* 2nd ed., Chicago: University of Chicago Press, 1964.

Lewis, O. J. "The hominid wrist joint," in S. L. Washburn, and P. Dolinow, (eds.), *Perspectives on Human Evolution Two,* New York: Holt, Rinehart and Winston, 1972, pp. 169–91.

Napier, J. R., and Davis, P. R. *The Forelimb Skeleton and Associated Remains of Proconsul africanus,* London: British Museum (Nat. Hist.), 1959.

Napier, J. R., and Napier, P. H. *Handbook of Living Primates,* New York: Academic Press, 1966.

Napier, J. R., and Napier, P. H. (eds.) *Old World Monkeys,* New York: Academic Press, 1970.

Newell, N. D. "Crises in the history of life," *Scientific American,* 208(2): 76–92, 1963.

Pilbeam, D. "*Gigantopithecus* and the origins of Hominidae," *Nature,* 225: 516–19, 1970.

Rosenblum, L. A., and Cooper, R. W. (eds.) *The Squirrel Monkey,* New York: Academic Press, 1968.

Sarich, W. M. "A protein perspective," in J. R. Napier, and P. N. Napier (eds.), *Old World Monkeys,* New York: Academic Press, 1970, pp. 175–226.

Schultz, A. H. "Some factors influencing the social life of primates in general and early man in particular," in S. L. Washburn (ed.), *Social Life of Early Man,* Chicago: Aldine, 1961, pp. 58–90.

Simons, E. L. "On the mandible of *Ramapithecus,*" *Proc. Nat. Acad. Sci. U.S.A.,* 51: 528–36, 1964.

Tattersall, I. "More on the ecology of North Indian *Ramapithecus,*" *Nature,* 224: 821–22, 1969.

Tobias, P. V. "Early man in East Africa," *Science,* 149: 22–23, 1965.

Van Valen, L., and Sloan, R. E. "The earliest primates," *Science,* 150: 743–45, 1965.

Wood, A. E. "Eocene radiation and phylogeny of the rodents," *Evolution,* 13: 354–61, 1959.

3

Pliocene and Pleistocene Hominids

In Chapter 2 we considered the coevolution of plants and animals in the Late Mesozoic and Early Cenozoic through adaptive radiation in both the primates and the flowering plants of the evolving tropical rain forest ecosystem. Increasing aridity led to the development in this ecosystem of valley grasslands, on which there evolved pongid and hominid forms, differentiating from a common anthropoid stock. Despite the long geological interval that has since elapsed, certain characteristics of these early hominoid forms, and in particular of *Ramapithecus,* appear to be identifiable in our own contemporary species. This chapter considers the evolution of Pliocene and Pleistocene hominids from this point on, and their continuing divergence and differentiation, which resulted in their acquisition of still further human features.

The Pliocene-Pleistocene Boundary

The duration of the Pliocene epoch has not yet been accurately determined. The Miocene is believed to have ended some 12 to 13 million years ago, and the Pleistocene until comparatively recently was thought to have begun about 1 million years ago. However, with the increasing availability of relevant absolute dates for both biological and inanimate material, a growing consensus now places the Pliocene-Pleistocene boundary at somewhere between 3½ and 4 million years ago. As this postdating has not generally affected the time estimates of the onset of glaciation (currently the first traces of significant glaciation are placed between 2.5 and 3.0 million BP), the interval of the Pleistocene that preceded the ice ages, generally known as the Villafranchian, has consequently automatically become greatly extended. The situation is further confused by the fact that "Villafranchian" is sometimes applied to the last epoch of the Pliocene. If this system is followed, the Pleistocene epoch commences with the first evidence of the onset of cold conditions and may be assessed as having lasted 3.0 million years.

The Villafranchian

The Villafranchian is essentially a stratigraphically defined geological interval recognized by the contemporary occurrence of the remains of the four genera *Equus* (horse), *Elephas* (elephant), *Camelus* (camel), and *Bison* (bison). As these genera are believed to have originated in the New World and subsequently to have spread from there to the Old World, their occurrence in Eurasia implies some continuity of the Bering land bridge, and therefore a significant alteration of the relative levels of sea and land in that region.

The periods of continuity between Eurasia and North America appear to have become successively shorter as the Cenozoic progressed, and the opportunities for transcontinental migration accordingly became more restricted. After the last of the four genera to evolve, *Elephas*, had migrated, there was apparently no major faunal movement across this Bering land bridge apart from that of *Homo* himself.

Many authorities now consider that the start of the Villafranchian, and therefore of the Pleistocene period, can be dated from the time when *Elephas* is first noted in Old World deposits. This time has yet to be confirmed by potassium/argon dating, but it appears likely that it was somewhere around 4 million BP (Table 3-1).

Dating Techniques

The radiometric measurement of age by the potassium/argon method utilizes the estimated decay rate of the radioactive isotope of potassium, K40, to form argon (described in Appendix I). In addition, there is now a dating method for this period employing an analysis of deep-sea sediments. This is based on the circumstance that climatic fluctuations are

TABLE 3-1 *Estimated Chronological Range of Principal Hominid Grades*

There is still some uncertainty as to the precise dating of these geological periods, which is really of no overriding concern. As seen from this table, there is even more uncertainty of the precise dates of appearance and extinction of particular hominid "grades."

Period	Grade	Approximate life span of "grade" or species
Holocene 0 to 30,000 BP	*Homo sapiens*	40,000 years
Pleistocene 30,000 to 4 million BP	*erectus-sapiens* (2 million to 40,000 BP)	? 2 million years
	Homo erectus (4 million to 2 million BP)	? 2 million years
	Paranthropus species *Homo africanus* (? continuing from Late Pliocene, relicts persisting perhaps to about 1 million BP) ↓	3 million years
Pliocene 4 million to 12–13 million BP	(? found in Late Pliocene)	
	(continuity into Early Pliocene) ↑	
Miocene	*Ramapithecus* (Late Miocene)	? 5 million years
	Proconsul (Middle Miocene)	? 5 million years

reflected in the planktonic composition of these sediments. If changes in the sediments can be correlated with a number of independently established dates, an absolute time sequence of sediments can be produced. Using such comparatively new methods of dating, which are further described in Appendix I, the time of the first major glaciation, the Günz (which ended the Villafranchian), is now placed at 1½ million years ago.

Tropical Time Sequences

The somewhat indefinite time schedule for the Pleistocene is further confused when one turns to the tropics, and correlations with tropical African sequences are still very incomplete (Table 3-2). It is therefore somewhat

TABLE 3-2 *Correlations Between Regional Climate Sequences in the Pleistocene*

These still are uncertain. The original basal sequence, which referred to glaciations in the Swiss Alps, now must be correlated with nonalpine continental and alpine glaciation occurrences in temperate North America and Eurasia, and with wet and dry periods in the tropics and subtropics. Correlations and absolute dating shown here for the several wet and dry episodes in African Pleistocene prehistory are still largely conjectural.

Pluvial sequences in tropical eastern Africa	Glacial intervals in Swiss Alps	Approximate absolute datings in 1000 years BP
Dry phase	Interstadial	10
Gamblian pluvial	Würm	70
Dry phase	Interstadial	
Kanjeran pluvial	Riss	300
Dry phase	Interstadial	
Kamasian pluvial	Mindel	500
Dry phase	Interstadial	
Kageran pluvial	Günz	750

difficult to integrate into a time sequence the next series of Pliocene and Early Pleistocene fossils to be discovered, which are for convenience commonly lumped together under the general term of *australopithecines*. This word means "southern apes" and implies no special relationships with either pongid (ape) or hominid (human) lines.

Australopithecines

The first of the australopithecines to be correctly so designated was found at Taung, in the northernmost part of the Cape Province of South Africa. It was the skull of a juvenile animal about 6 years old and was described by a professor of anatomy, R. A. Dart. With the amazing combination of skill and intuition characteristic of the several workers who have been associated with such hominid finds, Dart immediately recognized in this skull of a 6-year-old a form of ape entirely different from any previously discovered in early Pleistocene deposits, and he named it *Australopithecus africanus*. It is the first of a group of what are often called "slender" or "gracile" australopithecines found especially in the Transvaal Province of South Africa, which will probably come to be known as *Homo africanus*, although the nomenclature in this hominid line is excessively confused at the present time. Indeed, it seems cavalier to engage in taxonomic lumping and so casually to dismiss the arguments arising from the patient, scholarly, and skilled work devoted to each of these still rare hominid finds. However, some simplified theme must be identified. Although about 150 australopithecine sites have so far been discovered, combining all the bones together would still not produce a complete, composite skeleton. Only in about three sites are enough individuals represented to begin to get some idea of population variation. Under these circumstances some workers, such as Richard Leakey, consider it useless for the present to go further than to call all "slender" types *Homo* and all "robust" types australopithecines. It does seem possible to go a little further, and as Robinson has claimed that *Homo africanus* is already a legitimate name, it can be applied without creating further taxonomic problems to the earliest "slender" forms while placing all "robust" types in the equally legitimate genus *Paranthropus*. Alternatively the cut-off can be made later, as Pilbeam does, naming all Pleistocene hominids through *habilis* as *Australopithecus*. Or it is possible to follow Birdsell and call all the earlier hominids australopithecines, making the group pithecanthropines to accommodate the different forms of what used to be called *Homo erectus* and placing Neanderthals and modern humans as different subspecies of *Homo sapiens*.

Homo africanus

Subsequent findings of skulls and partial skeletons of mature animals reveal that the hominids included here within the aggregate species *Homo africanus*, which appear to have been extensively distributed in southern Africa from 2½ million to 3 million years ago, were about one half or one third the size of present-day Americans. Their weight is estimated to have varied from about 60 to 120 pounds. They had pronounced bony ridges above their eyes, and jaws that were massive in relation to the size of their skulls. Although these last two features would appear to suggest their placement in the pongid line, their brains were significantly and substantially larger than those of modern apes in comparison to their size; estimates of cranial capacity vary from 600 to 900 cubic centimeters. On this basis alone they are best regarded as hominids.

Examination of skulls of *H. africanus* shows that the spinal cord entered the bottom rather than the back of the skull (Figure 3-1), so that the head was carried erect. The implication that this hominid walked upright was confirmed when a complete human-like half pelvis was discovered in 1947 (Figure 3-1). Coupled with a strong reduction of the canines, which, while still large, scarcely projected beyond the incisors, this led to the deduction that *H. africanus* was an extensive tool user, or, in contemporary jargon, object user. (Figure 3-2).

The Earliest Occurrence of *H. africanus*

Although many of the earlier australopithecine finds were estimated to be about half a million years old, no absolute dates are available because the breechias in which this South African material occurs are not volcanic rocks, and the potassium/argon technique cannot be used. Some australopithecine remains at Olduvai Gorge and various finds around the shores of Lake Rudolf in Kenya were dated between 1.75 and 2.5 million years BP. Excavations in 1967 and 1968 by F. Clark Howell and his Chicago University group in the Omo Valley of Ethiopia, reported on by Richard Leakey, suggest that australopithecines of both *H. africanus* and *Paran-*

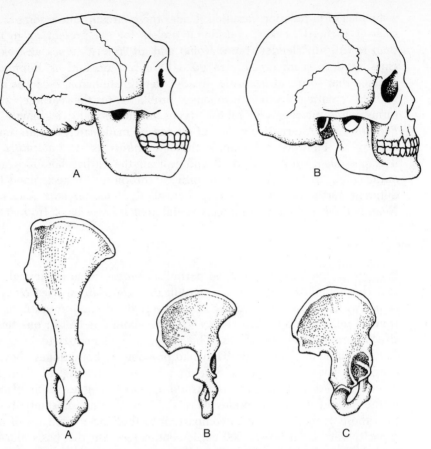

FIGURE 3-1. *The skull and pelvis of* **Homo** **africanus** reconstructed and drawn from a number of fragmental skeletal remains. Above: Skull of *H. africanus* (*A*) compared with that of *H. sapiens* (*B*). Below: Half pelvis of *H. africanus* (*B*) compared with that of a gorilla (*A*) and *H. sapiens* (*C*). This hominid form has been variously referred to as *Australopithecus transvaalensis, Homo transvaalensis,* and a number of other epithets. Its cranial capacity exceeded that of any known contemporary or fossil pongid and nearly overlapped that of contemporary man, so that it is now generally placed in the genus *Homo.* Although *H. africanus* was an extensive tool user, conclusive evidence of tool manufacture is lacking, and for the purposes of this text, tool-using forms such as this are all placed under this one species. Fossil remains have been found over a wide area of tropical Africa, and variously dated from about 4 million to some ½ million years BP.

thropus types (see p. 75) existed nearly 4 million years BP. Although the full significance of this postdating of australopithecine appearance relative to the revised estimates of the Pliocine-Pleistocene boundary has not been asessed, the broad hypotheses of australopithecine relationships discussed here seem valid. The one major difficulty remaining is that what some workers regard as a "gracile australopithecene" in this Omo-Rudolf area dates to approximately 5.5 million years BP. If it is a Pliocene *H. africanus* this is not disturbing. If on the other hand, as some specialists claim, these

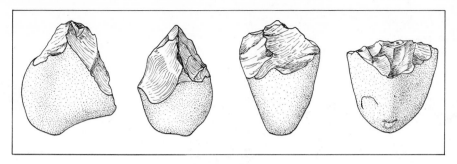

FIGURE 3-2. *Pebble tools* are primitive tools that could have been manufactured
artifacts made and used by *Homo africanus* or accidentally fractured pebbles
that were merely selected when needed. Pebbles such as this have been found
in various African localities in association with remains of several *Homo* species,
and also with those of *Paranthropus*. One-quarter natural size.

Pliocene finds represent the next form to be discussed, *Homo habilis,* then
relationships described here cannot be correct.

Olduvai Discoveries—*Homo habilis*

After the Transvaal finds, and before the recent Ethiopian and Kenyan
ones, discoveries were made elsewhere in Africa of Early Pleistocene
hominids. The most extensive were those unearthed in the Olduvai Gorge
in northern Tanzania beginning with *Zinjanthropus boisei* in 1959 by
L. S. B. Leakey, his wife, and his colleagues. Some of these were estimated to
be even older than the Transvaal material, potassium/argon dating of one
giving an age of nearly 2 million years. Unlike the South African material,
the remains were extensively and undeniably associated with pebble tools,
and gradually it was realized that they fell into two fairly distinct groups.

To state very simply what is coming to be a generally accepted pattern
for these varied finds, on the one hand is a hominid group "gracile" of
smaller body size but with a proportionally larger brain and less pronounced
jaws and brow ridge, which is generally classified as one or another species
of the genus *Homo*, creatures with an omnivorous diet. On the other hand
is a second hominid group ("robust") with a larger body size and more
massive skull, with proportionally a rather lower brain capacity, whose
diet was largely vegetarian; this group is placed in the genus *Paranthropus*
(Figure 3-3). The dentition of *Paranthropus* shows wear on the crowns
of the molar teeth, which indicates that these creatures could *chew*, a
distinction from modern pongids whose projecting peglike canines prevent
such a rotary action of the jaws, as noted previously.

Pebble Tools

In 1959 Mary Leakey found fragments of a skull, subsequently allocated
to *Zinjanthropus boisei,* lying on a working floor for the manufacture of

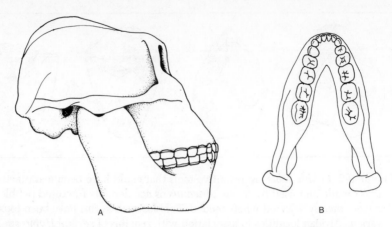

FIGURE 3-3. *Skull (A) and lower jaw (B) of* **Paranthropus** whose remains cover approximately the same time span as *Homo africanus,* 4 million to ½ million years BP. They are distributed more widely over the Old World, however, reaching from tropical Africa to as far east as central China and into Indonesia.

pebble tools. This association of pebble artifacts, which Mrs. Leakey now prefers to call *choppers,* is cited as the first acceptable proof of what archaeologists call a paleolithic or early stone-age culture that can be linked with any australopithecine. In 1960 less crude tools were found associated with another type of australopithecine, which because of this greater tool-making ability L. S. B. Leakey named *Homo habilis.* From essentially tool-using and perhaps occasionally tool-making *Homo africanus,* a cultural ritual of extensive tool manufacture had now been clearly established with *Homo habilis.* Unlike the Oldowan chopper, the cutting edges of this second class of artifacts had been shaped on both sides. The more sophisticated tool kit now included—besides hand axes—choppers, picks, cleavers, awls, and even spheroids that might have been used as bolas. The bolas are known to have been used for hunting in South America. Two or three stone balls are linked by leather thongs; thrown at a fleeing animal, they may become entangled around its legs and bring it down.

Asian Australopithecines and other Hominids

The crude pebble tools that provide evidence of the earliest forms of human culture have now been sequentially arranged from artifacts discovered at Olduvai, and the culture has been termed *Oldowan.* The adaptive success of even the crudest of these tools, coupled with anatomical adaptions freeing the forelimbs for their manufacture, transport, and use—and perhaps associated with the development of other behavioral traits such as speech—is evidenced by their extensive occurrence from the Mediterranean coast to the Cape of Good Hope. Later and less crude forms are found even more extensively in many of the tropical and subtropical regions of the Old World. Australopithecine remains also have been discovered there.

Peking Man

Some Chinese hominid finds were made at the same time, a few even previously to that of the African forms, but they were mostly too scattered to permit the construction of any substantial hypothesis before the African story had been partially pieced together. The best-known of these Chinese forms is "Peking man."

In 1927 several "human" teeth were discovered in a cave near the town of Choukoutien, about 30 miles southwest of Peking. After studying one of these teeth, Davidson Black, a Canadian anatomist at the Peking Union Medical College, stated that it belonged to a human species, which he named *Sinanthropus pekinensis* (Figure 3-4). Peking man is thus not an australopithecine, but belongs to the group for which some workers have established the genus *Pithecanthropus*. Excavations over the next 10 years in the same area revealed not only a rich series of fossil hominoid remains but also hearths with which were associated butchered and roasted animals. These various forms were painstakingly and accurately described and illustrated by a German refugee, Franz Weidenreich. This was fortunate, because virtually all the original fossil material crated for transport to the United States for safer storage disappeared at the time of the attack on Pearl Harbor and has never been recovered. This circumstance was almost as disastrous as the experience of J. T. Robinson in South Africa, who, returning to begin a new season's excavations, discovered that the cave limestone he was working for hominid remains had been sold to a toothpaste manufacturer.

Java Man

In 1890, many years before the work in China, a Dutch anatomist, Eugene Dubois, inspired by reading Darwin's *Descent of Man*, had searched for and found the first remains of what came to be called "Java man." Dubois recognized the hominid nature of his finds and placed them in a new species, *Pithecanthropus erectus* (erect ape-man). The significance of his finds was lost because European "authorities" attacked Dubois' ideas so heatedly that he eventually retreated into silence for 30 years. Java man was not regarded in its true light until similar fossils were discovered, some close to the site of Dubois' original find. Most of the Indonesian material is judged to

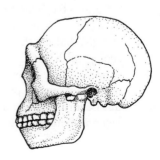

FIGURE 3-4. *Skull of the earliest form of Peking man.* Although later material from the same deposits seems to belong definitely to *Homo erectus,* the earliest remains from the Choukoutien series can probably be considered as referable to *Paranthropus.* They are tentatively dated to about 600,000 BP.

fall within the species *Pithecanthropus* (*Homo*) *erectus,* as also does Peking man. Several more recent finds may possibly be australopithecines of the "gracile" type, and resemble *Homo habilis.* Too few remains have yet been found to be sure this earlier hominid reached Indonesia, but there are no present theoretical grounds for supposing such a dispersal impossible.

European Australopithecines

It appears that the split-pebble culture, which hominids adopted toward the close of the Lower Pleistocene and which persisted at least 600,000 years, conferred a great ecological advantage and led to extensive hominid dispersal over the Old World. Such a wide distribution is not common among animals, but it was achieved by other carnivores, for example, *Panthera* (lion) and very large herbivores such as *Elephas* (elephant).

To date no hominid remains in Europe have been clearly related to the pebble-tool cultures encountered there, but one fragmentary find received attention that history may consider beyond due proportion to its significance. In 1907 a hominid lower jaw was found in a sand pit in the village of Mauer, close to Heidelberg in West Germany. It was associated with an extensive megafauna including elephant, rhinoceros, horse, sabertoothed tiger, hippopotamus, and giant beaver. Even without modern methods of radiometric dating, this permitted placement of the find in the interglacial interval between the Mindel (second) and Riss (third) glaciations of the middle Pleistocene. A new species of *Homo, H. heidelbergensis,* was defined by O. Schoetensack at Heidelberg University on the basis of characteristics exhibited by this jaw. Like Peking man, *H. heidelbergensis* is far from being an australopithecine and is also now placed in *Homo erectus,* discussed in the next chapter, which deals with the later stages of human evolution.

Trophic Displacement Among Early Hominids

From the fragmentary evidence of the scattered African remains discussed here, it would appear that over a wide geographical area of central and southern Africa, as perhaps in certain other regions of the Old World, various hominid species, first assigned by their discoverers to many genera but here reduced to either *Paranthropus* or *Homo,* were sympatric (Figures 3-5 and 3-6). This being so, it may be surmised that there was competition between at least some of these coexisting hominid species, because their upright habit, their tool-using capacities, their occupation of a similar ecological habitat (valley grasslands scattered in wooded or forested landscapes), their more or less omnivorous diet, and probably their similar troop behavioral patterns, would bring them into competition. Whenever this occurs, ecologists have postulated that there must be *character displacement* if both species are to survive, as was described in the introductory chapter. That is to say, where sympatry develops, and two or more populations live in the same habitat, there has to be modification of a particular character or characters that causes further niche diversification between the competing species. These then come to occupy slightly different ecological niches, and the species populations are thus able to avoid direct and annihilating competition.

The concepts of competitive exclusion and niche diversification can be applied in the case of *Homo* and *Paranthropus*, if the sweeping assumption is made that all these Pliocene and Early Pleistocene hominids can be allocated to one genus or the other. W. M. Schaffer has suggested that *trophic displacement* or a behavioral diversification occurred. Whereas the species of both genera originally had an omnivorous diet, members of the genus *Homo* became progressively more specialized in object use and eventually in tool manufacture. They used this greater technology to maintain their omnivorous diet and remained essentially both primary and secondary consumers (Figure 3-7). Contrariwise, species of the genus *Paranthropus* became increasingly and eventually exclusively vegetarian, i.e., primary consumers, and did not greatly extend whatever object-use ability they originally possessed.

Put in slightly different terms, there was at first a jack-of-all-trades (*Homo africanus*) who was tool using and omnivorous. This jack came to be sympatric, i.e., coexisted, with a somewhat similar jack of a different taxonomic origin (*Paranthropus*). The latter was therefore forced by competition to adopt an exclusively vegetarian diet, becoming *P. robustus*, while the first jack became a tool maker and accentuated the carnivorous element of its diet. This hominid jack (*H. habilis*) then evolved into a specialized tool-making carnivore (*H. erectus*), while *Paranthropus*, with competition thus reduced, survived and evolved before passing to extinction about half a million years ago into the larger form known as *Australopithecus* (*Paranthropus*) *boisei*. The development of large molars with thick enamel in this "robust" line of australopithecines could be explained if it is accepted that a significant portion of their diet was grass and other small seeds that they chewed.

The Aggression Theory

A rival theory explaining the origin of forms attributable to the genus *Homo* must be mentioned; it was presented by Robert Ardrey after reviewing the work of R. A. Dart. This theory is based on the discovery by Dart in South Africa of many fossil baboon skulls, 42 in fact, that had been battered in. The holes in them fit the indentations that could be made using the distal end of the femur of a small antelope. Ardrey and Dart maintained that such femurs were present in significantly larger proportions than they should have been if they were not used as weapons in this way. They presented the thesis that cultural behavior attributable to the genus *Homo* evolved when australopithecines discovered how to use a tool to destroy other sympatric animals, and especially other and competing an-

FIGURE 3-5. Paranthropus *male and female.* The stature is upright and locomotion bipedal, teeth large, causing prognathy, but canines not exceeding the length of the other teeth, brains about the size of modern gorilla brains. Total weight probably averaged about 140 pounds in mature males, somewhat less in females; skin color black. Time period about 2 million years BP, distribution Old World tropics.

thropoid species. Thus, they claimed, originated the aggressive behavioral feature that has since characterized all hominid evolution. Ardrey subsequently extended this theory by incorporating it into a *territorial* concept. There are two interesting side issues arising from the evidence utilized in this aggression theory. The first is that the position of the holes in the baboon skulls suggested the blows causing them had been struck by a usually right-handed creature. All modern apes and monkeys are ambidextrous. Second, nocturnal baboon hunts are known to have been extensively practiced at one time or another by various human groups. In fact they represent the one exception to the otherwise diurnal predation habits of all known prehistoric human societies.

Whether the emergence of *H. erectus* resulted from trophic displacement, the development of aggressive or territorial behavior, or some combination of these, it would have been impossible unless associated with the further development of speech and language.

Figure 3-6. Australopithecus (Homo) africanus male, female, and infant, contemporary and sympatric with *Paranthropus* illustrated in Figure 3-5. Similar upright stature and bipedalism, black skin pigment, hairiness, prognathy and large teeth, but with less massive jaws and slighter stature, averaging about 100 pounds. Tool using; shown here bashing bones with stones so as to permit extraction of bone marrow, but little different from *Paranthropus* in the respect. Time period the same, about 2 million BP, distribution also Old World.

Evolution of Language

Hockett and Ascher have described how language may have evolved from a call system such as it seems reasonable to suppose persisted from the *Proconsul* grade through *Homo africanus* and *H. habilis*. They presume that many generations of chattering in an increasingly elaborate call system favored selection of variations which increased the sensitivity of the vocal tract. This greater sensitivity eventually made possible the insertion of minor variations within the sequence matrix of the call system. It became necessary to listen for these variations and it was no longer possible to identify a message completely from the opening cadences of the call. In this way true language developed from within a call system, still retaining features such as intensity, pitch, and variation.

The basic similarities of all modern languages, Hockett and Ascher argue, is evidence of their common origin. They believe language originated

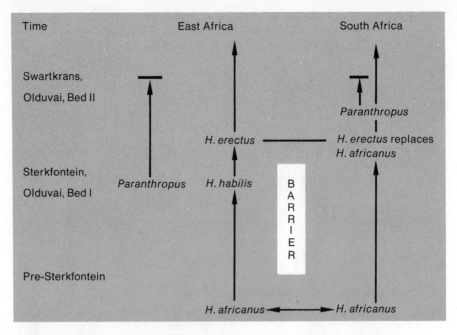

FIGURE 3-7. *Trophic displacement in the genus* Homo illustrating the time sequence and stratigraphical relationships of the several lower Pleistocene hominid species. This is one of several possible schemes that can be prepared to fit the still scanty information on these early human forms.

but once, about 2 million years ago, and the increase in cranial capacity in hominids, observed from then until about 40,000 years ago, is associated with the gradual elaboration of language and other cultural features. Two other researchers whose work will be discussed later, P. Lieberman and E. S. Crelin, have emphasized that anatomical modifications of the pharyngeal area were also a necessary prerequisite for the full development of speech.

An interesting observation made by A. H. Schultz is that the development of language provided for continuing selection of a sense of hearing. This was necessary not for the detection of *external* sounds, the approach of possible predators, but to sense *social* sounds emitted by other members of the band. This suggestion is in agreement with more recent observations on the maintenance of spatial patterns in primates such as were discussed in the previous chapter. The sense of smell, having no such selective advantage, continuously declined from prosimian grades, although the custom of rubbing noses still practiced among some human groups may be a ritual survival of this vestigial sense.

Significance of Language

There is great significance in the creation of language in the early hominids and its gradual development and perfection as a precise means of communication in the various forms of *Homo erectus*. Perhaps because of the immense archeological effort devoted to the study of artifacts, the effects of evolution and use of various types of tools have been somewhat over-emphasized. Such expressions as "man the toolmaker" constantly underline this latter aspect of human behavioral evolution, and there is little or no reference to "man the language user." As will be discussed in later chapters, some recent investigations on the capacity of the pharynx in various hominid forms, taken together with psychological studies of brain functions, are directing more attention to this aspect of human evolution.

It is possible today to recognize between 5,000 and 10,000 different languages, the actual number depending on the distinctions made between a language and a dialect. Whatever this total, all known modern languages and dialects are exceedingly complicated, and none could be labeled "primitive." For example, the language group used by Australian Aborigines embodies one of the most formally complex grammatical structures known. All languages are structured to impart accurate information on the past, the present, and the future. This is in direct contrast to the evolution of tool manufacture, where even in the contemporary world examples may still be found of peoples with paleolithic, mesolithic, and neolithic industries.

Language leaves little trace of its existence unless it is transcribed in a written form, but it seems likely that in early hominids the development of languages was an essential precursor to the further evolution of social structure beyond that described for nonhuman primates and to the more material aspects of cultural evolution such as tool making. To illustrate this probability rather bluntly, in modern society a deaf and dumb child is as much handicapped in acquiring social learning as a blind child. Until the invention of written forms of language, we learned primarily by listening and asking questions. This is still the basic form of instruction followed in our early schooling.

Historical Linguistics

The first modern studies of languages took the form of what is called *historical* or *comparative linguistics*. Developed toward the middle of the nineteenth century, this discipline explored the relationship between languages by identification of *loan words*, together with phonetic evolution and divergence. Loan words are those transferred from one language to another. This is done more particularly by bilingual individuals whose parents spoke different languages, but also by travelers, traders, and invaders.

Studies on comparative linguistics were able to trace modern languages back some 2,000 or 3,000 years, relating them to the common sources from which they derived, and sometimes identifying the base languages, but usually hypothesizing the existence of a common language no longer spoken.

Because the discipline was studied especially by European scholars, the first language family to be thus explored was the Indo-European group, including ancient languages such as Hittite, which is extinct, and Sanskrit, which is known in written form only. Others were Persian, Greek, Italic dialects, which survive in the Romance languages, the Celtic languages, and the Germanic and various Slavic languages as well as the Semitic,

Aramaic, Arabic, and Ethiopian group. To illustrate how certain features are common to all these languages, the word for "mother" essentially has an "m" sound as the French *mère*, Spanish *madre*, German *mutter*, Latin *mater*, Arabic *um*, and so forth.

Language Transmission

It must be remembered, however, that language is a ritual of human behavior. Like all rituals, it has to be learned, and between one individual and another this learning process is imperfect. Language also becomes modified in use so that it is always changing from generation to generation. There is no such thing as *correct speech* or *correct language;* there is only *contemporary speech* and *contemporary language.* Meanings, words, pronunciation, and form change continuously.

Until written languages were developed some 4,000 years ago, beginning with the Sumerian pictographical writing or about 5,000 BP, these modifications were not recorded. Now that they are recorded, it has become apparent how rapidly languages change. Biblical English to us appears formal and stilted; the language of Chaucer is alien even to individuals whose mother tongue is English. Classical Arabic as expressed in the writings of the Koran employs many words no longer used in, for example, colloquial Egyptian or Syrian.

Modern communications have generally hastened the rate of loan word introduction into many languages. They have also greatly stressed a limited number of widely used languages. What appears likely to happen within the next few centuries is the emergence of three or four languages—say, Russian, Spanish, English, Mandarin—as world tongues with which all scholars will be familiar. Other languages will probably gradually disappear, except in illiterate pockets of the world, either through disuse or from being swamped with loan words.

The Origin of Speech

An excellent presentation of the origin of speech has been provided by C. F. Hockett. He suggests that in order to investigate the origin of speech and language, it is necessary to study the basic design features of all communication systems whether they relate to humans, animals, or machines. Using this approach, it is possible to postulate the methods of communication of ancestral hominids and the evolutionary steps by which their communication system developed into language in successive hominid stages. The thirteen design features of any communication system that Hockett identifies are

1. Vocal-auditory channels
2. Broadcast, transmission, and directional reception
3. Rapid-fading transitoriness
4. Interchangeability
5. Total feedback
6. Specialization
7. Semanticity
8. Arbitrariness
9. Discreteness
10. Displacement
11. Productivity
12. Traditional transmission
13. Duality of patterning

The first five of these design features are common to the communication systems of all land mammals. Use of the *vocal-auditory* channel, for example, leaves the limbs free for other activities, thereby contrasting with the "bee dance," described by von Frisch, which requires the whole body. The second feature, *broadcast transmission,* means that the communication can be picked up by any auditory system within range. The *rapid fading* of the signal contrasts, for example, with the mapping out of a territory with excreta or some persistent glandular secretion. *Interchangeability* differs, for example, from communication during courtship displays, where the male can only transmit male signals and receive female transmissions, and vice versa. In rather the same way, *feedback* also differs from such communications as courtship displays, because the transmitter can listen to what he is transmitting, whereas the animal performing a courtship display is unable to view his actions.

To these five design features a *call system,* such as that found in primates like gibbons, adds three other features. The first of these is *specialization,* use of the transmission as a signal triggering a particular feeling. *Semanticity* still further defines this triggering and restricts it to a particular range of feelings; an example is a gibbon danger call. Call systems also show *arbitrariness* in that the sounds have no obligatory relation to the feeling being triggered. This contrasts with the bee dance, in which the movements are faster if the source of nectar is close, slower if it is farther away.

Discreteness, which is a feature of call systems as well as of language, refers to the use of a discontinuous rather than a continuous series of sounds. In the bee dance, the variation of signals is continuous rather than discrete. A given signal in call systems and languages does not merge continuously into another one, but is always discrete from it; if the signal is not made clearly it may be confusing and depend on the discretion of the receiver to interpret in one way or the other.

Early hominids, or protohominids as they are sometimes described, appear to have incorporated these nine features in a call system such as is now exemplified by modern gibbons. The problem of the origin of speech can therefore be defined as the determination of how and when the remaining four features of language were inserted into or superimposed on such a hominid call system.

Blending

The feature labeled *productivity* distinguishes what linguists call a *closed* system from an *open* one. Hockett and Ascher describe how *blending* can convert the former to the latter. They postulate the representation of acoustic contours of two calls, for example, by representing them with a series of letters ABCD for one call and EFGH for another. ABCD might mean "food is here," whereas EFGH conveys "danger is coming." The calls are so familiar that hearing only a portion can lead to a correct interpretation.

An early hominid encountering both food and danger might then stumble on a blending of these calls, using the first part of the one and the last part of the other in the form ABGH. Should this new call become established in use, then both the two old calls and the new one will have become *composite*. That is, in ABCD the first part AB comes to mean "food" and the second part CD to convey "no danger"; in the second call EFGH, EF says "no food" and GH communicates "danger."

Open Systems

If, as Hockett and Ascher suppose, the original closed call system had ten calls, when each of these blended once with each of the others, there would be a total of 100 calls. The system is still a closed one, but each of the 100 calls now has two parts, some of which recur in other calls. This is the basis for building composite signals, and is the essential feature of an open system that transforms it from a call system into a language. Once this openness has been achieved, the characteristic known as productivity— with its great potentiality—has been added. This potentiality cannot, however, be completely realized without some form of *traditional transmission*.

This pattern of learning may be observed in children, who are believed initially to acquire language in the form of a closed system. The child first appears to learn whole sentences as inflexible units; only later does he perceive the blending possibilities of these sentences and begin to develop the productivity of the language. This is also the basis of the modern method of foreign language teaching employed in schools. Foreign languages are now taught first through conversation, which provides the framework of whole sentences in the form of a closed system. As familiarity with the language proceeds, these set sentences are blended with others to provide a more refined and open communication system in the language.

Selection Pressures

In many accounts of human evolution much stress is placed on the strong selective pressures for manipulative skills in hominids that provided for more efficient and more effective tool making and tool using. It is possible that a comparable or even greater selection pressure developed for the characteristics of traditional transmission and productivity in speech. Genotypes that produced phenotypes eager to acquire by traditional transmission the full language insofar as it had been developed, and innovative in increasing its productivity, must have been more "fit" than genotypes whose phenotypical expression did not display these characteristics. They therefore would have been subject to strong positive selection pressures.

Duality of Patterning

The *duality of patterning* feature of language is necessitated by the small number of discontinuous sounds available for language construction, or the building of *morphemes*, which is the linguist's term for words or other units of meaning. To illustrate this, the meaningless sounds "n," "t," and "e" can be arranged in two ways in English to make meaningful morphemes, as "net," and "ten." Combinations of four or more sounds can be made into even larger numbers of morphemes.

Call systems do not have this language characteristic. Nor do they have the feature of *displacement,* which can trigger communication of events and objects in a different space or time. This is, however, a feature of some other known language communication systems, for example, the bee dance.

Language Sophistication

Various unsuccessful attempts have been made during this century to teach human language to chimpanzees. It has been assumed from these failures that brains at the ape grade are incapable of comprehending an "open" language system such as has just been described. Recent studies at Santa Barbara, California, by a man and wife team, the Premacks, working with a 7-year-old female chimpanzee, Sarah, suggest to the contrary that even at this ape level the primate brain can handle the symbolism of human language. The Premacks report that Sarah has a reading and writing vocabulary of about 130 words and that she understands the syntactic concepts of word class and sentence structure. Of course, the terms "reading" and "writing" are not used here in their normal connotation. Sarah "reads" from a vertical sequence of variously shaped and colored plastic symbols of different sizes. She "writes" by selecting from a collection of these symbols and arranging them sequentially. Thus a vertical sequence of symbols for the individual words "Sarah insert apple pail banana dish" produced appropriate actions such as a human would perform if given the written instruction "Sarah, put the apple in the pail and the banana in the dish."

In language use as in object use, it seems that our human species is not unique. As in our object use, which is unique only by virtue of the enormous expansion of our tool-making industries, so the uniqueness of our language use must lie in our enormous development of language. Speech patterns commensurate with this language developed only slowly; we did not fully evolve them until we reached our *Homo sapiens sapiens* stage, as will be discussed later.

Through the 3 or 4 million years of development of *Homo erectus* it may be imagined that these four features of *displacement, productivity, duality of patterning,* and *traditional transmission* were developing and refining an open communication system, in the form of what we know as speech or language, from a closed call system. The uniqueness of the genus *Homo* and its final evolutionary development is thus not solely related to progress in the manufacture and manipulation of tools. This by itself would probably have been insufficient to permit the rise of hominids appreciably beyond the pongid stage, if no better method of communication had been devised than a closed call system, even if supplemented with a sophisticated range of grimaces (Figure 3-8). The adaptive acquisition of an effective lan-

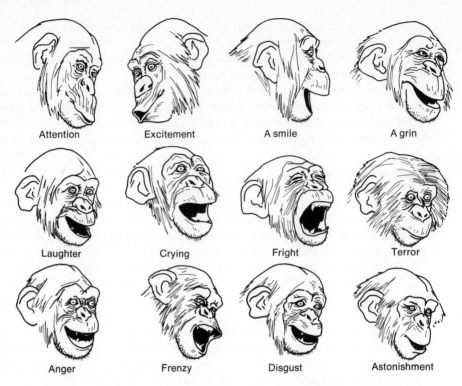

Attention	Excitement	A smile	A grin
Laughter	Crying	Fright	Terror
Anger	Frenzy	Disgust	Astonishment

FIGURE 3-8. *Facial expressions of young chimpanzees.* Before the evolution of an effective language, an intricate series of facial expressions probably evolved among early hominid grades, more sophisticated than that of chimpanzees illustrated here, but basically similar. The use of such facial expression was of significance in mother-infant relationships even after language was perfected and may also have caused the change from dorsal to the ventral copulation that probably characterized all hominids.

guage permitted the development, storage, and transmission of ritualistic behavior and so provided the opportunity for cultural evolution, which quantitatively is so uniquely human.

The previously noted absence of an *elementary* form of language in any living human group suggests that speech patterns had already achieved considerable sophistication in all populations of *Homo erectus* by Middle Pleistocene times at the latest. Further quantum cultural evolution in the form of permanent settlements permitted a further specialization of activity, which is evident in modern industrial societies and which could not have been achieved without an even more effective communication system. Nor apparently could our speech patterns have been made so effective without biological changes in the nature of the pharynx as our *H. erectus* grades evolved into *H. sapiens*.

Simultaneous with the perfection of speech and language in evolving Pleistocene hominids, other variations must have been occurring; one of the more important of these would have been the degree of skin pigmentation.

Skin Color and Its Significance

Variation in the skin color presently found in human groups has occasionally led to much speculation as to its causative factors. The most plausible explanation for the diversity in this characteristic has been given by W. F. Loomis, who believes that different types of skin color are adaptations to maximize ultraviolet light penetration in high latitudes and minimize it in low ones (Figure 3-9).

A world map of the extent of pigmentation among major groups of aborigine peoples, before the onset of the current waves of migration in the last few centuries, shows a marked correlation between skin pigmentation and latitude. The reversible summer pigmentation and keratinization known as *sun tan* is seen by Loomis as a means of maintaining a physiologically constant rate of vitamin D synthesis despite the great seasonal variation in the ultraviolet radiation experienced in high latitudes.

Vitamin D

The principal function of vitamin D in human metabolism is considered to be the regulation of calcium absorption from the intestine and control of the rate of deposition of inorganic minerals in unossified bone. This vitamin, sometimes known as the "sunshine vitamin," is produced in the skin, where very short ultraviolet radiation of wavelengths from 290 to 320 millimicrons transforms the provitamin 7-dehydrocholesterol into vitamin D.

This vitamin cannot be obtained in significant quantity in the more usual human diets. In high latitudes it is not present in any foodstuffs in winter; elsewhere it occurs principally in the liver of bony fishes, most familiarly as cod liver oil.

FIGURE 3-9. *Skin reflectance* as measured, using a reflectance spectrophotometer and expressed for a range of light wavelengths in European (Caucasoid); Chinese (Mongoloid); Burmese (Mongoloid); and Nigerian (Negroid) skins. The steep rise in reflection with longer wavelengths is due to reflectance from hemoglobin in subcutaneous blood vessels.

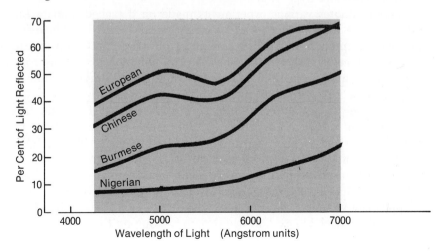

Absence of vitamin D synthesis for any significantly long period causes the disease known as *rickets*. In modern societies sufficient quantity of vitamin D to permit normal bone calcification is obtained through artificial fortification of milk and other foods.

Hypervitaminosis D

Whereas too little vitamin D causes the typical rickets symptoms of bowlegs, knock-knees, and twisted spines, especially in infants, amounts of vitamin D in excess of 2.5 milligrams per day result in symptoms of what is called *hypervitaminosis D*. In this pathological condition arising from an excess of vitamin D in the body, the levels of both calcium and phosphorous in the blood are raised. The condition is fatal once it affects the kidneys and kidney stones have developed.

Apparently human metabolic processes have no ability to regulate the amount of vitamin D in the bloodstream or to selectively secrete this substance to obtain the physiological regulation of its concentration in the body. Loomis and workers before him have proposed that the rate of vitamin D synthesis in the stratum granulosum of the skin is regulated by pigmentation and keratinization of the overlying stratum corneum. This controls the amount of solar ultraviolet radiation penetrating the outer layers of the skin and therefore determines the amount of vitamin D synthesized in the underlying region.

Experimental Evidence on Ultraviolet Reflectance

Loomis quotes the work of Beckmeier, who in 1958 estimated that 1 square centimeter of white skin could synthesize up to 18 IU (international units) of vitamin D in 3 hours. This rate would provide a sufficient daily dose of vitamin D for northern European infants who daily have no more than their faces exposed to "some fresh air and sunshine." Loomis proceeds to calculate that such an unpigmented individual would synthesize in 6 hours in the tropics up to 800,000 IU of vitamin D, eight times the amount sufficient to produce hypervitaminosis D.

Loomis quotes various experiments made on detached human skin that demonstrate that the degree of blackness as measured by skin reflectance is inversely proportional to the amount of ultraviolet radiation transmitted. It has further been shown that the common mutation for albinism in black human groups produces an individual whose skin ultraviolet transmission is within the range of that of the skin of nonpigmented groups.

From the various figures available, it is apparent that if only the face of a Negro infant is exposed in a high latitude winter climate, too little vitamin D is synthesized to meet its physiological requirement. It is therefore not surprising that evidence from medical experience finds Negroes the most susceptible of all peoples to rickets. These observations are now becoming theoretical in view of the general availability and use of cod liver oil to supplement vitamin D deficiencies in the diet.

The data available explain why Negroes living on the equator synthesize

5–10 per cent vitamin D as compared with unpigmented individuals under these same conditions, and why Negroes living in tropical climates do not suffer from kidney stones and other signs of overcalcification. Not all workers accept this explanation of skin pigmentation; some, like H. F. Blum, place a greater emphasis on negative correlations that have been demonstrated between pigment intensity and the susceptibility to skin cancer. There are certain other objections to what could be termed the Loomis theory of skin pigmenation; an extensive consideration of these, together with a full presentation of alternative hypotheses, has recently been published by W. J. Hamilton.

Origin of Depigmented Groups

Loomis proceeds to extend his observations and theories to account for the origin of white skin. He considers that because of their tropical primate beginnings, early hominids were probably both deeply pigmented and fur covered. (This is the supposition made in the previous chapter in discussing the movement of primates from a forest to a savanna environment.) He supposes that the northward migratory expansion of early hominids would have been halted somewhere near the Mediterranean because of the increasing tendency of infants to develop the bent legs and twisted spines of rickets. This factor would also limit the penetration of fully pigmented hominids to about latitude 40° S. The one modern exception to his rule is found in the Eskimo and Aleut peoples, who are moderately pigmented despite their occurrence in extremely high latitudes. Loomis accounts for this anomaly on the basis of the fish oil and meat diet of these groups, which provides an adequate but not excessive dose of vitamin D.

As melanin granules (which are black or brown) and yellowish keratohyaline granules (which produce nails) both reduce the amount of ultraviolet penetration and are controlled by polygenic systems, it may be supposed that any gene mutation in these systems would produce a phenotype with a somewhat lighter skin, which if inherited would permit still further permanent penetration of high latitudes. Thus, by slow increases in the gene frequencies of such mutant alleles, there would gradually evolve peoples with a demelanized and dekeratinized skin of maximum transparency who were capable of living in the highest latitudes without suffering from underproduction of vitamin D.

Morphological Evolution

So far we have been concerned mainly with the behavioral (language), physiological (melanin production), and cultural evolution of Pleistocene hominids, but the morphological evolution is also of interest, particularly because the changes that took place represented the last major morphological adaptations to occur in any *Homo* grade. Moreover, some morphological adaptations, it has already been noted, were associated with the improvement of speech and the greater sophistication of tool making.

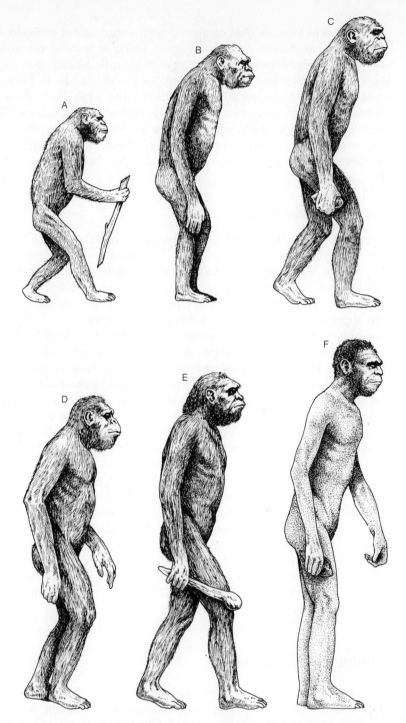

FIGURE 3-10. *Conjectural reconstruction of stages in the evolution of Pleistocene hominids* all found in tropical Africa. *A. Proconsul* circa 20 million BP. *B. Ramapithecus* circa 15 million BP. *C. Homo africanus* circa 1.8 million BP. *D. Paranthropus* circa 1.7 million BP. *E. Homo habilis* circa 1.5 million BP. *F. Homo erectus* circa 1 million BP.

The series of conjectural reconstructions of the body form in Figure 3-10 illustrate the comparatively small change in morphology that occurred during the Pliocene and Pleistocene. The Miocene mold, as shown by forms such as *Proconsul* and *Ramapithecus*, remained essentially the same, although substantial variations occurred in overall body size.

The most significant morphological changes occurred in the skull (Figure 3-11). The massive jaws, large teeth, and brow ridges of Miocene hominoids became reduced in later hominids whereas the cranial capacity increased. This produced a change in the center of gravity of the head and permitted a reduction in the size of both the nuchal (neck) and temporal (lower jaw) muscles (Figure 3-12).

These changes in the balance of the head were associated with the completely upright stature attained by final adaptations of the locomotor apparatus. The pelvis broadened to provide a greater attachment surface for the muscles that provide forward movement. The effect of this broadening was to shorten the ilium and bring close together the articulation points of the leg, the *sacral articulation,* and the *acetabulum* (Figure 3-13). From this came a bipedal striding walk unique among animals and a behavioral characteristic of the genus *Homo.* The morphological adaptation to bi-

FIGURE 3-11. *Evolution of the skull, dentition, and brain in anthropoids. A. Gorilla (Gorilla gorilla). B. Paranthropus. C. Homo africanus,* grade. *D. Homo erectus* grade. *E.* Contemporary man (*Homo sapiens*). In all the hominids the teeth, although still varying in size, have crowns attaining a common level, and tend to become smaller. The brain case and brain enlarge proportionally, and the sagittal crest disappears as the jaws are reduced. Simultaneously prognathy is also reduced.

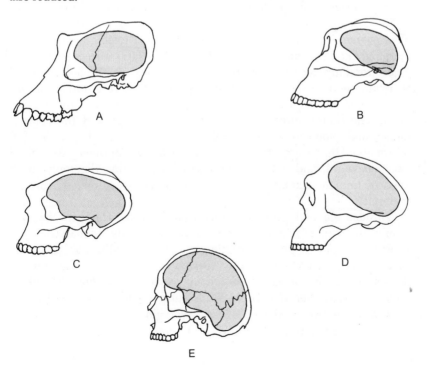

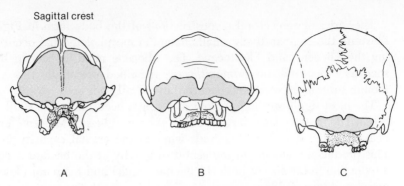

Sagittal crest

A   B   C

FIGURE 3-12. *Occipital view of some anthropoid skulls* showing how the neck muscles (the nuchal musculature) became reduced during *africanus-erectus-sapiens* evolution as the center of gravity of the head moved nearer the point where the head pivots on the occipital condyles with the top of the spine. This change developed from reductions in the at-first massive jaws and teeth, the thickness of orbital ridges and of the facial bones in general, through the adoption of a more upright carriage made possible by other relatively minor skeletal changes. The size of the nuchal muscles as judged by their area of attachment indicated here by stippling on the skulls of (A) gorilla (*Gorilla gorilla*), an orthograde primate whose spine still enters from the back of the skull. (B) *Homo erectus* grade. (C) *H. sapiens* grade. This provides evidence for archeologists and anthropologists as to the nature of the rest of the head, its muscles, and the posture and hominid and pongid forms. The sagittal crest, to which are attached the temporal muscles supporting the lower jaws, furnishes evidence of this same kind.

pedalism was probably completed, apart from some minor modifications, by the beginning of the Pleistocene, 4 million years ago (Figure 3-14).

The forelimb of hominids underwent even less change (Figure 3-15). The *scapula,* by which each forelimb is attached to the thorax, is very little modified in all primates. The *clavicle,* which permits a freer maneuverability of the forelimb, also is scarcely modified in any anthropoid. With all such skeletal structures, as described in Appendix II, a vast store of information on minutiae has been accumulated by long and careful study and measurement. The bones, through their articulations, origins of muscular attachments, and blood and nerve supplies, provide a great wealth of factual information that supports speculation on other morphological, physiological, behavioral, and culture features. It is probable that only in the science of human anatomy is so much information extracted from so little material.

Considerable and probably somewhat fruitless argument has revolved around the issue of whether *Homo habilis* was really "human" or not. There is no substantial weight of objection among paleoanthropologists to placing the hominid represented by the Leakeys *habilis* skull in the genus *Homo.* This specimen has a measured cranial capacity of about 657 cubic centimeters. Three other *H. habilis* skulls subsequently discovered at Olduvai have a range from 600 to 684 ccs., averaging about 642 ccs. Previously, "human" forms were held to have a minimum cranial capacity of 700 ccs.,

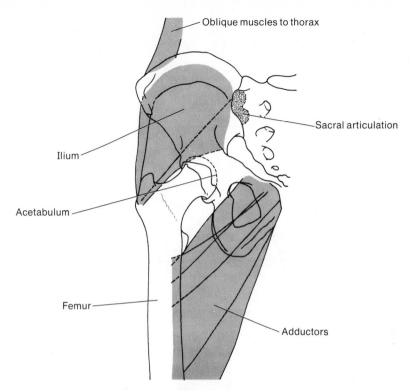

FIGURE 3-13. *The principal thigh muscles controlling lateral movement* in man (*Homo sapiens*) where all the weight is borne on one leg at a time at each step, the abductor muscles (upper shaded area) pulling the body forward in this action and the abductors balancing it.

and that of modern *H. sapiens* is known to range from approximately 800 to 2,000 ccs.

The Leakeys believed that at Olduvai *H. habilis* slowly evolved into *H. erectus,* the earliest forms of which had cranial capacities in the region of 750–800 ccs., thus overlapping with the lowest range found in comtemporary humans. Although *H. habilis* on this basis appears to have an undeniable claim to be regarded as "human," there remains the question of whether or not to exclude from this category the "slender" type of australophithecine, with cranial capacities ranging from 435 to 562 ccs., and a quite sophisticated tool kit. If this form is called, as has been done here, *Homo africanus,* it leaves the extensive and possibly aggregate *Ramapithecus* population as the only hominid form excluded from the genus *Homo* and not regarded as "human."

Mental Capacity

Some contemporary experimental work is available that is relevant to these deductions as to mental capacity of hominids made from morphological observations. Various experiments have been conducted to determine the

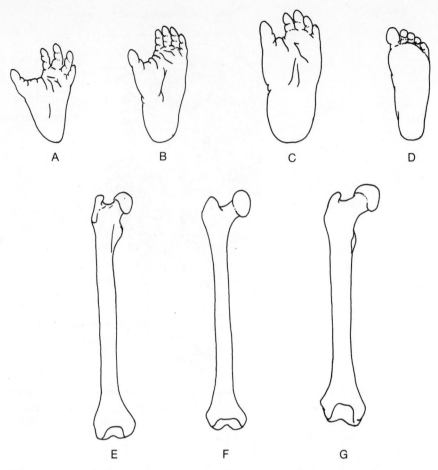

FIGURE 3-14. *Evolution of locomotor equipment* in hominids involved only minor changes over the past 15–20 million years once bipedalism replaced brachiation. Above: The outline of the feet of (A) the chimpanzee (*Pan troglodytes*), (B) the lowland gorilla (*Gorilla gorilla*), (C) the mountain gorilla (*Gorilla beringei*), (D) contemporary man (*Homo sapiens*). Below: The leg bone (femur) of (E) *Proconsul*, (F) lowland gorilla, (G) modern man drawn to the same size and therefore not to the same scale.

reasoning powers of monkeys, no species of which even begins to approach the cranial capacity of an australopithecine. The most ambitious of these studies, reported by Weinstein, tested monkeys' ability to identify both similarities and differences. Nine objects of different shapes and colors were placed on a tray. The monkey being tested was handed one of these as a symbol and had to select others that resembled it. Thus a triangle required all *red* objects to be selected, a circle all *blue* ones. The ability of the test monkeys to perform the necessary symbolic association revealed a relatively high power of reasoning.

Comparison of such performances at the monkey grade with those of contemporary man permits a crude correlation between hominid cranial

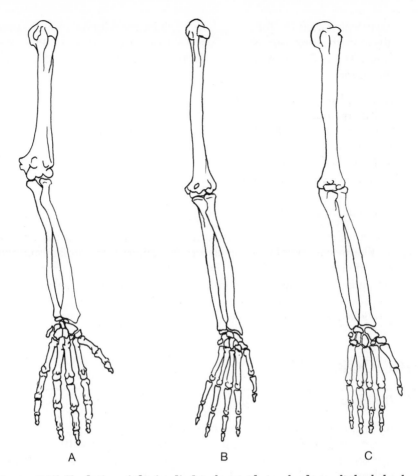

A   B   C

FIGURE 3-15. *Evolution of the forelimb* in hominids involved very little skeletal adaptation over the past 15–20 million years once bipedalism replaced brachiation as the commoner mode of locomotion. The forelimb skeletons of (*A*) *Proconsul,* (*B*) chimpanzee (*Pan troglodytes*), (*C*) contemporary man (*Homo sapiens*), drawn so as to appear the same length, show very little variation. Evolution, which has proceeded from a clumsy tool user such as *Proconsul* to form exquisitely delicate performers such as the violinist Yehudi Menuhin, has scarcely involved the basic skeleton structure while enormously refining the sensory perception and the muscular coordination.

capacity and probable mental ability. On this basis a reasoning capacity, but *not* a cultural knowledge, as high as that of contemporary man had probably been developed in hominids by 2 million years BP, for some cranial capacities of *Homo erectus* from that period, as has been noted overlap those observed in contemporary humans. The further evolution of *Homo erectus* and the reasons why it must be considered the direct ancestor of *H. sapiens*—or the changes occurring as the *erectus* grade evolved into the *sapiens grade*—are considered in the next chapter.

The earliest stages of the *H. erectus* grade, which could well have originated at least as long ago as 4 million years may be considered to have contributed the following human characteristics over and beyond those given in the previous chapter as persisting from Miocene-Pliocene hominoids. These characteristics evolved further in *H. erectus* and persisted into the *H. sapiens* grade:

1. Completely bipedal habit
2. Enlarged brain
3. Geographical variations in degree of pigmentation
4. Reduced canines
5. Basic evolution of speech and language
6. Improved tool-making dexterity
7. Social division of labor
8. Extended infant and juvenile dependence

It must be emphasized that with the appearance of the *H. erectus* grade in Pleistocene hominids, much of the observable human *morphological* evolution had been achieved. Of the above list, only the first four features are entirely morphological; the rest are at least partially behavioral. Further evolution was more concerned with behavioral than with morphological and physiological characteristics.

BIBLIOGRAPHY

References

Ardrey, R. *African Genesis,* New York: Delta, 1961.

Ardrey, R. *The Territorial Imperative,* New York: Delta, 1966.

Birdsell, J. B. *Human Evolution: An Introduction to the New Physical Anthropology,* Chicago: Rand McNally, 1972.

Blum, H. F. "Does the melanin pigment of human skin have adaptive value?" *Quart. Rev. Biol.,* **36:** 50–63, 1961.

Campbell, B. G. "Conceptual progress in physical anthropology—fossil man," *Ann. Rev. Anthrop.,* **1:** 27–54, 1972.

Clarke, R. J., Howell, F. C., and Brain, C. R. "More evidence of an advanced hominid at Swartkrans," *Nature,* **225:** 1219–22, 1970.

Dart, R. "*Australopithecus africanus*: the man-ape of South Africa," *Nature,* **115:** 195, 1925.

De Villiers, H. "The first fossil skeleton from South West Africa," *Trans. Royal Soc. S. Africa,* **40** (Part III): 187–96, 1972.

Greenberg, J. H. "Language universals: a research frontier," *Science,* **166:** 473–78, 1969.

Guilcher, A. "Pleistocene and Holocene sea level changes," *Earth-Science Rev.,* **5:** 69–97, 1969.

Hamilton, W. J., III. *Life's Color Code,* New York: McGraw-Hill, 1973.

Hockett, C. F. "The origin of speech," *Scientific American,* **203**(3): 88–96, 1960.

Hockett, C. F., and Ascher, R. "The human revolution," *Curr. Anthrop.* **5:** 135–68, 1964.

Howell, F. C. "Remains of Hominidae from Pliocene/Pleistocene formations in the lower Omo Basin, Ethiopia," *Nature,* **223:** 1234–39, 1969.

Isaac, G. "The diet of early Man," *World Archeology,* **2**(3): 278–99, 1971.

Jolly, C. "The seed-eaters," *Man,* **5**(1): 5–26, 1970.

Leakey, R. E. F. "Early *Homo sapiens* from the Omo River region of southwest Ethiopia," *Nature,* **222:** 1132–38, 1969.

Leakey, R. E. F. "Skull 1470," *National Geographic,* **143:** 818–29, 1973.

Loomis, W. F. "Skin-pigment regulation of vitamin-D biosynthesis in man," *Science,* **157:** 501–6, 1967.

Pilbeam, D. *The Ascent of Man,* New York: Macmillan, 1972.

Premack, A. J., and Premack, D. "Teaching language to an ape," *Scientific American,* **227**(4): 92–99, 1972.

Robinson, J. T. "Homo 'habilis' and the Australopithecines," *Nature* **205:** 121–24, 1965.

Savage , D. E., and Curtis, G. H. "The Villafranchian stage—age and its radiometric dating," *Geol. Soc. Amer.,* Special Paper, 1970.

Schaffer, W. M. "Character displacement and the evolution of the Hominidae," *American Naturalist,* **102:** 559–71, 1968.

Schaller, G. B., and Lowther, G. "The relevance of carnivore behavior to the study of early hominids," *Southwest. J. Anthrop.,* **25**(4): 307–41, 1969.

Schultz, A. H. "Some factors influencing the social life of primates in general and early man in particular," in S. L. Washburn (ed.), *Social Life of Early Man,* Chicago: Aldine, 1961, pp. 58–90.

Simons, E. L. "Assessment of a fossil hominid," *Science,* **160:** 672–75, 1968.

Simons, E. L., Pilbeam, D., and Ettel, P. C. "Controversial taxonomy of fossil hominids," *Science,* **166:** 258–59, 1969.

Tobias, P. V. *The Brain in Hominid Evolution,* New York: Columbia University Press, 1971.

Tobias, P. V. "The distinctiveness of Homo habilis," *Nature* **209:** 953–57, 1966.

Van Lawick-Goodall, J. *In the Shadow of Man,* Boston, Mass.: Houghton Mifflin, 1971.

Further Readings

Altman, S. A. "Social behavior of anthropoid primates: analysis of recent concepts," in E. L. Bliss, *Roots of Behavior,* New York: Harper, 1962, pp. 277–85.

Bishop, W. W., and Clark, J. D. (eds.) *Background to Evolution in Africa,* Chicago: University of Chicago Press, 1969.

Bishop, W. W., and Miller, J. A. (eds.) *Calibration of Hominoid Evolution* Edinburgh: Scottish Academic Press, 1971.

Campbell, B. G. *Human Evolution,* Chicago: Aldine, 1967.

Carpenter, C. R. "A field study of the behavior and social relations of the gibbon," Comparative Psychology Monographs No. 16, 1940 (republished in C. R. Carpenter, *Naturalistic Behavior of Nonhuman Primates,* University Park, Pa.: Pennsylvania State University Press, 1964).

Caspari, E. "Selective forces in the evolution of man," *American Naturalist*, **97**: 5–14, 1963.

Coppens, Y. "Les restes d'hominidés supérieures des formations Plio-Villafranchiennes de l'Omo en Éthiopie," *C. R. Acad. Sci. Paris*, **272-D**: 36–39, 1971.

Denes, P. B., and Pinson, E. N. *The Speech Chain: The Physics and Biology of Spoken Language*, New York: Doubleday, 1973.

Emden, J. M. "Natural selection and human behavior," *J. Theoret, Biol.*, **12**: 410–18, 1966.

Goodall, J. M. "Tool-using and aimed throwing in a community of free-living chimpanzees," *Nature*, **201**: 1264, 1964.

Harrisson, B. *Orang-utan*, London: Collins, 1962.

Holloway, R. L. "New endrocrinal values for the australopithecines," *Nature*, **227**: 199–200, 1970.

Klein, R. G. *Man and Culture in the Late Pleistocene*, Scranton, Pa.: Chandler, 1969.

Korn, N., and Thompson, F. (eds.) *Human Evolution, Readings in Physical Anthropology*, New York: Holt, Rinehart and Winston, 1967.

Kurtén, B. *The Ice Age*, New York: Putnam, 1972.

Kurth, G. (ed.) *Evolution and Hominisation*, 2nd ed., Stuttgart: Fischer, 1968.

Leakey, L. S. B., and Goodall, V. M. *Unveiling Man's Origins*, Cambridge, Mass.: Shenckman, 1969.

Le Gros Clark, W. E. *Man-Apes*, New York: Holt, Rinehart and Winston, 1967.

Loring, B. C. *The Stages of Human Evolution*, Englewood Cliffs, N.J.: Prentice-Hall, 1967.

Medawar, Sir Peter. "What's human about man is his technology," *Smithsonian*, **4**(2): 22–29, 1973.

Morton, D. J. *The Human Foot*, New York: Hafner, 1964.

Napier, J. *The Roots of Mankind*, Washington, D.C.: Smithsonian Institution Press, 1970.

Oakley, K. P. *Man the Tool Maker*, 6th ed., London: British Museum (Nat. Hist.), 1972.

Reynolds, V. *Budongo, an African Forest and Its Chimpanzees*, New York: Natural History Press, 1965.

Schaller, G. B. *The Year of the Gorilla*, Chicago: University of Chicago Press, 1964.

Spuhler, J. N. (ed.) *The Evolution of Man's Capacity for Culture*, Detroit: Wayne State University Press, 1961.

Washburn, S. L. (ed.) *The Social Life of Early Man*, Viking Fund Publications in Anthropology No. 31, Chicago: Aldine, 1961.

Washburn, S. L., and Jay, P. C. (eds.) *Perspectives in Human Evolution*, New York: Holt, Rinehart and Winston, 1968.

West, R. G. *Pleistocene Geology and Biology*, New York: Wiley, 1968.

Wolpoff, M. H. "The evidence for multiple hominid taxa at Swartkrans," *American Anthropologist*, **72**: 576–607, 1970.

4

Homo Erectus

With the emergence of *Homo erectus* the long story of human biological evolution is almost over, but as regards *cultural* achievement it is only just beginning. It took perhaps 1½ billion years for life to evolve on this planet, and another 3 billion years before animals appeared. Almost a further half billion years passed before primates developed, but then after only a further 57 million years forms referable to the genus *Homo* had evolved. From the first known culture (Oldowan) to the second (Acheulean), there was an interval of a mere half million years, which is approximately the time that then elapsed before the industrial revolution. From that point, a mere one or two centuries ago in most regions, quantum advances in human culture proceeded very rapidly, through the "atomic age" to the "computer era." The progress of cultural evolution has so accelerated, that from time intervals recorded in billions of years we are reduced to expressing the passage of time first in centuries, then in years, and now even in months.

Therein lies the uniqueness of man. Other forms of life have evolved, undergone adaptive radiation, been modified by character displacement from jacks to specialists in the process of niche diversification, and passed to extinction with the inevitable onset of environmental change to which they had become too specialized to adapt. Human populations have evolved essentially by *cultural* speciation. Trophic displacement and niche diversification therefore affect mainly cultural practices. As one culture moves to extinction, the morphological and behavioral characteristics of the jack remain, largely unmodified, to give rise to another cultural variant able to survive in the changed environment.

Our anatomical, morphological, physiological, and behavioral characteristics are not entirely immutable. The influence of selection pressures has undoubtedly produced modifications of gene frequencies and resulted in adaptation in these type of features, as have certain independently operating genetic phenomena. But such change is slow. From *Homo erectus*, through *Homo sapiens*, and on into the future, we really have but one biological population, slowly but continuously evolving under selection pressure; eventually behavior rituals will supply the knowledge to control these pressures. Only then shall we be able, if we wish, to stabilize gene frequencies and fully control further physical and behavioral evolution.

Physical Characteristics of *H. erectus*

It seems likely that most earlier forms of *H. erectus* were about half the size of the average modern American, weighing approximately 60 pounds. The carriage, as judged by the form of the pelvis, was just as upright, and the rest of the skeleton, apart from its small size, varied little (Figure 4-1). Estimates of cranial capacity range from 800 to 1,100 cubic centimeters, which overlaps that calculated for normal functioning in modern man (800–2,000 ccs). The large teeth also fall within the modern human range. Such proportionally large teeth would make the face somewhat prognathous (protruding), with a back-sloping chin and a receding forehead. For several reasons, *H. erectus,* at least in its later forms, must have been effectively naked.

The Naked Ape

This naked condition of man, unique among the 193 surviving species of monkeys and apes (Anthropoidea), which are otherwise covered in hairy fur, has inspired a variety of theories. One of the most plausible attributes our "naked" condition to our almost exclusively diurnal vision, which was almost certainly the case with *H. erectus*. Most predation by top carnivores is completed at dusk or dawn, and these larger carnivores all have some degree of night vision. Forced to operate, because of lack of night vision and a relatively slow running pace, in tropical or subtropical temperatures in full daylight at a speed more likely to wear down rather than overtake a prey, *H. erectus* would soon have built up too much metabolic heat to be dissipated without a pause in activity. With this *coursing* habit there would have been strong selection pressure for any mutants producing a reduced body fur covering. Supposedly there was a parallel and com-

FIGURE 4-1. *Confrontation between* Homo erectus *and* Paranthropus *at Olduvai in Tanzania.* The two *erectus* males have no body hair, and have a basic social division of labor that permits them to be away from their family base on a hunting expedition. *Paranthropus* retains body hair and lacks this division of labor; females and young all move with the band, each member of which gathers food only for itself. *H. erectus* would probably increasingly in geological time attack, kill, and eat young or disabled *Paranthropus* when opportunity arose, but not the contrary. While hunting, *H. erectus* would have the huge advantage of a greater facility in dispersing metabolic heat through his sweating and naked skin during a prolonged midday chase. Time period circa 600,000 BP.

pensatory selection of mutants for shivering, which is estimated to provide up to three times the resting heat production; otherwise the nights may have proved a little chilly before the cultural acquisition of fire and clothing.

Modern game animals normally rest in the shade during the heat of the day, and soon become exhausted if forced to run at this time. Any predator that has adapted to effective metabolic heat dispersal by loss of a furry covering and the development of sweat glands will obviously have a high rate of successful kills, even of animals many times larger, provided they can be stampeded into running. Nakedness would, however, have one other drawback in addition to the possibility of excessive chilling when at rest. Because it would permit still further exposure of the skin to ultraviolet rays, the possibility of *hypervitaminosis D* would be increased. For this reason it has been assumed that all the hominids considered here were heavily pigmented, even more so than the fur-covered hominoids such as *Proconsul*.

The fact that none of the other 192 species of monkeys and apes are hairless is explained according to this predation theory by the circumstance that none of them is so exclusively carnivorous. Although many may have a significant secondary consumer element in their omnivorous diet, this comes from eating insects, small rodents, scorpions, worms, and other small animals rather than larger ones that must be run down on foot by coursing.

The diverse opinions regarding many salient features of human evolution are well illustrated by contrary views on the origin of hairless skin. J. B. Bresler reproduced excerpts from eight letters to the editor of the journal *Science* prompted by a provocative statement of H. Bentley Glass referring to "loss of certain unnecessary structures, such as bodily hair once clothing was invented. . . ." Alternative suggestions accounting for the loss of hair ranged from that described here, through a nutritional explanation, to one involving ticks and body lice. As noted in the last chapter when considering skin pigmentation, such problems of what may be termed ethological physiology have recently been reviewed at some length by W. J. Hamilton.

Cultural Characteristics of *H. Erectus*

Desmond Clark has provided a graphic description of the life of a small group of *H. habilis* on the bare open mud flats bordering the Lower Pleistocene Olduvai Lake. The culture he describes persisted for more than 600,000 years from its first potassium/argon confirmed date of occurrence in 1,700,000 BP before being superseded by that of *H. erectus*. The essential features of the life of *H. habilis* are considered first to have been a hunting and gathering economy, that is, meat eating supplementing a vegetable diet, and second the collection in a central place of the results of foraging. Stone and more rarely bone equipment were developed to deal with this food supply. According to Clark, the temporary camp at Olduvai was made by several individuals only. The accompanying animal remains were juvenile or small mammals, frogs, lizards, and other creatures of similar size. Virtually every bone that could have contained marrow was broken (Clark notes that a modern African hunter-gatherer group is known to use bone marrow as the first solid food for infants). Artifacts used as implements in the camp were of five types: natural cobbles, "bashing stones" pebbles flaked to make choppers, and small nodules, which could represent the end product when fashioning the choppers and obtaining the last category, flakes, used in cutting and scraping (Figure 4-2).

Acheulian Cultures

Desmond Clark proceeds from this human culture of *H. habilis* to examine its successor, that of the more advanced and complex *H. erectus,* the "hand-axe culture" known collectively as Acheulian (Figure 4-2). It is believed that the African hunter-gatherer paleolithic culture of the Middle and Upper Pleistocene had appeared by 500,000 BP and in some places persisted until 50,000 years ago. By at least 60,000 BP, as determined by radiocarbon dating, *H. erectus* lived in medium-sized groups in open country marginal to water supplies, camping around open hearths. Wooden spears, clubs, and stones were used for hunting quite large animals (Figure 4-3), although dead ones were still scavenged, fruit and honey were gathered (Figure 4-4), and wooden digging sticks were used to unearth plant storage material. Large animals were probably driven into boggy ground, dispatched by stoning and spearing, dismembered, and eaten on the site (Figure 4-26). There had to be some form of language to permit coordination of these varied activities.

Behavioral Features of *H. erectus*

Archeologists such as Clark base their assessments of behavioral patterns on factual evidence. Behavioralists such as Desmond Morris must be more speculative. From the latter's viewpoint, it can be supposed that *H. erectus* exhibited some major behavioral features, the most important being *pair bonding.*

In *H. erectus* the sexual behavioral pattern probably changed as a result of selection pressures favoring greater parental care for the young. Pair bonding between the male and the female parent would provide this; an extension of female receptivity both into pregnancy and into times of the menstrual cycle other than the few days of potential fertility would be a factor strengthening the pair bond. It would need no new triggering device, for the original physiological and behavioral stimuli that initiated copulation—what we call "falling in love"—would also establish the arrangement.

Among surviving social anthropoids there are many forms of behavior during periods of sexual activity. As noted in Chapter 2, such behavior appears to be closely correlated with the nature of the environment occupied by a particular species. Pair bonding would appear to have selective advantages for a terrestrial social group in which a primary division of labor and establishment of food-sharing habits separated males and females for days at a time. It would permit the lengthening of the female's period of receptivity without precipitating fighting among the males. This in turn would facilitate further elaboration of a division of labor necessitated by the demands of increasingly long periods of juvenile dependence. The females would be able to specialize in child rearing and food gathering

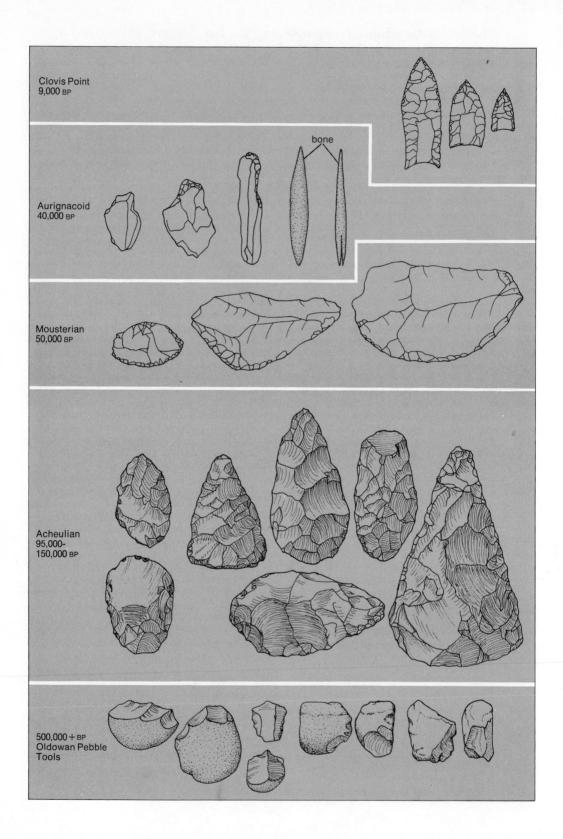

Clovis Point
9,000 BP

Aurignacoid
40,000 BP

bone

Mousterian
50,000 BP

Acheulian
95,000-
150,000 BP

500,000+ BP
Oldowan Pebble
Tools

FIGURE 4-3. *Hunting techniques of* erectus-sapiens *populations in Africa.* A hunting group of Broken Hill men dismembering a rhinoceros caught in a pit trap. Lower Gamblian pluvial phase, about 70,000 BP.

around a home base while the males foraged for animal protein, some of which would be brought back to the now dependent pair-bond family.

Exogamous Mating

The practice of exogamous mating within bands or hordes may have been an early social development after the establishment of pair bonding. M. D. Sahlins suggests that such behavior would assist in reducing sexual competition; it would also have implications for the rate of further social and physical evolution. A theoretical example of how such exogamous behavior would operate is illustrated in Figure 4-5. Such formalized ritual patterns

FIGURE 4-2. *Some of the major groupings of artifact types* characterizing particular cultural stages of human social evolution. The use of tools for dating purposes tends to produce circular arguments, so that absolute dating procedures, such as radiometric techniques, are preferable. The geographical variation in the times at which various cultures appeared, as revealed when absolute dates are available, illustrates the wide range in the chronological sequences of earlier cultures.

FIGURE 4-4. *An* **erectus-sapiens** *group in Africa* at an Acheulean phase about 70,000 BP gathering wild honey. Note the bark platter hacked out on the spot using a hand axe. Honey is still an important element of the diet of Pygmy peoples in Africa, and a simple type of beehive is commonly hung in natural vegetation by many other contemporary African groups. A bird group known as honey guides will direct any animal that will take notice (including man) to a wild bees' nest by repeated calls and flutterings.

for mate selection are described from pre-Columbian Indian societies and elsewhere.

Incest taboo rituals frequently reinforce exogamous mating behavior in surviving stone-age societies. At the same time, they establish kinship ties that must considerably modify territorial behavior and would tend to establish a concept of tribal territory shared by intermarrying bands or hordes. Moreover, exogamy would increase the introduction of new loan words evolved by individual bands, and therefore increase the rate of language change and evolution.

Division of Labor

Pair bonding may be considered as reinforcing an increase in the division of labor. The adventurous male can scavenge and hunt unimpeded by the lesser mobility of his pregnant or child-encumbered mate. The patient

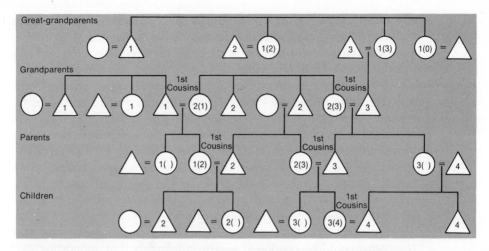

FIGURE 4-5. *Diagram of exogamous mating* by hypothetical contemporary hunter-gatherer bands. Group selection for such behavior probably occurred early in the history of *erectus-sapiens* populations. Selection pressure favoring such behavior would develop because of the need to avoid intraband rivalry over mates and to prevent possible genetic consequences of close inbreeding would produce potentially disadvantages homozygosity for many recessive genes. This system can result in frequent first-cousin matings, which could have this same result. △ = male ○ = female; four different bands indicated by numbers 1, 2, 3, and 4. Mated females transfer to the band of male mate, indicated by the number in parentheses. Blanks instead of numbers indicate bands other and additional to those labeled 1, 2, 3, 4.

female can concentrate on the more time-consuming labor of digging up edible portions of plants, collecting seeds, and breaking open bones for marrow extraction, while caring for the infant and juvenile offspring of the pair bond.

The comparatively equal loss of body hair and equal performance in walking, running, jumping, and climbing ability between the sexes in contemporary humans strongly suggest that these adaptations were made before any such division of labor occurred. A marked lack of throwing ability among modern human females implies that from the time of *H. erectus*, when temporary camps had evolved, women no longer participated in hunting expeditions, so were not further selected over the next 250,000 years for greater projectile-throwing. At the same time their dexterity at manipulating small objects demonstrates that they continued to use, and perhaps manufactured, particular ranges of preparatory tools. Once fire was associated with the temporary camp, women would have fire tending and cooking added to the tasks of child care and plant-food gathering.

It is interesting to speculate that with such divisions of labor in *H. erectus* there would have been selection pressures operating on mental characteristics such as a better-than-average aptitude for adventure and for short-term strategic planning, aggression, and a command language in males, and longer-term policy planning, patience, and a descriptive language in females. The now established differences in reaction to crowding between human males and females may have developed at this time. The generally thicker head hair covering of modern females compared to males suggests that even in *H. erectus* there was not always complete accord in a pair bond. Blows may occasionally have been traded, which afforded a selective value to the thicker hair protecting the portion of the body most susceptible to fatal damage. By the same token, the lesser head covering of the male suggests that blows were rarely exchanged between males temporarily associated in hunting parties. This implies the probable existence in such bands of some kind of accepted social hierarchy system, perhaps based on known hunting skills. The facial hair on the jowls and upper lip has been surmised by some authors to represent a physical reinforcement of such male dominance patterns.

Speculation like this as to selection for specific behavioral traits in *H. erectus* could be continued almost indefinitely, but supporting evidence is so nebulous as to make such speculation unprofitable. What appears certain is that changes in gene frequencies of particular alleles were more rapidly made in *H. erectus* populations than can be achieved in populations of *H. sapiens* at this present time. Selection pressures favoring particular alleles probably resulted in more extensive adaptation during the earliest periods of the almost 4 million years of existence of *H. erectus* and intermediate *erectus-sapiens* populations than has been possible at any subsequent time in *H. sapiens* history. For one thing, the time span for the present *H. sapiens* grade, perhaps a mere 20,000–25,000 years, has been very much shorter. For another, selection can operate only through a difficult survival rate or differential production of offspring, and the social structure of *H. sapiens sapiens* grade populations has increasingly buffered this effect. The old adage that "human nature does not change" is now substantially correct, although the vast numbers of modern man may eventually promote some exceptions to this generalization.

Fire and Cooking

The only positive evidence for the earliest association between *H. erectus* and fire in Africa is from deposits as recent as 50,000 years BP. This can be interpreted two ways: either African populations of *H. erectus* were very late in acquiring the use of fire, or, alternatively, no evidence of earlier use has so far come to light.

In the case of Peking man the association of *H. erectus*, burned bones, and charcoal in caves is very much older, dating back to 400,000 BP (Figure 4–6). This does not imply any more than the *capturing* of fire from lightning strikes, volcanoes, or other such natural occurrences. It was probably a very, very long time before the making of fire was discovered.

The importance of *preserving* fire still retains its significance in present-day rituals, where an "eternal flame" is an important symbol in many cultures. The Vestal Virgins of Rome were responsible for keeping a fire going. In some modern tropical forest communities, women may be seen

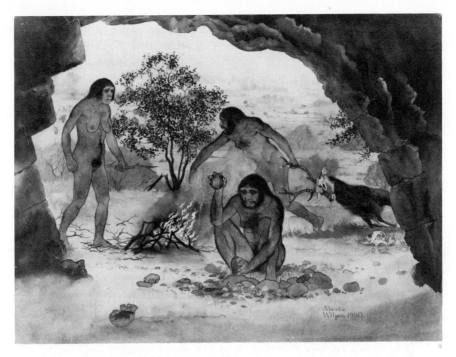

FIGURE 4-6. *Tableau depicting life of Asiatic* **Homo erectus** (Peking Man) approximately 300,000 BP. Division of labor in the family unit is presumed to have required females to remain around the cave shelter as fire tenders (fire may have been captured rather than created at this time), vegetable food gatherers, and attendants for the young. Males hunted, butchered, made tools, and maintained the territory, while keeping in contact with other similar bands of the tribe to facilitate exogamous mating arrangements as daughters became nubile or sons matured.

walking from one site to another bearing burning brands on their heads (protected by several layers of leaves). Probably fire was often transported during diurnal wanderings by carrying glowing embers in clumps of damp clay.

It is sometimes argued that cooking meat in a fire made it more digestible for *H. erectus.* Whether this is true or not, the meat would certainly be easier to consume, especially by the young and by the middle-aged whose teeth had begun to wear down or decay. The significance of cooking might therefore lie more in its effects in reducing morality in the young and the old than in providing a greater digestibility. Less time would be spent in eating cooked meat. If what has been presumed regarding the granivorous diet of hominids is true, the greatest dietary effect of the use of fire would be in the preparation of grains. Even before the invention of containers in which grains could be boiled to make a gruel or porridge, ground wild grain could be baked into a form of bread, as for example is the practice among some modern Australian Aborigines.

There is circumstantial evidence that *H. erectus* in Europe (Figure 4-7) was also using fire for cooking meat by about 300,000 BP. This evidence comes from a site at Farralba in Spain, where bones of many specimens of megafauna, including deer, horses, aurochs, and elephants, were found in association with charcoal and showed signs of being broken. However, no hominid skulls or skeletons have been located at this site.

Perhaps this early employment of fire for cooking and probably for heating in Europe and Asia, together with the use of animal skins for clothing, was prerequisite to extension of *H. erectus* into colder latitudes. Populations remaining in equatorial Africa required no such protection against cold, so only adopted fire for cooking much later, and clothing often not at all. It may be possible that the use of fire for cooking had to be discovered accidentally through its initial use for heating—so groups that did not need the extra warmth did not have an opportunity to discover cooking.

Geographical Forms of *H. erectus*

Extrapolating from the studies of Loomis on pigmentation discussed in the previous chapter, it must be assumed that what ecologists are accustomed to call geographical races, which will here be termed *geographical*

FIGURE 4-7. *Tableau depicting life of European* **erectus-sapiens** (Swanscombe man) approximately 200,000 BP during the Riss-Würm interglacial. A group of males from the same tribe, but representing several bands, has driven a reindeer (caribou) into a bog, where it can be killed more readily with sharpened stakes and shafted tools. This would be one of the known localities within the tribal territory where such a hunting technique could be executed; animals previously dispatched on this spot included the aurochs. Butchering must have been effected with tools either left around or manufactured as the occasion demanded.

forms to avoid confusion, had developed in *H. erectus* contemporaneously as migration carried these hominids into the temperate zones. Although "Peking man" was in terms of pigmentation marginal at 40° N, some lightening of the skin had probably occurred. "Heidelberg man," at nearly 50° N latitude, must have been *white*, although developing a marked sun tan in summer. Such depigmented north temperate groups of *H. erectus* would be paralleled by depigmented southern groups in temperate Africa. It is not surprising to find there in modern times white groups, represented by the virtually extinct Hottentots, and brown groups, as the Bushmen of the Kalahari.

Genetic Continuity

Although such geographical groups would be established concurrently with migration north or south of the tropical zone, the uniformity of cultural artifacts throughout the area of dispersal of *H. erectus* demonstrates that cultural contact remained between the scattered populations. Where there was cultural contact it can confidently be assumed that breeding contact remained. Either mutual admiration or the vicissitudes of local fighting would be sufficient to overcome to at least some extent any cultural barriers against intergroup mating. In the case of highly selected characteristics such as skin color, the progeny of such crosses would sometimes tend to be less fit than the parents and would not significantly modify the gene pool of either group by introgression. In other traits less obviously of selective value, such as wavy versus straight hair, intergroup crossing must have ensured that mutant alleles eventually spread into all groups despite any comparative geographical isolation.

That some measure of breeding isolation did exist between major groups is evidenced by the present distribution of certain traits that appear to have no selective value. The most extensively studied series of such traits are *blood groups*, details of whose global distribution are known more fully than for any other genetically determined system.

Blood Groups

The blood serum of some individuals can clump the red cells of certain others. From this reaction, three basic types of blood groups were initially recognized, A, B, and O. A fourth, AB, was later added to this early classification. The differences are believed to be caused by slight variations in the mucopolysaccharide materials at the blood corpuscle's surface, and there is some indication now that diet as well as heredity may influence their occurrence.

It is interesting to look again at the same three geographical areas referred to when discussing skin color groups, where it may be supposed that East Africa had black-skinned forms of *H. erectus*, Heidelberg had white, and Peking brown or yellow. The maps in Figures 4-8 and 4-9 ignore for this purpose the blood groups of some other parts of the world. With regard to the gene G for blood group O, equatorial Africa has typically the highest frequency (70–80 per cent), Europe is lower (60–70 per cent), with tongues reaching south in Africa, and Asia is the lowest (40–50 per cent), with higher frequencies in coastal regions.

With gene G^B for blood group B, Asia has the highest frequency (20–30 per cent), again lower on the coasts, equatorial Africa is lower (10–15 per cent), and Europe is the lowest (5–10 per cent).

In the case of gene G^A for blood group A, Europe has the highest frequency (25–30 per cent), Asia a lower figure (15–20 per cent), lower in coastal areas, and equatorial Africa the lowest (10–15 per cent), with some central intrusion of 15–20 per cent frequencies.

Putting these figures together in another way, with the area's most common blood group first, the next most frequent second, this gives

Europe	A	(O)
Equatorial Africa	O	(B)
Asia	B	(A)

W. Boyd examined the question of a more detailed range of blood groups in relation to major groupings of human populations, arriving at a scheme similar to the above. It differs only in identifying as separate entities two splinter groups, Australian Aborigines and Ameridians, the New World aborigines. The scheme he arrived at, omitting these last two, may be summarized as

FIGURE 4-8. *The distribution among contemporary aboriginal human populations of the gene I^A which produces the A blood type.* The frequency of this gene is highest in populations of the Caucasoid group, lowest in Early Mongoloids of the Americas.

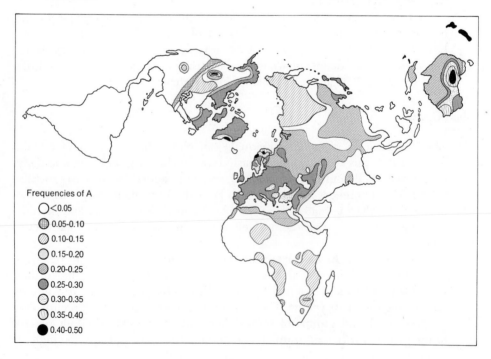

Frequencies of A
- ○ <0.05
- ⊕ 0.05-0.10
- ◐ 0.10-0.15
- ◯ 0.15-0.20
- ◍ 0.20-0.25
- ● 0.25-0.30
- ◒ 0.30-0.35
- ◑ 0.35-0.40
- ● 0.40-0.50

Caucasoid High frequency of Rh *cde* and *CDe*, moderate of other blood group genes. *M* usually above, *N* below 50 per cent

Negroid Very high frequency of Rh *cDe,* moderate frequencies of other blood group genes

Mongoloid High frequency of *B*, little if any *cde*

It is tempting to argue that such groupings of an apparently nonselective character in modern human populations result from genetic differences that arose between *H. erectus* groups as mutations for skin color permitted some of them to migrate from the parent stock in equatorial Africa and form distinct geographical groups in higher latitudes. Possible migration routes have been considered during studies of an interrelated problem, the origin of the human populations of the New World. To anticipate ideas presented later as this account of human evolution develops, it will be surmised in this text that geographical groups of *H. erectus* evolved *in situ* into Caucasoid, Negroid, and Mongoloid forms of contemporary human populations. The process was a continuous one, with no natural breaks, but for convenience populations intermediate between *H. erectus* and *H. sapiens sapiens* are here called *erectus-sapiens* populations. Again to anticipate later passages, there is a general recent tendency to call these *erectus-sapiens* populations *Neanderthals.* It is also now clear that blood groups are not

FIGURE 4-9. *The distribution among contemporary aboriginal human populations of the gene* G^B *which produces the B blood type.* The frequency of this gene is highest among Recent Mongoloids, but it is quite absent from the Early Mongoloid populations of the Americas, probably because of the operation of the "founder principle."

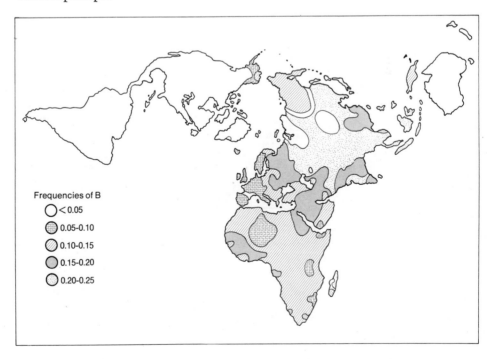

Frequencies of B
- < 0.05
- 0.05–0.10
- 0.10–0.15
- 0.15–0.20
- 0.20–0.25

even presently the neutral traits they were originally thought to be. They have been shown to be associated with selective resistance to some infectious diseases such as smallpox and tuberculosis.

Early Migrations

The possible restrictions to population mobility imposed by the spread of ice sheets in the Old World over territory suitable for hominid occupation have been considered by a number of workers. There appears a high probability that during an ice age most of Europe would have been quite uninhabitable, although because of lowered sea levels there may have been routes across the Mediterranean Sea. Populations in Asia would have been isolated there by a southern extension of Caucasus ice sheets to the Persian Gulf (Figure 4-10). The same almost complete isolation would have been true for Asian populations during an interglacial period because of the substitution in this Caucasus-Persian region of a dry zone for an ice-sheet barrier (Figure 4-11).

As the climate moderated after a glacial interval, migration from the south into western Europe would be possible, but factors of time and distance probably prevented European hominid groups from extending sufficiently eastward to encounter similarly migrating Asiatic populations spreading

FIGURE 4-10. *Eurasia during a glacial episode of the Pleistocene.* Mountain glaciers effectively isolate any *erectus-sapiens* populations that have penetrated east of the Caucasus Mountains. In Europe alpine glaciers may similarly have isolated any European populations, although lowering of sea levels may have permitted dry crossings of the Mediterranean Sea area.

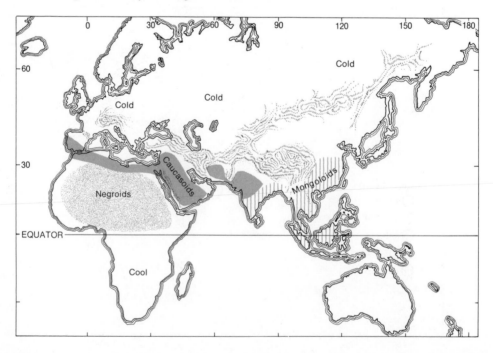

westward. The formidable barrier of the massive Himalaya range would under all conditions tend to restrict the southward migration of Asian populations.

Pleistocene barriers to migration therefore appear to confirm the deductions from examination of such traits as blood group differences, that essentially white European groups, black African groups, and yellow or brown Asiatic groups of *erectus-sapiens* populations evolved as distinct geographical entities in comparative breeding isolation (Figures 4-12 to 4-15).

Neanderthal Man

The progress of human evolution in this chapter as developed in this chapter indicates the existence of three geographical forms of *erectus-sapiens* populations that developed independently in response to their differing environments. However, all possessed a common Acheulian culture and undoubtedly were still undergoing some gene exchange throughout their dispersal area.

So far we have ignored the complications that arise from the undisputable existence of another hominid form whose discovery antedates that of *H. erectus* historically, and whose remains and associated artifacts quantita-

FIGURE 4-11. *Eurasia during an interglacial episode of the Pleistocene.* The warmer and drier conditions would impose a desert barrier tending to isolate any *erectus-sapiens* populations that had succeeded in penetrating east of the Caucasus Mountains. The Mediterranean Sea and desert conditions in North Africa would similarly tend to isolate any European populations.

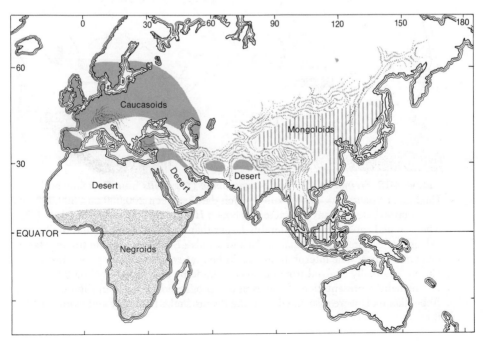

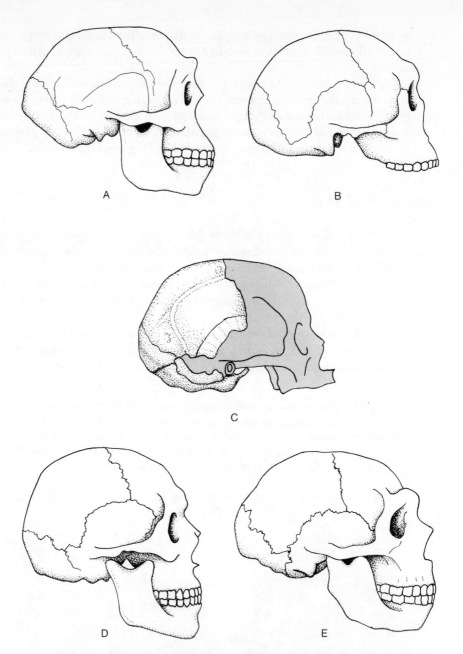

FIGURE 4-12. *Steinheim (B) and Swanscombe (C) skulls* from Germany and England, respectively. These skulls, which date between 200,000 and 300,000 BP, are intermediate in many characters between *H. erectus* (A) and *H. sapiens* (D). Their cranial capacity is considerably larger than in *Homo erectus,* and the brow ridges are less pronounced than in Neanderthals (E). They provide fragmentary evidence at best; one interpretation could be that *erectus-sapiens* populations in Europe gradually evolved from an *erectus* grade by approximately 70,000 BP. A schematic representation of stages in this progression is given in Figure 4-14. Acheulian tools were associated with the Swanscombe fragment and evidence of fire.

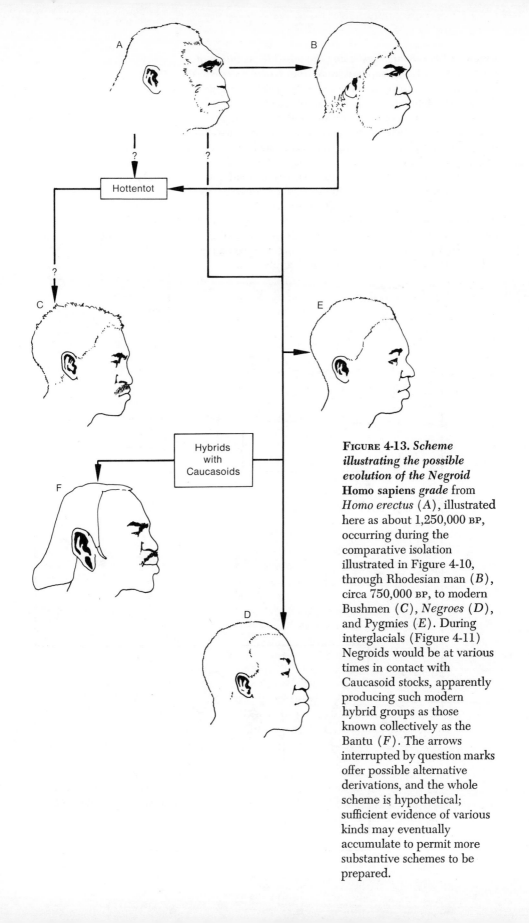

FIGURE 4-13. *Scheme illustrating the possible evolution of the Negroid* **Homo sapiens** *grade* from *Homo erectus* (*A*), illustrated here as about 1,250,000 BP, occurring during the comparative isolation illustrated in Figure 4-10, through Rhodesian man (*B*), circa 750,000 BP, to modern Bushmen (*C*), *Negroes* (*D*), and Pygmies (*E*). During interglacials (Figure 4-11) Negroids would be at various times in contact with Caucasoid stocks, apparently producing such modern hybrid groups as those known collectively as the Bantu (*F*). The arrows interrupted by question marks offer possible alternative derivations, and the whole scheme is hypothetical; sufficient evidence of various kinds may eventually accumulate to permit more substantive schemes to be prepared.

FIGURE 4-14. *Scheme illustrating the possible evolution of the Caucasoid* **Homo sapiens** *grade* from *Homo erectus* (A), as in Figure 4-13. Migration during interglacials (Figure 4-11) and comparative isolation during glacials (Figure 4-10) led to the evolution of such forms as Solo man (B), Indonesia circa 1,000,000 BP, and the now extinct Tasmanian man (C), as well as to modern Caucasoids such as the Ainu (E) of Japan, Dravidians (F) of India, and Australian Aborigines (G). Modern Libyans (D) may be related to Swanscombe man (circa 250,000 BP), who could have been ancestral to Cro-Magnon man (H) (circa 50,000 BP) and to Neanderthals (I) (circa 50,000 BP), also probably such forms as the modern Mideast Caucasoids (J). All males except D. The same remarks about the highly speculative nature of this scheme as in Figure 4-13 apply here.

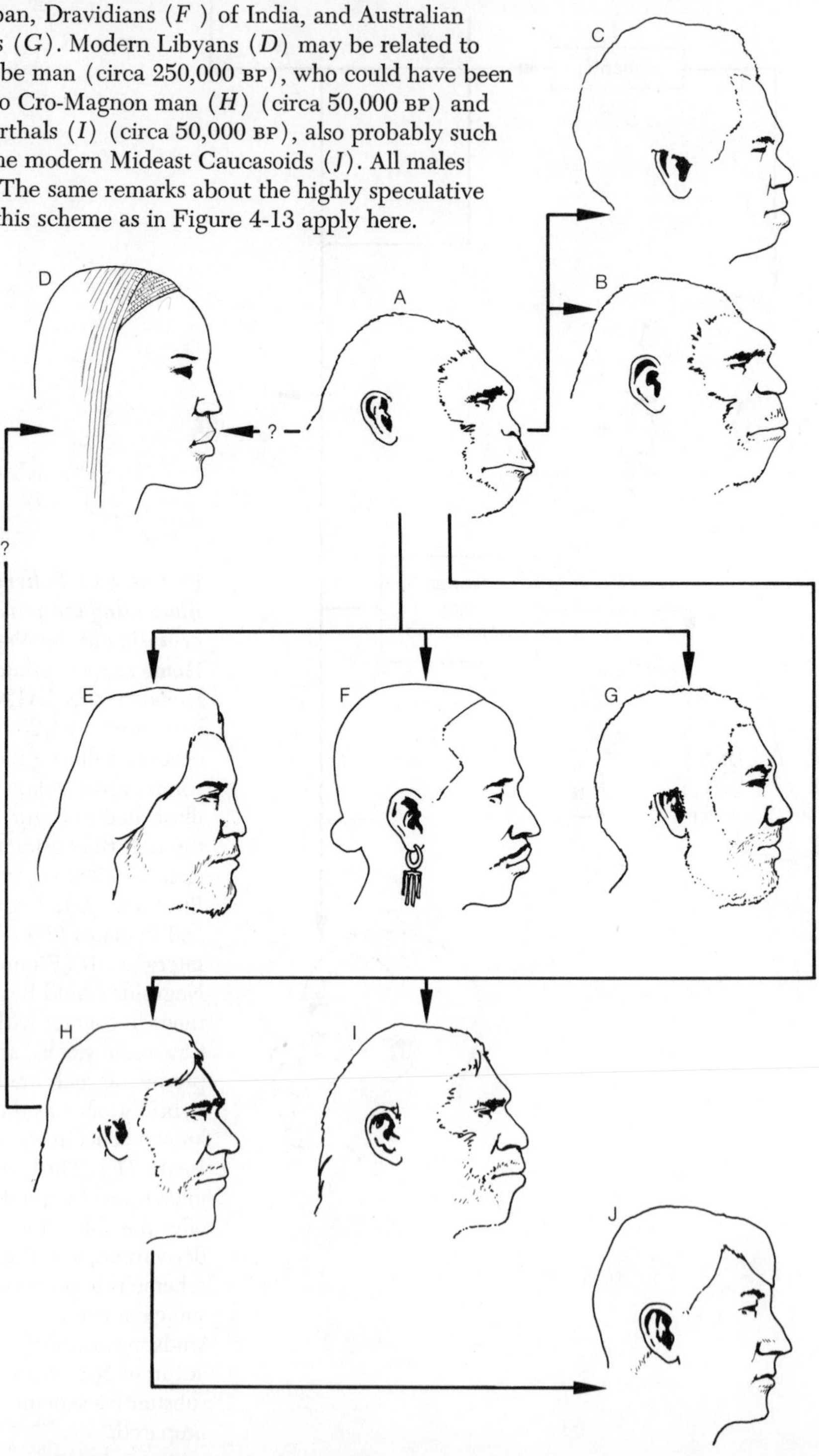

tively far exceed those of any other hominid yet considered. The first report of artifacts of this form appeared in 1849, 10 years before the publication of Darwin's *Origin of Species,* when M. Boucher de Perthes described some fashioned flints found in Somme River gravels in France. In an imperious manner not entirely unfamiliar to us today, the French establishment pronounced its judgment on this report: *"L'homme fossile n'existe pas."*

That this was not to be the last word on the subject was first conclusively demonstrated with the unearthing in 1856 of thigh and arm bones, ribs, clavicle, pelvis, and skull cap sections from a cave in Germany. It was situated in a valley a little to the east of Dusseldorf, which had been named "Neanderthal" after the nom de plume *Neander* formed by translating into Greek the name of a seventeenth-century poet that by sheer coincidence was Neumann (new man). Confusion and controversy surrounded these finds for many years, as they did more scanty but earlier remains of "Gibralter man." Although appropriately named by an astute Irish professor of anatomy, William King, as a new species of man, *Homo neanderthalensis,* support for this view was not generally received until after a dozen or so similar finds were unearthed in Europe between 1870 and 1880.

Controversy and doubt persist, and it is still not clear beyond question whether modern man evolved through a Neanderthal stage or whether this

FIGURE 4-15. *Reconstructions illustrating possible evolution of mongoloid group of* Homo sapiens *grade A.* H. erectus *(Africa) circa 1,250,000* BP. *B. Peking man (China) circa 1,000,000* BP. *C. Paleo-Indian. D. Recent Mongoloid female (China). E. Bering Sea Mongoloid (Alaska). All males unless otherwise stated.*

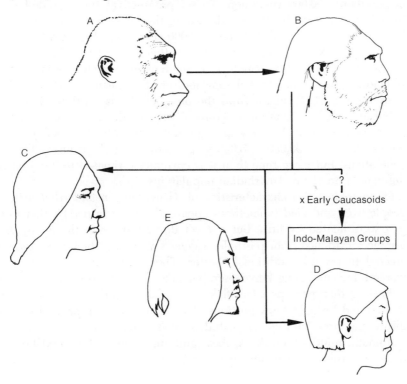

was a specialized evolutionary form, selected from the European geographical group of *H. erectus,* which eventually died out without contributing any new alleles to the gene pool of modern man.

Presently a consensus of opinion holds that in the warmest parts of Eurasia, and in Africa, Neanderthals represented an evolutionary stage through which regional human populations passed in the course of evolving from *Homo erectus* to contemporary humans. It is generally believed the Neanderthal population of cold northwestern Europe passed to extinction before the establishment there of modern *Homo sapiens.* A similar consensus regards Neanderthals as referable to *H. sapiens,* and usually named as a subspecies, *H. sapiens neanderthalensis.*

Neanderthal Characteristics

It is now usually accepted that remains of Neanderthal man in Eurasia are associated with a particular stone-axe culture known as *Mousterian* (Figure 4-16). Apart from a limited number of sites such as Mount Carmel in Israel and some still controversial ones like Broken Hill in Zambia, all Neanderthal remains have been found in Eurasia, especially in western continental Europe. There are no significant anatomical differences between these Neanderthals and modern man except in the face, pharynx, and dentition. The teeth were very large, which partially accounts for the characteristic prognathy. The cranial capacity was the largest known in hominids, ranging from 1,300 to 1,600 cubic centimeters. Earlier or "progressive" types of European Neanderthals date from the last interglacial, the Riss-Würm, which began about 100,000 years ago. These "progressive" types graded into the "classical" types of the Würm glaciation, the latest remains known having been dated to an interstadial about 40,000 years ago (Figure 4-16). Perhaps "progressive" Neanderthals arose even earlier, toward the close of the Riss glaciation about 150,000 years ago. Fossil remains, which now include about 100 more or less complete skeletons, indicate European Neanderthal populations were most abundant from the onset of the last (Würm) glaciation, dated at 75,000 ± 5,000 BP radio-metrically, to the temperate interstadial, which began about 42,000 years ago and lasted about 10,000 years. This represents a mere 30,000 ± 5,000 years, allowing perhaps only 1,000 ± 250 generations, but more time than it is currently estimated to have taken for differentiation of the Amerindian populations.

Summarizing the characteristics of *H. sapiens neanderthalensis,* these people lived and died (sometimes apparently being buried) either in caves that were heated by fires, but *not* yet decorated, or in the open. In both instances evidence from postholes suggests that skin shelters or tents were erected to provide additional shelter. Their stone tools were principally worked flint, including knives, scrapers, cleavers, borers, and saws, the last indicating that perhaps some flakes may have been hafted. Neanderthals hunted and brought back portions of the kill to their hearths. Remains indicate the prey in Eurasia included such animals as woolly rhinoceros, mammoth, reindeer, musk ox, ibex, glutton, chamois, fox, wild horses and oxen of various species, and marmot. Neanderthals may have killed large

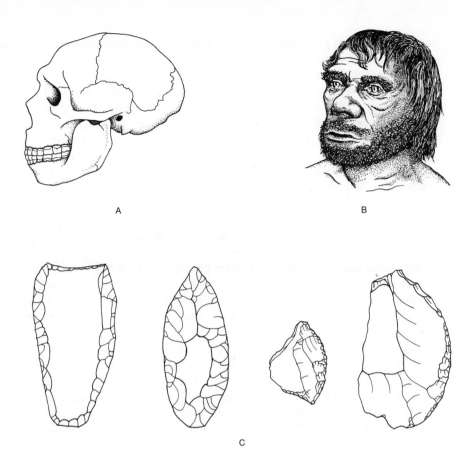

A   B

C

FIGURE 4-16. *A "classical" Neanderthal skull and Mousterian tools,* both typical of
forms dating to about 60,000 BP. Cultural artifacts identified as Mousterian are
widespread over Eurasia, suggesting that this typically Neanderthal flake culture
was adopted by other *erectus-sapiens* populations now often referred to as
Neanderthals. Such cultural spread indicates some degree of cultural contact
between populations located in widely scattered regions of the Old World.

predators such as cave lions, bears, and hyenas, either because they were
competitors or perhaps in the case of the bear for ceremonial purposes.

Like modern Bushmen, European Neanderthals were probably not averse
to taking small animals such as hares, tortoises, snakes, or birds on an opportunity basis, but they were primarily hunters of large herd animals.
Although their hunting skills had increased, they were still too inefficient
to support their population on the relatively lighter production that would
result from chance encounters with small prey or solitary large animals.
Thus typically in Neanderthal sites bones of reindeer, oxen, horses, rhi-
noceroses, and mammoths predominate. The hunting of such herd animals
places a premium on two interrelated factors. A fair number of adult males
are required to form a hunting party, a moderately sophisticated means of
communication is necessary to work out the attack strategy among its
members.

These two prerequisites must have strongly influenced the further biological and cultural evolution of *erectus-sapiens* populations wherever they occurred. Several workers, among them P. Lieberman and E. S. Crelin, have recently reconstructed the vocal tracts of some Neanderthal populations. They conclude that owing to the lack of a well-developed pharynx, a deficiency revealed by these reconstructions, the resonating chamber at the back of the throat was restrictive. Neanderthals of western Europe may consequently have possessed only about 10 per cent of the speaking ability of contemporary humans. On such evidence it seems likely African Neanderthals had a better ability, and European Neanderthals in warmer areas may have had approximately the same capacity as contemporary humans.

However this may have been, and the matter is still controversial, an even more interesting area of speculation is currently developing from a rather different source. It has been known for some time that the unique and general right-handedness of humans is the result of a partial separation of functions between the two cerebral hemispheres of the brain. Even great apes are ambidextrous and do not exhibit right-handedness. The first convincing evidence for its development appears in Neanderthals. It was apparently a practice to stuff chunks of meat into the mouth, then, gripping them with the teeth, sever the still free portion with a stone knife. The evidence for this practice resides in the frequent appearance of fine scratches on the tooth enamel. Invariably these are oriented from top right to bottom left, the direction in which a right-handed person would naturally cut. Right-handed dexterity is controlled primarily by the left hemisphere. Another principal function of the left hemisphere concerns language use. It is tempting to surmise that in the biological evolution of Neanderthals, dexterity of the left hand was traded off against the potential advantage of devoting more of the right hemisphere to this activity and to language development and instead, using the right hemisphere to enhance conceptualization. While selection for further manual dexterity of the right hand would continue strongly, biological adaptation would also be marked as regards the size of the pharynx and the further specialization of the hemisphere with respect to language functions.

Cultural evolution of Neanderthals would lie more especially in the development and sophistication of group hunting techniques through improved language and better weapons. In fact, the transition from Neanderthal to modern human populations is generally accepted as represented by the replacement of middle paleolithic with upper paleolithic industries. Some workers suggest putting a date on this, and saying every population after that date was *Homo sapiens sapiens* and before that was *Homo sapiens neanderthalis* or an earlier hominid. However, this works only for the moment, because there is a dearth of fossils from this *erectus-sapiens* transition period. When more are found, they will probably invalidate this criterion, because biological evolution like cultural evolution proceeds at different rates in different areas.

Because the survival of a hunter-gatherer band would relate primarily to the number of able-bodied *men* in a band available for group hunting, women and their gathering activities were of lesser significance. There is a strong suggestion that female infanticide was practiced at this time, for

Neanderthal remains display a sex ratio favoring males by about 10 per cent. This is very high taking into account the fact that far more males would die on hunting expeditions away from their native hearths, near which most females would by contrast have a high chance of being interred and subsequently located by archeologists. The slight tooth wear of many Neanderthals supports the predominance of the meat diet, which is easier on the teeth than the mastication of plant food gathered by the women. This is another contrast with modern hunter-gatherers, who may have only 20–40 per cent meat in their diet. Neanderthals used fire, but it is believed they could not make it. They probably carried hot embers in a dab of wet mud to kindle new fires as required. Although they could dress animal skins, they laced them together with leather thongs to form garments. Stitching probably remained unknown to most of them, nor did they discover how to make pottery. Although there is ample evidence of extensive ceremonial behavior, no musical instruments have been found. Pigments may have been applied to the body, but no graphic art developed otherwise.

Positive material evidence as to Neanderthal behavior and beliefs of a most exciting nature has recently been revealed by R. S. Solecki, who organized a major cave exploration at Shanidar, in the Kurdistan region of northern Iraq. This excavation not only confirmed that Neanderthals had a belief in survival after death that caused them to bury their dead formally and with selected artifacts but that, in one instance at least, *local wild flowers had been strewn around the corpse*. This major excavation work at Shanidar also confirmed that Neanderthals sometimes at least cared for their aged and infirm rather than abandoned them. Less praiseworthy practices have been noted in some finds, where unchallengeable evidence of cannibalism of Neanderthals by Neanderthals has been obtained.

Trophic Position of Neanderthals

While they were still very incompletely known, it was tempting to dismiss populations of *H. neanderthalensis* in Eurasia as variants of the white geographical group of *H. erectus* that had specialized in an exclusively hunting economy (Figure 4-17). Neanderthals differed morphologically from *H. erectus* principally in the larger brain size and smaller posterior teeth, both features that are consistent with a greater concentration on predation. As top carnivores, the Neanderthals could never have had a very high population density; development of new hunting techniques might always have resulted in overkill of particular prey species and permanent disturbance of the tundra ecosystem in which these populations lived. The facial characteristics, which became progressively more pronounced, are consistent with the need for such a top carnivore (as in the case of the modern Eskimo) to chew not only meat but also hides in order to prepare skins for use as clothing.

Trophic isolation of this exclusive white hunter group would isolate it in Europe, away from the still unspecialized and possibly neanderthaloid hunter-gatherer populations of the Mediterranean region, Asia, and Africa. The climate amelioration of the Würm interstadial may have destroyed the game species on which the specialized *H. neanderthalensis* populations depended exclusively for survival in Eurasia, an event to which overkill may have contributed, driving these populations to extinction.

This view of Neanderthals is not in conflict with two major finds already noted. The first is the discovery of possible intermediates on Mount Carmel in Israel, the second the so-called Rhodesian or Broken Hill man, exhibiting "boskopoid" tendencies—the African version of Neanderthals, according to some workers (Figure 4-13). One plausible explanation for these apparent anomalies is that early Eurasian Neanderthals, which the Mount Carmel

FIGURE 4-17. *A Neanderthal band setting out on its northward spring migration.* After overwintering in a wooded area of central Europe, this small band is heading north with all its possessions, tools, skins, and small game provisions to exploit the rich summer hunting resources of the more northerly tundras. Time period toward the close of the Riss-Würm interglacial about 80,000 BP, and relatively warm.

finds could represent, "Rhodesian man" (or more likely woman), and "Solo man" represent parallel stages in the progression of Caucasoid, Negroid, and Mongoloid *erectus-sapiens* populations, respectively, to modern man. The classical Neanderthals would then be a European subspecies that passed to extinction about 40,000 BP. They would have coexisted with other evolving Eurasian *erectus-sapiens* (Neanderthal) populations, but their essentially carnivorous diet of large herbivores on the European periglacial tundra and taiga would have separated them geographically from the omnivorous human populations of more temperate and tropical areas.

Cro-Magnon Man

The possible initial overemphasis on European Neanderthals arising from the circumstances of their location in western Europe may equally apply to the last major fossil series to be discussed. This has come to be called *Cro-Magnon man*, based on a rather imperfect skull found in a cave shelter near a great rock (*cro-magnon*) in the village of Les Eyzies in southwestern France in 1868. Although the teeth are missing, all the skull measurements fall within the range of those of living Europeans. Indeed, photographs of contemporary individuals closely resembling this type have been published.

This first Cro-Magnon skull was that of a man about 50 years old. Subsequently in the same locality remains of two other adult men, a woman, and an unborn child were found. All date from the Gottweiger interstadial, aproximately 30,000 years BP, and related artifacts represent upper paleolithic tools. Over the last 100 years or so since the first Cro-Magnon discoveries, archeological activity over the 3,500 or so square miles of the Dordogne district, where these finds were made, has been so intense as to yield a greater amount of prehistorical human material than any other known area of the earth. It is now believed that these Cro-Magnon populations of the Dordogne are essentially referable to and indistinguishable from our own species *Homo sapiens sapiens*. The oldest remains date back to about 40,000 BP, forming a continuous and unbroken line to the present time, and relating without any chronological discontinuity to the last Neanderthals. Strictly speaking, Cro-Magnon populations were restricted in their distribution to northwestern Europe. Some authorities hold to this distinction, others apply the name to any remains that fall clearly within the physical range of diversity of contemporary human groups, regardless of where the remains have been found. As will be discussed in Chapter 6, it might be advisable to retain the more restricted use until further evidence is available on human migration in Cro-Magnon times and more general agreement is reached as to the mode of progression from *H. sapiens neanderthalis* to *H. sapiens sapiens*.

It is thus considered that by 30,000 BP the European group of *erectus-sapiens* had assumed the anatomical form of now-living descendants classifiable as *H. sapiens sapiens* (Figures 4-12 and 4-14). The same may well have been true of the African *erectus-sapiens* group (Figure 4-13), and it was true for the Asiatic forms, as subsequent migration into the Americas from Beringia and into Australia from further north revealed (Figure 4-15). Despite the comparative absence of any further *physical* evolution from this time, the immense range of late paleolithic, mesolithic, and neolithic cultures still have not evolved, and many migrations had yet to occur. Some of these were undoubtedly prompted by further climatic change. In considering the diversity of modern men, a number of these points will be amplified. Cultural changes other than those exhibited in worked stone tools will now briefly be examined.

Migration of *H. erectus-sapiens* Populations

As has been noted several times, dispersal of the hominid stages about the world was remarkably wide (Figures 4-18). It is true that as a top carnivores, which most species of the genus *Homo* proved to be, hominids could be expected to have the widespread distribution that fossil remains indicate; even the largely vegetarian species of the genus *Paranthropus* seem to have been widely dispersed. Several factors may have been responsible for this. The greater intelligence resulting from the proportionately larger brain and the possession of early forms of cultural behavioral patterns would permit more ready adaptation to somewhat differing habitats and ecosystems. This adaptability in turn would facilitate the accommodation of popu-

FIGURE 4-18. *Distribution of finds of skeletal remains attributable to* Homo erectus. Shaded area indicates position of polar ice sheet and European alpine glaciation about 500,000 BP, which is the approximate date for many of these finds. By contrast with *Paranthropus, Homo africanus,* and *H. habilis,* remains of which have not been found outside tropical areas, *H. erectus* could exist in a wide range of climates, from tropical to arctic.

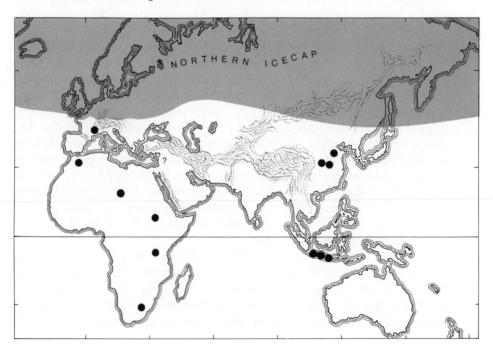

lation surpluses to any new territories that they might invade. Feedback mechanisms, which in other animals might provide for regulation of population numbers by restricting births or increasing deaths, or both therefore may not have restricted population growth because outmigration prevented overexploitation of any particular habitat or ecosystem. Emigration and outmigration have been a constant feature of the development of all human societies, although in modern times restrictive laws have sometimes reduced this flow to a level where its selective nature causes it to be known as the "brain drain."

As far as early *H. sapiens* populations are concerned, we are rapidly discovering more about one major migration, the occupation during the closing stages of the Pleistocene of the largest remaining virgin land mass, the New World.

Occupation of the New World

No remains of any hominid, or of any hominoid other than *Homo sapiens* ever have been found in the New World. It has therefore been concluded that until the appearance of *Homo sapiens* pongid and hominid evolution were restricted entirely to the Old World. This unexplained circumstance is somewhat curious in view of the fact that other forms of mammalian evolution were proceeding in the same period and about as extensively in both the eastern and western hemispheres. Moreover, camels, horses, bison, and elephants actually appear to have originated in the New World and migrated to the Old. Such migrations would not have been possible except across what is generally known as the *Bering land bridge*. More precisely, these mammals migrated by way of *Beringia*, for during the period when readjustments of sea level relative to land exposed a continuous land mass between North America and Asia, this was at least 400 miles wide, which would in no way constrict animals into a narrow migration passage (Figure 4-19). Any movements through Beringia would therefore have been a normal pattern of territorial dispersal similar to what would occur on any continental mass.

The Bering Land Bridge

If human migration into the Americas was by land—and there is presently no substantial evidence to suggest that it was by sea or occurred more than once—it could have taken place only at three possible dates in the Upper Pleistocene (Figures 4-19 and 4-20). The first was from about 70,000 to 50,000 BP, the second from about 32,000 to 28,000 BP, the third from about 22,000 to 15,000 BP. These are the dates determined by the stratigraphical, palynological, and radiometric evidence of when sufficient areas of Beringia would have been exposed to permit migration from Asia to North America. A few years ago considerable weight of archeological evidence suggested that the last date was probably the most likely one for this spread of *Homo sapiens* populations into the western hemisphere. It agreed with the oldest radiometric dating of any presently known Paleo-Indian remains of *Homo*

PLATE 4. *Evidence of Paleo-Indian cultures in the American continents* is being assembled slowly by carefully and scientifically planned digs, such as this one in Wyoming. Archeological evidence from such digs is more readily accessible when modern earth-moving machinery can be used to remove the overburden. Stratigraphical and other spatial relationships are now more precisely recorded, and a wide range of techniques from other disciplines can be employed to convert what might otherwise be pure conjecture into quantified probabilities.

sapiens in North America, that of "Laguna woman" dated to 17,000 BP. Claims of earlier finds predating this period have since been substantiated. Although a considerable body of archeological opinion supported the original dating, it did not preclude the movement of other cultures at an earlier date.

That earlier cultures were present in North America is now an inescapable conclusion from the still-accumulating evidence. A human skull found in the Los Angeles area during construction work there in 1936 was dated in 1971, using the "protein clock" technique, to at least 23,600 BP. A caribou bone scraper found in the Yukon in 1972 yielded a C14 date of 27,000 BP. Dwarf mammoth bones that had been split and burned have been found on Santa Rosa Island, off the southern California coast. C14 dating places their age at 29,000 BP. Artifacts associated wtih mammoth bones from central Mexico

PLATE 5. *Evidence of Paleo-Indian cultures in North America* frequently
demonstrates that large animals were butchered, if not killed. The four teeth of
this fully grown mammoth, the last of its three sets, and the rest of its skeleton
were littered with several hundred Clovis tools broken, dropped, or discarded
when it was butchered. This skeleton, provisionally dated to about 12,000 BP,
was found recently in an arroyo some miles east of Tucson, Arizona. (One-fifth
natural size.)

date to about 20,000 BP. New methods of dating, described in Appendix 1,
may indeed require that the occupation of the New World be regarded as
occurring in the first of the suitable periods, 50,000 to 70,000 BP.

Further finds of datable remains or artifacts may place human entry into
the New World in a more certain time frame. Although there is every expectation the date will be pushed backward, paleoanthropologists are currently
reluctant to consider a period earlier than 40,000 BP, which rules out the late
Dr. Leakey's finds at Barstow, in the Mohave Desert, which he alleged were
artifacts dating to 50,000 BP. The evidence for this claim of an entry not earlier
than 40,000 BP has recently been summarized by C. Vance Haynes. Part of
the reluctance to admit earlier dates for the invasion of North America is that,
as was seen previously, Neanderthals persisted until about 40,000 BP, and
transition to modern humans was not apparent until about 25,000 BP. Neander-

thals had adapted well to periglacial environments, but paleoanthropologists do not believe that their cold-weather technology could have coped with the arctic winter of Beringia. For example, they joined skin garments together by inserting thongs and did not know how to *stitch* them, for needles have been found in only one Neanderthal site.

There is not only the question of the human occupation of Beringia, there is also the problem of when elements of this population could have moved southward into the American continents. As the diagrams in Figure 4-22 show, an unscalable glacial wall at various times would prevent any migration southward from the ice-free Beringia tundra. Haynes has recently discussed these problems, and the illustration in Figure 4-20 schematically summarizes the present position.

Early Mongoloids

From about 18,000 BP, hunter-gatherer groups of a nomadic pre-Mongolian type known as *Early Mongoloids* can be demonstrated to have been spreading northward through the pacific hinterland of northeastern Asia. The movement could have begun earlier, as indeed it must have if the datings for North America just discussed are correct. These populations are termed Early Mongoloids because the flattening of the face, the loss of brow ridges, and certain other characteristics of Recent Mongoloids had not yet reached their fullest development. The eyes were lengthened horizontally, but most probably no fatty eyelids were present. Fossil remains of these people are always associated with a tool culture known as the "projectible tip," in which the spear heads, supposedly hafted, have been secondarily flaked to provide a sharp apex (Figure 4-21). It is thought that such bands were hunting the large game animals that characterized these Upper Pleistocene times, and the projectile tip was necessary to provide penetration to the thick hides of these animals.

Remains of projectile-tip peoples have been found as early as 18,000 BP in the northeastern section of Asia adjoining the now-inundated land of Beringia. The earliest such remains in Alaska are dated at 12,000 BP. No remains have yet been located from the submerged areas of the Bering Straits.

However, similar projectile tips of about the same date have been found much farther south in the North American continent. For them to have reached such a distance, it is presumed that the occupation of Beringia must have been completed at the latest by 15,000 BP. It probably occurred considerably earlier, as already noted. The term *occupation* is perhaps a misnomer. One, or at most several, bands of a projectile-tip hunter-gatherer population were probably all that ever occupied this territory, and it is doubtful whether they even realized they were crossing from one continent to another. If it was indeed only one band, it could have numbered as few as 100 individuals. The so-called *founder principle*, which is of general application about the world, appears to apply to certain genetic traits of Amerindians. It provides circumstantial support for the supposition of both

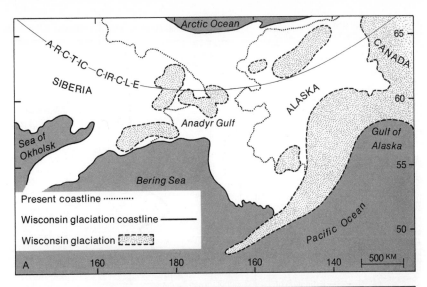

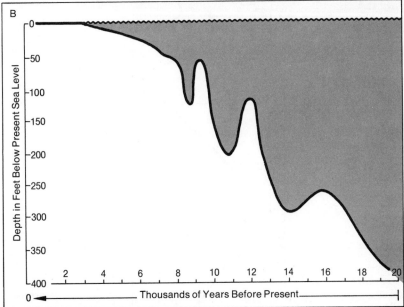

FIGURE 4-19. *The Bering land bridge* showing (A) the land area exposed during the maximum marine regression and (B) the fluctuating Pleistocene sea levels. See text for further explanation.

the unique event of migration to the Americas and the very small size of the migrating group.

Migrations in the Americas

The area of the land in North America available for colonization in this final major marine regression of the Upper Pleistocene would be compara-

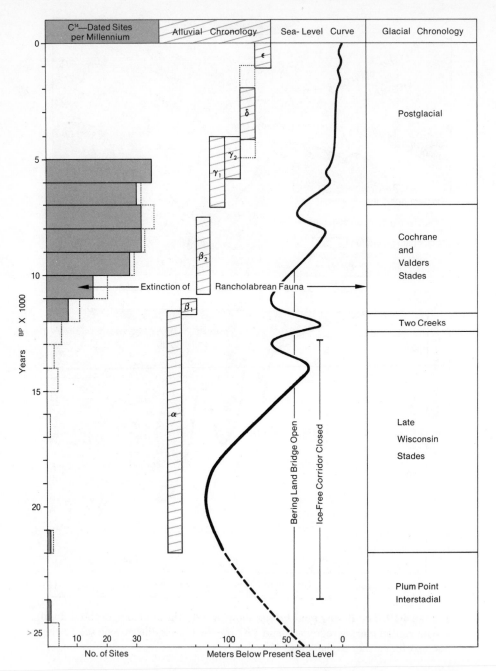

FIGURE 4-20. *Correlations between Pleistocene fluctuations in sea level, glacial chronology, alluvial deposition, and Paleo-Indian sites in North America.*

tively restricted. The great ice sheets of the Wisconsin still stretched over the whole of Canada, apart from the so-called *nunatak,* or exposed ice-free tundra, covering much of Alaska. The Cascade and Sierra Nevada mountain

A   B

FIGURE 4-21. *Tools of the general type known as projectile tip.* These occur extensively in northeastern Asia, from about 18,000 BP, and forms shown in *A* are found later over a wide area of North America. The tools believed to have been hafted, the point being used to penetrate the hide of large mammals. Folsom (*B*) is the name given to a common North American type of the general projectile tip class occurring from about 12,000 to 7,000 BP.

chains were only just beginning to be elevated. Between them and the Pacific Ocean would lie most of the exposed land surface north of the more southerly extension of the Wisconsin ice sheets, which in the central part of the country would probably be at least as far south as Colorado and Kansas (Figure 4-22).

Although the nomadic projectile-tip hunter-gatherers would be living in tundra in Beringia, as they penetrated farther and farther south through the Pacific Northwest they would encounter increasingly luxuriant ecosystems. These would mainly be of a coniferous forest type, interspersed with grasslands on which an extensive megafauna had evolved. Many remains of this megafauna have been found in the tar pits around Los Angeles, and it has been named the *Rancholabrea fauna* after the largest of these tar pits to be examined to date, on the old Spanish land holding of Rancho La Brea (Figure 4-23). Radiocarbon dates show that the remains form a sequence from about 40,000 to 10,000 years BP.

FIGURE 4-22. *Glaciation in North America at various times* as correlated with absolute dates obtained from radiocarbon measurements. Even during the several glacial episodes illustrated here, a western corridor and portions of Alaska remained clear of glaciers.

11,000-12,000 BP   10-000-11,000 BP   9,000-10,000 BP   8,000-9,000 BP

FIGURE 4-23. *The Rancholabrea fauna:* a few of the commoner large mammals typical of this community. *A. Megalonyx* (giant sloth). *B. Arctodus* (giant bear). *C. Smilodon* (sabertoothed cat). *D. Camelops* (camel). *E. Titanotylopus* (protogiraffe). *F. Preptoceras* (giant ox).

The Rancholabrea Fauna

The Rancholabrea fauna has been extensively described and illustrated. It was composed of horses, camels, bison, elephants, mammoths, several species of ground sloth, sabertoothed cats, lions, wild dogs, dire-wolves, a number of large bird species, and many smaller genera and species of modern rodents. The projectile-tip nomadic bands would already have had a long cultural history of hunting such animals, many of which (or their ecological equivalents) are known from Asia at this time. The American populations of these animals would, however, never have been exposed previously to any hominid-type predator, and they had no prior selection for predator-prey relationships in connection with hominids. The larger animal species would thus be easily killed by the invading Monogoloids (Figure 4-24).

Pleistocene Overkill

It could be surmised that with such an abundance of easily obtainable food, there was a population explosion among these hunter-gatherer bands, which would rapidly increase in number and in territory occupied. There is still much controversy as to the large animal–man interrelationships. One school of thought holds that what has come to be called "pleistocene over-kill" occurred between about 15,000 and 12,000 BP, when many of the genera of the Rancholabrea megafauna passed to extinction (Figure 4-25). Simultaneously, remains of hunter-gatherer Early Mongoloid projectile-tip or "Clovis-tip" peoples began to appear in scattered areas of western and southern North America. At this time camels, horses, saber-toothed cats, elephants, mammoths, all species of sloth, lions, hunting dogs, dire wolves, and several forms of deer and antelope became extinct. The only forms of larger animals surviving to the present time from this rich fauna are bison and pronghorn antelopes.

P. S. Martin was the first to draw attention to the phenomenon that he named "Pleistocene overkill." He examined the disapperance rate of genera of game animals that it might be supposed early *H. sapiens* populations were hunting and killing. He found that although during various epochs of the Tertiary period such extinction had proceeded at a relatively smooth and steady rate, the Late Pleistocene was characterized to coincide in some instances with new migrations or major cultural advances in early *H. sapiens* populations.

African Overkill

In Africa the megafauna had been in contact with hominids for at least 4 million years, perhaps a little longer if hominid evolution can be extended backward into the Pliocene for any significant length of time. During this long period a predator-prey relationship would have developed in which any evolutionary advance permitting hominid predators to kill prey more readily would be paralleled by selection of the prey for a greater ability to escape this enhanced predation.

FIGURE 4-24. *A Paleo-Indian hunt.* Folsom man attacking a camel near Tule Springs, Nevada.

Such evolutionary balances are essential for the maintenance of stability in any predator-prey relationship. What would seriously disturb this balance would be cultural rather than morphological or physiological evolution occurring in the hominid predators. The speed at which cultural selection and adaptation could take place in the predator would be entirely beyond the prey's capacity to match by selection and adaptation of morphological and physiological features alone.

Martin notes that with the appearance of the utilization of fire and hafted weapons in Africa, apparently about 50,000 years ago, 40 per cent of the megafauna genera passed to extinction (Table 4-1).

Hunting Innovations

There is a supposition that the general method of game hunting practiced by *erectus-sapiens* populations in Africa prior to this time was to drive animals into swampy areas, where their mobility would be greatly restricted (Figure 4-26). They were slaughtered before they could extricate themselves. A line of fire across a savanna would be far more effective in driving

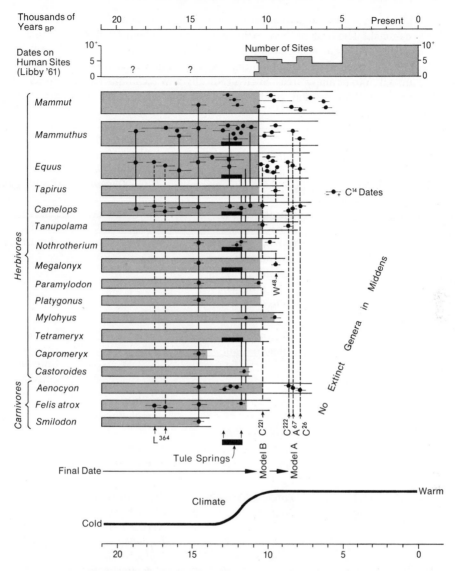

Figure 4-25. *Pleistocene overkill in North America.* See text for further explanation.

game downwind than a thin line of shouting, waving men. Likewise, a hafted weapon would be more effective and far less dangerous than a hand-held one in killing a bogged-down animal.

Pleistocene Overkill Elsewhere

If Martin is correct, Africa lost 40 per cent of its game genera as a result of these two cultural "mutations of fire and weapon hafting," but North

TABLE 4-1 *Pleistocene Overkill in Tropical Africa*

This table shows the date of extinction of various genera of large African mammals. The peak of these extinctions occurred about 50,000 BP, when it is believed fire using was first introduced to this continent. This could have led to a fundamental change in hunting techniques, in which fire was used to drive animals into swampy areas. Such a radically new technique could have seriously unbalanced the previous predator-prey relationships and led to the extinction of some species.

Order	Villafranchian and Early Middle Pleistocene (1.0–2.0 million years)	Late Middle Pleistocene extinction (last 100,000 years)	Living genera
Primates	Gorgopithecus Dinopithecus Cercopithecoides Australopithecus Paranthropus Telanthropus Parapapio	Simopithecus	Pan Gorilla Mandrillus
Carnivora	Lycyaena Meganteron Homotherium	Machairodus	Actinonyx Panthera Hyaena Crocuta
Tubulidentata			Orycteropus
Proboscidea	Anancus Stegodon Deinotherium	Archidiskodon Gomphotherium	Loxodonta
Perissodactyla	Metaschizogherium Serengeticeros	Stylohipparion Eurygnathohippus	Equus Diceros Ceratotherium
Artiodactyla (Suidae)	Potamochoerops Omochoerus	Potamochoeroides Mesochoerus Notochoerus Tapinochoerus Stylochoerus Metridiochoerus "Kolpochoerus" Orthostonyx	Potamochoerus Sus Phacochoerus Hylochoerus
Hippopotamidae			Hippopotamus Choeropis
Camedidae		Camelus*	
Cervidae		Megaceroides	Cervus
Giraffidae		Libytherium	Giraffa
Bovidae	Pultipkagonides Numidocapra	Homoioceras Bularchus Pelorovis Lunatoceras	Okapia Tragelaphus Boccercus Taurotragus

TABLE 4-1 *Pleistocene Overkill in Tropical Africa (cont.)*

Order	Villafranchian and Early Middle Pleistocene (1.0–2.0 million years)	Late Middle Pleistocene extinction (last 100,000 years)	Living genera
		Megalotragus (Gen. nov. 1) (Gen. nov. 2) *Makapania* *Phenacotragus*	*Syncerus* *Cephalophus* *Kobus* *Redunca* *Hippotragus* *Oryx* *Addax* *Damaliscus* *Alcelaphus* *Beatragus* *Connochaetus* *Aepyceros* *Litocranius* *Gazella* *Capra* *Ammotragus*
Total	19	26	40

* Living species surviving in Eurasia.

FIGURE 4-26. *Hunting techniques of* Homo erectus *in Africa.* Chelles Acheu men driving a now-extinct giant antelope into a swamp, about 500,000 BP. Fire was not apparently used at this time in Africa to drive animals or indeed for any purpose.

America lost nearly all its larger mammalian genera contemporaneously with migration of Early Mongoloid bands into the area.

By 1,200 BP virtually all the large animal species had disappeared, as judged from the skeletal remains of this age encountered in the Los Angeles area tar pits. Although there were climatic changes during this interval, which have been carefully and critically considered by D. I. Axelrod, they appear insufficient to account for such a large-scale and dramatic extinction (Table 4-2).

TABLE 4-2 *Summary of Mammalian Megafaunal Extinction and Survival*

	Africa	U.S.A. + Canada
1. Living genera (50 kg)	40	14
2. Later Pleistocene extinction	26+	35
3. Earlier Pleistocene extinction	19	13
4. Normal Pleistocene megafauna	66+	49
5. Later Pleistocene extinction intensity	39 per cent	71 per cent

As has already been mentioned, Early Monogoloid hordes using projectile-tip weapons had been migrating northward in northeast Asia from at least about 30,000 BP, and perhaps earlier. From their weapons and the animal bones associated with them, it is apparent that they were nomadic hunter-gatherers experienced in hunting a megafauna very similar in general composition to that of the Rancholabrea. Geomorphologists have estimated that one or more of these hordes could have penetrated to Beringia from 28,000 to 32,000 BP, when it was an exposed tundra-covered land mass. Finally, after one more exposure of Beringia, the end of the Wisconsin glacial period and a further melting of glacial ice inundated this area to open the Bering Straits, and the human horde or hordes of Beringia would have been edged out into Alaska proper. From there, migrations would have occurred in the normal manner for establishing territory; the first confirmed date of 23,600 BP for penetration to the Los Angeles area fits well with this theory, as do the several others recently determined dates already mentioned.

Nevertheless, many workers still object to the hypothesis that within some 10,000 years or so Pleistocene overkill could have exterminated such an abundant megafauna (Table 4-3). Yet it was not in this instance merely a case of a cultural change unbalancing a predator-prey relationship. This North American megafauna had never before had to cope in any way with predation from bands of cunning hominids armed with effective weapons and perhaps 500,000 years of cultural experience with this type of hunting.

Australia had only two carnivorous animals other than insectivores, the Tasmanian wolf and the Tasmanian devil—neither of them a social animal—so that again the megafauna, in this instance essentially a marsupial one,

TABLE 4-3 *Extinction of Larger Animals in North America in the Late Pleistocene*

This seems to have reached a peak about 12,000 BP and to have virtually ceased by 10,000 BP.

Order	Irvingtonian + Blancan extinction	Rancholabrea extinction (last 15,000 years)	Living genera (1.5–2 million) (years)
Edentata	Glyptotherium Glyptodon	Megalonyx Nothrotherium Paramylodon Eremotherium Boreostracon Brachyostracon Chlamytherium	
Carnivora	Borophagus Ischyrosmilus Chasmaporthetes	Aretodus Similodon Dinobastis Aenocyon Tremarctos*	Euarctos Ursus Felis Panthera Canis
Proboscidea	Rhyncotherium Stegomastodon	Mammut Cuvieronius Mammuthus	
Artiodactyla	Pliauchenia Titanotylopus Hayoceros Platycerabos	Platygonus Mylohyus Camelops Tanupolama Sangamona Cervalces Capromeryx Stockoceros Tetrameryx Bootherium Symbos Euceratherium Preptoceras Saiga* Bos	Cervus Odocoileus Oreamnos Ovibos Ovis Rangifer Antilocapra Bison Alces
Perissodactyla	Nannippus Plesippus	Equus* Tapirus*	
Rodentia		Castoroides Neochoerus Hydrochoerus*	
Reptilia (Testudinata)		Geochelone*	
Total	13	36	14

* Living species surviving south of the United States or in Eurasia.

TABLE 4-4 *List of the Larger Animals Recorded from the Rancholabrea Fauna*

Tetrameryx		*Chlamytherium*	Armadillos
Capromeryx	} Pronghorns	*Dinobastis*	
Stockorcerus		*Smilodon*	} Sabertoothed cats
Mammut		*Neochaerus*	
Curieronius	} Mastodons	*Hydrochoerus*	} Capybaras
Mammuthus	Mammoth	*Euceratherium*	
Megalonyx		*Preptoceras*	} Shrub oxen
Paramylodon		*Tapirus*	Tapir
Northrotherium	} Ground sloths	*Bootherium*	Bovid
Eremotherium		*Cervalces*	Moose
Platygonus	} Peccaries	*Brachyostracon*	Glyptodon
Mylohyus		*Saiga*	Antelope
Camelops	Camel	*Symbos*	Musk-ox
Equus	Horse	*Bison antiquus*	Long-horned bison
Castoroides	Giant beavers		
Arctodus	Bear		

would be unused to quick response to predation by morphological, physiological, or behavioral adaptation. Many marsupial genera also disappeared about this time, contemporaneously with the arrival of human populations.

The most recent extinctions to occur in the world were not so much of whole megafaunas, but rather of one particular element, large, flightless birds. The Maoris maintain that moas had disappeared from New Zealand before their arrival (Figure 4-27). European sailors have no such alibi for Mauritius, where the last dodo was killed about 1680. The arrival of a Polynesian group to populate Madagascar for the first time in 1,000 BP coincides with the date of disappearance of the giant flightless bird *Aepyrornis maximus*.

With the reduction in numbers and species of megafauna in these extensive game ecosystems, whether from climatic change or Pleistocene overkill or both, the hunter-gatherer bands must have undergone cultural evolution directed more to the gathering aspects of the economy than had been true when game was abundant. This change of emphasis may have helped to promote the so-called agricultural revolution that originated in various parts of the Old World and at least once and perhaps several times in the New World, where agricultural groups were clearly established by 5,000 BP.

Migration of Aleuts and Eskimos

The movement of a projectile-tip people through Asia and into the Americas was followed by the eastern and northern progression of later groups, with physical characteristics more closely approaching those of present-day

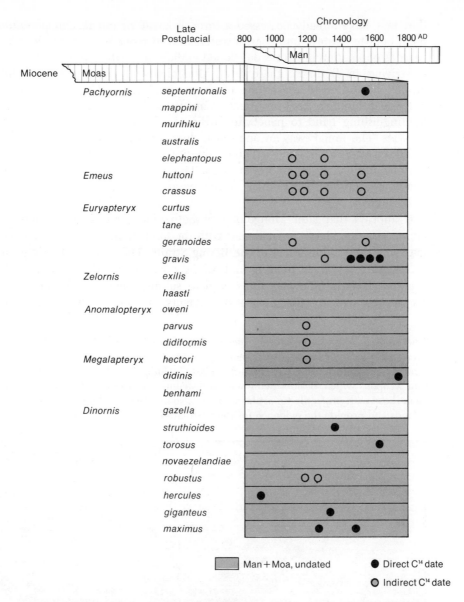

FIGURE 4-27. *Extinction of the moas in New Zealand* correlated with absolute dates obtained from radiocarbon dating.

Recent Mongoloids. Invariably associated with these later Mongoloids is a type of culture known as Aurignacoid, in which extensive use was made of bone and bone tools (Figure 4-28). The comparatively rapid migration of these peoples along the Pacific Coast of Asia may have been facilitated by two related factors. First, they were both hunters and fishermen, and the consumption of fish livers may well have provided the vitamin D necessary for a fairly deeply pigmented group to penetrate into northern latitudes without undergoing mutation toward lighter skin color. Second, they may have already possessed a form of kayak or umiak that permitted them both to venture into coastal waters and to move northward along the coast as population pressures developed in their original territories.

Arriving in Beringia, these later Mongoloids apparently divided into groups, according to the nature of the boats utilized. The kayak was an individual hunter's boat of great maneuverability, which permitted the hunting-fishing band to penetrate farther north into even more inclement habitats. The umiak was an open boat, less seaworthy, less maneuverable, and handled by a small crew. Those using it would need to supplement their hunting and fishing by land operations. This would include the catching of anadromous fishes like salmon, which periodically returned to the rivers of Beringia.

It appears that later Mongoloids of both kinds arrived in Beringia by about 5,000 BP, almost coincidental with the final submergence of that land mass and the establishment of the Bering Straits. The more southerly group continued their mixed hunting and fishing operations and persisted on the Aleutian Islands following the complete submergence of Beringia. They are represented by the present-day Aleuts. The other group followed

FIGURE 4-28. *Migration routes of Recent Mongoloids in North America.* The areas occupied by Aleuts and Eskimos can be identified from artifacts of the general Auragnicoid type (Figure 4-2). These include a number of bone tools and weapons which, besides being widely distributed in Eurasia at sites younger than those with projectile-tip artifacts, occur also in Alaska and the Aleutian Islands and provide the evidence for the postulated migration routes.

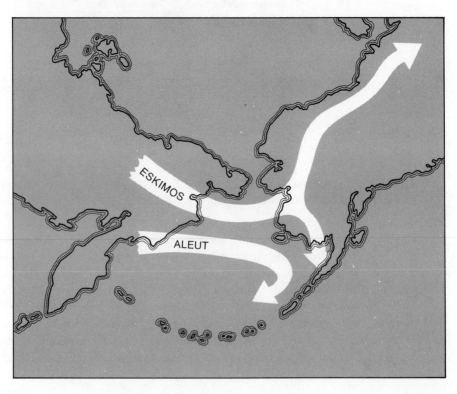

the retreating ice across Canada and along the coast of Greenland with their highly specialized and exclusively hunting-fishing operations and became the present-day Eskimos.

Other Prehistoric Migrations

The Americas were not the only continental masses excluded from early hominid and all pongid evolution. Australia, although a part of the Old World and located not too far from Indonesia with its quite extensive and well-established hominid and pongid occupations, has so far produced no records of hominid remains earlier than 18,000 BP.

Australasia

Like North America, the island of Australia appears to have been invaded in prehistoric times by land, when the accumulation of polar and continental ice sheets lowered sea levels to expose traversable portions of the continental shelf. However, there is one major difference between this Australian invasion and the American one. The Torres Strait is not deep, and intermittent crossing would have been possible during the Pleistocene whenever New Guinea and Australia formed a common land mass. But even at such times the depths of the Timor Sea isolated this enlarged land mass from the rest of southwest Asia. This is why the faunas of New Guinea and Australia are unique. Humans, the only land animals to migrate over this so-called Wallace line, must at some stage somehow have crossed the waters of the Timor Sea. Perhaps this migration was achieved by island hopping on rafts, a process the modern voyages of Thor Heyerdahl have shown to be eminently feasible. Before the introduction of canoes, such "line of sight" island hopping to cross open sea straits up to 25 miles across would nevertheless be hazardous. Some work by N. B. Tindale has shown that off the coast of northern Australia only about half the people who attempted to cross an 8-mile strait actually made it.

There is some evidence that Australia was occupied by at least two separate groups, the Tasmanians, who traveled down the east coast, and the Murrayians, who moved into the more arid regions. A third and northern group, the Carpentarians, may represent the fishermen from islands to the north who began to visit the continent from about 7,000 years ago, with the development of ocean-going canoes. There is one record from Indonesia of an Australian-like skull dating to 50,000 BP, which suggests that an ancestral group originated in this region. The date for the domestication of dogs by *Homo sapiens* in Eurasia appears to fall between 50,000 and 100,000 BP. The dingo is the only eutherian predator known from the Upper Pleistocene of Australia, and it seems a reasonable conclusion that the dingo was a domesticated dog that moved into Australia with the Carpentarians, who had obtained this animal through cultural contact with Eurasia.

Aboriginal Australians with the earliest forms of culture still used non-hafted tools and weapons. The effect on extractive efficiency of mounting weapons on wooden shafts is very apparent. It seems that the island of Tasmania was finally separated from the Australian mainland at a date that separated the people with nonhafted tools from those who later developed them. Tasmanians, the last of whom were exterminated in the nineteenth century, never developed hafted tools. Graphic eyewitness descriptions from early European travelers tell how Tasmanians used hand axes, for example, in climbing trees. The story of Aborigine history in Australia is fascinating and can hardly be summarized in a few paragraphs. Some idea of the many problems still unsolved may be obtained from a collection of recent papers edited by D. J. Mulvaney and J. Golson.

Island Migrations

Besides these major movements into the continental masses of the Americas and Australasia, many migrations have occurred by sea into smaller and sometimes larger islands. For example, the island of Madagascar appears to have been colonized by Polynesian groups about 1000 AD. Other contemporary movements of Polynesians by sea into the islands of Oceania have been suggested and are discussed further in Chapter 6.

Migrations in Historical Times

Migrations of human groups have not yet ceased entirely, although modern national frontier regulations tend to prohibit mass movement. When considering human diversity it should be remembered that, with few exceptions, all human societies have resulted from the eventual incorporation into one territorial group of population elements from a series of migrations, conquests, slave captures, colonizations, and other kinds of movement that have not only been continuously proceeding in *Homo sapiens* grade communities but also among other species of our genus and among even earlier hominid forms.

In this account of migration, no consideration has been given to the question of what form of *Homo sapiens* grade was involved, or whether the migrating groups are indeed referable to this species grade. It must not be assumed that *H. sapiens sapiens* evolved only once as Cro-Magnon man, and radiated out from Europe over the whole globe. There is considerable evidence for the supposition that each of the three main geographical groups already discussed, independently evolved from an *H. erectus* grade to an *H. sapiens* grade. Support for such speculation is further discussed in Chapter 6.

BIBLIOGRAPHY

References

Axlerod, D. I. "Quaternary extinction of large mammals," *Univ. Calif. Publ. Geol. Sci.*, **75**: 1–42, 1967.

Baker, J. R. "Cro-Magnon man 1868–1968," *Endeavour*, **27**: 87–90, 1968.

Binford, S. R. "Late middle Paleolithic adaptations and their possible consequences," *Bioscience,* **20:** 280–83, 1970.

Birdsell, J. B. "Preliminary data on the trihybrid origin of the Australian Aborigines," *Archeol. Phys. Anthrop. Oceania,* **2:** 100–55, 1967.

Bordaz, J. *Tools of the Old and New Stone Age,* New York: Natural History Press, 1970.

Boyd, W. "The contributions of genetics to anthropology," in A. Kroeber's *Anthropology Today,* Chicago: University of Chicago Press, 1953, pp. 488–506.

Brace, C. L., "Ridiculed, rejected, but still our ancestor, Neanderthal," *Natural History,* **77**(5): 39–45, 1968.

Bresler, J. B. (ed.) *Environments of Man,* Reading, Mass.: Addison-Wesley, 1968 ("An exchange of views," pp. 169–76).

Brose, D. S., and Wolpoff, M. H. "Early upper paleolithic man and late middle paleolithic tools," *American Anthropologist,* **73:** 1156–94, 1971.

Butzer, K. W. *Environment and Archeology, An Ecological Approach to Prehistory,* Chicago: Aldine-Atherton, 1971.

Chang, *The Archeology of Ancient China,* New Haven: Yale University Press, 1968.

Clark, J. D. *The Prehistory of Africa,* New York: Praeger, 1970.

Cole, S. "A spanish camp of stone age elephant hunters," *New Scientist,* **16:** 160–62, 1962.

Coles, J. M., and Higgs, E. S. *The Archeology of Early Man,* New York: Faber and Faber, 1969.

Coon, C. S. *The Story of Man,* New York: Knopf, 1970.

Giddings, J. L. *Ancient Men of the Arctic,* New York: Knopf, 1967.

Glass, H. B. "The ethical basis of science," *Science,* **150:** 1245–61, 1965.

Greene, D. L. "Environmental influences on Pleistocene hominid dental evolution," *Bioscience,* **20:** 276–79, 1970.

Harrison, G. A., Weiner, J. S. Tanner, J. M., and Barnicot, N. A. *Human Biology,* New York and Oxford, England: Oxford University Press, 1964.

Haynes, C. V. "The earliest Americans," *Science,* **166:** 709–15, 1969.

Haynes, C. V. "The Calico site: artifacts or geofacts?" *Science,* **181:** 305–10, 1973.

Hester, J. J. "Ecology of the North American Paleo-Indian," *Bioscience,* **20:** 213–17, 1970.

Ho, T. Y., Marcus, L. F., and Berger, R. "Radiation dating of petroleum-impregnated bone from tar pits at Rancho La Brea, California," *Science,* **164:** 1051–52, 1969.

Howell, F. C. "Upper Pleistocene men of the southwest Asian Mousterian," in G. H. R. Van Koenigswald (ed.), *Neanderthal Centenary,* Utrecht: Kemink en Zoon N. V., 1958, pp. 185–98.

Howells, W. W. "*Homo erectus,*" *Scientific American,* **215**(5): 46–53, 1966.

Isaac, G. L. "Studies of early culture in East Africa," *World Archeology,* **1** (1): 1–28, 1969.

Krantz, G. S. "Brain size and hunting ability in earliest man," *Curr. Anthrop.* **9**(5): 450–51, 1966.

Liebermann, P., and Crelin, E. S. "On the speech of Neanderthal man," *Linguistic Inquiry*, **2**(2): 203–22, 1971.

Lumley, H. "A paleolithic camp at Nice," *Scientific American*, **220**(5): 42–50, 1969.

Martin, P. S. "Africa and Pleistocene overkill," in P. S. Martin and H. E. Wright (eds.), *Pleistocene Extinctions*, New Haven: Yale University Press, 1967, pp. 75–120.

Morris, D. *The Naked Ape*, New York: McGraw-Hill, 1967.

Mourant, A. E. *The Distribution of Human Blood Groups*, Springfield, Ill.: Thomas, 1954.

Mulvaney, D. J. "The prehistory of the Australian Aborigines," *Scientific American*, **214**(3): 84–93, 1966.

Mulvaney, D. J., and Golson, J. (eds.) *Aboriginal Man and Environment in Australia*, Canberra, Australian National University Press, 1971.

Reed, C. A. "Extinction of mammalian megafauna in the Old World late Quaternary," *Bioscience*, **20**: 284–88, 1970.

Rideaux, T. *Cro-Magnon Man*, New York: Time-Life, 1973.

Sahlins, M. D. "The origin of society," *Scientific American*, **203**(3): 76–87, 1960.

Tindale, N. B. "Some population changes among the Kaladilt people of Bentinck Island, Queensland," in *Records of the South Australian Museum 1962*, Vol. 14, pp. 259–96.

Vayda, A. P. (ed.) *Environment and Cultural Behavior*, New York: Natural History Press, 1969.

Washburn, S. L., and DeVore, I. "The social life of baboons," *Scientific American*, **204**(6): 62–71, 1961.

Further Readings

Baker, P. T., and Weiner, J. S. *The Biology of Human Adaptability*, Oxford, England: Clarendon Press, 1966.

Binford, S. R., and Binford, L. R. *New Perspectives in Archeology*, Chicago: Aldine, 1968.

Binford, S. R., and Binford, L. R. "Stone tools and human behavior," *Scientific American*, **220**(4): 70–84, 1969.

Brace, C. L. "The origin of man," *Natural History* **79**: 46–49, 1970.

Brace, C. L., Nelson, H., and Korn, N. *Atlas of Fossil Man*, New York: Holt, Rinehart and Winston, 1971.

Braidwood, R. I. *Prehistoric Men*, 7th ed., Glenview, Ill.: Scott, Foresman, 1967.

Breuil, H. *Four Hundred Centuries of Cave Art*, trans. by M. E. Boyle, Montignac: Centre d'Etude et de Documentation Prehistorique, 1952.

Buffington, J. D. "Predation, competition, and Pleistocene megafauna extinction," *Bioscience* **21**: 167–70, 1971.

Campbell, B. G. "The roots of language," in S. Morton (ed.), *Biological and Social Factors in Psycholinguistics*, Logos Press, 10–23 1971.

Campbell, B. G. (ed.) *Sexual Selection and the Descent of Man*, Chicago: Aldine, 1972.

Chang, "Archeology of ancient China," *Science,* **162:** 519–26, 1968.

Clark, J. G. D. "Radio carbon dating and the expansion of farming cultures from the Near East," *Proc. Prehistoric Soc.,* **31:** 58–73, 1965.

Clark, J. D. "Acheulian occupation sites in the Middle East and Africa: a study in cultural variability," in J. D. Clark and F. C. Howell (eds.), Recent Studies in Paleoanthropology, *American Anthropologist,* **68**(2): Part 2, 394, 1966.

Clark, J. G. D. *The Stone Age Hunters,* New York: McGraw-Hill, 1967.

Clark, J. G. D., and Piggott, S. *Prehistoric Societies,* London: Hutchinson, 1965.

Constable, G. *The Neanderthals,* New York: Time-Life, 1973.

Crabtree, D. E., and Davis, E. L. "Experimental manufacture of wooden implements with tools of flaked stone," *Science,* **159:** 426–28, 1968.

Crow, J. F. "The quality of people: human evolutionary changes," *Bioscience,* **16:** 863–67, 1966.

Darwin, C. *The Descent of Man,* London: Murray, 1909.

Dixon, J. E., Cann, J. R., and Renfrew, C. "Obsidian and the origin of trade," *Scientific American,* **218**(3): 38–44, 1968.

Gould, R. A. "Chipping stones in the outback," *Natural History,* **77**(2): 42–49, 1968.

Haynes C. V. "Elephant-hunting in North America," *Scientific American,* **214**(6): 104–12, 1966.

Howell, F. C. *Early Man,* New York: Time-Life, 1965.

Howells, W. W. *Mankind in the Making: The Story of Human Evolution,* rev. ed., New York: Doubleday, 1967.

Hundert Jahre Neanderthaler/Neanderthal Centenary, Gedenkbuch der Int. Neanderthal Feier Düsseldorf, 26–30 August 1956, Utrecht: Kemink en Zoon, 1958.

Ju-Kong, W. "The skull of Lantian man," *Curr. Anthrop.,* **7**(1): 83–86, 1966.

Kurtén, B. "The cave bear," *Scientific American,* **226**(3): 60–62, 1972.

Klein, R. G. "Mousterian cultures in European Russia," *Science,* **165:** 257–65, 1969.

Leakey, L. B. S. *Olduvai Gorge 1951–1961, Fauna and Background,* Cambridge, England: Cambridge University Press, 1965.

Lee, R. B., and DeVore, I. (eds.) *Man the Hunter,* Chicago: Aldine, 1968.

Liebermann, P., Crelin, E. S., and Klatt, D. H. "Phonetic ability and related anatomy of the newborn and adult human, Neanderthal man, and the chimpanzee," *American Anthropologist,* **74**(3): 287–307, 1972.

Martin, P. S. "Pleistocene niches for alien animals," *Bioscience,* **20:** 218–21, 1970.

Race, R. R., and Sanger, R. *Blood Groups in Man,* 4th ed., Oxford, England: Blackwell, 1962.

Sheppard, P. M. "Blood groups and natural selection," *Brit. Med. Bull.,* **15:** 134–39, 1959.

Simons, E. L. "Some fallacies in study of hominoid phylogeny," *Science,* **141:** 879–89, 1963.

Solecki, R. S. *Shanidar, The First Flower People,* New York: Knopf, 1971.

Sonnenfeld, J. "Interpreting the function of primitive implements," *American Antiquity,* 28(1): 56–65, 1962.

Spuhler, J. N. (ed.) *The Evolution of Man's Capacity for Culture,* Detroit: Wayne State University Press, 1961.

Steward, J. H. *Theory of Culture Change: The Methodology of Multilinear Evolution,* Urbana: University of Illinois Press, 1969.

Ucko, P. J., and Rosenfeld, A. *Paleolithic Cave Art,* London: Weidenfeld and Nicholson, 1967.

Warner, W. L. *A Black Civilization,* New York: Harper and Row, 1964.

Washburn, S. L. "Tools and human evolution," *Scientific American,* **203** (3): 62–75, 1960.

Washburn, S. L. (ed.) *Social Life of Early Man,* Chicago: Aldine, 1961.

Weidenreich, F. *Anthropological Papers,* Memorial Volume, compiled by S. L. Washburn and D. Wolffson, New York: The Viking Fund, 1949.

Wilmsen, E. N. "Lithic analysis in paleoanthropology," *Science,* **161:** 982–87, 1968.

Young, J. Z. *An Introduction to the Study of Man,* Oxford, England: Oxford University Press, 1971.

Zeuner, F. E. *A History of Domesticated Animals,* London: Hutchinson, 1964.

5

The Origin of Urban Civilization

The previous three chapters have considered the ecological evolution of human populations from the first living organisms on this earth, through primate and ground-ape stages, to various *Homo erectus* grades, and finally to *Homo sapiens sapiens*. By the time they had reached this last level, human populations had a sophisticated communication system and social order, were equipped with considerable manual dexterity enabling them to fashion a range of artifacts entirely sufficient for their *ecosystem* roles, and had covered most of the earth with a mosaic of nucleated territories more or less stable and basically adequate for the support of the total global population, estimated as falling between 5 and 6 million persons.

The further cultural evolution of these first *sapiens* populations is closely correlated with selection for a "concentration" or "agglomeration" characteristic. That is, a selection factor emerged that favored behavioral activities directed toward the formation of interacting groups of ever-increasing size and density. The most obvious advantage provided by this selected characteristic is that it enhances the accumulation and exchange of information. Hunter-gatherer bands could make individual discoveries, but much time would elapse before these could be universally incorporated into the ritualistic behavioral patterns of all extant societies. Greater concentrations of individuals, and still further improved verbal communication systems, would lead to the accelerated dissemination of new ideas, until with modern methods of communication this process of cultural exchange mushroomed into what is often called the "information explosion." This "concentration" trait would have to be selected for concurrently with whatever traits promoted the establishment of permanent settlements.

The Earliest Settlements

Hunter-gatherer cultures of all hominid and human populations were essentially nomadic, although they might seasonally or cyclically occupy a specific section of territory. As Birdsell has demonstrated, the size of the territory occupied was determined by limiting environmental factors (Figure 5-1). This obligatory nomadism, dictated by both the local and the seasonal depletion of resources, still is exhibited by a few survivors of this type of culture such as the Bushmen of South Africa, or the Australian Aborigine hordes. In some instances, however, a chance combination of microecosystems could have provided on the same site a succession of over-

FIGURE 5-1. *Correlation between the mean annual rainfall and the territory size* for Australian Aborigine tribes. The area needed by the tribe is larger with a low rainfall, smaller with a high rainfall, as would be anticipated.

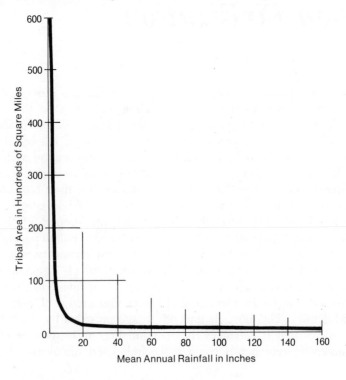

lapping resource productivity peaks sufficient to permit continuous occupation of the site (Figures 5-2 and 5-3).

Drinking Water

The first requirement for such a permanent site is an assured supply of drinking water. This is found, for example, on a seacoast, where shallow wells can tap the fresh water that floats on the salt-water table below the sand dunes. At the close of a Middle Pleistocene glacial, or the beginning of an interstadial, there would also be many suitable freshwater lake and river sites.

Food Supplies

All such permanent sites, whether coastal or interior, would also have readily accessible supplies of invertebrate food at any time of year. Supplemental gathering activities might be seasonal, like gathering acorns in the Pacific hinterland of North America, or picking fruits of various *Grewia* species along southern African rivers, or harvesting wild grains in many regions. Hunting similarly might have been restricted to a breeding season, when the more readily taken juvenile animals were available. In suitable areas, for example, southwestern France, game may have been so plentiful that cave sites could be occupied all year.

Such village settlement patterns must have arisen many times in these kinds of favored locations long before the so-called agricultural revolution.

FIGURE 5-2. *Schematic representation of a microenvironmental series* sufficient to permit *permanent* human occupation of the area. The transect illustrates conditions across a 20-kilometer section of the Tehuacan Valley, Puebla, Mexico.

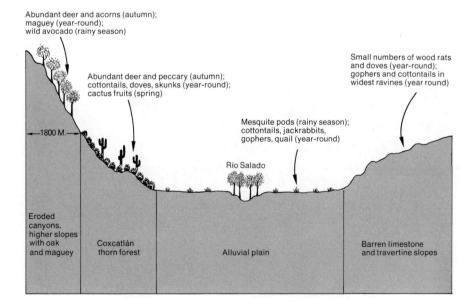

Abundant deer and acorns (autumn); maguey (year-round); wild avocado (rainy season)

Small numbers of wood rats and doves (year-round); gophers and cottontails in widest ravines (year round)

Abundant deer and peccary (autumn); cottontails, doves, skunks (year-round); cactus fruits (spring)

Mesquite pods (rainy season); cottontails, jackrabbits, gophers, quail (year-round)

—1800 M.—

Río Salado

Eroded canyons, higher slopes with oak and maguey

Coxcatlán thorn forest

Alluvial plain

Barren limestone and travertine slopes

Indeed a number of these permanent sites are known from the Middle East dating between 8,000 and 11,000 BP. The Kom Obo Plain in southern Egypt, some 30 miles downstream from the modern Aswan Dam on the Nile, seems to have been permanently occupied by hunter-gatherers for about 5,000 years from as early as 17,000 BP. The total human population of the approximately 250 square mile plain is estimated to have been around 150–200 persons, possibly about a half dozen extended family bands. The permanent nature of the settlement would facilitate the accumulation of cultural equipment, both material and behavioral (Figure 5-4). At the foot of sandstone cliffs on the Kom Obo Plain have been found the grinding stones used for hand milling wild grains. The provision of shelter around each pair-bonded hearth, to cite a simple example, must have been among the earliest innovations here as elsewhere. Once its effectiveness had been demonstrated, it could rapidly be copied and improved.

Early Settlement Life

It is tempting to conjecture as to the life of such preagricultural early *sapiens* settlements. From contemporary survivals persisting in out-of-the-way places and from archeological evidence such as is provided from the Nelson Cave site on the Cape Province Coast in South Africa, first occupied about 18,000 BP, some educated guesses can be made. Along many tropical shore lines small coastal villages can be encountered that have no access by road and very little connection by boat with neighboring settlements. Judging from the size of the kitchen middens, the village sites have often been

FIGURE 5-3. *A coastal site with sufficient microecosystems to support a permanent settlement.* A transect of the Ocoo area of coastal Guatemala about 15 kilometers long. This is taken at the site of a pre-Columbian settlement now known as Salinas la Blanca.

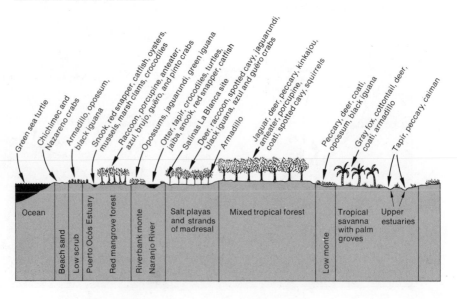

occupied for a considerable length of time. Proceeding backward in time, the life of such an early village as it was at first may have been somewhat as follows.

The settlement included fewer than 100 persons, and little more than a dozen hut clusters. The huts were built from fronds and trunks of coconut palms, with coconut fiber constituting the main tying, binding, and knotting material, as it did in many other such operations. Fishing was a continuous year-round occupation for the men and youths, but was carried out on an individual basis, using spears, baited hooked or wedged lines, and a ring or cast net. The women collected shellfish of various kinds from beach areas, and discarded shells piled up in middens near the village huts. Domestic water came from shallow wells 6 to 10 feet deep dug through the sand immediately behind the shoreline. Toward the end of the dry season this water became more brackish and rather unpleasant to drink.

Besides collecting shellfish, drawing water, gathering dead coconut fronds and flotsam cast up on the beach for domestic fuel, the women might have rather casually tended small overgrown beds of nut grass (*Cypersu esculentus*), which has a small underground rhizome, edible when boiled. Even today this is one of the few crops that will grow in close proximity to the beach, because of the driving salt spray that blows over coastal villages. Nut grass probably represents a very early selection from some naturally occurring ubiquitous tropical sedge species such as *Cyperus rotundus*. Women also foraged out on food-gathering expeditions behind the shoreline (Figure 5-4).

Diet

The diet of fish, shellfish, nut grass, berries, nuts, and coconut was varied periodically with game meat, such as when, at differing times of the year, the several local species of turtle came ashore to breed. The telltale tracks the female turtles left in the sand led to their detection and slaughter, and to the unearthing of the eggs they had concealed in the sand above the tide line. Other dietary bonanzas occurred when a whale or other large marine animal became stranded on the shore.

Domestic Animals

Dogs roamed around the village, but as scavengers; they were not fed or watered. At some time the settlement acquired a herd of swine. Like the dogs, the pigs were not fed, but lived especially on sand crabs from the beach and by scavenging on the middens. A few chickens, likewise unfed, also were obtained at some later stage. They were killed and eaten only on ceremonial occasions.

Division of Labor

In such an early village there was very limited communal sharing or division of labor, although the population had the social structure of an

FIGURE 5-4. *A hypothetical tropical African coastal settlement* in a mesolithic and pre-agricultural phase. This kind of settlement, although formed by Negroids, is very similar in structure and activity to that of Salinas la Blanca established by Early Mongoloid Paleo-Indians in Guatemala and illustrated in Figure 5-3. Hunter-gatherer bands probably quite independently established such permanent settlements along all tropical coasts, wherever the diversity and productivity of microecosystems were sufficient to support them, several millennia previous to the agricultural revolution.

extended family. The women of the pair bonds would associate in groups during their various gathering activities, but each returned with the fruits of her labors to her own hearth. Her daughters might assist her when they reached puberty and before they were married. Female activities were essentially gathering ones. Males fished individually, and likewise returned their catch to their own huts. The only times food might be shared were the few occasions when turtles or marine mammals were killed or washed ashore. Some fish would be smoked or dried and used to barter for tools and other artifacts, clothing materials, and probably brides; marriage would be exogamous, and wives would be obtained by trading with neighboring settlements.

Cultural Activities

Life in such a village would not be so continuously exhausting that there would be no time for leisure, and some crafts would develop as would communal dancing and storytelling. Perhaps the earliest specialization, with some women more proficient than others, would be weaving coconut fronds into baskets, fans, and other articles. Among the men, specialization may have taken the form of net making, or carving further bone tools, which eventually progressed to boat building. Thus both sexes probably had some early division of labor, but there would be no communal schooling, and special skills would pass on by social learning and some degree of inherited adaptation from mother to daughter, father to son. Population regulation would result from mortalities following spasmodic fights over wells and coconut grove areas, by accidental drownings, and perhaps by outbreaks of food poisoning from eating infected shellfish. There could also have been cultural practices for population regulation, as will be described in later chapters.

Origin of Cultivated Plants

Inland settlements of similar complexity, size, and activity such as that just described must have been scattered along river and lake banks throughout much of the area first occupied by *Homo erectus* grade peoples, then later by evolving *erectus-sapiens* populations. Such favorable occupation sites would have continued to be inhabited as *erectus-sapiens* groups graded by imperceptible stages into *Homo sapiens,* whenever this might have occurred in such inland settlements. Perhaps occasional waves of a more highly evolved intermediate form swept over the settlements of a given region, just as one tribe frequently was observed to be in the process of overrunning another in eighteenth- and nineteenth-century Africa and elsewhere.

The First Cultigens

Food gathering would continue to be a major occupation of the women in the early *sapiens* populations and a wide variety of gathered seeds would be carried to the village. Reaped wild grasses might have been brought back for threshing, winnowing, and grinding. It is not difficult to imagine that some of these grass seeds would fall on the piles of debris that comprised the kitchen middens, the first urban garbage dumps. Essential characteristics of the soil of these middens would be a high mineral content, from the decomposing bones and seashells (of snails in inland settlements) that were thrown there, and also from fecal matter deposited on them. The disturbed nature of the midden soil would be coupled with good drainage on these loosely packed dumps, which would in any case be near the huts and so on raised ground, above the general flood level. It is not surprising that grain cereal crops like wheat and barley, which appear to have been those first cultivated, require soils with good drainage, high nutrient content supplied from artificial or natural manures, and an open soil structure attained by plowing and harrowing or digging and hoeing the soil in seed-bed preparation. Because of their heavy requirements for soluble nitrogenous salts, such cultigens are technically described as *nitrophilous* plants, although their phosphate requirements frequently are equally great.

The Fertile Crescent

According to classical evidence, the first cultivated plant appears to have been wheat. The area where this occurred probably was in the so-called Fertile Crescent of Iraq and Iran, roughly the upland area separating the valleys of the Tigris and the Euphrates from Turkey to the north and Iran to the east (Figure 5-5). In the upland areas (averaging about 1,000 meters altitude) of this Fertile Crescent there was adequate winter and spring rainfall, and herbaceous plants were abundant between and within the open canopy of the woodland ecosystems. As early at least as 50,000 years BP, grinding stones were employed in preparing food, presumably the harvested seeds of such herbaceous plants. Flint sickle blades have been found together with these milling and grinding stones, suggesting that wild cereals actually were reaped and brought to settlements for processing, as may be observed in the case of modern Australian Aborigines.

Early Farming Villages

The two earliest Middle East village-farming communities so far extensively explored date from 7,000 and 6,500 BP, although recently a site dating to at least 9,500 BP has been described. Jarmo, the more ancient of these two, and Tepe Sarab are approximately 120 miles apart in modern Kurdistan in northeastern Iraq. Jarmo was a permanent settlement with about two dozen mud-walled houses. It has been estimated that about 150 people lived there with dogs and other animals. Some tools were made from obsidian, the nearest source of which was 200 miles away.

Two kinds of cultivated wheat and one of barley have been found at Jarmo, together with lentils, peas, and a kind of vetch, although it is not certain that the last three were actually cultivated. Goats, dogs, and perhaps sheep were domesticated animals in the village, while wild species of goats, sheep, cattle, pigs, horses, asses, and dogs were all present in the surrounding areas.

Archeologists have considered that Jarmo and other settlements like it represent a relatively sophisticated stage of the first agricultural societies and that the actual origins of agriculture should be sought still earlier, a view now substantiated by the work of H. Cambel and R. J. Braidwood at Cayonu Tepesi. Some further support for this view has come with the

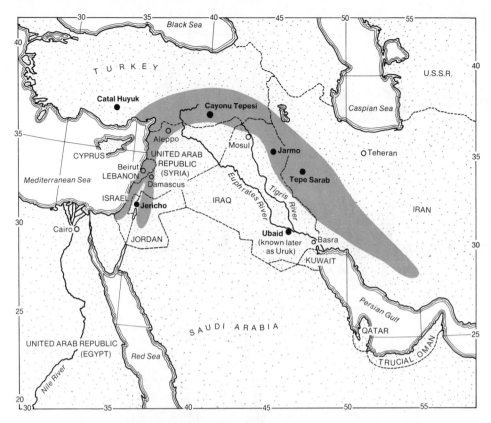

FIGURE 5-5. *The geographical area covered by the Fertile Crescent,* which lies in the region known at the beginning of this century as Mesopotamia, is now allocated among Iraq, Iran, Turkey, Jordan, and some other national territories. Although the cultivation of plants and domestication of animals developed independently in other areas, the Fertile Crescent is most significant in that cereal crops appeared earlier here on an extensive scale than anywhere else, and more animal species were domesticated here than in any other region.

excavation of a city known as Catal Huyuk, on the Anatolian Peninsula of Turkey, by British archeologist J. A. Mellaart.

Trading Settlements

Catal Huyuk was an early neolithic settlement, whose first occupation levels can be dated to 9,000 BP. It was initially a prepottery settlement, as evidenced from its oldest occupation level. It appears to have been established along a river bank and to have covered about 32 acres, although not all of this was necessarily occupied at one time. Evidence that it was more than a village of farmers comes from the variety of cultural materials encountered—especially the abundance and range of tools made from obsidian—together with the quality and consistent planning of the mud-walled houses (Figures 5-6).

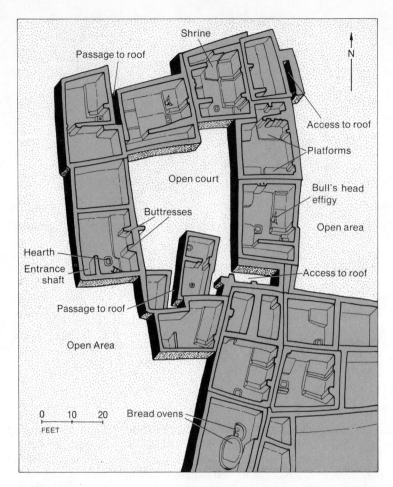

FIGURE 5-6. *A plan of the condominium accommodations in the neolithic city of Catal Huyuk* reconstructed to show how the house walls probably stood, but without including any roof structures. This Turkish city, at this time about 8,000 BP, had remarkably advanced agricultural and commercial activities. (From James Mellaart, "A Neolithic City in Turkey," *Scientific American,* **210**[4]: 95, 1964. Copyright © 1964 by Scientific American, Inc. All rights reserved.)

Agricultural Activities

Among the cereal grains identified from the oldest levels are six-row barley and a hexaploid type free-threshing wheat. Grain bins were found in every house, together with mortars for husking grain and querns for grinding it. Domestic animal remains include sheep, cows, dogs, and goats, but no pigs; wild animal bones indicate wild cattle, reindeer, wild asses, wild sheep, boars, and leopards. The dead were buried underneath the houses, and the disproportionate number of females found interred suggests that many men died away from the settlement, on foraging, hunting, or trading expeditions. Clothing worn by both sexes was made of leopard skin, fastened with bone pins.

An Early City

Mellaart considers that because of the obvious prosperity of this settlement of Catal Huyuk, which must be called a city because of the many activities that took place in it as compared with those of an agricultural village, it is clear that obsidian was traded there extensively. The source of obsidian nearest to the city is only 50 miles to the east, and other active volcanoes that would have produced this natural glass were not much farther distant. Obsidian is not commonly encountered in this region of Asia Minor or Europe; it was superior material from which to fashion the flake cutting tools of the neolithic period, a use first encountered in the mesolithic period.

Urban Origins

Jane Jacobs has speculated about the influence which this obsidian trade had on the development of the city. She hypothesizes that about 12,000 BP a hunter-gatherer group located deposits of obsidian at a site at first unintentionally incorporated within its territory. As this material was discovered and worked into cutting tools by the group, it was found that the material, the tools, or both could be traded with adjoining hunter-gatherer groups in return for products of the hunt such as live or dead animals, and for gathered food such as berries, nuts, and wild cereal grains.

Barter

Such would be the demand for this obsidian trade that a permanent area would be established within the group territory where barter could be carried on. This would not be at the site of the obsidian deposits themselves, which would be kept secret and protected, but rather some distance from them, probably on the perimeter of the group's territory. This obsidian-trading settlement would rapidly grow in size and prosperity as more representatives came in to offer the products of their hunting and gathering in trade.

Jacobs supposes this was the situation several millennia previous to the first known occupation of Catal Huyuk, perhaps as early as 12,000 BP, at a time before there were cultivated plants and domestic animals to be traded. She postulates that some of the wild animals brought in for barter were alive and that grain bins were filled to overflowing with all kinds of gathered wild plants.

Appearance of Cultivars

Seeds of all these wild plants would inevitably be scattered around the settlement accidentally. Some would germinate to become established on kitchen middens, at first being ignored. The plants more successful in propagating themselves on such sites would occasionally hybridize with other forms, a process facilitated by the admixture of species resulting from the obsidian trade. Eventually the more prolific of these natural hybrids, which may also have meanwhile mutated to give a more vigorous form or one with a higher yield, would come to the attention of one or more of the inhabitants, who might hit on the idea of gathering its seed and scattering it over the midden to extend the plant's area.

Origin of Domestic Animals

In the case of the live wild animals received in trade, the quieter ones would be left alive in preference to the more obstreperous beasts, which would be killed first because they were such a nuisance to keep or to feed. The former, which might forage for themselves around the settlement and be no great trouble, might have been allowed to remain alive even long enough to produce offspring. It would be an easy transition from this to domestication, although there seems little doubt that cattle were first domesticated for religious purposes rather than for actual consumption.

According to Jacob's suppositions, the agricultural revolution resulted from the development of at most a few trading settlements, *not* from the activities of the small permanent occupation sites that would have become generally established over a wide area during the early *sapiens* phase of human evolution. In support of this contention there is increasing evidence that in Mesopotamia from about 9,500 to 8,500 BP agricultural villages and hunting villages coexisted.

The Agricultural Revolution

Whatever the origins of agriculture, there can be no doubt that once cultivated plants and domestic animals had been developed, their use spread rapidly. The pattern of permanent settlement was already laid down, social contacts had been established through exogamous mating, and knowledge of the techniques of this major new culture could disseminate quickly. In the 2 millennia from 9,000 to 7,000 BP, neolithic agricultural cultures became widespread throughout the Middle East, and especialy in the valleys of the Tigris and the Euphrates and across the Sinai Peninsula in the Nile Delta. H. G. Baker estimates that by the year 3,800 BP Assyrian and Babylonian irrigation systems using brick-lined canals were irrigating an area of 110,000 acres with water from the Tigris and the Euphrates.

Parallel Development of Agriculture

The origins of agriculture in other parts of the world are now being more extensively explored, although they likewise are still the subject of much speculation.

The New World

In the New World many efforts have been made to trace the origin of corn and of various legumes and cucurbits. Despite much research, the real origin of corn (maize), which archeologists estimate dates back some 7,000 years, has not yet been clearly established as to time, parentage, or place—although the last probably was somewhere in Central America or in the northern part of the Andes. Both areas seem to have been possible centers of origin for another major crop, the "Irish" potato, while manioc (cassava), peanut, sweet potato, and a number of lesser crops such as amaranth, tomato, chili, *zapote*, avocado, cocoa, rubber, cocaine, cotton, and tainus yam (*Xanthosoma*) also originated in the New World (Figure 5-7). These plants were

FIGURE 5-7. *Some of the major food crops developed in the New World.* A. Groundnut (*Arachis hypogea*). B. Potato (*Solanum tuberosum*). C. Cassava (*Manihot utilissima*). D. Tomato (*Lycopersicum esculentum*). E. Tainus or yautia (*Xanthosoma sagittifolium*). F. Sweet potato (*Ipomea batatas*). G. Corn (*Zea mays*). H. Cocoa (*Theobroma cacao*).

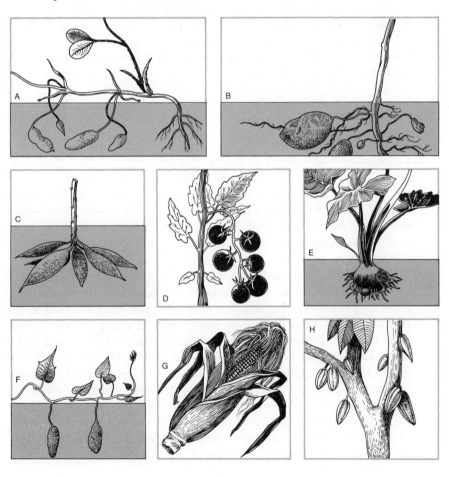

well established as cultigens in Mexico by what archeologists call the Middle Formative or Middle Preclassic Period commencing about 3,500 BP. It is easy to imagine that the establishment of great temple cities like Teotihuacan about a thousand years later would have greatly assisted in the rapid spread of the use of these various cultigens and any new cultivars of them that were selected.

Asia

In Asia, paddy rice, which is now the staple food of half the world's population, must have been among the first crops cultivated. There are a number of surviving wild species of this genus (*Oryza*), some of which are still gathered for food. Recent work at sites in Burma and Thailand, described by W. G. Solheim, places the development of cultivated paddy rice at least as early as 12,000 BP. Indeed, Solheim believes that rice and perhaps other plants were first cultivated farther to the north in southeast Asia about 15,000 BP. Solheim considers also that pigs may have been domesticated in northern Thailand by 12,000 BP, but that zebu cattle and water buffalo were not domesticated anywhere until about 9,000 BP. These dates for cultivated plants and domestic animals are thus earlier by several millennia than those so far provided for the Fertile Crescent. However, there is no evidence, or indeed any speculation, that cultivation and domestication practices spread from the one area to the other.

Other Tropical Areas

In other tropical areas there was probably an independent discovery of various forms of cultivated plants. For example, in West Africa, the oil palm (*Elaeis guineensis*) and various forms of *Diosorea* yam were probably among the earliest cultivated crops. Both groups occur on the margins between forest and savanna, which was probably the most favored occupation site for the temporary camps of hunter-gatherer people. It would be merely a question of extending the area of clearing around naturally occurring wild plants of these two forms in order to create something of a garden.

C. O. Sauer has argued for a wild root origin of cultural plants between 11 and 15 millennia BP, maintaining that hunter-gatherer bands could discover they could dig up a root, use most of it, and put some back to grow to sufficient size to harvest again on a later visit. This may explain the development of such crop plants as taro (*Colocasia esculenta*) in Asia and tainus yam (*Xanthosoma*) in South America, but cereals like barley and wheat in the Middle East appear both to antedate the appearance of any major root crop and to exert the strongest cultural influence (Figure 5-8). Curiously, some major clusters of hunter-gatherers, such as Australian Aborigines, never developed any cultivated plants, cereals, or roots—perhaps in the case of Australia, because this continent provided an insufficient number of sites suitable for permanent settlement.

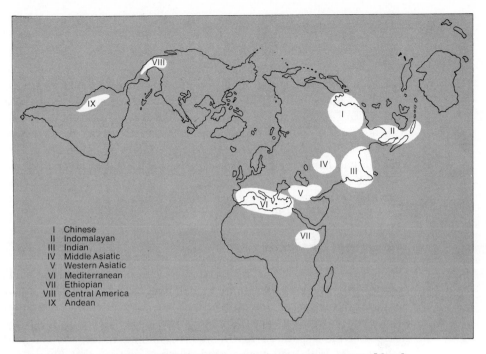

FIGURE 5-8. *The centers of origin of cultivated plants* as demonstrated by the Russian geneticist Vavilov. Some of these coincide with the known early centers of urban civilizations, for example V, the western Asiatic center, and VI, the Mediterranean center. On the other hand, the Ethiopian center is not associated with any known early urbanized society. The Eurasian belt from China to Spain, which includes centers I, III, IV, V, and VI, is the primary area for grain crop origins. Only two major cereals, sorghum (Africa) and maize (tropical America), originated outside this area, and it is possible that early cities developed only in such grain-raising areas. Root crops, which more particularly evolved in center II, were too bulky and sometimes too perishable to be transported any distance for purposes of barter; city evolution was not associated with them.

Domestication of Animals

Archeologists have developed a technique that enables them to decide on a quantitative basis when an animal species has been domesticated. By analysis of the fossils of a megafauna found at various levels of an occupation site over a given time interval, it is possible to detect a shift to a higher consumption of a predomestic wild species. A second shift is detectable by changes in the age at which these predomestic animals were killed. If these two shifts are associated with a continuing increase in the proportional representation of the predomestic species in the total animal remains found in the area, the shifts and the increase together demonstrate that cultural control over the species has been obtained, that the animal has become domesticated.

Animal Domestication in the Middle East

By use of this method of analysis it is apparent that in the Near East domestication of animals had taken place generally by the beginning of the seventh millennium BP, although, as already noted, domestic animals occurred at Catal Huyuk 2,000 years previously. It did not develop in Europe until at least a thousand years later, and in central, eastern, and southwest Asia not until 2 millennia later, i.e., 5,000 BP. However, as already noted, Solheim has presented evidence for an Asian date of 12,000 BP. In the New World it had hardly begun by historic times with the llama and its relatives, and Australian Aborigines have only the dingo, which they may have taken from some other group before reaching Australia.

Domestication of Cattle

Of all the problems that have interested the cultural historian, the ethologist, and the geographer in relation to animal domestication, the most fascinating has been the origin of domestic cattle. The concept that nomadic hunters first domesticated cattle for religious purposes has generally been rejected in favor of the hypothesis that it was sedentary farmers who did so. Some archeologists, with well-documented support, contend that *human* sacrifice preceded domestic animal sacrifice, and this seems possible in regard to some areas of the New World.

Bos primigenius

It is generally believed that all domestic cattle are derived from one wild species, *Bos primigenius,* the wild urus or aurochs (Figure 5-9). This species at one time ranged widely from the Pacific coast of Asia across Eurasia into Europe and North America. Although the last member of this wild population is believed to have died in 1627, a synthetic population has since been recreated by the selective breeding of cattle in domestic herds with urus or aurochs characteristics. With the wide distribution of urus was linked a considerable range in form, size, and probably also coat color, from the time when the animal first appeared in the fossil record during the Riss glacial period.

The possibility has been raised that a second wild cattle species, *Bos africanus,* occurred in North Africa and that this also was domesticated, being now represented only in relict populations in more remote areas. If this is so, the genes of this species would have contributed to further variation in the domestic cattle from Asia, which were eventually introduced into Africa.

Religious Motivation of Cattle Domestication

That the domestication of the urus may have been undertaken in the first instance for religious reasons is suggested by the resemblance between the huge curved horns of the species and the young moon. In fact, historical

FIGURE 5-9. *The aurochs or urus* (Bos primigenius), the wild cattle species that was widely distributed over temperate and subtropical Eurasia in the early Holocene and survived in a wild form until the seventeenth century. All breeds of domestic cattle are believed to have been developed from this ancestral stock.

studies suggest that the urus was regarded as a symbol of the moon goddess and was ritually killed in a charade symbolizing death and resurrection. It is supposed that captured wild animals were kept in corrals for sacrificial use at such ceremonies and that here some young would be born and receive more protection than they would in the wild. Infant mortality would be lower, and there would be a tendency for somewhat unusual forms to survive in greater numbers. Similarly, castration of bulls probably developed as a ritual before the effect this operation has in taming animals, and in improving the amount and quality of meat, was discovered.

It is significant that the religious cult of the city of Catal Huyuk was apparently based on cattle; representations and remains indicate a cattle altar in every dwelling (Figure 5-10).

Early Forms of Domestic Cattle

Quite early there seems to have been a selection for large horns in the domesticated form that has been known as "primigenius" (*Bos taurus primigenius*) and also for shorter-horned or polled types (*Bos taurus longifrons*). Although it is somewhat difficult to distinguish between the wild urus and early domesticated "primigenius" cattle probably maintained as a breed for ritual purposes, there is no problem with the short-horned or polled "longifrons" type, which appears to have been developed more for draft, beef, and milk production.

The religious aspects of the herding of cattle persisted into dynastic Egypt in the tombs of the bulls, which were ceremoniously buried under the Sakkara pyramid near Cairo, and even into modern times with the traditional Hindu sacred Brahman. Perhaps it is reasonable to suggest that

FIGURE 5-10. *A reconstructed shrine from the neolithic city of Catal Huyuk* showing the prominence given to animals and especially cattle heads. (From James Mellaart, "A Neolithic City in Turkey," *Scientific American,* **210**[4]: 100, 1964. Copyright © 1964 by Scientific American, Inc. All rights reserved.)

this ritual has penetrated into and persisted in the New World with the rodeo and "ride the Brahman bull" event and with the Latin American bullfights.

Although domestication of cattle was at first for religious purposes, it may be concluded that carcasses of the animals were normally eaten if they were not given any form of ritual burial. Just when the religious significance of the killing was abandoned, and the cropping of domestic cattle became solely an agricultural operation to provide food, is difficult to determine. It is possible that in some areas, simultaneously with the decline in the sacrificial motive, the milking possibilities rather than the beef potentialities received the greater attention.

There is some confirmation of this chronological account of animal domestication in the recorded use of cattle as the first draft animals. Wheeled vehicles have been found among the items buried during royal funerals in the Middle East dating back to 5,000 BP. In addition either the representations or the remains of sledges, plows, and oxen yokes are commonly encountered in the area at one time called Mesopotamia. In Mesopotamia cattle must have been used as draft animals for at least 7,000 years, and perhaps the use of the ox-drawn sledge antedated even this.

The Spread of Agriculture

During the 12,000 years from the fifteenth to the third millennia BP, it is probable that all the major plants and animals that have at one time or another been used in agriculture were domesticated or cultivated (Table 5-1). It is indeed erroneous to talk of *the* agricultural revolution, because although this process of domestication and cultivation probably first developed in only one area, either the Middle East or southeast Asia, it was repeated a number of times, by peoples apparently at least temporarily in complete cultural isolation, for example, the Amerindians. Agriculture probably had similar independent origins in yet other relatively isolated human groups.

TABLE 5-1 *Possible Origin and Date of Domestication of the Major Domestic Animals*

Domestication of animals was associated with the earlier stages of the urban revolution, and was likewise concentrated in the Eurasian belt where cereal growing developed. A number of domestic pets subsequently appeared in many parts of the world, such as peacocks (India), budgerigars (Australia), canaries (Canary Islands), white mice and hamsters (Middle East), as well as a few more recently domesticated animals kept for food such as the eland (Africa). All dates given here are estimates, often based on slight circumstantial evidence; much additional work must be done before more precise statements can be made.

Area where first domesticated	Date of domestication (year BP)
EUROPE AND WESTERN ASIA	
Dog (*Canis familiaris*)	20–50,000
Cattle (*Bos taurus, B. indicus*)	11,000
Sheep (*Ovis aries*)	11,000
Goat (*Capra hircus*)	11,000
Pig (*Sus scrofa*)	? 11,000
Horse (*Equus caballus*)	6,000
Ass (*Equus asinus*)	6,000
Cat (*Felis maniculata*)	4,000
Dromedary (*Camelus dromedarius*)	3–4,000
Rabbit (*Oryctolagus cuniculus*)	3,000
Goose (*Anser anser*)	3,000
Pigeon (*Columba livia*)	? 3,000
EASTERN ASIA	
Chicken (*Gallus domesticus*)	4–5,000
Elephant (*Elephas indicus*)	5,000
Water buffalo (*Bos bubalus*)	3–4,000
Yak (*Poephagus grunniens*)	3–4,000
Bactrian camel (*Camelus bactrianus*)	3–4,000
Duck (*Anas platyrhynchos*)	3–4,000
NEW WORLD	
Llama (*Lama huancus*)	4–5,000
Turkey (*Meleagris gallopavo*)	2–3,000
AFRICA	
Guinea fowl (*Numida numida*)	2–3,000

The selection of cultigens during these several millennia certainly involved a wide range of plants. Something like 3,000 plant species have been utilized as cultivated plants throughout history. Partly because of increasing mono- culture and mechanization, perhaps only about 300 of these now survive in sufficient acreage to be regarded as anything but botanical curiosities or "backyard vegetables." No more than a dozen of these provide the staple foods for 90 per cent of the world's population.

Much attention has been given in this century to the improvement of existing crop plants by concentrating on their center of origin and search- ing out related plants growing there for use in crop-improvement breeding programs. Such activities have greatly increased our understanding of the various stages of development of many cultivars. One of the most imagina- tive workers in this area was the brilliant Russian geneticist N. I. Vavilov. Among his many fundamental contributions are the concepts of primary and secondary crop plants and of centers of origin of cultivated plants (Figure 5-8).

Primary crop plants are those such as wheat and barley that were deliber- ately cultivated and distributed for their own merits. *Secondary crop plants* developed incidentally as *weeds* adulterating the primary crop. The spread of wheat from Asia Minor into northwestern Europe, for example, also carried along some seeds of rye, and possibly of oats, as weed contaminants. In the colder and wetter areas to which they were introduced, both these species tended to grow and yield better than the wheat. In many such areas these secondary crop plants therefore replaced the wheat that had been the intended introduction.

Wheat

Wheat is not only one of the most important food crops of the world and a primary crop but it also illustrates very well the kind of evolutionary history characteristic of many cultigens. In the wheat genus *Triticum* there are some 14 different species, some wild, others known only in cultivation, but all originating in the Old World. These species can be arranged in three groups according to the number of chromosomes their cells contain: *diploids* (14), *tetraploids* (28), and *hexaploids* (42). Their interrelationships are shown in Figure 5-11.

FIGURE 5-11. *Cultivated wheats and their interrelationships.* A. Cultivated einkorn, *Triticum monococcum.* B. Wild einkorn, *T. boeoticum,* which is believed to have hybridized with (*C*) a diploid wild grass, *Aegilops speltoides.* Following doubling of the chromosome number of this hybrid, it produced a tetraploid series including (*E*) macaroni wheat, *T. durum,* (*F*) wild emmer, *T. dicoccoides,* (*D*) emmer, *T. dicoccum,* and (*G*) another tetraploid type, Persian wheat. The wild emmer is believed to have hybridized with (*H*) a second wild grass species, *Aegilops squarrosa,* to produce a hybrid that, following chromosome doubling, further hybridized with Persian wheat to produce the bread wheat series in *K.*

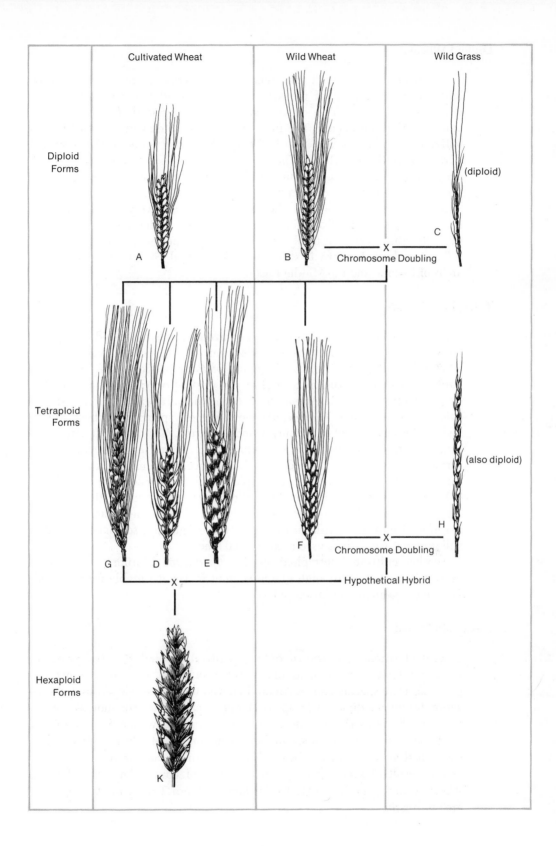

Cultivated Wheat

Wild Wheat

Wild Grass

Diploid
Forms

A

B

C

(diploid)

X

Chromosome Doubling

Tetraploid
Forms

G   D   E

F

H

(also diploid)

X

Chromosome Doubling

X ———————————————— Hypothetical Hybrid

Hexaploid
Forms

K

Diploid Wheats

The diploid wheats, with two sets of chromosomes, are assumed to be both the most ancient of the series and the foundation from which this crop was developed. Two diploid species are recognized, *Triticum boeoticum* and *T. monococcum.* The first, known as wild einkorn, is a wild grass native to southern Europe and the Middle East, and is one of the supposed parents of cultivated wheats. The second, einkorn (the name referring to the occurrence of one seed per spikelet), is thought to be a cultivated form of the first, with slightly larger seeds and less fragile heads.

In both species, however, the flower stalks are very brittle and the heads break so easily that there is inevitable loss during harvesting. Moreover, the glumes remain firmly attached around the seed, making the grain difficult to hull. Einkorn is still grown for animal consumption in certain areas of Europe and the Middle East.

Tetraploid Wheats

The seven wheat species usually recognized in the tetraploid group are considered to have arisen by hybridization of wild einkorn with another wheatlike Mideast wild grass, *Aegilops speltoides,* followed, as frequently happens in such crosses, by a doubling of the hybrid chromosome number to give a tetraploid form. Among these tetraploids is a wild emmer, *T. dicoccoides,* and an emmer, *T. dicoccum.* Both of these species have been identified in the charred wheat grains found at Jarmo. The fact that tetraploid rather than diploid wheats are represented there suggests either that Jarmo was indeed, as surmised, a fairly advanced agricultural settlement or that cultivation and gathering of tetraploid wheats preceded that of diploids.

Emmer was grown extensively in the early civilizations of the Mediterranean region to as far north as Britain. It retains the disadvantageous features of einkorn, brittle heads and attached glumes, so that where it is still grown it similarly is used as an animal feed.

Another cultivated tetraploid species, *T. durum* (durum), also still is grown; its exceptionally high gluten content makes it suitable for the preparation of spaghetti and macaroni.

Hexaploid Wheats

Hexaploid wheats are known only in cultivation, and the five principal forms sometimes are lumped under one species, *T. aestivum,* sometimes given separate specific names. All are believed to have developed by hybridization between a diploid wild grass, *Aegilops squarrosa*—common as a weed in wheat fields in the Middle East—and tetraploid wheat, followed by a doubling of the chromosome number. Common bread wheat, which if distinguished as a species is labeled *T. sativum* or *T. vulgare,* is supposed to be the result of such a cross with a variety of emmer known as Persian wheat occurring in the region that is now northern Turkey or the adjoining area of Russia.

Varieties of bread wheat have firm stems that do not tend to shatter when reaped and glumes that open readily to release the grain when threshed. It may be found necessary to predate the time of development of such free-threshing hexaploid wheats, because, as already noted, grains of one have now been identified in an early occupational layer of Catal Huyuk.

Preparation of Wheat

Because of the difficulty of hulling diploid and tetraploid wheats, whose clasping glumes retain the grain, it seems probable that after light roasting these wheats were soaked in water and made into gruel. If this preparation were left around for several days it would become contaminated with wild yeasts. The working of these yeasts may have prompted the subsequent development of leavened bread on the one hand and the brewing of cereal grains on the other, but the concept of alcoholic beverages appears to have had many independent origins.

The preparation of wheat flour by grinding grain, and the baking of bread from this, seems to have developed only with the selection of free-threshing hexaploid wheat varieties, at least in the Middle East. Elsewhere, where water supplies were less plentiful, grinding grain and baking the moistened flour may have been the original practice.

Domestic Animals

Considering the abundance of wild species available, the number of animals domesticated is comparatively few (Table 5-1). The early domestic forms developed from naturally occurring species in the Middle East included dogs, cattle, horses, donkeys, goats, sheep, cats, two species of camel, poultry, ducks, and geese. The New World contributed only the llama, its relative the alpaca, and the turkey. Europe provided the reindeer, and Asia the yak, the water buffalo, and the elephant. Two continents, Australasia and Africa, yielded no domestic animals at all, except the ostrich and guinea fowl. The African elephant is not counted because the one center in which it was domesticated has not continued its operations following the independence of the Congo. The absence of domestic animals originating in Africa is exceedingly curious. The common eland, for example, is readily domesticated; after an animal has been corralled for a day or so, it is difficult to drive it from the homestead. In South Africa herds of springbok and blessbok are commonly seen grazing with domestic animals on farms in the central regions. They are kept in this semiwild condition for game cropping purposes.

The Urban Revolution

It is little more satisfactory to speak of the urban revolution than it is to speak of the agricultural revolution. It certainly is not correct that the agricultural revolution preceded and precipitated the urban revolution. As has been discussed already, recent archeological work and the writings of such individuals as Jane Jacobs indicate that the so-called urban revolution may have developed in some instances at least, directly from trading settlements. In such cases, the urban revolution actually preceded the agricultural revolution which the urban revolution made possible, and from the basis of which the urban revolution then proceeded to unfold further.

Farming Villages

In a number of regions, the agricultural revolution did not lead to or was not associated with any further urban development. This seems to be true of, for example, some civilizations in the New World. Despite the existence of many fine Mayan temples and other stoneworks, a Mayan city—city being defined as a center of multifarious manufacturing industries as opposed to marketing activities—never evolved, only ceremonial centers that eventually were abandoned. Nor did true cities first arise in Egypt, despite the similar existence of temples. In dynastic Egypt, civilization probably developed under the Pharaohs for 2 millennia before true cities began to appear.

In the Fertile Crescent of southwestern Asia, village farming communities had become fully organized by 7,500 BP, and for the next 1,500 years at least they continued to prosper and to spread from the upland areas into the alluvial plain between the Tigris and the Euphrates rivers. It is in this valley that archeologists have excavated the early Mesopotamian cities that seem to have been the world's first real urban civilizations.

Urban Development

One reason for this urban development may have been the increased productivity made possible by the use of irrigation water in agriculture; another may have been the need to develop some kind of control organization to limit and direct the utilization of irrigation water. The city would also provide the organization necessary for storage and barter of the additional productivity of such civilizations.

It appears very reasonable to suppose that regions having cereal grains as their principal food crop would enjoy a much higher productivity than those mostly forest areas where root crops were in vogue. Cereal grains like wheat, maize, and rice require full sunlight for growth, thus necessitating complete clearing of the land. Unlike the majority of root and fruit crops, which are grown as part of a still more or less intact forest ecosystem, cultivated grains do not have to compete with other plants of their ecosystem, and provide a maximum return in terms of the incident sunlight. Early Mesopotamian cities would thus enjoy a number of advantages, the high productivity and reliable yields provided by irrigation agriculture, and the civic discipline needed to organize this high productivity.

Middle East Cities

A number of settlements known to have been at the city level by 6,000 BP have been excavated. Archeologists recognize four periods in the development of these cities, the earliest of which is named after Ubaid, the first site to be found. This first stage lasted for a couple of centuries, to give way to the even briefer Walda. The third period, the Protoliterate, produced the first written record and occupied the rest of the fourth millennium. The last and final phase is known as Early Dynastic. During this period, between 5,000 and 4,500 BP, independent city-states reached the height of their prosperity in the Tigris and Euphrates valleys and in lower Egypt. The Early Dynastic period is characterized by the replacement of a ritual priest class by a hereditary ruler group. Before this, temples and their priestly hierarchies had supervised a multitude of economic and ritual activities. Afterward, short-term war leaders had their political authority continued into succeeding periods of peace. Temple organizations nevertheless continued to prosper and acquired or held great estates of their own. The palace was concerned with raising and supplying its army, and constructing and maintaining the defensive wall of the city.

Uruk

This Early Dynastic series may be illustrated by the city of Uruk in southern Mesopotamia, which extended over 100 acres and probably had a population of about 50,000. Records from the time indicate that one of the temples in Uruk had a total congregation of 1,200 people, including 90 herdsmen, 80 soldier-laborers, 100 fishermen, 125 sailors, pilots, and oarsmen, 25 scribes, 20 or 25 craftsmen (such as carpenters, smiths, potters), and 250–300 slaves. Presumably the rest of the inhabitants, the cultivators and women and children, were not listed. Although this city and others like it in southern Mesopotamia, in the Nile Delta, and in the Lower and Upper Nile regions were based on military organizations and engaged in military activities, parallel developments in the Indus Valley in Pakistan illustrate that this military emphasis was not an essential feature of the evolution of the city. The one constant observation is that division of labor among craftsmen had apparently already by this time become extensive.

The Further Evolution of Human Societies

If we assume for the sake of the present review that the agricultural and urban revolutions were elaborations of the ritual behavior of human societies, there has been comparatively little time for their spread and evolution through all the widely scattered and sometimes partially isolated segments of the human population. Despite the ever-accelerating pace of exchange of cultural information, it is inevitable that some areas are still occupied by populations unable for one reason or another either to appreciate the significance of this information or to utilize it. A contemporary review of human societies on a global basis thus reveals a mosaic of varying degrees of social evolution, with numerous stages of adoption of the ritual and cultural behavioral patterns that have evolved. The many elements of this mosaic can be arranged to indicate the existence of a continuum of social progress, or rather a series of continua. Alternately, if a very broad classification is adopted, certain distinct grades or stages may conveniently be distinguished, and the mosaic can then be segmented into six somewhat arbitrary but convenient categories as follows:

1. Hunter-gatherer groups
2. Early Agricultural groups
3. Advanced Agricultural societies
4. Industrializing societies
5. Colonial societies
6. Advanced Industrial societies

This arrangement of categories is developmental and on an evolutionary ecological basis. The groups become larger as the series advances, which is why the term *societies* replaces *groups* in the later categories. Examples can be cited from the contemporary world of all these categories. It is also possible to maintain that any category except the first has had an evolutionary history that involved proceeding through the earlier categories. Special circumstances permitted some Advanced Industrial societies to internalize the fifth stage. The United States and Australia, for example, had undeveloped frontier areas within their own territories that absorbed the surplus population. Some small countries were able to achieve demographic changes that also enabled them to avoid passing through a colonizing phase.

Hunter-Gatherer Groups

Hunter-gatherer groups still have a few representatives, the best known of which are the Bushmen of Southwest Africa and the Australian Aborigine groups, although these are no longer in a mesolithic or any other form of stone-age culture. Typically they now occupy habitats of marginal productivity, although there is ample evidence that this was not always the case.

Early Agricultural Groups

Early Agricultural groups are still found, and some from New Guinea have been described recently by R. A. Rappaport in terms that explain the ecological significance of the feedback mechanisms regulating their population size. By contrast with hunter-gatherers, surviving groups of this societal level are found in areas of relatively high productivity. Their survival seems to be related to geographical isolation from other peoples.

Advanced Agricultural Societies

Neither of the first two categories is associated with a permanent settlement of any size or of any greater significance than a village of farmers. In Advanced Agricultural societies the villages become in many cases market towns, with a primary significance as barter-trading settlements. Often the market's importance may be judged from the frequency with which it is held, whether it is a 3-day, a 4-day, or a 5-day market, and so forth. When it becomes of sufficient importance to be a daily affair, it may become associated with streets of various kinds of artisans, such as leatherworkers, metalworkers, potters, weavers, and dyers. In the city of Kano in central Nigeria, representing such a market, these various activities appear to have been established for at least a thousand years.

Industrializing Societies

The Advanced Agricultural societies constitute many of the now-independent nations of the tropical world in Africa, southwest Asia, and southern Central America. They have ceased only recently to be dominated politically by the western powers, which experienced the last of the major revolutions, the so-called industrial revolution, in the eighteenth century, as they evolved into the fourth category of Industrializing societies. They then underwent a population explosion, becoming Colonial societies in the nineteenth century.

Advanced Industrial Societies

This development carried most of the western nations into the last category, Advanced Industrial societies, by the beginning of the twentieth century. Japan reached the stage of an Industrializing society appreciably later, and thus did not enter the colonizing phase until the beginning of this century. Some nations, Egypt, for example, may only now be evolving from the fourth category of an Industrializing society, and trying to move toward a Colonial society, although they no longer have any area for expansion in which to accommodate the surplus of individuals resulting from their population explosion.

The Ecological Development of Human Societies

This series or sequence of societal development can be described as an ecological succession. The characteristics of an ecological succession—increasing diversity, competition, and structural complexity, and decreasing dominance and net productivity—are displayed on a *society* instead of a *community* basis in the progression through the six stages arbitrarily recognized. Such an application of the succession concept represents an extension of its original use. Previously it has been applied to a series of communities, each with differing groups of dominants, rather than to a series of social organizations dominated by the same species at different cultural levels. Further examination does, however, seem to confirm the parallel.

Diversity

There is, for example, an increasing diversity of the various elements of these societies. This is illustrated by the division of labor, which is essentially lacking except between sexes in the pair bonds of a hunter-gatherer group, but which begins to develop in the agricultural categories and is extensive in the guild systems of Industrializing societies. Advanced Industrial societies have an even more diversified division of labor, or degree of specialization, as is documented in "DOT," the *Dictionary of Occupational Titles*, prepared by the U.S. Department of Labor. Expressed in ecological language, this extreme subdivision of tasks into many specialties is comparable to *niche diversification*. The division of labor does not necessarily follow the same route in every instance, but many proceed from branching at different points. This leads to the evolution of *ecological equivalents* in various cultures.

Competition

Increasing specialization results from adaptation to a steady heightening of competition between the various elements that develop in each successive societal stage. This further specialization is facilitated by the increasing structural complexity of the habitat. The jack-of-all trades rarely survives the evolution of society to the next grade. In the severe competition between specialists in the Advanced Industrial societies, redundancy is a very real threat to many trades and professions. It is now quite common to see magazine advertisements that begin, "Is your husband obsolete. . . ." For management and labor the constant movement toward redundancy of many specialists because of ever-intensifying competition presents both practical and moral problems.

Dominance

Whereas the first three features of evolving societies tend to increase in degree, there is a parallel trend toward a reduction of dominance by one or more elements over the rest. As has been noted, early civilizations were dominated by priests joined by, or sometimes superseded by, hereditary rulers. Generally in more modern societies priests and hereditary rulers no longer dominate; power has passed to smaller and more numerous circles of dominance that build around various types of individuals such as sportsmen, entertainers, politicians, scientists, and sometimes professional figures.

Succession

What causes succession, or what directs the movement from one stage in this series to the next, is an *energy differential* known in ecological terms as the net primary productivity. It is the difference between the total gross productivity of an ecosystem and the amount of energy the ecosystem disperses in maintaining itself, mostly in the form of heat and respiration, some as stored energy. In a hunting group such as Eskimos, or a hunter-gatherer group such as the Bushmen or Australian Aborigines, there appears to be too little difference between the amount of energy absorbed and the amount dispersed to move the group into a higher societal category. Early agricultural groups in the tropics appear to have sufficient net primary productivity to cause this further evolution, however, a number of feedback mechanisms limit the gross productivity, as described by Rappaport. When these feed-back mechanisms are interrupted, gross and net productivity rise, and the group moves forward into the category of an Advanced Agricultural society.

In the past such societies frequently were exploited by colonial powers, who skimmed off the surplus portion of the gross production in the form of raw agricultural and mineral products, which they then transferred for processing in their metropolitan factories. After attaining independence, Advanced Agricultural societies progressed into the category of Industrializing societies by utilizing their considerable net primary productivity.

There is another indicator of the productivity of a society, an economic one, the gross national product (GNP). This per capita statistic is somewhat nebulous in that it measures neither a rate of energy expenditure nor tangible industrial production. It provides only a guide to the amount of money expended each year on a per capita basis in a given nation. This figure varies up to a little over $4,000. Only two nations, Sweden and the United States, are in the $4,000-plus category. In the $1,701–4,000 class are 29 nations, from Australia to the Virgin Islands. With certain exceptions these are all what are defined here as Advanced Industrial societies. The exceptions are Kuwait, Libya, Qatar, and the United Arab emirates, which enjoy enormous oil royalties, and the Bahamas, Bermuda, and the Virgin Islands, which have succeeded in exploiting an equally enormous potentiality for tourism.

Some 90 nations, from Afghanistan to Zambia, fall into the $0–400 category. Without exception these are Advanced Agricultural societies. In this category they will remain until they obtain substantial sources of auxiliary energy to develop the production to power them on to later stages of societal succession. Fourteen nations, from Albania to Surinam, are just embarking along this route and can be called Industrializing societies. These nations have a GNP of $401–600, and include Latin America countries such as Costa Rica and one European country, Albania. Further along this industrializing process are 11 nations, Bulgaria to Yugoslavia, with a GNP of $601–800. A century or so ago, these nations would have been on the verge of becoming Colonial societies. Indeed one of them, Portugal, was until very recently the most aggressive of the remaining colonial powers. Jamaica and Mexico, also in this GNP category, are producing surplus population spilling out into other territories, one of the main characteristics of this colonizing phase; another one, South Africa, is developing an internalized colonizing structure.

In the final stages before attaining the category of an Advanced Industrial society are some 17 nations, from Argentina to Venezuela, with a GNP of $801–1,700. They include a number of Latin America countries and several European ones such as Greece, Hungary, and Spain, which is terminating its colonizing phase. Some independent countries that were once colonized areas of Colonial societies are now in this category, including Malta and Singapore.

On the evidence of the GNP, and allowing for the anomalies noted, of a total of 161 nations, about a fifth have entered the present final stage of societal succession, over a half are still stranded near the beginning. These include the most populous nation on earth, the Peoples Republic of China, and the second most populous, India. The remaining quarter of the nations are somewhere in passage from an Advanced Agricultural to an Advanced Industrial society. One of the most important problems of human ecology is to formulate theoretical explanations of why nations are strung out in this progression, and from this theoretical base to formulate practical schemes for the advancement of those that elect to move further toward the present climax of this societal succession. As yet no nation has chosen to do otherwise.

The Metal Ages and the Industrial Revolution

This essentially ecological approach to the evolution of human societies largely ignores the sequences obtained when this process is considered in relation to other criteria. The progress of civilization can be examined, for example, in terms of the metals used for tool manufacture, the type of fuel energy utilized, and the degree of automation.

An extensive literature has developed concerning the successive use of stone, bone, copper; bronze, and iron, into the "alloy age." Recent works have suggested that the transition from stoneworking to metalworking occurred in the obsidian-trading cities of western Asia that have been described here. The progress from wood to coal, to oil, to atomic energy, with a side turn to water and wind power, provides a different but equally instructive view of civilization. The industrial socioeconomic theory may well recede somewhat in importance in what Hardin calls the synthetic *computer-slave-automated-atomic* society. Many books can be and have been written about such aspects of societal evolution, but further consideration of them is incidental to the main purpose of this text. It is necessary, however, to examine further certain aspects of urban ecology.

The Ecology of the City

Considering the central importance of cities in contemporary civilization, it is almost incredible that so little has been made of the ecological processes that control their establishment, growth, and survival. During this century various attempts have been made to establish model cities, including a few modern capitals such as New Delhi and Brasilia, but the basic treatment of these appears to be architectural rather than ecological, a point that only a very few schools, such as that developed by Ian McHarg at the University of Pennsylvania, have appreciated. Generally speaking, cities, even capital cities, have just happened. Moreover, the reason why migration to them from rural environments occurs is largely unexplored. The one major established concept which has survived from classical human ecology is Burgess's concentric zone hypothesis. This supposes that economic determiners separate a city into concentric zones, each replacing in time the next outermost as the city grows, in a process comparable with ecological succession. Starting from the central downtown commercial area, Burgess recognized industrial, transitional degraded, multifamily high density, single family low density, and commuter zones.

Migration from Rural to Urban Environments

In the United States, where agricultural production in terms of per capita output is more efficient than in any other country, there has been a continuous migration from rural to urban environments (Figure 5-12). This has created city centers in various regions, of which 26 now have populations of over 500,000. Such city centers are by no means assured of perpetual growth by this migration process; some have already, according to their own projections, entered a phase of population decline (Table 5-2).

TABLE 5-2 *Actual and Projected Population in Selected Cities*

Statistics are rounded to the nearest 10,000 and given in thousands. The largest cities, New York, Chicago, and Los Angeles, will continue the steady rise in population through this generation, as will a number of smaller cities such as Oakland, Calif. Some smaller cities such as Nashville, Tenn., Tampa, Fla., and San Antonio, Tex, will undergo a more spectacular increase in size, whereas the population of a few large cities, such as St. Louis, Mo., will actually decline.

	1950	1960	1965 (Thousands)	1970	1980
New York	7,890	7,780	8,100	8,240	8,550
Los Angeles	1,980	2,480	2,740	3,000	3,670
Chicago	3,600	3,550	3,680	3,600	3,770
Nashville	170	170	450	470	520
Oakland	380	370	390	400	420
Tampa	120	270	300	350	440
San Antonio	410	690	680	780	950
St. Louis	860	750	700	670	660

There has been much speculation as to the causes of this urban decline. Some of the most stimulating ideas have been presented by Jane Jacobs, who relates it to loss of industrial diversity. She contrasts the economic growth of two English cities, Manchester and Birmingham, over the past century. Manchester at the beginning of this period was a great textile center, but its industry was specalized in this one activity. Birmingham was a diversified jumble of much smaller enterprises, but now its industries are expanding on this diversified basis. Manchester has stagnated with the relative decline of its textile industry as international competition heightened.

Jacobs identifies the same kinds of factors operating in the United States. Los Angeles, she notes, made its remarkable recovery from the loss of its aircraft industry, which had mushroomed during World War II, by diversifying into the electronics and other sophisticated industrial fields, thereby conforming to the Birmingham type of expanding city. Detroit, with its automotive industry, and Pittsburgh, with its steel mills, conform to the Manchester type.

This actually is only a portion of the argument that Jacobs develops for the expansion and decline of cities. There is a considerable similarity between her full thesis and the process of ecological succession, just as was noted in the case of societal development. There are, however, other aspects of ecology that must be considered besides economic diversification and population size, of which transportation represents one of the most intriguing. C. A. Doxiadis has evolved a theory relating the expansion of a city to its varying modes of transportation.

Urban Transportation

According to Doxiadis, the city limits are determined by a given time interval from its center, usually from 10 to 15 minutes. Urban evolution occurs in such a way that the suburbs are never more than 15 minutes away from the city center. When the only form of transportation was by foot, 15 minutes represented roughly 1 mile. Medieval walled cities such as Sienna in Italy, Kano in Nigeria, or the City of London therefore attained a maximum size of 1-mile radius. As various forms of city transportation developed, this radius could be expanded. On the outer perimeter subsidiary city centers would be developed, again related to transportation but to less rapid forms. Finally, with a municipal tram or bus service, a city could sprawl over an area with a radius of up to 10 miles, and contain up to a

FIGURE 5-12. *Growth of the urban population of the United States* as represented in census figures, 1790 to 1970. The rural population needed to maintain agricultural production has increased very little in size during the twentieth century, despite a virtual doubling of the total population, because of increases in per capita agricultural productivity. At the same time there has been a continuous migration of population from rural to urban areas.

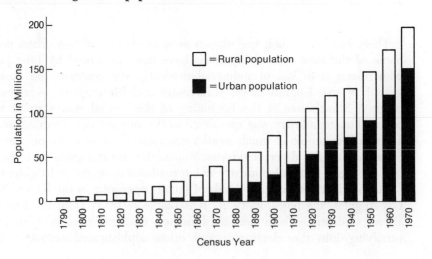

dozen or so suburban centers on its periphery, which people would proceed to on foot from areas 10–15 minutes' walk away.

In the megalopolis, a flight of 10–15 minutes by air sets a radius of about 50–100 miles. High-speed nonstop rapid transit may achieve the same magnitude of city size.

The Flight to the Suburbs

Alvin Toffler has recently developed the idea that we are moving toward *superindustrial* societies. This would be the new climax of the societal succession discussed in this chapter. In these superindustrial societies, urban concentrations lose their significance in the promotion of information exchange. In Toffler's view, the modern media for information transmission, in particular the electronic ones such as radio and television, have removed the necessity for face-to-face confrontation. The suburbanite can be as well informed and up-to-date as the dweller in the inner city. By implication Toffler forecasts a decline in city life and an increasing localization of industry and domiciles in suburbia.

Cities stagnate or decay for many reasons, and this extraordinarily complex subject is now being investigated extensively for the first time. Economy and transportation are just two of the many factors that may be responsible for bringing about a decline. Another way of investigating this problem is to build special experimental cities, a procedure now being explored.

The Experimental City

Proposals to build experimental cities have developed from a number of directions, and a further approach is discussed in Chapter 13. One particular proposal, the Experimental City project in Minnesota, was presented by a group representing federal, state, university, industrial, and private interests. This group started from the premise that if the present population of the United States were accommodated in 800 cities of approximately 250,000 inhabitants each, evenly dispersed over the country, environmental crises such as traffic congestion, riots, and pollution would be nonexistent. They further considered that the future emphasis in cities will be on the trading of *mentifacts* rather than *artifacts*, the exchange of ideas rather than goods. Basically, as this chapter has tried to explain, this opportunity for cultural exchange is the city's primary attraction for people and the essential causal ecological factor promoting migration from rural to urban environments.

The Experimental City group projected into a theoretical future and noted that the approximately 2–3 billion acres of land surface on the earth will, by 2068 AD, have to accommodate an estimated 15 billion people. These could be dispersed in 60,000 cities of a quarter of a million inhabitants, each surrounded by 40,000 acres (64 square miles) of open land.

Having thus established the theoretical global feasibility of such an approach to urban life, the group proceeded to outline details of the controlled city. Essentially it would be a well-contained, self-perpetuating ecosystem, in which all energy and material transfers are internalized, so that cycling processes prevent the accumulation of unused wastes at any point. It is beyond the scope of this text to pursue details of the experimental city further, but the opinion may be hazarded that if the Minnesota group had succeeded in founding such a city, they would have experienced no difficulty in finding a quarter of a million individuals willing to move in. The same is probably true for the Compact City, an even more recent proposal by G. B. Dantzig and T. L. Saaty for the structuring of new cities.

Environmental Confrontations

The surface of this earth is now about almost completely occupied by human groups and societies at the different stages of evolution that have been specified here. One of the elemental problems presented by this circumstance arises from the fact that no feedback mechanisms such as were imposed by the environment in hunter-gatherer groups, or by behavioral ritual in Early Agricultural groups, are now operating to restrict population growth in any of the societal categories except the last, Advanced Industrial societies. This is the basic cause of the present environmental crises that have been too slowly noticed, at least in the Advanced Industrial societies. For some of the other societies, the consequences of this fundamental but inescapable situation do not have to be understood; they are only too apparent. Famine and undernourishment are found only in very local areas in the so-called Advanced Industrial nations, but they are a way of life for the great majority in most of the Advanced Agricultural societies. In attempts to remedy this situation, wholesale use is made of pesticides without reference to possible longer-term and more widespread consequences. There also has been 25 years of such use in the Advanced Industrial societies, and this has many manifestations. These Advanced Industrial societies deliberately release huge quantities of wastes as well as pesticides into the air and water of their environment, indeed into the whole biosphere.

There is a tendency to attribute all present difficulties to the "population explosion." Although this is essentially true as regards the occurrence of famine, disease, poverty, lack of education, lack of opportunity, and low standards of living, it is not the sole cause of our misery. A second major contributing factor is the inexorable and seemingly inevitable expansion of urban life and activity. In later chapters the pathological and auto-toxic by-products of the urban revolution will be considered individually. Essentially technological problems of human ecology, they have technological solutions. They are the particular aspects of human ecology that presently cause most concern because disregard of the effects of any single one now threatens to bring almost immediate disaster. The threat from any of them is just as real as what would follow the outbreak of atomic warfare, and the chances for survival of even a small segment of our global human population would be at about the same level. If such an environmental disaster were to befall the human race, and if there were accidentally to be a small segment of survivors, there is some question as to whether even the present levels of Western civilization could ever again be attained. The total resources of the earth have been so far exhausted already that the succession of societal categories as described here, which was made possible by the attainment of a particular level of net primary productivity, could not be repeated. Because of this despoiling of resources, it might never again be possible to achieve a sufficient net primary productivity in a critical mass adequate to permit the succession from a surviving pioneering segment to the full flowering of Advanced Industrial societies that has been obtained by the western world in the middle of this century.

To control population growth is the first and most urgent step that must be taken, but many other measures must be instigated simultaneously. What must be done, and the ecological basis for such actions, will be described after an examination in the next chapter of the diversity of contemporary human populations that parallels this review of the origin and development of contemporary societies.

BIBLIOGRAPHY

References

Adams, R. M. "The origin of cities," *Scientific American,* **203**(3): 153–68, 1960.

Baker, H. G. *Plants and Civilization,* Belmont, Calif.: Wadsworth, 1965.

Bartlett, A. S., Baghoorn, E. S. and Berger, R. "Fossil maize from Panama," *Science,* **165**: 389–90, 1969.

Birdsell, J. B. "Some environmental and cultural factors influencing the structuring of Australian aboriginal populations," *American Naturalist,* **87**: 171–207, 1953.

Braidwood, R. J. "The agricultural revolution," *Scientific American,* **203**(3): 130–48, 1960.

Braidwood, R. J., Cambel, H., and Watson, P. J. "Prehistoric investigations in southeastern Turkey," *Science,* **164**: 1275, 1969.

Burgess, E. W. "The growth of the city," in *The City,* Park, R. E., Burgess, E. W. and Mackenzie, R. D., Chicago: University of Chicago Press, 1925, pp. 47–62.

Cambel, H., and Braidwood, R. J. "An early farming village in Turkey," *Scientific American,* **222**(3): 50–56, 1970.

Coe, M. D., and Flannery, K. V. "Microenvironment and Mesoamerican prehistory," *Science,* **164**: 650–54, 1964.

Dantzig, G. B. and Saaty, T. L. *Compact City,* San Francisco, Cal.: Freeman, 1973.

Deevey, E. S. "The human population," *Scientific American,* **203**(3): 194–204, 1960.

Doxiadis, C. A. "Man's movement and his city," *Science,* **162**: 326–34, 1968.

Hammond, N. "The planning of a Maya ceremonial center," *Scientific American*, **226**(5): 82–91, 1972.

Harlan, J. R., and Zohari, D. "Distribution of wild wheats and barley," *Science*, **153**: 1074–80, 1966.

Isaac, E. "On the domestication of cattle," *Science*, **137**: 195–204, 1962.

Jacobs, J. *The Economy of Cities*, New York: Random House, 1969.

Mangelsdorf, P. C., MacNeish, R. R., and Galinat, W. C. "Domestication of corn," *Science*, **143**: 538–45, 1964.

Martin, P. S., and Wright, H. F. (eds.) *Pleistocene Extinctions: Search for a Cause*, New Haven: Yale University Press, 1967.

Mellaart, J. "Deities and shrines of neolithic Anatolia: excavations at Catal Hüyük, 1962," *Archeology*, **16**: 28–38, 1963.

Mellaart, J. "A neolithic city in Turkey," *Scientific American*, **210**(4): 94–104, 1964.

Perkins, D., Jr., and Daly, P. "A hunter's village in neolithic Turkey," *Scientific American*, **219**(5): 98–106, 1968.

Protsch, R., and Berger, R. "Earliest radiocarbon dates for domesticated animals," *Science*, **179**: 235–39, 1973.

Rappaport, R. A. *Pigs for the Ancestors*, New Haven: Yale University Press, 1967.

Sauer, C. O. *Agricultural Origins and Dispersals*, New York: American Geographical Society, 1952.

Sauer, C. O. "Seashore—primitive home of man?" *Amer. Philosoph. Soc. Proc.*, **106**: 41–7, 1962.

Solheim, W. G., II. "An earlier agricultural revolution," *Scientific American*, **226**(4): 34–41, 1972.

Spilhaus, A. "The experimental city," *Science*, **159**: 710–15, 1968.

Toffler, A. *Future Shock*, New York: Random House, 1970.

Vavilov, N. I. "Studies on the origins of cultivated plants," *Bull. Appl. Bot., Genet., Plant Breed.*, **16**: 139–248 (English summary), 1926.

Further Readings

Burkhill, I. H. "Habits of man and the origins of cultivated plants of the Old World," *Proc. Linn. Soc. London*, **164**: 12–42, 1953.

Byers, D. S. (ed.) *The Prehistory of the Tehuacan Valley*, 2 vols., Austin: Peabody Foundation and University of Texas Press, 1968.

Cain, S. A. "Man and his environment," *Population Bulletin*, **22**: 96–103, 1966.

Cockrill, W. R. "The water buffalo," *Scientific American*, **217**(6): 118–25, 1967.

Coon, C. S. *The Hunting Peoples*, Boston: Little Brown, 1971.

Flannery, K. V. "The ecology of early food production in Mesopotamia," *Science*, **147**: 1247–56, 1965.

Gorman, C. "The Hoabinhian and after: subsistence patterns in southeast Asia during the Late Pleistocene and Early Recent Periods," *World Archeology*, **2**(3): 300–20, 1971.

Grist, D. H. *Rice*, London: Longmans Green, 1953.

Harlan, J. R., and Zachary, D. "Distribution of wild wheats and barley," *Science,* **153:** 1074–80, 1966.

Harris, M. "The cultural ecology of India's sacred cattle," *Curr. Anthropol.,* **7:** 51–56, 1966.

Heiser, C. R. "Some considerations of early plant domestication," *Bioscience,* **19:** 228–31, 1969.

Helback, H. "Ecological effects of irrigation in ancient Mesopotamia," *Iraq,* **22:** 186–96, 1960.

Helbark, R. "Domestication of food plants in the Old World," *Science,* **130:** 365–72, 1959.

Higham, C. F. W., and Leach, B. F. "An early center of bovine husbandry in Southeast Asia," *Science,* **172:** 54–56, 1971.

Hutchinson, J. (ed.) *Crop Plant Evolution,* Cambridge, England: Cambridge University Press, 1965.

Leeds, A., and Vayda, A. P. (eds.) *Man, Culture, and Animals: The Role of Animals in Human Ecological Adjustments,* Washington, D.C.: American Association for the Advancement of Science, Pub. 78, 1965.

Lorenz, K. Z. *Man Meets Dog,* London: Pan Books, 1959.

Lowry, W. P. "The climate of cities," *Scientific American,* **217**(2): 15–23, 1967.

Marshack, A. *The Roots of Civilization,* New York: McGraw-Hill, 1972.

Mason, I. L. A. *World Dictionary of Breed Types and Varieties of Livestock,* Commonwealth Bureau of Animal Breeding Genetics, Techn. Commun. No. 7, 1951.

Mumford, L. *The City in History: Its Origins, Its Transformations and Its Prospects,* New York: Harcourt, Brace and World, 1961.

O'Flaherty, C. A. "People, transport systems, and the urban scene: On overview," *Intern. J. Environ. Studies,* **3:** 265–85, 1972.

Reed, C. A. "Animal domestication in the prehistoric Near East," *Science,* **130:** 1629–39, 1958.

Tindale, N. B. "Ecology of primitive man in Australia," in A. Keast et al. (eds.), *Biogeography and Ecology in Australia,* Monographiae Biologicae No. 8, The Hague: Junk, 1959, pp. 36–51.

Triestman, J. M. *The Prehistory of China: An Archeological Exploration,* New York: Doubleday, 1972.

Ucko, P. J., and Dimbleby, G. W. (eds.) *The Domestication and Exploitation of Plants and Animals,* London: Duckworth, 1969.

Wright, H. E., Jr. "Environmental changes and the origin of agriculture in the Near East," *Bioscience,* **20:** 210–12, 1970.

Human Diversity

The emergence of man has been a continuous evolutionary progress, first from our ancestral anthropoid stock, then from one hominid grade to another. As this progress has been charted over the last four chapters, it has been noted how our contemporary human populations in many biological and cultural features reflect characteristics inherited from one or other of our previous ancestral forms. A critical look at the immense diversity displayed by our species in the contemporary world suggests that it is possible to find many traits, both biological and cultural, whose incidence still lags behind or has accelerated ahead along this evolutionary path in one segment of our populations as compared with another.

Familiar examples of this have already been noted. One of our most stable characteristics, dentition, is undergoing change. The third set of molars, the "wisdom teeth," is being lost. In some human populations these molars still emerge before maturity, in others they may never erupt at all. The overall size of the teeth has been greatly reduced in some populations, whereas it remains large in others. In such respects we may consider that some of our populations have not yet fully progressed from *Homo sapiens neanderthalis* or, alternatively, that some already may have moved beyond *H. sapiens sapiens*. Likewise, culturally some may still not have evolved beyond the *H. sapiens neanderthalis* human grade, persisting as paleolithic hunter-gatherers, the characteristic culture of this grade. Other segments of our population, by contrast, may already culturally have evolved beyond the *H. sapiens sapiens* grade; certain biological traits may already have been adapted to this further cultural evolution. Our second 12,000 years of exposure to the selection pressures of urban ecosystems are certain to produce further cultural and biological adaptations to these ecosystems that will make twentieth-century man appear in retrospect as ill-adjusted as a chimpanzee riding a bicycle. Our descendants of these futuristic days, geologically but a moment away, undoubtedly will regard with tolerant amusement our clumsy attempts to elevate to a pedestal of uniqueness an evolutionary stage of the *H. sapiens sapiens* grade they doubtless will classify with neolithic man.

The purpose of this chapter is primarily to examine the ecological features of the morphological and physiological diversity of contemporary human populations in these terms and to relate them to any environmental factors with which there appear to be adaptive evolutionary relationships. In the course of such a review it will be apparent that biologists have had to base their theories solely on observational analytical experience. The amount of experimental work (as compared for example with that which has been carried out in order to obtain a similar understanding of the fruit fly *Drosophila*, the mold *Neurospora*, or the bacterium *Escherichia coli*) is negligible. At the same time, although ethical considerations always will impose a limitation on studies of human adaptations, in no other organism is there such a huge population available for study, such an accumulation of precisely recorded data, or such strenuous efforts made to preserve all offspring of all matings. We have only just begun to exploit the opportunity that these unique circumstances present.

The nature of the more obvious differences between the various modern populations we classify as *H. sapiens sapiens* has awakened as much interest and promoted as much controversy as the question of man's origin. Such differences also have been the source of considerable misunderstanding, great bitterness, and some conflict. On the one hand there have been the well-intentioned but overenthusiastic protagonists of the point of view that discernible differences are so slight and the situation so variable as to make it impossible to recognize any valid group patterns. At the other extreme are dogmatists who insist that appreciable differential adaptation has occurred, sufficient to produce significant differences in mental as well as physical attributes.

Continuous Versus Discontinuous Variation

In human populations as in all species it is sometimes convenient to group discernible differences of form, function, and behavior into either *continuous* or *discontinuous* variations. Continuous variations, sometimes described as *quantitative* differences, are measurable in some abstract way, and there are no discrete breaks between one expression and another of the diversity. There are tall persons and short persons in a given human population, but between these two extremes is a continuous series of persons of intermediate height. Continuous variation is polygenic; it is controlled by a complex system of genes and is an expression of the total effect of all the genes involved, modified normally by interaction with the environment.

Discontinuous variation, providing *qualitative* differences, has no such gradations. Although it also can be modified somewhat by interaction with environmental factors, the differences never intergrade into one another. For example, circus dwarfs usually exhibit a variant known as *achondroplasiac dwarfism*. They may vary somewhat in height, but there is a discrete upper limit to this variation compared with the variation in height of individuals not exhibiting this condition. Achondroplasiac dwarfism results from a single gene difference in the usual genotype, as do many other examples of discontinuous variation.

Another obvious discontinuous variation in human populations is the distinction between male and female individuals, in this instance produced by differences in a whole chromosome rather than a single gene. In other animals, and more particularly in plants, study of the chromosome complements has frequently led to a greater understanding of the causes of diversity. In human populations such studies have been undertaken only within the last decade or so. Meanwhile both continuous and discontinuous variations have been investigated using methods developed from environmental ecology or from genetic-physiological studies.

Human Cytogenetics

The normal diploid chromosome number in *Homo sapiens* is 46. This karyotype or chromosome complement has been classified by size on the "Denver system," which numbers and groups the 44 autosomes into "large" pairs, 1–5; "medium-sized," 1–12; and "small," 13–22. Of the two sex chromosomes, X can be classified as medium-sized, but its length varies both among individuals and among cells of the same individual. Y is small but usually distinguishable by its shape. In numbering the autosomes, the longest is placed first, the shortest last, and the others sequentially in order of length (Figure 6-1).

Other Primate Karyotypes

Diploid chromosome numbers vary considerably in other primates, as may be seen from Table 6-1. Prosimians range from 38 to 80, Old World monkeys from 42 to 72, and gibbons from 44 to 50; gorillas and chimpanzees have 48. Morphologically, the karyotype of the chimpanzee, apart from an extra pair of chromosomes and differences in pair 22, appears quite similar to that of humans (Figure 6-1). For the present, any such similarities provide no evidence supporting particular theories of affinities among the primates in

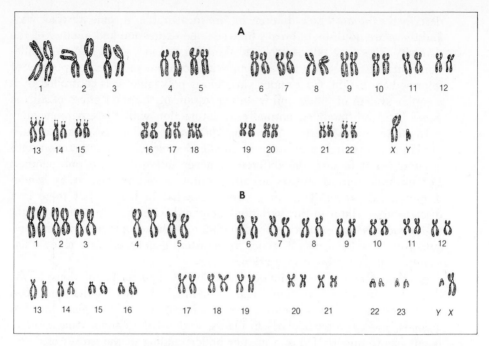

FIGURE 6-1. *Human karyotype contrasted with that of a pongid.* Apart from the additional homologous pair of chromosomes (pair 16) in the male chimpanzee (*Pan troglodytes*) karyotype shown in B, the chromosomes are morphologically very similar to those of the male human (*Homo sapiens*) karyotype shown in A. This close similarity is further emphasized by the occurrence of an aneuploid, trisomic 21, causing Down's syndrome (mongolism) in man, and a similar syndrome in chimpanzee, as reported by McClure and colleagues. Because of the extra pair of chromosomes in the chimpanzee karyotype, this syndrome is designated as trisomic 22 in chimpanzee, but it is apparently in each case the same morphologically similar homologous pair of chromosomes that are involved.

their relationship to the Hominoidea, any more than they can be used to identify group diversity in man.

Chromosomal Anomalies

Observations on human karyotypes are now made by the culture in vitro of blood lymphocytes stimulated to divide. This technique has permitted the identification of many chromosomal abnormalities with particular clinical conditions, but it has not yet been applied extensively to investigations of population cytogenetics. W. M. Court Brown suggested that when this is done, at least 1 per cent of newborn children will be found to show chromosomal abnormalities. About half of these will be structural rearrangements, one quarter autosomal aneuploids, the rest abnormalities in the sex chromosomes (Table 6-2).

Individual abnormalities in chromosome numbers in human karyotypes have been correlated with a number of congenital defects, but there is so far no suggestion of intergroup variation. There is, however, an apparent

TABLE 6-1   *The Range of Primate Karyotypes*

Primate group	Diploid number
Prosimii	
Tupaia glis	62
Tarsius bansanus	80
Lemur nongaz (mongoose lemur)	60
Lemur nacaco (black lemur)	44
Galago crassicaudatus	62
Galago senegalensis	38
Nycticebus coucang (slow loris)	50
Perodictus potto (potto)	62
Ceboidea	
Callithrix chrysoliucos	46
Ateles paniscus	34
Cebus apella	54
Cercopithecoidea	
Cercopithecus mitis	72
Cercopithecus mona (mona monkey)	66
Cercopithecus neglectus	60
Ceropithecus patas (patas monkey)	54
Macao fuscata	42
Papio gelada (gelada baboon)	42
Hominoidea	
Pongidae	
Hylobates moloch	44
Hylobates syndactylus	50
Gorilla gorilla (gorilla)	48
Pongo pygmaeus (orangutan)	48
Pan troglodytes (chimpanzee)	48
Hominidae	
Homo sapiens (man)	46

TABLE 6-2   *Frequency of Chromosomal Aberrations in Newborn Human Infants*

Type of abnormality	Per cent occurrence in population
Trisomy 21 (Down's syndrome: mongolism)	0.15
Trisomy 13/15	0.07
Trisomy 17/18	0.50
Aneuploidy in sex chromosomes (XXY:XXX:XXXY:XX:XYZ, etc.)	0.18
Total anomalies	0.90

correlation between the occurrence of XYY forms and height. More than half the males with this aneuploidy who have been measured are over 6 feet tall. Court Brown also reported that an additional Y chromosome somewhat increases the chance of a male developing psychopathic symptoms.

Polymorphic Genes

Although some variation in human traits is thus accountable on a cytogenetical basis, by far the greatest amount of variation is due to genetic polymorphism. The genetic analysis of human populations based on polymorphic variation observed in blood group characters is one of the earliest examples of investigations dependent on this phenomenon. Electrophoretic examination of the enzymes and other proteins possessed by one individual inevitably reveals differences when these bands are compared with those obtained from other individuals. When individuals are found to have differences in the electrophoretic band that a given enzyme produces, such differences normally have a simple genetic origin. In the majority of cases it seems to make little difference what particular simple variant of an enzyme or other protein we possess as a result of this polymorphy in our genes. That is to say, most mutations which have caused this extensive polymorphy have been neutral in their effects on the human beings involved. A few, however, bestow considerable selective values, either positive or negative, for the carrier. Sickle-cell anemia provides an extensively investigated example of such an advantageous mutation. Its effects are discussed later in this chapter.

Polymorphisms may be grouped either as *balanced* or as *transient*, but in theory at least there can also be *neutral* polymorphisms which, as indicated in the previous paragraph, are not subject to selection pressures. Balanced polymorphisms, for one of several possible genetical reasons, provide up to a particular point a net selective value for the heterozygote. Sickle-cell anemia is an example. In transient polymorphisms the polymorphic pattern is apparently not maintained by selection pressures either, but this negative feature is very difficult to prove experimentally. Transient polymorphisms result from several different situations.

One implication, theoretically, of the existence of polymorphic variation in gene composition is that it should be possible by the use of electrophoretic examination of human enzymes to determine whether or not race theories are confirmed. To date, all examinations have produced somewhat inconclusive results, as recently described by Cavalli-Sforza. Differences in genes have been detected between one human group and another, but these have not greatly exceeded the difference found between two individuals chosen at random from the *same* group.

Anthropometry

The first and to date the only class of scientists to embark on extensive mensuration of group differences in contemporary human populations in an impartial manner were the physical anthropologists, who concentrated essentially on morphological characteristics such as height, weight, sitting height, head length, skin color, nature of hair, and width of nose. The result of their labors was the compilation of a huge mass of physical measurements, whose highest potential value probably never will be fully exploited, because such work generally neglected to record simultaneously the precise environmental parameters relating to each group studied. Unfortunately also for *anthropometry*, the science embracing such studies, the more dimensions measured, the less likely they are to be independent. Moreover, ecologists now tend to minimize correlations with macroenvironmental factors, emphasizing instead considerations of the microenvironment. It is likely that for some time attention will have to be concentrated on reactions between the microenvironment and specific polymorphisms.

Before examining some conclusions of the physical anthropologists based on an examination of anthropometric data, it is pertinent to review the geographical situation in respect to the Old World distribution of *Homo erectus* in the Middle Pleistocene, and to look again at any possible connections between this and the distribution of distinct groups of contemporary man.

Major Groupings of *Homo sapiens*

As was discussed in earlier chapters, it seems reasonable to postulate that populations of a *Homo erectus* grade evolved gradually into a *Homo sapiens* grade level, undergoing, as far as physical characteristics are concerned, changes in only a few features such as cranial capacity, pharynx position, and tooth size. As a result of these changes, facial and skull characteristics became somewhat modified. Because we are still uncertain of when this evolution occurred, and because there could not really be any abrupt transition, it seems better to refer to a *H. erectus* grade and a *H. sapiens* grade. The term *erectus-sapiens population* has been used here to refer to transition groups between these two grades. Currently there is, as already noted, a growing tendency to call these transitional forms Neanderthals.

In regard to *behavioral characteristics*, by contrast with physical features, there was an extensive further evolution in developing *erectus-sapiens* cultures. As the number of individual hordes or bands increased, territoriality would begin to impose a restriction on further wide migration of these evolving populations. There would also be an additional stabilizing territorial effect with the appearance of permanent settlements in favored microenvironments. Therefore, it seems most likely that the gradual evolution from the *H. erectus* grade to the *H. sapiens* grade proceeded, albeit at somewhat varying rates, in all geographical areas, developing from the existing groups already occupying those areas.

The skulls of all Early and Recent Mongoloids are characterized by certain features such as a more or less flattened face, a rather large cranial capacity, and so forth, and are thereby distinguishable at all stages from the skulls of Caucasoids and Negroids occurring in the two other main geographical areas. In the same way the characteristics of the Cro-Magnon skulls and associated skeletal remains are Caucasoid, clearly not Early or Recent Mongoloid, and recognizably different from Negroid.

Some major movements of populations over long distances surely occurred as *erectus-sapiens* populations approached the last few millennia preceding historical times. The most spectacular of these movements resulted in the first occupation of the continents of North and South America and of Australasia. Despite such important exceptions it seems legitimate to suppose that by the time human populations entered recorded history, three major geographical groups of *sapiens* grade existed that can be recognized, described, and diagnosed as different from one another on a number of criteria. These include the cranial characteristics that permit physical anthropologists to work from both contemporary and extinct populations by comparing skeletal remains. It also appears reasonable to postulate that evolution from the *erectus* to the *sapiens* grade took place at approximately the same geological time in all three groupings.

The ancient geographical groups that emerge on this hypothesis have been labeled *Caucasoid, Negroid,* and *Mongoloid;* they characteristically occur in western Europe, tropical Africa, and eastern Asia, respectively. As already noted, there is every indication that genetic and cultural contact among these geographical groups was never lost for any significant period of time. Their isolation was therefore only a matter of a greater or lesser degree of spatial separation, varying with the conditions prevailing in the macroenvironment and with the particular ritualistic behavioral patterns involved.

Distribution

When examining the distribution of *erectus-sapiens* populations in previous chapters, it was noted that hunter-gatherer assemblages with relatively stabilized territories had become established over most of the habitable regions of the Eurasian and African continents by, at the latest, about 200,000 BP. Major land migrations, notably into the "empty" continents of the Americas and Australasia, occurred subsequent to this date, followed by extensive and repeated migrations into Oceania by sea. In occupied territories, however, migratory movements of this kind must have been more difficult, as they would involve fighting and possibly a temporary depletion of resources. It can be assumed, therefore, that for at least 200,000 to 300,000 years ancestral groups of contemporary man adapted and evolved, each in its own territorial area and within the particular parameters of a great variety of differing ecosystems.

It would then be reasonable, by extrapolation from experience with other animal species under these circumstances, to expect the following:

1. Varying gene frequencies among various major geographic human groups
2. Some mutant alleles unique to particular major geographical groups

3. Phenotypic acclimation to particular environmental factors
4. Genotypic variation as a result of the selection of particular morphological, physiological, and behavioral characteristics, sometimes in the same direction as changes resulting from acclimation
5. Examples of genetic drift
6. Examples of the founder principle

In this chapter, as we review the diversity of contemporary human populations, numerous instances will be observed where one or another of these phenomena is in operation. In each case it has produced a new form "fitter" in relation to some circumstance at some time. The characteristic need no longer provide a "fitter" phenotype, indeed rather the opposite. "Fitter" is not in any case to be equated with "superior."

It would be intellectually dishonest to deny or minimize diversity in human populations arising from such ecological causes. It is most likely that the selection for "fitter" phenotypes at some time in the 4-million-year history of *erectus-sapiens* populations has produced diversity not only in morphological and physiological but also in behavioral and mental characteristics. In its time and place each of these selections would be uniquely superior to all other variations subject to the same selection pressures. They are therefore of ecological interest in this respect, and also in regard to their present representation in contemporary populations. A number of recent books on the potentially inflammatory subject of the "races" of mankind, which discuss human diversity at some length, are listed in the items for further reading at the end of this chapter. This subject becomes inflammatory only if it is taken out of its ecological context and given emotional overtones.

In Chapter 4 the salient features of the evolution and continental migrations of *Homo erectus* were reviewed. It was noted that major topographical discontinuities—mountain barriers, seas, ice sheets—had at different times isolated various sections of global populations. It was this *abiotically* imposed isolation within which the several geographical sections of the *erectus-sapiens* populations evolved before reaching the *sapiens* grade. It is not surprising to find that virtually all workers who have examined the diversity of contemporary human populations are agreed that a Negroid group can be defined in the continent of Africa, a Caucasoid group in Europe, and a Mongoloid group in Asia. Other continents have presented a greater confusion of opinion. For this reason, Amerindians and Australian Aborigines usually are separated as fourth and fifth groups, and care is taken to avoid discussion of the present inhabitants of the diverse islands commonly grouped together under the heading Oceania.

The speculations on evolution and migrations presented in Chapter 4 can be extended logically to include these other groups, and this subject is therefore taken up again here. In doing so, one basic assumption has to be made about an issue most previous texts have ignored. "Modern man" is tacitly assumed in most descriptions to be a type of Cro-Magnon man or a development from this form. This cannot possibly be so if there has been the degree of continental isolation postulated in these pages. The last common ancestor of all men was *Homo erectus*. Because the subsequent adaptation of *erectus-sapiens* populations has been basically the result of selection leading to *cultural* evolution, *mental* characteristics have developed very much along parallel evolutionary lines. Because *physical* characteristics have not been changed extensively since Early Pleistocene times, no breeding barriers have developed. Our various geographical forms have therefore remained interfertile both in theory and occasionally in actuality. This is an implication that cannot be sidestepped or glossed over; it must be subjected to scientific scrutiny.

The concept of an independent parallel evolution of several geographical human subgroups from a *H. erectus* to a *H. sapiens* grade was first extensively presented by C. S. Coon about 10 years ago. In a world suddenly and belatedly highly sensitive concerning human rights, Coon's ideas were labeled racist. Once a label has been attached, it is often very difficult to remove, but later work is tending to substantiate some of the ideas on races that Coon presented. In a carefully prepared and extensive paper, D. S. Brose and M. H. Walpoff have examined the hypothesis that there was a discontinuity between *H. sapiens* and earlier human forms in all areas except the one in which *H. sapiens* evolved and from which the new populations radiated over the earth. True *H. sapiens* remains are invariably the only human form associated with *industries* of the kind archeologists call *Upper Paleolithic* assemblages. Earlier *erectus-sapiens* (Neanderthal) population remains are never associated with anything more advanced than *Middle Paleolithic* assemblages. Brose and Walpoff demonstrate that the distinction between these two industries is artificial. They have found overwhelming evidence everywhere that Middle Paleolithic assemblages occur of local industries transitional to Upper Paleolithic. In other words, modern *H. sapiens* populations, with Upper Paleolithic industries, have not abruptly displaced Neanderthals with Middle Paleolithic assemblages.

It might still be argued that modern humans with their advanced tool kits only slowly infiltrated into regional Neanderthal populations and only slowly taught these backward locals how to make more advanced tools. Brose and Walpoff have countered this argument by demonstrating that early modern *H. sapiens* remains can be found associated with *Middle Paleolithic* industries. In such transitional populations, the local morphological form and range of the preceding Neanderthal populations are reflected by a similar local morphological form and range in the succeeding *H. sapiens* populations. In other words, local genetically produced variations are continuous through the transition. This would not be so if migrating *H. sapiens* individuals were hybridizing with the local population, for further genetic variation would incidentally be introduced.

Finally, Brose and Walpoff demonstrate that the still limited number of remains classifiable as transitional between Neanderthal and modern human morphological forms are associated always with Middle Paleolithic industries. If modern humans with their Upper Paleolithic industries were radiating out from a center of origin, it is unlikely they would have implanted their genes so successfully as to produce intermediate skeletal morphologies without also having some influence on associated industrial assemblages.

Clearly, further dispassionate work along these and the many lines now available can eventually firmly establish or finally reject this still controversial view of the current stage of human evolution. In this chapter it is assumed to be substantially correct.

Negroids

Africa south of the Sahara contained during the Late Pleistocene the most homogeneous group of hominids of any continent. This area seems to have been both the center of origin of australopithecines and the region of sympatric interaction between *Homo* and *Paranthropus* that eventually led to the selection of *H. erectus* that had adapted to a human ecological niche. Whereas several waves of hominid migration left the continent of Africa, none apparently returned, at least until the Holocene, or postglacial times. African *erectus-sapiens* populations therefore were little influenced by mixture with other groups and had a longer period of undisturbed regional evolutionary history than any other. This is reflected in the vast range of cultural behavior, as described by social anthropologists, which exceeds that of any other geographical area, and the considerable morphological variation, again more extensive than in any other region of comparable size.

The possible interrelationships of major Negroid groups are illustrated in Figures 6-2 and 4-13. These schemes are based partly on archeological finds illustrated in Figure 6-2, again arranged in a sequential evolutionary pattern. Further intensified work could substantiate, disprove, or modify such very tentative hypotheses of Negroid interrelationships. In any case, these became much obscured by waves of southerly migrations of various Caucasoid stocks from the Mediterranean area, which developed more especially during and after the last glacial period. This led to the establishment in tropical Africa of pure Caucasoid groups such as the Rea Sea Hamitics including tribes such as the Beja, Beni Amer, and Hadendoa (Figure 6-3), together with synthetic Caucasoid-Negroid populations such as the Bantu, who have come to occupy much of the tropical African savanna regions. Negroid groups remained generally pure in forested areas because these Holocene Caucasoid invaders were inseparable from their domestic grazing animals, for which there were no suitable fodder plants in the tropical rain forest.

That Negroid stocks evolved from forms very close to the earliest form of *H. erectus* is also evidenced from the occurrence elsewhere in the tropics of such fossil forms as *H. soloensis* or Solo man. All these *erectus* forms must have had hair that grew very slowly, and was "kinky" or "woolly." Negroes, Bushmen, Hottentots, and Congo Pygmies display this feature, as do Pacific groups regarded as "primitive"—the Kanaks of New Caledonia, the Papuans, and the Negrito pygmies of New Guinea. The now-extinct Tasmanians possessed it also. It could be an environmentally determined feature like the black skin that characterized all trop-

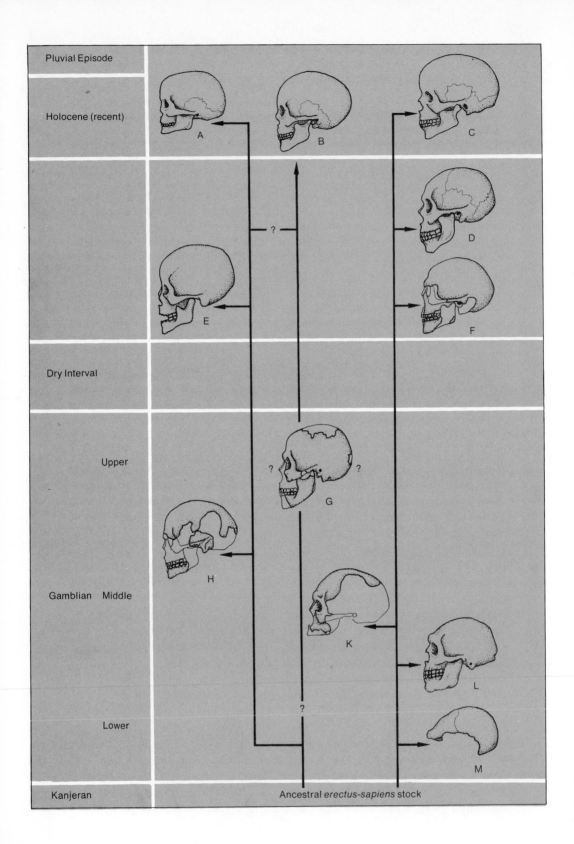

Pluvial Episode

Holocene (recent)

A

B

C

D

E

F

Dry Interval

Upper

G

H

Gamblian Middle

K

L

Lower

M

Kanjeran

Ancestral *erectus-sapiens* stock

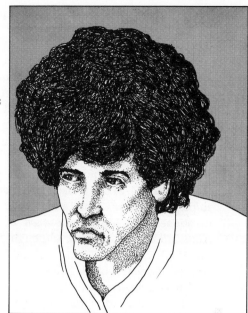

FIGURE 6-3. *Hadendoa male* showing the caucasoid features characterizing several allied tribes collectively classed as Red Sea Hamites, occupying the Red Sea Hills of the Sudan and the former territory of Eritrea, on the western shores of the Red Sea. These pastoral Caucasoids are believed to have emigrated from the central portion of the southern Mediterranean coast at a time in the Late Pleistocene or Early Holocene, when the Saharan ecosystem was more productive, partly at least because of a higher rainfall.

ical inhabitants whatever their origin, but the balance of evidence is against this, especially as it seems to be linked with small external ears and perhaps other as-yet uninvestigated traits.

Caucasoids

As *H. erectus* produced mutant less-pigmented forms, so populations were able to penetrate from the tropics into northern latitudes. Simultaneously there seem to have been other morphological changes—for example, in the method and rate of growth of the hair, which became longer and

FIGURE 6-2. *Interrelationships of Negroid erectus-sapiens populations* as conjectured from some recent and fossil material representing such forms as (A) Bushman (contemporary), (B) *Hottentot* (South Africa, extinct twentieth century), (C) Negro (contemporary) (D) Elmenteita A (Kenya), (E) Matjes (South Africa), (F) Cape Flats (South Africa), (G) Tuinplaats (South Africa), (H) Boskop (South Africa), (K) Florisbad (South Africa), (L) Rhodesian man (Zambia), (M) Hopefield (South Africa). An ever-increasing accumulation of such skeletal material may eventually make possible a statistical evaluation of the significance of any observed and measured variation in it. Improvements in dating techniques are gradually providing an absolute chronology, and increasing paleoclimatic knowledge is supplying an ever more reliable paleoecological background. Combined with biochemical, genetic, and morphometric studies of contemporary populations, these should eventually permit the construction of a conjectural scheme depicting the evolution of the *sapiens* grade in Africa and elsewhere. For the sake of simplicity here, prehistoric (Hamitic) and historic (Semitic) intrusions into Africa from the north are ignored.

more or less wavy or curly. Perhaps this was an adaptation to provide better head protection during the colder winter season at the higher latitudes.

Although Caucasoids, like Negroids, appear to have migrated until historical times only by land, subtropical forms must have penetrated the northly margins of the *H. erectus* Asiatic areas of dispersal. Such groups as the Dravidians of India, Vedda of Ceylon, Ainu of Japan, and especially the "Murrayian" types of Australian Aborigines, with Caucasoid features such as long wavy hair and medium large ears, can be accounted relics of such a former distribution, except that the Australians must have headed south across the tropics.

In Figure 4-14 is a diagram of some of these Caucasoid interrelationships. Australians, Ainu, and so on, not only have the Caucasoid features mentioned but also an *absence* of features of other main groups. The Ainu are not Mongoloid, although now isolated in a sea of Mongoloids; the Murrayian Australians are not Negroid, although some live within the tropics and have black skins.

Caucasoids occupied Europe for a long time, long enough for two events to occur that did not happen in any other primary group. The first was that a major Late Pleistocene section of *sapiens* grade, the Neanderthals, differentiated, became very abundant, and passed to extinction. In other primary geographical groups major Late Pleistocene subsections of *sapiens* grade did not apparently pass to extinction, at least not until the twentieth century. The second unique event was that Caucasoids occupied the higher latitude regions with weaker sunlight long enough for depigmented mutations to spread through the population; this affected hair, eye, and skin color. Depigmentation of the skin has occurred in all groups, but only in Caucasoids did this extend to other epidermal structures.

Mongoloids

It is difficult to avoid the conclusion that Mongoloids evolved parallel with Caucasoids and in the northerly area of distribution of the earliest *H. erectus* form. No prehistoric Mongoloid forms are found in Europe or Africa. Later Mongoloid forms are reminiscent of late Neanderthals in that they appear to be specialized forms further adapted to a particular mode of existence. In the case of Recent Mongoloids, these forms appear to have evolved adaptations to a particularly rigorous winter period such as is encountered on the Asiatic steppes. This adaptation enabled the Eskimo groups to spread, for example, along the northern shore of Canada and occupy parts of the Greenland coast. Some of these Mongoloid migrations were by sea, and either this habit was copied or human populations had now simultaneously in many areas evolved a *sea-exploration grade*. In either case, in the Late Pleistocene and persisting into historical times, many previously unoccupied islands became settled, often by people of synthetic populations apparently resulting from Recent Mongoloid admixture with Early Mongoloid or even with direct descendants of early

H. erectus stocks. These explorations by sea involved especially the smaller islands of Oceania, the last virgin territories uninvaded by man. All the islands on continental shelves had by now been occupied during one interval or another of marine regression during a glacial period.

Oceanic Groups

The problem of classifying the contemporary populations of Oceania on the basis of three primary geographical groups involves even more speculation; authorities such as S. M. Garn take the line of least resistance by designating them as distinct populations. A plausible hypothesis which involves some confirmable deductions would run something as follows: *Homo erectus* was widely distributed throughout the Old World. An early wave of tropical forms of this species spread rapidly through the Old World tropics and is represented by Solo man in Java and Rhodesian man in Africa. Pygmy types developed by parallel evolution from these forms, represented by the contemporary Congo Pygmies in Africa and Negritos in New Guinea (Figure 6-4). Papuans, according to this hypothesis, are the

FIGURE 6-4. *Past and present distribution of Pygmies.* The shaded area represents the conjectural distribution at the close of the Late Pleistocene, the solid area the modern distribution of Pygmy groups in Africa and Negrito peoples in southeast Asia. There are at least two hypotheses as to the origin of such groups, one proposes that they arose independently as an adaptation to high-temperature regimes, the other that their stature has varied little from the *africanus-erectus* stock from which it is surmised they evolved. The second hypothesis receives additional support from the circumstance that Negrito-like forms were the first human invaders of Australia, and the *only* prehistoric occupants of Tasmania.

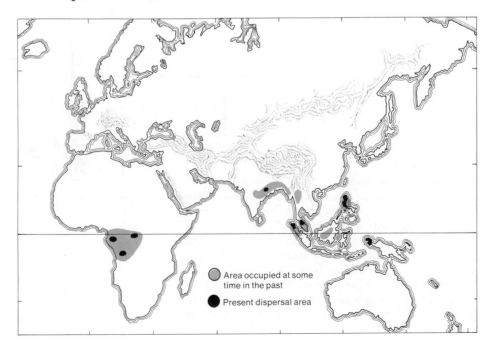

○ Area occupied at some time in the past

● Present dispersal area

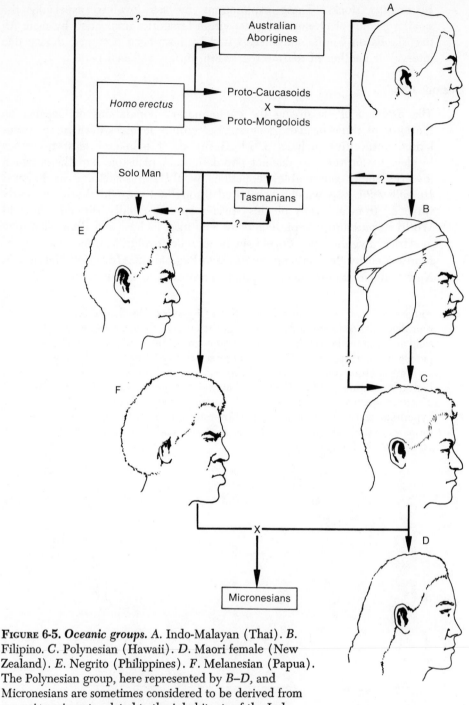

FIGURE 6-5. *Oceanic groups.* *A.* Indo-Malayan (Thai). *B.*
Filipino. *C.* Polynesian (Hawaii). *D.* Maori female (New
Zealand). *E.* Negrito (Philippines). *F.* Melanesian (Papua).
The Polynesian group, here represented by *B–D*, and
Micronesians are sometimes considered to be derived from
sea-going migrants related to the inhabitants of the Indo-
Malay peninsula (*A*), derived from admixture between
Early Mongoloid stocks and a partially differentiating
Caucasoid group. Melanesians (*F*) and Negritos (*E*) may
have differentiated directly from a Pacific group of *erectus-
sapiens* stock established in Oceania before any of the later
migrant group arrived, as may the southern group
(Tasmanians) and more northerly groups of Australian
Aborigines.

Homo sapiens grade evolved from Solo man (Figure 6-5). They have some Negroid features, notably slow-growing kinky hair and small ears—which, as already noted, are not specifically Negroid but rather traits of the earliest *erectus-types*. Papuans and similar groups form what are often called the Melanesians, and have no Mongoloid features.

Another wave of the early *H. erectus* stock moved eastward into Asia after evolving into proto-Caucasoid stocks. These eastern migrants continued east to form the Ainu of Japan, left a remnant behind as the Vedda of Ceylon and the Dravidians of India, and penetrated by land routes into Australia to form the Murrayian groups of Australian Aborigines. They also had no Mongoloid features. In all these migrations into Australia, Ceylon, and Japan, these early groups moved across the continental shelves when such land was exposed during a glacial period (Figure 6-6).

Some of these migrating proto-Caucasoid stocks mingled with Early Mongoloid stocks in southeast Asia. About the third or fourth millennium BP they became sufficiently skilled in boat making to succeed in long ocean crossings. Sailing originally from the Malayan peninsula, they penetrated the whole of Oceania in successive *Polynesian* migrations, of which New Zealand, for example, probably received two before the main Maori invasion of the fifteenth century AD. (Figure 6-5). Madagascar was reached by the Polynesians about the tenth century AD. Some of the Polynesian groups fused with Melanesians to invade the small islands that comprise Micronesia.

The distribution of the main language groups in Oceania (shown in Figure 6-7) does not conflict with this broad hypothesis. Language relationships are, however, noted for their unreliability as evidence of ancestral relationships. Until much more archeological and anthropological work has been undertaken in Oceania, it is not really profitable to speculate even this far as to group origins and relationships. M. Levison and collaborators have recently constructed a computer model that simulates the settlement of Polynesia.

FIGURE 6-6. *Evidence of Australian Aborigine origin,* which is very fragmentary at present, is provided by these three skulls. *A.* This skull, found near Melbourne and dated to 18,000 BP, represents the oldest human remains found in Australia. *B.* Closely resembling the skull in *A,* this comes from Wadyak in Java, and its age has not yet been positively determined. *C.* An adolescent skull from Sarawak believed to be approximately 40,000 years old and to belong to the same population group as the other two skulls.

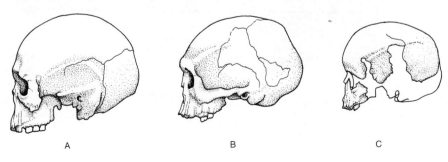

A   B   C

Because of this partial segmentation of global human populations into distinct interbreeding units, an ecologist would attempt to recognize and establish two kinds of diagnostic characters typifying particular units. The first class of characters would be genetically determined, the second the result of environmental response. Moreover, some environmental modification of the expression of particular genetically determined characters would be expected. Stated in ecological terms, there should be detectable differences of a *genotypic* nature, which arose because of differences in genetic inheritance, for such reasons as are listed on p. 192, and variations of *phenotypic* origin resulting from the response of the same genotypes to different environments, or through the operation of the same environment on different genotypes. In making their careful and extensive measurements on the morphological characteristics of various groups, physical anthropologists generally neglected to distinguish between

FIGURE 6-7. *The distribution of Oceanic languages* agrees with the major migrations through Oceania suggested in the text. A Polynesian group (*D*) essentially coincides with the area occupied during the principal Polynesian movements by sea and contrasts with A, the language group of the Australian Aborigines who moved in earlier by land. The Papuan forms (*B*) are distinct from A and D, as would be expected if such peoples are ancient relict groups. Melanesian languages (*C*), used in this sense to exclude Papuan, would be expected to differ both from the languages of such ancient stocks and from the language of Polynesian peoples, as, unlike the latter, they contain no Mongoloid admixture.

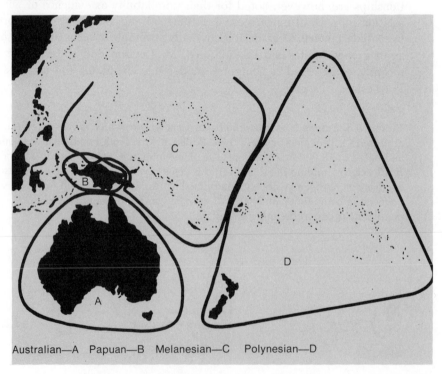

Australian—A   Papuan—B   Melanesian—C   Polynesian—D

diversity arising from these two different sources. Numerous attempts nonetheless have been made to establish relationships between these measurements and features of the macroenvironment.

Ecological Rules

Two rules were early established regarding temperature responses. These classical ecological laws, which have been known generally as Bergmann's rule and Allen's rule, hold that variations in *size* in the first, and *body form* in the second, over a given geographical range of a homeotherm group are related to thermal gradients in that range (Table 6-3). Traditionally animal ecologists have considered the size relationships, and anthropologists have attempted to apply the body-form correlations (Figure 6-8).

Temperature is one of the limiting ecological factors operating on a global scale. It broadly determines the nature of the biotypes and ecosystems of particular latitudes and altitudes, and may be shown to be correlated with such ecological phenomena as species diversity and productivity. The relationship between temperature and body-heat regulation in homeotherms has been extensively explored in recent years, and the validity of these two rules has been seriously questioned. Physiologists in particular have challenged both on the grounds that there are too many exceptions. Some also maintain that adaptations to particular ranges of environmental temperature are physiological rather than anatomical. There are those human ecologists who would admit the general applicability of the rules to animals, but who also maintain that man's culture enables him to modify the effects of temperature without any biological adaptation. This last view largely ignores the circumstance that if the *sapiens* grade is regarded as a species population, *Homo sapiens* has probably been in existence for an estimated minimum of 200,000 years. Cultural adaptations that significantly render man independent of external temperatures, and especially the exploitation of fire, have been operating universally under extensive control for approximately one quarter of this time, and in more sophisticated cultural patterns for perhaps only 2,000 or 3,000 years. Even in modern times, many economically structured societies still expose their less privileged members to considerable extremes of heat and cold, and selection pressures promoting anatomical and physiological adaptations in relation to temperature cannot be ignored.

Biometric Studies of Body Form

E. Schreider analyzed body build in order to quantify it for mensuration purposes and permit a more scientific determination of possible correlations with environmental parameters such as temperature. He studied ten metric characters in eleven human populations and discovered that the volume to surface ratio is very similar in groups as distinct as Parisian workmen and Somali nomads. More specifically, he found that this volume to surface ratio figure fluctuates around 0.5 liter per square decimeter, and concluded that

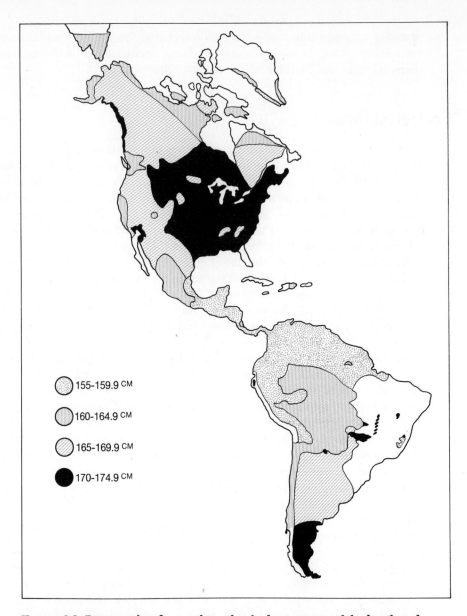

Figure 6-8. *Bergmann's rule* postulates that body size in a widely distributed homeotherm *increases* with *decrease* in temperature. This holds true for Amerindian populations, which show clines of increasing stature with increasing latitude north and south of the equator. Environmental selection for this characteristic cannot on present certain evidence have been operating for more than 20,000 years. Similar correlations hold for sitting height, head size, and other anthropometric statistics. In the far north these correlations are not significant, because in this case the measurements are taken from Eskimos, who have probably had less time in which to adapt to prevailing temperatures.

despite the obvious imperfection of his methods, the ratio appears higher in European than in tropical populations. The body weight to estimated trunk

TABLE 6-3 *Environmental Adaptations of Body Form*

Each of the three main contemporary groups listed here is graded downward in terms of increasing mean ambient temperature. There are associated clines of decreasing height and weight in each group, as would be anticipated according to Bergman's rule. There are also clines in the ratio of height to weight, as would be expected from Allen's rule. See text for further explanation.

Population	Height (centimeters)	Weight (kilograms)	Ratio
CAUCASOID			
Finland	171.0	70.0	2.44
United States (Army)	173.9	70.2	2.48
Iceland	173.6	68.1	2.55
France	172.5	67.0	2.57
England	166.3	64.5	2.58
Sicily	169.1	65.0	2.60
Morocco	168.9	63.8	2.65
Scotland	170.4	61.8	2.76
Tunisia	173.4	62.3	2.78
Berbers	169.8	59.5	2.85
Mahratta (India)	168.8	55.7	2.94
Bengal (India)	165.8	52.7	3.15
NEGROID			
Yambasa	169.0	62.0	2.78
Kirdi	166.5	57.3	2.90
Baya	163.0	53.9	3.02
Batutsi	176.0	57.0	3.09
Kikuyu	164.5	51.9	3.17
Pygmies	142.2	39.9	3.56
Efe	143.8	39.8	3.61
Bushmen	155.8	40.4	3.86
MONGOLOID			
Kazakh (Turkestan)	163.1	69.7	2.34
Eskimos	161.2	62.9	2.56
North China	168.0	61.0	2.75
Korea	161.1	55.5	2.90
Central China	163.0	54.7	2.98
Japan	160.9	53.0	3.04
Sudanese	159.8	51.9	3.08
Annamites	158.7	51.3	3.09
Hong Kong	166.2	52.2	3.18

surface ratio likewise revealed two distinct groups, one in the temperate, the other in the tropical zone (Table 6-4).

Schreider supposed from these analyses that a geographical gradient exists for the body weight to body surface ratio in human populations. He found

TABLE 6-4 *Limb Length/Body Weight Ratios (Centimeters/Kilograms) for Adult Males*

Allen's rule postulates that in a widely distributed homeotherm the length of the extremities tends to increase with increasing temperature. These figures relate arm and leg length to body weight, so that the greater the limb length relative to the body weight, the higher the ratio. Limb length in the anthropometric statistics used in this table increases from cooler regions (Europe) through subtropical (Mexico, Mideast) to hot regions (equatorial Africa, India).

	Limbs to weight ratio average	Body height average
73 Parisian workers	4.88	168.8
47 Finns	4.89	169.5
80 French soldiers	4.91	168.2
50 French students	4.94	174.6
300 British soldiers	5.00	170.9
113 French soldiers	5.02	168.9
504 Ukrainians	5.08	167.3
120 Sicilian soldiers	5.09	169.1
100 Tonkinese	5.37	159.9
82 Otomis (Mexico)	5.51	157.6
31 Arabs (Yemen)	5.63	162.2
18 Asheraf (Somalia)	5.64	170.9
119 Nhungues (Mozambique)	5.66	167.9
123 Darod (Somalia)	5.74	172.2
119 Rahanoween (Somalia)	5.83	169.4
51 Gobaween (Somalia)	5.90	168.4
47 Dir (Somalia)	6.01	172.9
26 Antumba (Mozambique)	6.06	164.9
87 Hawyah (Somalia)	6.21	170.0
18 Korana (South Africa)	6.21	159.8
35 Indians (Madras)	6.61	168.4
95 Aka Pygmies (Congo)	6.98	144.1
115 Basua Pygmies (Congo)	7.03	144.3

it impossible to develop these particular analyses further because of a dearth of measurements using these criteria. He therefore had to utilize the more extensively recorded parameters of body height and weight. These generally conformed to the geographical gradient of the weight to body surface ratio, but he discovered a more complex situation than was originally supposed. From this complexity it was concluded that either there were exceptions to these ecological rules, or that a number of distinct gradients exist for each of the groups examined.

These and other considerations led Schreider to restate Bergmann's rule: "In races of closely related homeotherm species, the relative value of the body surface, expressed as a volume of the mass, increases in climates which at least during part of the year subject the thermolitic mechanisms to stress. The inverse tendency appears in climates which over-facilitate the elimination of heat."

Examining the assumptions of Bergmann's rule, Schreider considers that the universally quoted examples of the Eskimo and the Nilotic anatomy are in some way fallacious and may lead to error (Figure 6-9). He supposes that there is a convergence for these biologically significant ratios in human groups, but considers that corrected formulas may, when applied, still reveal some sort of relationship supporting the original hypothesis. He concludes that the limb length to body weight ratio does play a part in thermoregulation, that with an environmental temperature of 35–36.9°C the negative correlation between the body temperature and the ratio is significant. However, Schreider admits this correlation is not very strong and that there is no reason to expect it to be so, because the anatomical conditions of thermoregulation are not the only casual factors influencing this ratio. The plurality of cross-factors partly explains why physiological correlations are generally low.

FIGURE 6-9. *Correlations between body size and form and prevailing temperature.* The attenuated limbs of the Dinka woman from the southern Sudan on the left have been related by Allen's rule to a direct correlation between limb length and temperature. This same rule requires that the Eskimo (right) have relatively foreshortened limbs, as seems to be the case.

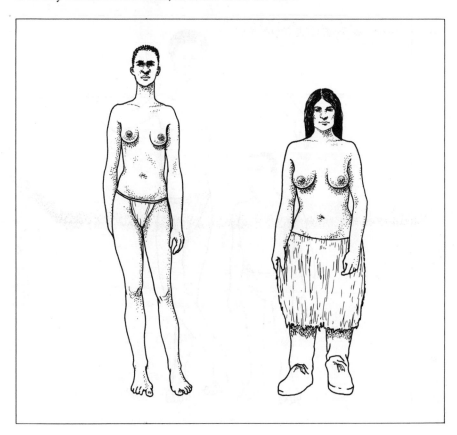

Extreme Variations in Body Form

The greatest extremes of body form in contemporary human populations are encountered on the African continent (Figure 6-10). They are the several groups of Pygmies, where the Ituri group, for example, averages about 1.44 meters (4 feet 8 inches) in height, and such other groups as the Nilotics (Dinka, Nuer, Shilluk) of the Sudan, Masai of East Africa, and Watusi of central Africa, the first group of which averages approximately 1.78 meters (5 feet 10 inches).

Current explanations for the evolution of such contrasting morphological forms in relatively close geographical proximity suppose that both are selections for increasing the rate of heat dispersal into the environment. Pygmies achieve this because of their increased surface area to bulk ratio. The very tall groups are tall by virtue of attenuation of the limbs and trunk associated with little if any increase in bulk. In other words Pygmies are conforming to Bergmann's rule, Nilotics to Allen's. Pygmies are moist-

FIGURE 6-10. *Extremes of body form* as shown by the Watusi (right) and the Pygmy (left) drawn to the same scale. These two African groups live in approximately the same temperature regimes, and within 500 miles of one another, although the Watusi prefer savanna, the Pygmy forest habitats. The usual explanation for this contrast is that both forms are adaptations to increase the rate of heat loss.

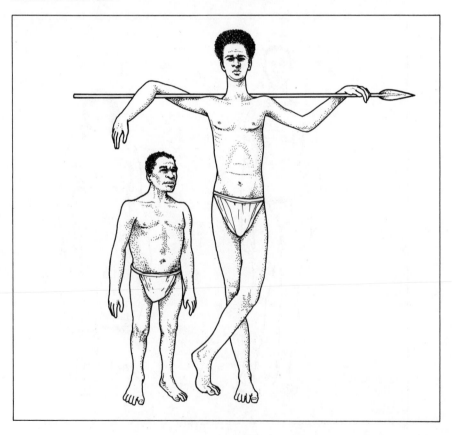

forest hunter-gatherers whose supply of protein is limited. Nilotics, Masai, and Watusi are pastoralists, with a relatively high-protein diet. The additional skeletal material can be added without the risk of attendant nutritional problems.

If these explanations are correct, there should be other Pygmy groups in the moist forests of other continents, and this is indeed so. Pygmies are found in the Andaman Islands and in New Guinea, and it seems more plausible to argue their appearance there as an example of parallel evolution rather than as long-distance dispersal from a common African ancestral Pygmy stock. However, Pygmy morphology and genetics could be read in a somewhat different way, if their close approximation to the height and weight of early forms of *Homo erectus* is stressed. Pygmies would not then be regarded as genetic variants of *erectus-sapiens* stock, but as retaining for nearly 4 million years the ancient body-form characteristics of *H. erectus*.

Pigmentation

Of all the differences among contemporary human groups, skin color has been the most obvious, the most controversial, and the most misunderstood. The term "race" has frequently been intended implicitly or explicitly to refer to a group with a particular range of skin color, and has provoked more agonistic behavioral reactions than any other human characteristic. The latest hypothesis accounting for the presence of pigments in the superficial layers of the human epidermis, presented in Chapter 2, affords a plausible explanation for most of the phenomena that have been observed in relation to this characteristic.

Distribution of Pigmentation Levels

Skin color is an inherited trait determined by a polygenic system whose individual phenotypic expression is modified within certain limits by environmental interactions. In other words, individuals inheriting a particular gene combination will be somewhat darker or lighter than the average individual with this particular gene combination if they have more or less than average exposure to the sun's ultraviolet rays.

Throughout history there has been, until modern artificially reinforced diets appeared, strong selection pressure with respect to this characteristic of skin color. The selection pressure was of this high intensity because, according to Loomis's theory discussed earlier, at a particular latitude individuals with too deep pigmentation would have too little vitamin D synthesis; individuals who were too pale would have too much. Neither group would be likely to survive to breeding age.

The original human population of the Americas is known to have been on this continent for at least some 23,000 years, perhaps much longer than this (Appendix 1). As noted earlier, it may have developed from a population very limited in number and almost completely depigmented. The occupation by this "white" founder band or bands of a region of as high latitudes as

Beringia could only have been achieved by depigmented populations, or by still pigmented people receiving an adequate supply of vitamin D through the consumption of fresh fish livers. It is generally agreed that the founder Paleo-Indian populations were not fishermen, they therefore must have been depigmented.

At the opening of the European era the range of pigmentation exhibited showed a close inverse correlation with latitude (Figure 6-11). Those occupying the tropical zone of Central America are almost as heavily pigmented as any black group encountered in the Old World. As this extent of pigmentation had to be produced by adaptations to the selection pressures developing as the originally white Early Mongoloid stock of the Americas penetrated further into low latitudes, it would be expected that the genetic system controlling pigmentation in Amerindians would differ from that encountered in the Old World. This possibility does not seem to have been investigated.

Other Clines of Skin Color

Each of the geographical groups Negroid, Caucasoid, and Mongoloid contains populations that show similar correlations between latitude and skin color. In the first group the greatest amount of pigmentation is found in West African Negroes of the forest areas. The Banta peoples of the upland plateaus outside the equatorial forest areas are somewhat lighter in skin color, becoming reduced to "brown" in the region of the Tropic of Capricorn with people such as the Zulus. Bushmen of the Kalahari Desert, generally classified as Negroid, are similarly brown-skinned. The Hottentots, also Negroid, who at one time occupied the southern areas of South Africa, were almost completely depigmented.

Mongoloids show the same inverse correlation with latitude, varying from the depigmented peoples of Mongolia to very dark-skinned Indonesians. Theoretically this dark skin would be controlled by the same polygenic system as in representatives of the Negroid stock, but again this does not appear to have been studied.

Caucasoids, like Mongoloids, appear to have been selected for depigmentation during a northward migration through the northern hemisphere, supplying the brown-skinned populations of the Mediterranean region and the white-skinned groups of northern Europe.

It is possible that one of the paler-skinned Caucasoid populations that had adapted to higher latitudes later migrated south and east through Eurasia, leaving relicit populations behind, and founded an Australian Aborigine group. If this is true, these early Caucasoid colonizers of Australia—like the Early Mongoloid settlers of the New World—would have had to adapt to lower latitudes again. Their pigmentation, which is very heavy in the northern Australian tribes occupying the tropical zone, would be expected to be controlled by a separate genetic system from that encountered in Negroes of Negroid stock. In this particular instance there is literature describing work which appears to confirm this theoretical expectation.

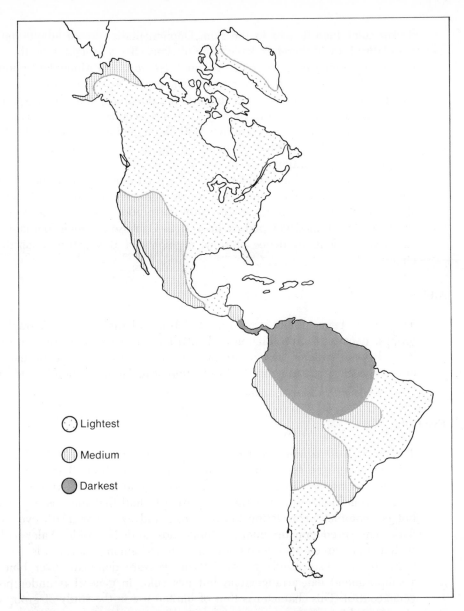

FIGURE 6-11. *Distribution of skin pigmentation in aboriginal Amerindian populations.* The darkest peoples occur in the tropical and subtropical areas of Central America. The lightest are the Fuegians (extreme south) and the Iroquois (extreme north). The Eskimos in the far north are anomalous, as explained in the text.

Hair Color

Because hair is derived from epidermal cells, hair color would not be expected to differ significantly from that of the skin itself. This holds true with one qualification, which is necessary because a relatively small amount of pigmentation in the epidermis will produce in the hair a much darker color than it does in the skin. Depigmentation as an adaptation to high latitudes will therefore have more obvious effects in lightening the skin than in lightening the hair color. Although Negroid and Mongoloid populations with depigmented skin are known, only Caucasoids are ever encountered with blond hair. This suggests the possibility that the adaptation of Caucasoid groups to the highest latitudes occurred very much earlier than these other two main groups. With Mongoloids and Negroids there has not been sufficient time for mutant forms to be selected that exhibit a complete depigmentation in the epidermal apparatus.

There is in some Caucasoid populations a curious red coloration of the hair, associated with the clumping of minor pigmentation into freckles on the skin, both features being correlated with the predominance of the pigment carotene. This trait developed exclusively in Caucasoids, again perhaps because of the long residence of some segments of this group in northern climes.

Albinos

There have been albino forms reported from all three main geographical groups. Because the production and distribution of melanin are controlled by polygenic systems, the occurrence of such mutants suggests that the enzyme pathways leading to the production of melanin can be blocked by a single mutational change.

Eye Color

The amount of pigment in the epidermis is reflected not only in skin and hair color, but also, by its presence in the iris, in eye color. The function of the pigment appears to be prevention of damage to the sensitive eye tissues by ultraviolet light. It is not then surprising to find that there is a general, but not complete, correlation between hair and eye color. Dark eyes, like dark hair, predominate among Mongoloids and Negroids. Paler colors such as blue, green, and yellow are to be found among Caucasoids.

In northern Caucasoid groups, brown eyes are dominant over blue in a simple mendelian arrangement, but eye color in general is under polygenic control. This character is thus of limited use in the study of group environmental interactions.

Hair Texture and Form

Considering hair texture and form, there appears less reason why this should show as marked environmental adaptations as skin color. Hair can be classified into three categories, long and straight, long and wavy or curly, short and crinkly or kinky. The distribution of these three hair types in the Old World conforms very closely with what would be expected if they characterized modern peoples derived from Mongoloid, Caucasoid, and Negroid groups, respectively. The most extreme form of crinkly hair is found in the tight spirals of the Bushman's head; this would indicate directional evolution that has proceeded to an extreme in this ancient Negroid stock.

In Mongoloids the hair is not only straight but has a greater cross-sectional area than in Negroids and Caucasoids. That is, individual hairs are thicker, although the total head covering is not.

Baldness

Even though the *form* of the hair may thus be less a matter of environmental adaptation than its color, its *persistence* during life appears to show adaptation as a result of environmental pressures or, in this case, the removal of environmental pressures. The various patches of hair still distributed over the surface of the human body are sometimes referred to as *ornamental* hair. If this were literally true, this relict hair covering would be solely decorative, and subject only to sexual selection pressures. However, the skull has the thinnest layer of subcutaneous fat of any part of the body, and the brain tissues are the most susceptible to damage of all body organs. An intelligent guess might be made that head hair therefore has at least an ancillary protective function. This being so, cultural developments that introduced the use of artificial covering such as helmets would be expected to reduce the selection pressures for maintenance of this protective covering and be associated with the random spread of mutations away from this character.

This is largely what has been found. In those areas of the Mideast where the "metal ages" appeared about 7,000 BP, several millennia before they developed in other parts of the world, the incidence of premature baldness is far higher than elsewhere. The genetic system producing premature baldness in man is located in the sex chromosomes, so that males heterozygous for this character will bald early, whereas a heterozygous female will be just as persistently covered as one who is a homozygous dominant. Only the homozygous recessive female will be prematurely bald, so this is a quite rare condition.

Effects of Removal of Selection Pressure

This tendency of hair characteristics to mutate away from the genes producing them when selection pressures no longer operate against such mutants may be expected eventually to affect all such characteristics of human morphology, physiology, and behavior. Because additives in the form of vitamin D are now extensively used in temperate countries, and clothing has been adopted in tropical ones, the interaction between latitude and skin pigmentation is no longer so important. Movement of populations to different latitudes is relatively (but not completely) free from selective disadvantage of skin color inappropriate to a different climate. Thus, after a lapse of time that will certainly run to many millennia, it is likely that skin pigmentation will disappear, and with it at least one cause of intergroup friction arising from human diversity. Long before this, however, it is equally certain that chemical control of genes will make possible the creation of any desired level of pigmentation or form of hair.

Facial Form

Facial form clearly is a very complicated characteristic, controlled by different but interconnected genetic systems. Nevertheless, terms have been applied that generalize observable differences. Two such terms are *flat-faced* and *prognathous*. Some of these forms are the result in particular of variation in another characteristic, the size of teeth.

Tooth Size

The presence of very large teeth, as in Neanderthals, necessitated a corresponding protrusion of the lower portions of the face in order to accommodate them and resulted in the appearance of what is sometimes called a receding chin. Reconstruction of Neanderthals (Figure 4-17) always show this marked prognathy and the receding chin feature.

Among living forms the largest teeth known are those of the Australian Aborigines (Table 6-5). These are associated similarly with a receding chin and a degree of prognathy, but not as pronounced as in Neanderthals (Figure 6-12).

The size of the teeth is considered an adaptation resulting from positive selection pressures of the environment. In prehistoric times the food of

FIGURE 6-12. *Correlation between tooth size and extent of prognathy.* Large teeth as in Neanderthals (*A*) require more facial volume for their accommodation than the small ones of a Mideast Caucasoid (*D*). Aboriginal Australians (*B*) have the largest teeth of any contemporary group; those of Eskimos (*C*) are almost as large.

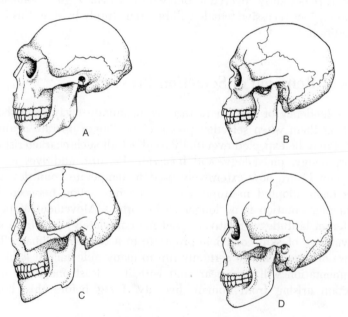

Australian Aborigines apparently was cooked in the ashes of a fire, then eaten, ashes and all. The abrasion resulting from the incorporation of gritty ash in the food contributed to the rapid wearing down of even these large teeth. By middle age the incisors were worn down to the gums, a circumstance that led to early death because of inability to consume enough food.

Another factor contributory to this abrasion was the use of teeth for gripping items while they were being worked with the hands. This wears the incisors in particular, unlike the chewing effects, which especially affect the molars. Among Australian Aborigines the incisors were the teeth to suffer the most excessive wear by middle age. The top and bottom sets came together instead of overlapping in what is now known as the overbite. Worn incisors would be prevented from meeting by the projection of the molar crowns and would no longer be able to grip anything small.

Selection for Large Teeth

In such a population as the pre-European aboriginal peoples of Australia, any random mutation resulting in smaller teeth, particularly in smaller incisors, would shorten the life span of the individual bearing it. There would thus be a reduction in the number of offspring with this trait in the next generation. By contrast, positive selection pressures would increase the size of the teeth gradually to the point where some other factor began to

TABLE 6-5   *Variation in Tooth Size*

Mean length in millimeters of selected teeth in various population samples. For each of the three tooth samples the measurement of the first tooth of each group is cited: in general the first tooth of each group varies least in size between one individual and another. Although the variation reflected here is usually greater between geographical populations than within them, it probably reflects differing diets and varying feeding behavior in the *Homo sapiens sapiens* stage rather than the results of selection operating from as far back as a *H. erectus* grade.

| | Upper Jaw | | Lower Jaw |
	Incisor	*Molar*	*Molar*
MONGOLOID			
Japanese	8.4	10.2	11.1
Aleut	8.3	10.2	11.4
Peruvian (Indian)	8.8	11.1	11.9
CAUCASOID			
European	8.4	10.4	11.2
Australian Aborigine	9.4	11.4	12.3
NEGROID			
Sub-Saharan	9.3	11.1	11.8
Bushman	8.3	9.9	10.9

interact that balanced the advantages of larger teeth against the disadvantages of further increases in their size.

Among extant populations, the next largest teeth to those of the Australian Aborigines are found in Eskimo populations, and also to a lesser extent in Aleuts (Table 6-5). The selection for this trait can again be explained on a cultural basis. Eskimos too use the incisors for gripping objects that are being worked with the hands; even more significantly, they chew skins to tan and soften them in preparing furs as garments. Eskimos also have no overbite, or at least did not until the present generation, which no longer is rigidly observing traditional cultural practices.

Evolution of Small Teeth

With two exceptions, Bushmen and Recent Mongoloids, European Caucasoids are the geographical group with the smallest teeth. Indeed European material from the middle ages already shows a reduction in tooth size as compared with mesolithic and paleolithic Europeans. When the relative size of contemporary human teeth in the Old World is mapped (Figure 6-13), it appears that populations with the generally smallest teeth are centered in an area where agricultural communities first developed extensively. This

FIGURE 6-13. *Distribution of relative tooth size* in aboriginal human populations of the Old World before the major migrations of the nineteenth and twentieth centuries. the smallest teeth occur in the Eurasian region where cereals, especially wheat, rice and barley, were first cultivated beginning some 11,000 years ago. The crude figures of tooth size used for this table were adjusted to *relative* figures by allowing for an obvious correlation with body size.

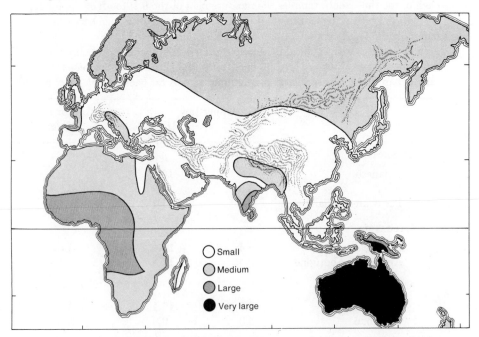

Small
Medium
Large
Very large

suggests that with the change from the consumption of wild plants and roots to that of gruels and other preparations made from the flour of ground cereals, lesser demands were placed on the teeth. Mutations resulting in the development of smaller teeth were no longer eliminated by the operation of strong negative selection pressures.

Distribution of Tooth Size

A belt of small-toothed people stretches from the Pacific coast of China to northwest Europe, including both Mongoloid and Caucasoid populations. Combined with the fact that the occurrence of large teeth is found equally in peoples derived from Negroid, Caucasoid, and Mongoloid stocks (Table 6-5), this strongly suggests that tooth size is determined by interaction with the cultural practices of individual peoples, both as to the preparation and nature of the food consumed and as to the extent teeth are used as a clamp.

There is a further complication in the diversity encountered in tooth form, the presence in some groups of shovel-shaped incisors. This also has been explained environmentally as another mutation away from large teeth in people with advanced food cultures.

Nasal Characteristics

The form and shape of the nose are, like the degree of prognathy of the face, partly correlated with the size of the teeth, rather than a specific indication of a particular ancestry. Large teeth increase the frontal width of the lower face. This increase widens the breadth of the base of the nose, and results in extended nostrils. Peoples as widely apart geographically as Australian Aborigines (Caucasoids), West African Negroes (Negroids), and Eskimos (Mongoloids) are found to have wide noses and flared nostrils, supposedly for this reason. Contemporary populations of the Middle East, with agricultural food cultures believed to precede most others by several millennia, not only have small teeth but also correspondingly thinner lower faces, narrow noses, and small nostrils.

Nose Form

Physical anthropologists record nose form, as opposed to the width and the size of the nostrils, by using the *nasal index*. This is arrived at by expressing the breadth of the nose as a percentage value of its length. A low nasal index (47) therefore is associated with a long, narrow type of nose, a high index (51) with a relatively short, wide one. Although correlations can be obtained between this index and certain environmental parameters, the nasal index generally is not considered a good character to use. Despite this lack of confidence in it, short, wide noses appear to be more characteristic of the tropics, and long narrow noses typical of more temperate areas (Figure 6-14 and Table 6-6).

TABLE 6-6 *Correlation Between Nasal Index and Climate*

A low index, a short, wide nose, is correlated with a warm moist climate; a high index, characterizing long, narrow noses, is correlated with a cool dry climate.

Nasal index of 146 populations correlated with	Correlation coefficient	Standard error
Dry-bulb temperature	0.63	0.050
Relative humidity	0.42	0.068
Dry-bulb temperature and relative humidity	0.72	0.040
Wet-bulb temperature	0.77	0.034
Vapor pressure of the air	0.82	0.027

Long noses and high bridges also are characteristic of peoples living in desert areas. It sometimes is suggested that such long noses are adaptations to selection pressures arising from dry or cold air or a combination of these environmental conditions in which the air requires dampening, warming, or both, in long nasal passages before it enters the lungs.

In the tropics, where the air is both moist and warm, there is no selection pressure for the development of longer nasal passages, nor is there any pressure to select against mutations resulting in somewhat shorter nasal passages. The short, low, squat nose with small but flared nostrils supposedly is the type which all people would develop eventually if we all lived in air-conditioned homes with effective humidifiers and the thermostat turned up rather on the warm side. Similar results would occur if everyone lived in moist tropical regions.

Mongoloid Adaptations

There are some objections to a purely environmental explanation of nose form, for the flattened faces of Mongoloids are associated with a relatively low and comparatively small nose, despite the usual explanation for this face form as an adaptation to extremely cold conditions. The broad fat-padded cheeks, eyelids, and forehead of Recent Mongoloids, if indeed they are genetic adaptations to the cold, are the latest major morphological adaptation in *erectus-sapiens* populations. Such adaptations must have occurred within the last 500 or 600 generations. They were not present in 25,000 BP when Early Mongoloids were entering Beringia, and climatic conditions had ameliorated throughout the world by 10,000 BP.

Mongoloid Eye Fold

The slitlike form of the eyes in Mongoloids is accentuated by the deposition of fatty tissues in an exterior fold of the upper eyelid (Figure 6-15). This is a characteristic exhibited in its most extreme form among modern Asiatic populations of northeast Asia. It is also present to a lesser extent

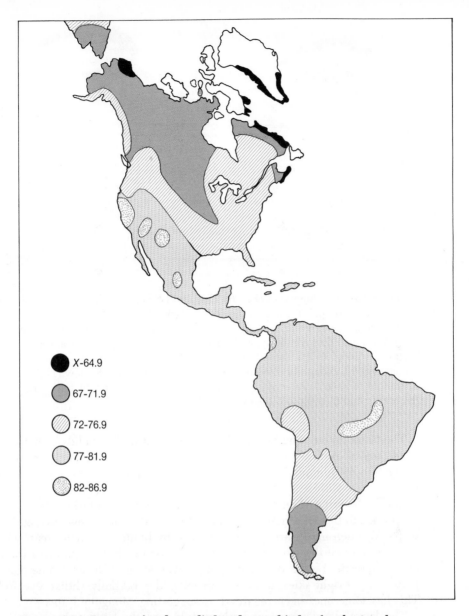

FIGURE 6-14. *Bergmann's rule applied to the nasal index,* for aboriginal Amerindian populations. A low index, indicating a short, wide nose, is correlated with warm moist climates; a high index, characterizing a long, narrow nose, with cold dry climates. As in the case of body form and skin pigmentation previously illustrated, these adaptations to environmental selection pressures have apparently occurred in the less than 20,000 years during which Amerindians could have been occupying their pre-Columbian territories. Even were very recent new dating figures to be accepted, the period for these adaptations is extended only to 50,000 years.

in Amerindians, and is therefore judged to have been a feature of the Early Mongoloids.

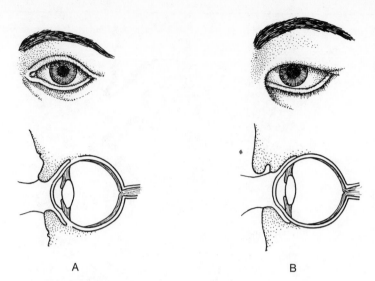

FIGURE 6-15. *The Mongoloid eye* (B) characteristic of the whole Mongoloid geographical group, but less pronounced in Early Mongoloids such as contemporary Amerindians, contrasted with a Caucasoid eye (A). The Mongoloid eye is narrowed dorsoventrally, and the additional fatty fold shown in the sectional view gives it a slanted appearance. The narrowing of the eye aperture is considered an adaptation to provide relief from sun glare off snow. The fold is thought to have a protective function against cold wind, to which the rather flat Mongoloid face exposes the eyes.

The usual explanation for this deposition of fat in the eyelids is that under conditions of extreme cold it protects against frost damage to this comparatively thin membrane. Likewise the general form of the eye, the tendency to assume a slitlike aperture, is considered to be the result of selection pressures favoring reduction of the amount of light entering the eye, an adaptation to the conditions of high light reflectance in a snow-covered landscape. Wooden masks are sometimes worn by hunters in such areas, which similarly reduce the glare by having only narrow slitlike apertures for the eyes. Interestingly, the Bushman groups, who stare at distant game against perpetually bright skies, have also developed a partially slitlike eye form. This has sometimes been regarded as suggesting Mongoloid ancestry.

The lesser development of both eyefold and the slitlike shape in Early Mongoloid forms, or in modern groups descended from Early Mongoloid forms, suggests that although these two characteristics may first have been adaptations to selection pressures exerted by cold winds and snow-covered landscapes, directional evolution through random drift has continued even when the selection pressures that initiated such adaptations diminished.

Negroid stocks essentially restricted to the African continent never experienced extremes of low temperatures and would not be expected to have developed these particular cold-protective adaptations. Caucasoid groups, however, in Late Pleistocene times could be expected to have been exposed to somewhat similar conditions. Considering their greater extent of depigmentation, the length of this exposure would have been for a longer period.

These particular eye characteristics do not appear in Caucasoid stocks, possibly because the ameliorating effect of the Gulf Stream somewhat reduced the severity of the winter conditions in northwestern Europe.

Thus the Mongoloid type of eye seems to provide an example of a characteristic that was environmentally determined, but was an adaptation that occurred in only one of the three major geographical groups of *erectus-sapiens* populations.

Other Morphological Variations

Two minor variations in the morphology of the head—small ears and thick lips—appear to characterize at least some Negroid stocks. For neither of these features would there appear to be any obvious selective value, either presently or in the past. Some Caucasoids have very large ears, but there is no record of the auditory mechanisms performing any differently with such large appendages as compared with small ones. A possible thermoregulatory function does not fit the circumstances either. It is possible that the ear flaps are in the process of reduction to a vestigial condition, and that the mutations leading to this reduction are more numerous in Negroids than in other groups.

The thick lips of Negroids may be the result of sexual selection. P. V. Tobias has argued that the *steatopygia* (Hottentot bulge) of Bushmen and Hottentot women, giving the hugely fattened buttocks, is another example of such a process of sexual selection (Figure 6-16). This might provide a better explanation of both phenomena than any theory of environmental selection, random drift, or mutation.

A comparatively neglected source of possible information on group differences is provided by *dermatoglyphics,* the study of skin ridge patterns observable from prints of fingers, palms, and soles. These ridge patterns, which are constant and unique in each individual, can be analyzed into whorls, loops, and arches. The frequencies of representation of each of these three types characterizes aborigines from different geographical areas. Cummins and Midlo, in a comprehensive work, provide data illustrating how with the use of such indexes as the frequency of arches and whorls, peoples of Caucasoid, Negroid, and Mongoloid stocks can be distinguished. Moreover, on this basis Bushmen, Hottentots, and African Pygmies emerge as a distinct Negroid subgroup, Ainu most closely resemble Caucasoid stocks, and Eskimos and American Indians form an isolated group.

Several other morphological variations await further detailed examination. Ear wax may be of a dry or wet type; the relative sizes of the first, second, and third molar teeth may vary on a consistent pattern.

Considerable attention was once given to the varying occurrence of sweat glands in different human groups. There are 2–5 million such glands on the body, and their density ranges from 150 to 350 per square centimeter. Figures were produced to support the view that acclimation to higher temperatures would increase the number in an individual and that there were also genetically fixed differences between various groups. Recent

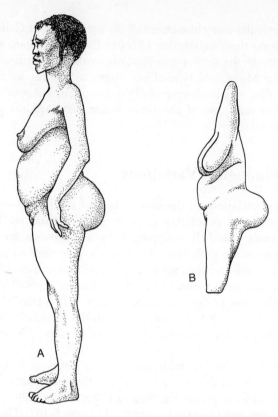

FIGURE 6-16. *Steatopygia.* **A.** Side view of female Bushman from the Kalahari Desert showing the outgrowth of fatty tissues in the buttocks to which the term *steatopygia* has been applied. This condition characterizes Bushmen and Hottentot groups. It is considerably less exaggerated in the males and is not found to occur in any other modern peoples. The relatively common occurrence in Europe from about the close of the Pleistocene of figurines such as that illustrated in *B* suggests that possibly steatopygia was a feature of early *Homo erectus* populations generally, but survived only in these two African groups. If this were substantiated, it would favor the food storage–heat exchange explanation of steatopygia rather than the sexual selection one.

work suggests, however, that the number of sweat glands has already been determined at birth. Possibly the differences previously recorded arose because observations were made on the number of glands actually functioning at a given time.

Biochemical Variations

Some reference has already been made to the existence of different blood groups in contemporary human populations. The biochemical study of such diversity was the natural successor to the physical anthropologists' activities. In this instance, although the observations are mostly qualitative rather than quantitative, they have a scientific basis and are both repeatable and suitable for the application of statistical tests of validity. Furthermore, a major portion of the observable diversity in biochemical characteristics is genetic and inherited, not a phenotypic expression of environmental interactions.

Such characters are grouped here as biochemical variations because the application of biochemical tests is usually required for their detection. It is a purely arbitrary grouping that has no other significance than convenience. Included within this definition are the blood groups and hemoglobin mutants already considered, many other blood groups such as the rhesus, P Lutheran, Kell, Duffy, Diego, and Gm systems; G-6PD deficiency; haptoglobins, transferrins; BAIB excretion; and other examples of genetic polymorphism. The impetus that human immunological work received from what are fast becoming regarded as classical studies on genetic coding, and more recently in response to demands for a deeper understanding of organ and tissue transplant phenomena, is rapidly extending this list. At the risk of appearing biased, it is possible to select for mention some of these polymorphic situations that affect blood group systems and secretory and taste functions.

The Rhesus System

Polymorphic situations arise when several mutant forms of an allele coexist in a population at such levels of gene frequencies that their persistence must be related to factors other than the mutation rate alone. The *rhesus system* at one time was believed to result from the existence of two alleles, *Rh* and *rh*, giving rhesus-positive genotypes *RhRh* and *Rhrh*, with the single recessive rhesus-negative genotype *rhrh*. The presence of these alleles is determined by agglutination tests using rhesus monkey red cells in rabbits.

The effect of rhesus incompatibilities between mother and offspring is lethal to newborn babies. The rhesus factor genetic system is, however, more complicated than was first thought. A more recent hypothesis postulates rhesus alleles in closely linked combinations among three loci, *C, D,* and *E*. Using this last interpretation, considerable variation has been found in the rhesus system throughout the world, and some theories as to possible ancestry and migration routes of particular population groups have been based on this (Table 6-7).

The Diego System

Some blood group systems, such as the Diego antigen, concide precisely with one of the three main *erectus-sapiens* groups. The Diego antigen is of widespread occurrence in Early Mongoloid populations of Amerindians in both Americas, and also in Asiatic, that is, Recent, Mongoloids such as Japanese and Koreans (Table 6-8).

Abnormal Hemoglobins

Generalizations such as this supposed confirmation, from blood group biochemistry, of the hypothetical three independently evolving geographical

TABLE 6-7 *Distribution of Rhesus Genes in Populations of Different Geographical Groups*

These figures appear to support the proposition that gene frequencies for the rhesus factor lie at different levels in Caucasoid, Mongoloid, and Negroid populations; Eskimos and Amerindians are classifiable most nearly as Mongoloids, Australian Aborigines as Caucasoids. However, a more critical examination of such statistics needs to be undertaken before indisputable conclusions can be drawn.

Geographical group	Genes							
	CDE	CDe	CdE	Cde	cDE	cdE	cDe	cde
CAUCASOID								
Danes	0.1	42.2	0	1.3	15.1	0.7	1.8	38.8
Italians	0.4	47.6	0.3	0.7	10.8	0.7	1.6	38.0
Spaniards	0.1	43.2	0	1.9	12.0	0	3.7	38.0
Australian Aborigines (Early Caucasoid)	2.1	56.4	0	12.9	20.1	0	8.5	0
MONGOLOID (RECENT)								
South Chinese	0.5	75.9	0	0	19.5	0	4.1	0
Japanese	0.4	60.2	0	0	30.8	3.3	0	5.3
MONGOLOID (EARLY)								
Eskimos (Greenland)	3.4	72.5	0	0	22.0	0	2.1	0
Navaho	1.3	43.1	0	0	27.7	0	28.0	0
Blood	4.1	47.8	0	0	34.8	3.4	0	9.9
Chippewa	2.0	33.7	0	0	53.0	3.2	0	8.0
NEGROID								
Bushmen (Early Negroid)	0	9.0	0	0	2.0	0	89.0	0
Shona (Rhodesia) (Mixed Negroid–Caucasoid)	0	6.9	0	0	6.4	0	62.7	23.9

groups of *erectus-sapiens* populations inevitably involve much oversimplification. If the oversimplification is substantially ignored, and subspecies or races are erected on this basis, then these entities may nevertheless be entirely artificial and have no real evolutionary significance. It is therefore interesting to examine the situation in regard to the occurrence of *abnormal* as opposed to normal hemoglobins in modern populations. The distribution of the genes producing these abnormalities is among the best known of the genetic systems operating in any naturally occurring population.

Hemoglobin molecules are formed by two different polypeptide chains known as the X and B chains, two of each occurring in each molecule. Formation of the X chains is controlled by a different gene from the one that

TABLE 6-8 *Frequency of Diego-Positive Phenotypes in the Major Geographical Groups of Contemporary Man*

The figures provided are estimated percentage frequencies of occurrence of Diego-positive (Di [a+]) individuals in the population specified.

Geographical group	Population	Percentage Di (a+)
MONGOLOID		
Early Mongoloid	Caingangs, Brazil	46
	Carajas, Brazil	36
	Guajiros, Venezuela	5
	Caribs, Venezuela	36
	Guahibos, Venezuela	15
	Maya Indians, Mexico	18
	Chippewa Indians, U.S.A.	11
	Apache Indians, U.S.A.	4
Recent Mongoloid	Japanese	12
	Koreans	6
	Alaskan Eskimos	1
CAUCASOID		
Early Caucasoid	Australian Aborigines	0
Caucasoid	U.S.A. whites	0
	Asiatic Indians	0
NEGROID	Liberia and Ivory Coast	0
	Bushmen	0

controls the B chains. Detectable mutations occur very rarely in these genes; they result in the replacement of a single amino acid by another at a particular position on one of the X or B polypeptide chains.

Other mutations resulting in the production of abnormal hemoglobin occur that involve more extensive changes than a single amino acid replacement; sometimes even a change in chromosome structure is involved. In the heterozygous condition with most mutations the condition is not lethal, but tends to reduce the individual's life span.

With international agreement in 1952 the various abnormal hemoglobins were named A, F, M, S, and so on; and it is relatively simple to recognize these using electrophoretic techniques. Some had already been characterized by their clinical expressions. *Sickle-cell hemoglobin,* or *sickle-cell anemia* as it has alternatively been named, had long been recognized from the sickle shape which some red cells in a fresh blood film from an affected person assume when they are deoxygenated, particularly when this is done by adding sodium metabisulfite as a reducing agent. This condition was believed to be present in peoples of Negroid ancestry, and it is now known to indicate the presence of the hemoglobin labeled S, which is the result of a single amino acid replacement. The frequency of heterozygotes in a population may sometimes exceed 15 per cent.

Another inherited clinical condition, *thalassemia,* received its name because it is most often recorded among Mediterranean stocks (Figure 6-17). When it results in severe anemia, usually fatal during childhood, it is known as *thalassemia major* or *Cooley's anemia.* Milder forms of this anemia are called *thalassemia minor.* These mutations are associated with a wide range of red cell distortions.

Thalassemia results from a number of changes in gene composition and is a relatively common mutation, sometimes attaining a gene frequency in excess of 15 per cent.

F. B. Livingstone has considered the global distribution of these hemoglobin polymorphisms in relation to that of both various forms of malaria and possible "breeding units." Thalassemia, which seems to convey some resistance to malaria caused by *Plasmodium malariae* and *P. vivax,* almost coincides in its distribution with that of endemic forms of this malaria (Figure 6-18). It appears to be encountered extensively in peoples of Negroid, Caucasoid, and Mongoloid stocks, all of which are represented within this endemic malarial area. Hemoglobin S, by contrast, appears to relate primarily to malaria resulting from infection by *Plasmodium falciparum.* This form of malaria is found especially in Africa, Greece, and India (Figure 6-18), and it is not therefore surprising that sickle-cell trait is noted primarily in peoples of Negroid stock, less in Caucasoids, little in Mongoloids.

FIGURE 6-17. *Distribution of thalassemia in the Old World* is independent of the distribution of the three main human geographical groups. It therefore differs from sickle-cell anemia in this respect, but may similarly be correlated with the distribution of a form of malaria (see Figure 6-18).

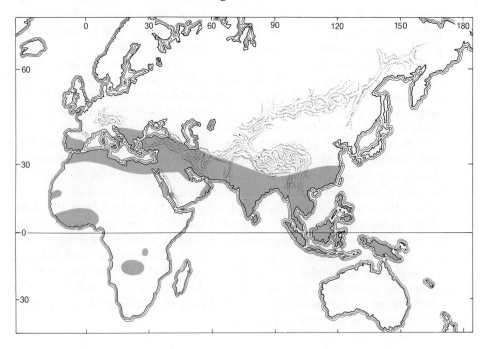

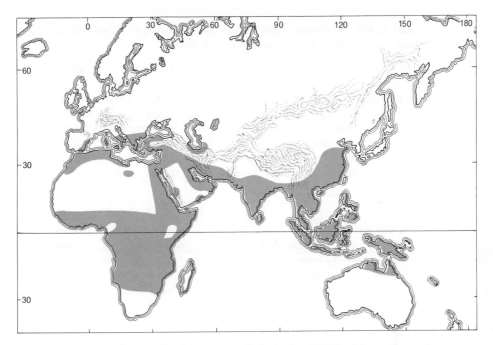

FIGURE 6-18. *Distribution of falciparum malaria in the Old World* is contrasted with the dispersal area of thalassemia shown in Figure 6-17. Falciparum malaria results from infection by *Plasmodium falciparum.*

Thus Livingstone has established that a high incidence of malaria is associated with a cultural change from hunting-gathering to sedentary agriculture, which has taken place during approximately the past 10,000 years, and that the distribution of hemoglobin polymorphisms cannot be used as evidence supporting the establishment of any human breeding units at this time. Nevertheless, although he does not note this, it does provide evidence of genetic contact among the various human populations of the Old World in recent times.

Twinning Rates

Before proceeding to examine some more complex correlations between human diversity and biotic or abiotic pressures, it would be possible to run through a whole gamut of apparently unrelated features that show such correlation in some measure. One of these is the rate of twinning in the three major geographical groups. The figures reproduced in Table 6-9 are comparatively old and very incomplete, but essentially similar figures have recently been published by K. F. Dyer. It should be possible to obtain considerably more data on this subject from various parts of the world. Despite the complication that the dizygotic twinning rate is related to age (Figure 6-19), these figures do suggest that monozygotic twinning rates are low in all contemporary human groups, whereas dizygotic twinning rates show

TABLE 6-9 *Twinning Rates Contrasted Among the Three Major Geographical Groups of Contemporary Man*

Figures for dizygotic twins (those arising from the separate fertilization of two individual ova) are separated from those of monozygotic twins (which have resulted from the division of the one fertilized egg to form two individuals). While the second type shows no significant differences in occurrence among groups and is always low, dizygotic twinning rates are very much higher in Negroids than in the other groups. The one figure for Mongoloids indicates this group may have a very low value. Perhaps the extent of dizygotic twinning was at one time subjected to negative selection correlated with the seasonal rigor of the environment.

Geographical group	Population sampled	Rate of dizygotic twinning per 1000 maternities	Rate of monozygotic twinning per 1000 maternities
NEGROIDS	Ibadan, Nigeria	39.9	5.0
	Kinshasha, Congo	18.7	3.1
	Negroes, Jamaica	13.4	3.8
	Salisbury, Rhodesia (Bantu)	26.6	2.3
CAUCASOIDS	Greece	10.9	2.9
	England and Wales	8.9	3.6
	Sweden	8.6	3.2
	Italy	8.6	3.7
	France	7.1	3.7
	Spain	5.9	3.2
MONGOLOIDS	Japan	2.7	3.8

a marked inverse correlation with predictable rigors of the environment in the three areas of the world where these primary population groups evolved.

Athletic Performance

Once considerations are extended from morphological characters, which can be precisely measured, to less tangible and less readily quantifiable traits, one enters the realm of controversy. Among the least controversial of such features are differences among the major human groups in athletic performance.

Records of the Olympic games performances of athletes from many contemporary societies are now sufficient to indicate that individuals of Negroid ancestry are far more successful in the short-distance races, the dashes or sprints (particularly those up to 400 meters), than individuals apparently belonging to the other two major groups. There is no immediately obvious morphological or physiological explanation for this. Nor is the reverse true, that members of one or the other major group are more successful in the longer-distance events. There is the added complication in these long-

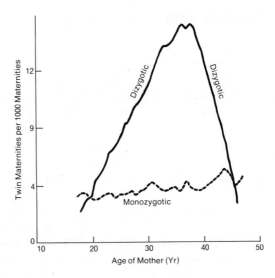

Figure 6-19. *Twinning rates* are affected by the age of the mother, insofar as dizygotic twins are concerned; the monozygotic twinning rate appears to be independent of the mother's age. This correlation will affect the figures on geographical differences in twinning rates shown in Table 6-9 only if it can be shown that the age periods of children are significantly different, which seems unlikely.

distance races that when games are held at sites near sea level the competitors who have been long adapted by residence to the more rarified atmosphere of higher altitudes are at an advantage. This is because they are conditioned to process oxygen at lower pressures. Through entirely unrelated circumstances there has apparently in recent Olympics been a greater proportion of such high-altitude residents among competitors who are of Negroid rather than Caucasoid or Mongoloid ancestry, and a corresponding bias in the competitive performance of these individuals. The suggestion of a positive correlation between ancestry and performance requires a more critical examination than space permits here.

Cultural Differences

Athletic performance is measurable in absolute terms. An extraterrestrial form of higher life would be able to comprehend varying abilities in the time taken to cover a specific distance of the earth's surface in a given time interval. However, *cultural* features such as music have no such absolute values and cannot be detached in time and space from the populations for whom they are an expression of ritualistic behavior.

Musical Ability

It is tempting to generalize that Negroids are more responsive to musical tones and rhythms than Caucasoids or Mongoloids. Certainly among peoples of Negroid ancestry there are more musical scales known than in the other two groups combined. Caucasoids now mostly utilize but a single eight-note musical scale and normally must be specially trained to adhere to that. "Tone deafness" commonly prevents many individuals of this group from detecting fluctuations in pitch of up to half a tone.

This comparative lack of musical response may be associated with the absence of tonal elements in Caucasoid languages that are common in those of the other two groups. It may also account for the comparatively indifferent performance of Caucasoids as linguists. Many Negroids even today cope effectively with a staggering range of languages. A Negroid college student, for example, may be simultaneously fluent in five tongues. First is a home language, which may be quite different from that used in primary education. For historical reasons secondary education will probably be in either French or English, but this is learned in a residential institution, often in an area where the local *lingua franca* also has to be learned, to make a fourth language. Meanwhile the parental home may have been moved to a new area where a fifth language is the dominant *lingua franca*. Before embarking on a college career the student may therefore be speaking five different languages, varying in form from Romance, Teutonic, Semitic, and Bantu. Mongoloids not infrequently are presented with the same kind of necessity, Caucasoids sometimes also—as with the Scandinavians or the Dutch—but rarely contend with more than four languages, and these have either a Romance or a Teutonic form.

Mental Traits

Next to diversity in skin color, possible variation in *mental ability* among the primary geographic groups has aroused more controversial interest than any other human attribute. Study of this aspect of diversity presents many other problems. First, it is impossible to provide a completely acceptable definition of intelligence. A simple statement such as "intelligence is essentially genetically determined, and heredity plays a dominant role in in-

TABLE 6-10   *Hereditary Influence on Intelligence*

The IQ test scores of approximately 40 adopted children compared with those of 40 natural (biological) children raised in homes at various social levels. Heredity rather than home influences appears to play the determining role in deciding the mean IQ of the various groups. Home influences may play a quite minor role; there is a strong suggestion that children made available for adoption may have had biological parents with a higher than average IQ.

Father's occupation	Adopted	Natural
Professional	113	119
Managerial	112	118
Clerical and skilled manual	111	107
Semiskilled	109	101
Unskilled	108	102

fluencing its level in a given individual" becomes meaningless, because it is not clear just what is alleged to be inherited.

A distinction has been accepted between *intelligence* and *aptitude*. In terms of this schism, intelligence has been defined as the ability to associate separate abstract ideas, and aptitude as the ability to perform particular specialized mental or physical activities. According to this definition, the musical and linguistic skills discussed above become aptitudes. At the same time it would appear that a certain level of intelligence is required before particular aptitudes may be expressed, as in mathematics. The most recent development of such a synergistic interaction is in the mathematical symbolism of computer programming. Although there are good and bad programmers, many people lack the aptitude or the intelligence or both and cannot even *begin* to do any programming. Considering the relatively high wages programmers can command in Advanced Industrial societies, it may be surmised that until the demographic transition is quite completed, selection pressures will tend to increase gene frequencies for this ability, whatever its origins. It is also apparent that "brain-drain" phenomena are tending to concentrate these abilities in regional centers.

Measurement of Intelligence

Doubtless many sincerely impartial attempts such as the recent controversial work of A. R. Jensen have been and will be made to define and measure intelligence, but for the present this seems to be an unnecessarily disturbing approach. The figures cited in Table 6-10 illustrate the kind of evidence about the relative effects on intelligence of nature versus nurture that is advanced in this controversy. Such figures seem to indicate that the heritable element in determining relative intelligence is substantially greater than the cultural one.

With the current stage of knowledge, "intelligence" could be regarded simply as an "aptitude" and measured as such, as for example J. P. Guilford suggested. That is, intelligence can be measured and rated as the ability to perform a particular type of mental exercise and *not* transferred from its specific context in relation to this type of exercise. It is unnecessary to attempt a measurement of the intelligence of computer programmers; it is sufficient to express a measure of their ability to program.

Aggression

Rather similar difficulties are encountered when considering diversity in another complex of traits sometimes labeled as *aggression*. There is an increasing tendency to include under this heading not only violent expressions of human behavior but also all the abilities to invent, innovate, "cope," and "drive." It would be a reasonable deduction that the environments described as having a low *predictability* select for genotypes producing more aggression than those with a higher predictability. In the north temperate zone, predictability shows a broad inverse correlation with latitude. Of the 22 or 23 successive civilizations that historians commonly recognize in this portion of the globe, all but one have been overrun, or at least seriously threatened, by barbarians from the north. The one exception is provided by the Moorish invasion of Europe following the rise of Islam. Northerners in Eurasian countries frequently are considered hardier or more aggressive than southerners. After a time, local migration of a brain-drain type may concentrate southern power in the hands of an inmigrant northern caucus, simultaneously depleting the north of its most aggressive genotypes.

During the long course of *erectus-sapiens* evolution, the persistence of less aggressive groups in more predictable environments, with the greater leisure to innovate in other cultural directions that such an existence would provide, may well have made most important contributions both to the diversity of the human gene pool and to the store of cultural knowledge. Such considerations reinforce the obvious need to conserve as great a diversity as possible in human groups and to resist cultural tendencies that would impose a greater conformity.

Ecological Influences on Human Diversity

At the beginning of this chapter it was noted that if human diversity is explicable entirely on the basis of evolutionary theory as developed from a study of other organisms, it should be possible to detect the same kinds of evolutionary phenomena as in these other populations. It now is possible to review the features previously outlined to see to what extent the diversity presented here may be attributed to such sources.

1. VARYING FREQUENCIES FOR THE SAME GENES AMONG THE HUMAN GROUPS CONSIDERED TO HAVE BEEN DIFFERENTIATED. This kind of variation has been illustrated here extensively, and is consistent with the primary geographical groupings. Genetic systems producing such features as basic blood groups and rhesus factors are distributed with consistently maintained variations in gene frequencies among groups, even though no selection pressures can be identified at present. Where selection pressures can be found, as in such traits as resistance to thalassemia, skin color, and tooth size, gene frequencies are clearly related to the environment rather than to ancestral groups.

2. UNIQUE ALLELES. A number of these have been mentioned among them coarse hair and fatty upper eyelids in Mongoloids, steatopygia in southern Negroids, thick protruding lips in equatorial African Negroids, Negroid Pygmy groups, and Caucasoid freckles. Not all are morphological by any means. The occurrence of blond hair solely in Caucasoids is of biochemical origin. Other less apparent unique biochemical traits are the Diego antigen in Mongoloids, the absence of *cdE* rhesus factor combinants in Negroids. and the presence of *Cde* combinants in Caucasoids.

3 and 4. PHENOTYPIC ACCLIMATION AND GENOTYPIC FIXATION. Waddington discussed the circumstance that individual phenotypic responses to environmental stress are similar to the adaptations made as a result of selection pressures resulting from such stress. This has been illustrated by the density of *active* sweat glands in human skin. A phenotypic response is demonstrated by the fact that long periods of tropical residence are found

to increase the number of active glands. A higher density of active sweat glands is also a genetic characteristic of equatorial Negroid populations. In a similar way, the amount of pigment in the skin increases following considerable exposure to a higher ultraviolet content in sunlight. This also is a genetically determined trait.

5. GENETIC DRIFT. This phenomenon has not been specifically identified here, because its major effects on human populations must have been far back in *H. sapiens* history. In these early times quite small migratory bands are thought to have existed. The Early Mongoloid group ancestral to all Amerindians could have been, as has been noted, as few as 100–500 persons strong. From recent studies on groups as small as this, for example, on the Dunkers sect in the United States, it is certain that genetic drift would have occurred in such a band. Observations on this point have not yet been fully assembled in the literature.

6. FOUNDER PRINCIPLE. The same applies to studies of the founder principle insofar as it must also have affected these early migratory bands. The complete absence of the B blood group in Amerindians is believed to be the result of its exclusion from the sample represented by the original Beringian founder band. This blood group does occur in other Mongoloids, including Eskimos and Aleuts.

Causes of Diversity

In a text such as this, there is space to mention only the salient features of human diversity, but insofar as these have been considered, they appear to conform to the general pattern of evolutionary development that any widespread biological group would be expected to exhibit. Although there is no difference in principle, there has been a considerable difference in emphasis. Evolution in *erectus-sapiens* populations, as is mentioned several times in this text, has related more especially to cultural evolution. It is the behavioral, i.e., mental, traits that have been especially and characteristically the material on which selection and adaptation have operated during the past 4 million years. We have no reason to suppose that such selection is not still a continuing process.

Cultural Evolution

The cultural adaptation that permits cultural evolution is very flexible indeed, and may occur one or several times per generation. In an attempt to define culture more closely, so that its evolution could be examined more critically, Julian Huxley recognized three separate but overlapping aspects:

1. *Sociofacts*   Direct behavioral acts such as rituals
2. *Artifacts*   Material results of sociofacts, e.g., tools or music
3. *Mentifacts*   Potential behavior, assumptions, ideas, values, intentions, etc., for example, "water flows downhill"

All higher animals possess sociofacts, but artifacts are rarely produced by animal populations other than members of the genus *Homo*. Nevertheless, there is no great difference between them, the latter being the physical product of the former. Both types of culture can be learned by imitation; neither requires either a society or a language (sociofact) for transmission.

Mentifacts, however, could not evolve before an effective communication language was developed. *Social learning* of mentifacts would also greatly accelerate the rate of familiarization of juveniles with accumulated mentifact information stores. Once effective language and mentifacts had become established in human culture, an extremely high selection value must have become attached to variants with a greater than average interest in them. Primates all show inquisitiveness, but intellectual curiosity in the form of *increasing educability* must have been and still is a primary adaptation in *erectus-sapiens* populations. Morphological and physiological adaptation lagged far behind in the 4 million years of *erectus-sapiens* cultural evolution, but this was not significant because mentifacts and artifacts provided the adjustments that "biological" selection had failed to produce. It is possible now that continuing pressures, coupled with the immense size of the human population, have resulted in the production of mentifacts at such a rate that they cannot all be integrated properly into global cultures. Some resultant artifacts are accumulating as wastes, and no natural reducer or artificial sociofact process in our ecosystem has the information necessary to decompose and recirculate them.

Finally, a crucial point must be made about cultural as compared with biological evolution. The latter can take place only in populations and is achieved by selection that leads to adaptation within a population. Leaving aside those special cases in microbial groups where certain adaptations can be transmitted genetically, only the population subject to the selection process will be able to adapt and thus evolve.

Cultural evolution is not restricted to the population in which it occurs; any other population may *mimic* the sociofact, *use* the artifact, *learn* the mentifact. The speed and ease with which this cultural exchange proceeds have attained such a high value that in the Advanced Industrial societies possessing the modern artifacts that facilitate information dissemination on cultural advances, cultural diversity is more evident between generations than between groups.

This has brought to our modern world another threat of extinction of a portion of its human diversity. Those populations lacking artifacts to receive and utilize all this new cultural information will be unable to adapt by cultural evolution to the adjusted parameters of the abiotic elements of the new ecosystems. The very speed of our cultural evolution therefore will reduce inevitably some of our population diversity. The nineteenth and twentieth centuries will have witnessed the complete destruction of all hunting and hunter-gatherer populations that were present when we entered this critical period. It is possible that the same inexorable fate may also extend to all agricultural societies, Early and Advanced. Conceivably only the urbanized Advanced Industrial societies, geared to the mass production of artifacts and the mass dissemination of mentifacts, will survive into the next millennium.

Individual Adaptation

Until now most reviews of human evolution have stressed population adaptions. W. S. Laughlin, for example, states: "The crucial importance of the concept of population and of population thinking becomes obvious when it is appreciated that *evolution takes place only in populations . . .*" [author's italics]. This is no longer true. *Cultural evolution now takes place in individuals.* The "fitter" individual has adapted to a changing environment, but not necessarily because of any trait that will be transmitted to a potentially larger number of offspring. The ability to adapt will depend on the individual's cultural history and cultural equipment. If an individual is surrounded by a suitable array of artifacts, has been provided with a sufficient store of mentifacts, and has been taught the sociofact tricks of applying these, that individual will respond by his own adaptation to cultural modification which would not even be perceived by someone deprived of appropriate artifacts. This fundamental change imposed by cultural evolution has such far-reaching implications that this point will be discussed further in later chapters.

Synthetic Societies

From this generalized and necessarily oversimplified account of human diversity, it may be deduced that during the approximately 4 million years of *erectus-sapiens* evolution there have been alternating periods of territorial stabilization and isolation succeeded by extensive migration, hybridization, and integration. The periods of comparative isolation in localized territories must have far exceeded in total time the almost cataclysmic episodes of migration and integration. They have been periods of stationary population growth and stable economic situations, the realization of the "S" phase or logistic mode of the ecologist. The cataclysmic episodes, the latest of which we are now immersed, have been the ecologist's "r" phase, the exponential growth of populations and economies.

The most recent of these episodes has resulted from the population explosions that were a consequence of the industrial revolution in northwestern Europe, which moved its previously largely agricultural societies into a colonizing phase. The subsequent migrations transferred Caucasoid groups into areas of the globe already inhabited by hunter-gatherers and Early Agricultural groups of early Caucasoids (Australia), Early Mongoloids (the Americas), and Negroids (equatorial and southern Africa). Reverse migrations were also promoted—Negroids into Caucasoid Europe, for example—as well as large-scale deportations, such as Negroids to Early Mongoloid areas (North and South America).

Modern Migration Rates

A resurgence of nationalism among previously politically dominated groups, which was accelerated in the 1950s after the end of World War II, has

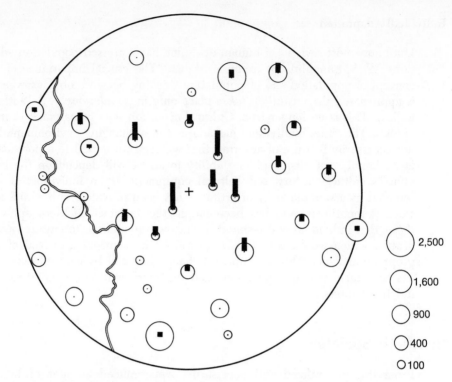

◯	2,500
◯	1,600
◯	900
◯	400
◯	100

FIGURE 6-20. *Village intermarriage.* Marriages in the village of Charlton-on-Otmoor, Oxfordshire, England, in 1861, when the population was recorded as 375 persons. The heavy vertical bars represent a comparative value for the number of marriages occurring between residents of that village and Charlton inhabitants. The position of the village of Charlton is marked with a cross; the Cherwell River, some 10 feet wide at this point, was obviously a major deterrent to courting couples. As the key indicates, different sized circles represent different sized village populations.

vastly reduced the level and rate of migration prevailing in the last few centuries. The position has once more become stabilized, each national group being within its own legally established territory. Some geographical regions have remained largely unaffected by this latest upheaval, and stabilization finds their populations little different in genetic composition from what they were in the last stable period. Figure 6-20 illustrates the stable rural conditions of nineteenth-century Europe, where it is calculated that the social units from which a mate was selected averaged about 500 persons. Even with the high mobility of the contemporary U.S. population a large percentage of individuals marry their high school or college sweethearts.

By contrast some islands, such as Hawaii received wave after wave of immigrants during this period. In Hawaii, an originally Polynesian people was infiltrated first by Caucasoids, then by Mongoloids from China and Japan, and finally by a complex of Early and Late Mongoloids, Caucasoids, and Negroids from the North American mainland (Table 6-11). Such a *syn-*

TABLE 6-11 *Ethnic Composition of the Synthetic Society of Honolulu County, Hawaii*

Estimated population averages for 1955–56 are arranged into primary groups and adjusted to the nearest thousand.

CAUCASOIDS		100,000
MONGOLOIDS		
Chinese	31,000	
Japanese	127,000	
Korean	6,000	164,000
OCEANIC GROUPS		
Filipino	36,000	
Hawaiian (or part)	63,000	99,000
OTHERS		12,000
Total		375,000

thetic society affords tremendous genetic opportunities for gene recombination followed by selection of "fitter" genotypes, better adapted to such an environment.

The United States

The largest contemporary synthetic society, indeed the largest universally diversified society ever known, is the more than 200-million-strong population of U.S. citizens (Table 6-12). There can be little doubt that despite modern medical care, welfare procedures, and tax differentials, natural selection is still operating on this massive gene pool to favor a greater reproduction or increased survival of offspring among the more "fit," whatever that term means as regards human traits.

The selection pressures in this instance have operated, at least until quite recently, through relative degrees of economic success. For example, the high financial remuneration now provided major league basketball players permits a very big man, with massive food consumption and special expenses for apparel, furniture, and other domestic appurtenances, to survive comfortably and afford, if he wishes, a large family. Previously such individuals existed miserably (e.g., as freak exhibits in a circus), and it can be supposed their fertility rate under these conditions was low.

The number of magnificently coordinated and fast-moving giants in the United States is visibly increasing. The same selection factors are providing larger and faster football players, in the case of quarterbacks, at least, correct and fast strategic problem solving is being subjected to vigorous selection pressures.

This is the way four linebackers were described in *Time* magazine several years ago: "Larsen (6 feet 5 inches, 255 lb); his forte is an explosive initial

TABLE 6-12 *Ethnic Composition of the Synthetic Society of the United States*

Ethnic origins are indicated by census statistics recorded in 1900. Figures expressed are thousands of individuals registered as foreign-born residents emigrating from the counties listed. It must be remembered that included in the other 66 million (87 per cent) of the population were the surviving *Negroid* descendants of some 10 million Negro slaves brought to the United States in the eighteenth and nineteenth centuries and, in the *Mongoloid* group, some 1½ million Amerindians who were the sole pre-Columbian inhabitants of the area.

Caucasoids		Mongoloids	
Austria	433	China	81
Great Britain	2,688	Mexico	103
Bohemia	157	Total	184
Canada			
(English language)	785		
(French language)	395		
France	104		
Germany	2,663	Others	117
Hungary	146	Grand total	10,046
Italy	484		
Netherlands	95		
Poland	383		
Russia	424		
Scandinavia	1,072		
Switzerland	116		
Total	9,745		

charge . . . that opens the way for Page (6 feet 4 inches, 250 lb) and Eller (6 feet 6 inches, 255 lb). Both are extremely quick and boast exceptional agility . . . . The iron man who makes it all work is Marshall (6 feet 5 inches, 250 lb), [who] has a quality of balance as great as any man I've ever seen." David might have had something more of a problem on his hands had he been faced with one of these modern-day Goliaths. Studies on the inheritance and number of offspring of individuals selected for these particular characteristics would be extremely interesting.

Net immigration figures for the United States presently run about 300,000 persons annually; the number of live births is approximately 1.8 million annually. Thus each year foreign genotypes are being added to the pool of genetic diversity in a ratio of 1 to 6. At this rate the population of the U.S. will continue to be a repository and test center for recombinants of all the human genes extant on this globe.

To botanical taxonomists this is a not unfamiliar situation. Several large tropical genera such as *Diospyros* (ebony) and *Terminalia,* which contain numerous species, have achieved this size because of niche diversification. Particular species have adapted to particular habitats in given ecosystems. This adaptation to particular ecological niches prevents extensive hybridization between the species, and taxonomists have little difficulty in establishing an acceptable grouping of taxa. However, if the natural ecosystems are degraded or destroyed, the position is entirely changed. Biotypes adapted to the original ecological niches are no longer the ones most favorably selected. Hybrids no longer perish because they are less "fit" in the environments of the new habitats created by degradation of the ecosystem. The taxonomist is presented with vast arrays of hybrid swarms aggressively colonizing these newly created dispersal areas and defying classification. Eventually, perhaps, natural selection will once more sort out the "fitter" genotypes and by niche diversification create a more orderly and thus a more readily classifiable complex. Meanwhile the original gene pool of the genus, which had been fragmented into individual taxa, has once more been reassembled. Hybridization has provided a vast reservoir of the genes of this pool, on which natural selection can operate to produce genotypes differing from those of the original species and better adapted to the new habitats.

Our new human habitats are the urban environments of our industrial world. New recombinants are needed if we are to better adapt biologically and culturally to these new situations. Synthetic societies such as those of the western hemisphere provide the necessary reassembly and scrambling of the human gene pool, which for several million years had been adapted to the narrow niches of more restricted environments.

Selection for Aggression

In these synthetic societies—whether of the past, like Britain or Rome, or contemporary, like the United States, Brazil, and Australia—one of the behavioral characteristics that must still be subjected to strong selection pressure is *aggression*. M. F. Gilula and D. N. Daniels provide a wide definition of aggression, including the entire spectrum of assertive, intrusive, and attacking behavior—sarcasm, dominance, and "coping." They point out the complex origins of aggression, which may arise from selection, frustration, or social learning. Creative aspects of aggression are usually described by the terms *drive, judgment,* and *intelligence.*

It would appear that considerable selective immigration into the synthetic society of the United States is concentrating individuals who express strongly aggressive characteristics. These are attracted especially to the three areas of developing megalopolises, the southern Pacific Coast, the northeastern seaboard, and the southern Great Lakes. This local concentration of U.S. citizens apparently is supplemented by a brain drain of individuals with similar aggressive characteristics from other parts of the world. The higher gene frequencies for aggressive polymorphisms in such megalopolies, it must be surmised, will lead to assortative mating and the partial separation of distant breeding units that may constitute the founders of a new species grade of *Homo,* which as suggested earlier could be called *H. sapiens innovatus.* This new grade will arise, or has already arisen, primarily as a result of cultural speciation. The new species grade will be characterized by a more stable and better accepted social hierarchy than in *Homo sapiens,* reproduction entirely separated from sexual intercourse, greater intelligence, and a harnessing of aggression into innovative and inventive directions. Individuals of this new species will possess the cultural equipment enabling them to adapt more rapidly to cultural change.

Although he does not say so specifically, Toffler has implied that *H. sapiens* has now reached the limits of adaptability to an ever-accelerating rate of cultural change. The pathological syndrome that develops from overstimulation by the swelling flood of new ideas he labels *future shock.* Victims of future shock, according to Toffler, show symptoms ranging from anxiety and hostility to authority through senseless violence to actual physical illness, depression, and apathy. They feel "bugged" or harassed and attempt various forms of withdrawal in order to reduce the number of decisions they must make. It may be conjectured that *H. sapiens innovatus* is being selected for biological combinations that provide a measure of resistance to future shock.

The gene frequencies of this new species grade will slowly introgress into the surviving populations of the *H. sapiens sapiens* grade until this grade has passed to extinction. Meanwhile, all human populations of whatever grade are imminently threatened by environmental crisis of immense, unprecedented proportions. These are a direct consequence of the removal of the feedback mechanisms that regulated human population growth. The ever-increasing depletion of natural resources, the discharge of externalized wastes into air and natural waters, and the deliberate or accidental discharge of persistent poisons into natural ecosystems are technical problems resulting directly either from failure to comprehend or from a deliberate flouting of ecosystem requirements. Their explanation, their effects, and their possible remedies are aspects of human ecology that will now be considered in the remainder of this text.

BIBLIOGRAPHY

References

Anonymous Extract from *Time*, October 17th, 1969, p. 62.

Baker, P. T. "Racial differences in heat tolerance," *Amer. Phys. Anthrop.*, **16:** 287–305, 1958.

Birket-Smith, K. *Eskimos*, New York: Crown, 1971.

Boyd, W. C. "Modern ideas on race, in the light of our knowledge of blood groups and other characters with known mode of inheritance," in C. A. Leone (ed.), *Taxonomic Biochemistry and Serology*, New York: Ronald Press, 1964, pp. 119–69.

Brose, D. S., and Walpoff, M. H. "Early upper paleolithic man and late middle paleolithic tools," *American Anthropologist*, **73:** 1156–94, 1971.

Buettner-Janusch, J. *Origins of Man*, New York: Wiley, 1966.

Bulmer, M. G. "The effects of parental age, parity and duration of marriage on the twinning rate,"*Ann. Hum. Genet.*, **23:** 454–58, 1959.

Cadien, J. D. "Dental variation in man," in S. L. Washburn, and P. Dolhinow (eds.), *Perspectives on Human Evolution*, Vol. 2, New York: Holt, Rinehart and Winston, 1972, pp. 199–222.

Cavalli-Sforza, L. L. "The genetics of human populations," *Scientific American*, 231(3): 80–89, 1974.

Coon, C. S. *The Origin of Races*, New York: Knopf, 1963.

Court Brown, W. M. *Human Population Cytogenetics*, New York: Wiley, 1967.

Court Brown, W. M. "Heredity and responsibility," *New Scientist*, 40: 235–36, 1968.

Cummins, H., and Midlo, C. *Finger Prints, Palms and Soles*, New York: Dover Publications, 1961.

Dyer, K. F. "Hidden variability in man," *New Scientist*, 44: 72–74, 1969.

Fermi, L. *Illustrious Immigrants, Chicago:* University of Chicago Press, 1968.

Garn, S. M. *Human Races*, Springfield, Ill.: Thomas, 1961.

Gilula, M. F., and Daniels, D. N. "Violence and man's struggle to adapt," *Science*, 164: 396–409, 1969.

Goodenough, W. "A problem in Malayo-Polynesian social organization," *American Anthropologist*, 57: 71–83, 1955.

Greene, D. L. "Environmental influences on Pleistocene hominid dental evolution," *Bioscience*, 20: 276–79, 1970.

Guilford, J. P. "Intelligence has three facets," *Science*, 160: 615–20, 1968.

Harrison, G. A., Weiner, J. S., Tanner, J. M., and Barnicot, N. A. *Human Biology*, New York and Oxford, England: Oxford University Press, 1964.

Jensen, A. R. "How much can we boost IQ and scholastic achievement?" *Harvard Educ. Rev.*, 39(1): 1–123, 1969.

Laughlin, W. S. "Race: a population concept," *Eugenics Quart.*, 13: 326–40, 1966.

Levison, M., Ward, R. G., and Webb, J. W. *The Settlement of Polynesia: A Computer Simulation*. Minneapolis: University of Minnesota Press, 1973.

Livingstone, F. B. "Aspects of the population dynamics of the abnormal hemoglobins and glucose-6-phosphate dehydrogenase deficiency genes," *Amer. J. Hum. Genet.*, 16: 435, 1964.

Lynn, R. "Genetic implications of the brain drain," *New Scientist*, 41: 622–25, 1969.

McClure, H. M., Belden, K. H., Pieper, W. A., and Jacobson, C. B. "Autosomal trisomy in a chimpanzee: resemblance to Down's syndrome," *Science*, 165: 1010–12, 1969.

Maricq, H. R. "Ethnic differences in the fingerprint data in an 'all white' control sample," *Human Heredity*, 22: 547–77, 1972.

Montagu, A. M. F. *A Handbook of Anthropometry*, Springfield, Ill.: Thomas, 1960.

Newman, M. T. "The application of ecological rules to the racial anthropology of the aboriginal New World," *American Anthropologist*, 55: 311–27, 1953.

Reed, T. E. "Caucasian genes in American Negroes," *Science*, 165: 762–68, 1969.

Schreider, E. "Ecological rules, body-heat regulation, and human evolution," *Evolution,* 18: 1–9, 1964.

Slobodkin, L. B., and Sanders, H. L. "On the contribution of environmental predictability to species diversity," in *Diversity and Stability in Ecological Systems,* Brookhaven Symposia in Biology No. 22, 1969, pp. 82–95.

Tjio, J. H., and Levan, A. "The chromosome number of man," *Hereditas,* 4: 1, 1956.

Tobias, P. V. "Bushmen hunter-gatherers: a study in human ecology," in D. H. S. Davis, (ed.), *Ecological Studies in Southern Africa,* The Hague: Junk, 1964, pp. 67–87.

Tobias, P. V. "Recent human biological studies in Southern Africa . . . ," *Trans. of the Royal Soc. S. Africa,* 40(3): 109–133, 1972.

Toffler, A. *Future Shock,* New York: Random House, 1970.

Waddington, C. H. *The Strategy of the Genes,* London: Allen and Unwin, 1957.

Further Readings

Ashley Montagu, M. F. *An Introduction to Physical Anthropology,* 3rd ed., Springfield, Ill.: Thomas, 1960.

Bloom, H. F. "Does the melanin pigment of human skin have adaptive value?" *Quart. Rev. Biol.,* 36: 50–63, 1961.

Boyd, W. C. *Genetics and the Races of Man,* Boston: Little, Brown, 1950.

Brace, C. L. "A non-racial approach towards the understanding of human diversity," in M. F. Ashley Montagu (ed.), *The Concept of Race,* New York: Free Press, 1962.

Brace, C. L., and Ashley Montagu, M. F. *Man's Evolution,* New York: Macmillan, 1965.

Campbell, B. G. *Human Evolution,* Chicago: Aldine, 1966.

Dobzhansky, T. *Mankind Evolving,* New Haven: Yale University Press, 1962.

Dobzhansky, T. "Genetics of race equality," *Eugenics Quart.,* 10: 151–60, 1963.

Harrison, G. (ed.) *Genetical Variations in Human Populations,* Oxford, England: Pergamon, 1961.

Hulse, F. S. "Technological advance and major racial stocks," *Human Biology,* 27: 184–92, 1955.

James, P. E. *A Geography of Man,* 2nd ed., London: Ginn, 1959.

Mather, K. *Human Diversity,* New York: Macmillan, 1964.

Mourant, A. E. *The Distribution of the Human Blood Groups,* Oxford, England: Blackwell, 1954.

Race, R. R., and Sanger, R. *Blood Groups in Man,* 4th ed., Oxford, England: Blackwell, 1962.

Roberts, D. F., and Bainbridge, D. R. "Nilotic physique," *Amer. J. Phys. Anthrop.,* 21: 341–70, 1963.

Sheldon, W. H., Dupertius, C. W., and McDermott, E. *Atlas of Men,* New York: Harper, 1954.

Washburn, S. L. *Classification and Human Evolution,* Chicago: Aldine, 1963.

Chapter

Population
Growth

Chapter

7

The first six chapters of this book were concerned with the origin and evolution of the human populations that have now universally achieved an exclusively *sapiens* grade occupying the whole of the habitable globe. This chapter and the succeeding two are devoted to the study of the size, structure, and fluctuations in numbers of these populations, that is, to *demography* or human population dynamics. Until the twentieth century the emphasis in such studies was on population fluctuations rather than population growth, if such matters were considered at all. Little account was paid to population structure. The emphasis on population growth is of quite recent occurrence, dating back no further in its present form than the early years after the Korean War. Even in literate societies many people still consider this emphasis misplaced; to illiterate societies the problem conceivably can be meaningless, despite the fact that they may already feel the impact of our present failure to resolve growth problems.

Until World War II any surplus population could always be accommodated somewhere, usually on the appropriate frontier—the American West, the Canadian Northwest Territories, the French Empire, the British Colonies,

Java, the Belgian Congo, Manchuria, "up-country"—almost anywhere. Now the frontier is no more, the empires and colonies no longer the happy hunting ground for minor sons seeking fame and fortune. Such up-country as remains is barren and uninviting.

To understand why human population growth has come so suddenly to demand such alarmed attention, we must examine the size of *habilis-erectus-sapiens* populations as they evolved through the various ecological stages described in Chapter 5.

Population Size in the Past

The occupational floor described for Olduvai *Homo habilis* may well have represented the base of activities of a single pair bond and their offspring. Among the Kung Bushmen of the Kalahari Desert Basin of Southwestern Africa, who represent one of the most primitive contemporary societies known, the band basically still is formed from such a family unit. The 1,000 persons who comprise the Kung group are divided up into 28 *bands,* each of which forms what anthropologists term an *extended family.*

A band might consist of perhaps 17 adults and 14 children, including an older leader with an older and a younger wife, their four sons and four daughters, three sons-in-law, three daughters-in-law, two unmarried children, and 12 grandchildren. Such a nomadic band would occupy a territory of approximately 30 square miles (20,000 acres), if territorial arrangements did not overlap.

Population Density

The derived Bushmen densities calculated from such figures (one person per square mile) are approximately the same as those of another surviving group of hunter-gatherers, the Australian Aborigines, who—at the time of the continent's historical discovery—were estimated to number approximately 300,000 persons. The Aborigine equivalent of the Bushman band was an extended family unit known as a *horde,* which usually was comprised of about 40 persons. A number of hordes together made up a *tribe,* the equivalent of the Bushman group, and an average tribe numbered about 500 persons. The size of the Aborigine tribal territory varied with rainfall, the ecological factor most commonly limiting ecosystem productivity (Figure 5-1). Availability of marine foods also increased the population density of coastal and insular tribes, but it was the additional food resources of riparian habitats that provided for the greatest increases in density.

The rainfall regimes of southwestern Africa and Australia are not comparable, especially as the latter shows variation on a continental scale. J. B. Birdsell quotes the widely varying densities as they relate to rainfall regimes in these two instances, and also those between the Shoshoni in the Great Basin of the United States and the Amerindians in central Baja California. Density of the last was more than 50 times that of the Australian Aborigines, despite generally comparable climatic conditions. Aside from deserts and the polar regions, no part of this earth still has such a low population density.

The carrying capacity for human populations has been forced upward in a progressive series of steps, as has been noted in earlier chapters. Each step represented some cultural advance, like elimination of some competitor or overexploitation of some resource, with a consequent overriding of the previous regulatory mechanisms restricting the population growth. As described in the first chapter, for most of our time we have held to a "K" phase at particular carrying capacity levels. Each major cultural achievement has permitted us, because of the larger extractive capacity it conferred, to change briefly to an "r" or exponential phase of growth, at the conclusion of which we assumed a higher carrying capacity.

Population Size in Hunter-Gatherer Populations

Using such figures for population densities as those just discussed, we can estimate that during the Middle Pleistocene, despite local overriding of regulatory mechanisms, the world's *erectus-sapiens* population did not exceed 3 million and could have been as low as 1 million (Table 7-1).

The reason the population density of these early human populations did not increase rapidly must always remain enigmatic. Comparison with remaining hunter-gatherers suggests a number of factors that may have prevented the survival of many children. First, environmental factors may have been so rigorous as to seriously affect longevity. If women, as well as men, generally

TABLE 7-1 *Increase in Human Population Density and Size over the Last 1 Million Years*

Improvements in extractive capacities resulting from cultural development of artifacts and sociofacts are correlated with a slow increase in population density and size, but only the urban, agricultural, and industrial revolutions of the last 6,000 years caused dramatic changes in both. Compare the estimated density for 2000 AD with those listed in Table 13-1.

Time BP	Cultural level	Average density in persons per square kilometer	World population in millions
1,000,000	Lower paleolithic	0.004	0.125
300,000	Middle paleolithic	0.012	1.0
25,000	Upper paleolithic	0.04	3.0
10,000	Mesolithic	0.05	8.0
6000	Neolithic/farming	0.05–1	80–150
2000	Farming	2	300
200	Farming	5	800
150	Farming/industrial	7	1,000
100	Farming/industrial	8	1,300
50	Farming/industrial	10	1,700
0–(1950 AD)	Farming/industrial	15	2,500
1975 AD	Farming/industrial	30	3,900
2000 AD	Farming/industrial	<50	7,000

did not usually survive beyond the age of 30 or so, their potential child-bearing years would be halved. If the environment included a cold winter, or a dry one, as many as one half of the infants could have perished during this time of severe environmental stress and not have survived to even a short adulthood. Eskimo populations at the beginning of this century illustrated both these features (Figure 7-1). Eskimo societies also had some social population regulation procedures, such as abandonment of unproductive dependents, infants, and ailing elderly individuals. Somewhat similar social practices probably operated among many hunter-gatherer groups, as was described in Chapter 4. The manner in which these communities existed, as well as their probable demographic statistics, is also becoming more clearly defined by extrapolation from remaining hunter-gatherer peoples.

Special Characteristics of *habilis-erectus-sapiens* Populations

Throughout the 4–5 million years of our evolution, the species of the genus *Homo* have been characterized by three features that have ensured, if not a high density, at least an always wide distribution. We have been what the zoo curator calls "opportunistic" in diet; that is, our food habits have never

FIGURE 7-1. *Demographic features of Eskimo populations.* This hypothetical population pyramid has been prepared from the actual statistics of a particular Eskimo population at the turn of this century. The conventional population pyramid, many examples of which are supplied later in this chapter, illustrates the proportion of the population contained in successive 5-year age groups (cohorts). These groups are subdivided into males (left) and females (right). Few individuals of either sex survive beyond the age of 35. Only about half the infants survive through their fifth birthday.

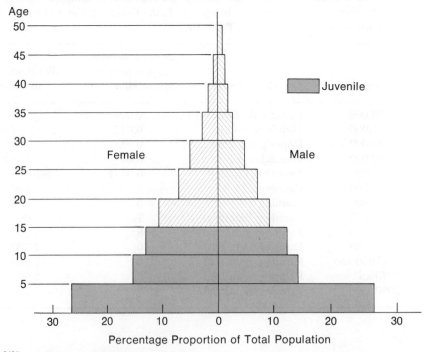

become highly specialized, and we have always possessed a considerable development of the behavioral rituals called "culture." These first two features have interacted synergistically, providing a ready facility for rapid extension into new and previously unoccupied territory. The third behavioral feature is a *social nature,* which has ensured the speedy dissemination and adoption of cultural invention and innovation.

Consequently, by the time evolution in the genus had attained what is here labeled the *erectus-sapiens* stages, beginning perhaps 2 to 3 million years ago, the whole of the habitable world with the exception of the Americas and Australasia was becoming occupied and divided into nucleated hunter-gatherer territories of varying size.

Trophic Relationships

J. B. Birdsell quotes a century-old work describing the food of Australian Aborigine tribes along the southwestern coastal region. This illustrates well the fact that Aborigine populations extracted all the food that could be economically obtained and prepared by the tools available at this stone-age cultural level from the several ecosystems they occupied. Arranged in categories according to the trophic level at which the human population had to function in utilizing each particular food source, the list of items eaten is shown in Table 7-2.

Because in an ecosystem only about one tenth of the energy input is transmitted from one trophic level to another (the rest is dissipated in metabolic and locomotor activity and as heat), only one hundredth of the biomass is available when a human population acts as a tertiary consumer or top

TABLE 7-2 *List of Food Items in an Australian Aborigine Diet Indicating the Trophic Level at Which the Aborigine Populations Would Be Operating in Each Instance*

As a primary consumer (herbivore)	As a secondary consumer (general carnivore)	As a tertiary consumer (top carnivore)
29 kinds of roots	6 sorts of kangaroos	2 species of opossum
4 kinds of fruit	5 medium-sized marsupials	dingos
2 species of cycad nuts	9 species of marsupial rats and mice	1 type of whale
Seeds of several legumes		2 species of seals
2 kinds of mesembry-anthemum	3 types of turtles	7 types of iguanas and lizards
7 types of fungus	11 kinds of frogs	8 types of snakes
4 sorts of gum	29 kinds of fish	
2 kinds of manna	All saltwater shellfish except oysters	
Flowers of several *Banksia* species	4 kinds of freshwater shellfish	
	4 kinds of grubs	

carnivore, compared with when it functions as a primary consumer or herbivore. This tends to make such items listed in Table 7-2 as whales, snakes, and dingos delicacies, relieving a fairly steady basic diet of roots interspersed with marsupials, fish, and other seafood.

Potential for Continuing Population Growth

Further increases in size of the global *erectus-sapiens* population could thus only be achieved by advances in what computer jargon labels "hardware" and "software." The "hardware" would be the physical equipment. Virtually no significant skeletal changes occurred in these last 2 to 3 million years of *erectus-sapiens* population development; it is only in somewhat larger cranial capacities that the physical equipment noticeably improved. Most further advances were achieved in the "software," the behavioral rituals that determined the cultural patterns, tools, and weapons of evolving societies, to which we have already applied the designations sociofacts, mentifacts, and artifacts. The simultaneous enlarging of the brain supposedly supplied some reinforcement to this cultural evolution.

Carrying Capacities

Each cultural advance would raise the level of the human secondary productivity in the individual group territory, thus increasing its *carrying capacity* for *erectus-sapiens* individuals. As the ecosystems contained in the territory could not receive any greater input of energy, this increased size of the *erectus-sapiens* population could only be achieved in one of two ways. It could be effected if there were a corresponding reduction in the biomass of competing species populations (for example, in the numbers of other top carnivores feeding on the same prey). Or it could be achieved by "mining" accumulated resources of the ecosystem, as when honey representing several years' production was collected from all the bees' nests in the territory or all the animals of a particular species were killed and eaten. Increases in human population size probably were achieved in both ways.

Global Population Size

At the beginning of this *erectus-sapiens* period 2 to 3 million years ago, the human world population size is estimated at about 125,000 (Table 7-1). Two major cultural advances were achieved before the end of this stage, probably between 50,000 and 300,000 BP, depending on the region. One was the *hafting* of tools and weapons, which previously had been held directly in the hand. The other was the use of *fire* not only in domestic cooking but also in hunting. It is not yet certain which development was the earlier, but it is likely that their cultural spread was uneven, whatever their point of origin. The use of fire, for example, is not recorded in Africa until comparatively late, 50,000 BP; and the use of hafted tools in Australia appears to have been introduced about 15,000 to 20,000 BP. Tasmanian man, as previously noted, was isolated from the human populations of the Australian mainland before the development of hafted tools and did not independently devise them. Both these major developments of hafting and fire use in hunting could have arisen independently in a number of territories; it is not essential for them to have had a unique origin. Wherever and however they did originate, their effect on extractive capacity and thus on carrying capacity for human populations had to be great.

Pleistocene Overkill

It is difficult to estimate the effect on population numbers of these two cultural advances, the use of fire and hafting, but no later than 50,000 BP there is notable evidence of their effects on at least some of the contemporary ecosystems. Indeed these might be regarded as the first evidence of the devastating and irreversibly destructive effects that human occupation has had subsequently and with increasing intensity on world ecosystems.

Overkill and Human Population Growth

During the time of this overkill in all continents, both food and water would have been in abundant supply, the ample protein diet would have reduced infant mortality and extended the life span. There would be for the time being no density-dependent controls on populaton growth. If food supplies were exhausted in any part of the territory, the nomadic band would move on to another part. If territories began to overlap, virgin lands within the potential dispersal area were still available for inmigration. It can be supposed that, in all the *erectus-sapiens* populations involved, family size was large and that the formation of new hordes and the occupation of virgin areas were at rates never again attained by hunter-gatherer populations. Birdsell believes such populations would double every generation. As shown in Table 7-1 and Figure 7-2, the world population is estimated to have tripled during this period.

Such a population explosion must have been associated with the migration of hunter-gatherer stock into Australia as well as the Americas. The earliest reliable date for any *erectus-sapiens* fossil remains from Australia is 18,000 BP. Access may have been gained to the continent before this, but not too early, for there is some evidence that the dingo (a semidomesticated dog) moved in about the same time. Early Mongoloids penetrating the Americas had no domestic dogs.

All these Pleistocene game animal extinctions can be interpreted as the result of overexploitation of a food resource, which would establish larger and more vigorous human populations than had existed previously. With the disappearance of much of the game, the same carrying capacity for the enlarged human population could be maintained only by assuming a primary consumer role and adopting a more vegetarian diet. Some authorities think that this enforced switch of trophic roles was the stimulus triggering the

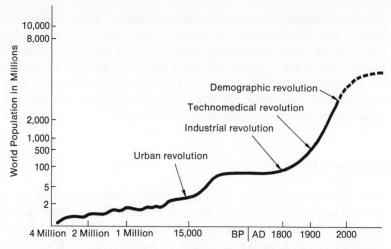

FIGURE 7-2. *Increases in the size of the total human population of the world.*
Plotted on a logarithmic scale on both axes, population increases are shown as
episodic and related to spectacular cultural advances permitting an increase in
the human carrying capacity of the environment. In prehistoric times such
advances would take such forms as hafting of weapons, invention of the spear
thrower, various uses of fire. The first spectacular population increase was
attained when permanent settlements developed, soon associated with agriculture.
The next was the result of the industrial revolution and the subsequent control
of mortality by health technology. The last revolution—the demographic one—
unlike all the previous ones, is assumed to have a stabilizing effect on population
growth, and to discourage further increase in our global population.

adoption of agricultural techniques in the Americas. Whether this is true
or not, density-dependent factors would again begin to influence further
population growth.

Population Growth in Early Groups

After these various Pleistocene overkill episodes, perhaps in some instances
concurrently with them, the peoples of various agricultural territories established from previous hunter-gatherers appear to have maintained balanced
populations whose growth was regulated by a number of feedback mechanisms. These were perhaps the first *human societies* that were permanently
density dependent and in which feedback mechanisms evolved that prevented utilization of ecosystem resources at a nonrenewable rate.

To illustrate the kind of feedback mechanisms in such Early Agricultural
groups, one tropical and predominantly vegetarian population is considered
here, contrasted with a temperate and almost exclusively carnivorous group.

Population Growth in an Early Agricultural Group

Small isolated populations in moist forest areas of New Guinea have been
known for some time to be head-hunters and have herds of domestic swine,

while operating essentially on a "swidden agriculture" pattern of shifting cultivation. Rappaport has reported studies of one particular group of about 200 persons in which there is a cyclic progress through swine herd increase to warfare, to extended feasting and truce, and back to swine herd increase. This effectively maintains both the human and domestic animal populations at a level where no irreversible damage is done to the microecosystems occupied.

The swine are not fed, but scavenge for themselves around kitchen middens, clearing up, among other items, human feces. They also feed in the moist forest. Whenever a sow farrows, as many males of the litter as can be caught are castrated. This reduces the number of boars to so low a figure that sows are most frequently impregnated by wild boars. The swine are not herded, killed, or controlled at this stage, and their number slowly builds until incidents of pig damage to gardens increase beyond a tolerable frequency. Formalized warfare of a highly ritualized form then breaks out between "offended" and "offending" parties, their relatives, and sympathizers. Some individuals are killed and others maimed in this warfare, which ceases when ritualized "honor" has been satisfied. Participants from both sides then join in an extended feast period during which virtually all the domestic pigs that can be caught are killed and eaten. When the swine herds have reached a very low density, the feast period is concluded, a truce is declared, and the group settles down to another slow build-up period for swine and human populations.

The feedback mechanisms in the group Rappaport studied were so effective in controlling both swine and human populations that a considerable area of virgin moist forest had remained uncut and excluded from the "swidden" mosaic.

Population Growth in a Nomadic Pastoral Group

In contrast, the Bedouin populations of the Mideast are pastoral peoples with an almost complete dependence on an animal rather than a vegetable diet. As recounted by G. H. Orians, in the north Arabian deserts of Saudi Arabia the main food supply of the Bedouin is camel's milk, together with meat from male calves, which are killed in order to conserve milk for human consumption. A single family unit requires from 15 to 20 milking camels for its maintenance. If cyclic drought reduces the forage in the territory of a particular tribe, this density of camels cannot be maintained. The tribe can therefore allow some of its camels to die and make good its needs by raiding or bartering with more fortunate neighbors, or it can resort to the same alternatives to obtain additional pasturage for its own camels. Either way a population control mechanism is likely to be invoked. Raiding will kill some male members of the groups involved. Bartering will lower productivity— or rather slow the rate of build-up of net productivity—and probably, through an enforced postponement of marriage, reduce population growth.

PLATE 6. *Evergreen riparian forest* fringing the bed of a seasonal river in Rhodesia. Rainfall in this area is low, averaging about 200 millimeters per annum. Crops such as sorghum, millet, and in heavier rainfall years corn, can often be grown, but nothing in years of below-average rainfall. Surface water is present in the shallow riverbed for three or four months of the year; after that, water is to be had only by digging in the riverbed. Nevertheless, this seemingly inhospitable microenvironment has supported permanent settlements of early agriculturalists for at least 1,000 years.

Advanced Agricultural Societies

It is possible, as was described in Chapter 5, that Advanced Agricultural societies arose in two ways. Primarily they may have been derived directly under the influence of a barter-trade city such as Catal Huyuk. Secondarily they may have developed through the superimposition of a trading function on a settlement in an Early Agricultural group. The founding of these Advanced Agricultural societies is estimated to have been associated with a rise in world population to somewhere in the neighborhood of 30 million. However these societies originated, it seems highly probable that, at least in their early history, Advanced Agricultural societies were independent of any feedback mechanisms regulating population growth.

Putting carrying capacity in terms of population density, R. J. Braidwood and C. A. Reed have estimated the increased carrying capacities that Advanced Agricultural societies were able to achieve in southwestern Asia as shown in Table 7-3.

Cultural monuments such as Mayan temples or Egyptian pyramids and numerous art works testify to the high level of net primary productivity

PLATE 7. *Palm-wine preparation* from the palm *Hyphaene crinita,* which occurs naturally in the riparian forest illustrated in Plate 6. The cut surface toward the tip of the stem of this palm exudes a sap that collects in the pot attached to receive it, as in this photograph. The sugar solution ferments because of contamination of the collecting pot by wild beasts. The vitamins obtained from drinking this fermented preparation probably help to keep mortality rates in the permanent riparian settlements of this area relatively low. When most of the palm trees, even small suckers as here, have been tapped, the whole settlement moves a few hundred yards up- or downstream, and the previously occupied area is allowed some years to recover. The early agricultural "Sutu" peoples of these riparian communities have remained in this region for at least ten centuries without overexploiting the riparian ecosystem or warring with other peoples.

that was available in this category of society for the expansion of cultural activities. It is nevertheless conceivable that this stage became a major hurdle for many evolving societies in all parts of the occupied world. Already some groups had remained at an earlier evolutionary stage, probably because environmental or behavioral limitations maintained a low net primary productivity. Thus Australian Aborigines and Kalahari Bushmen were held at a mesolithic grade in the first successional category, New Guinea settlements at the second. A study of West Africa illustrates the way in which many societies could be held at the third successional category of Advanced Agricultural societies.

TABLE 7-3 *Estimated Carrying Capacities of Various Early Societies*

These figures, for density in a given land-use pattern, do not compare with the overall figures of Table 7-1, but related to Table 13-1 they suggest that Sumeria might have looked much like modern Egypt in terms of population density.

Type of culture	Population density in persons per square kilometer
Pleistocene hunter-gatherers	0.01
Late Pleistocene hunter-gatherers	0.05
Early Agricultural communities (circa 7,000 BP)	10
Advanced Agricultural communities (Sumerian, circa 5,500 BP)	20

West African Societies

The pattern of ecosystem development in West Africa from the ocean to the borders of the Sahara is closely associated with the rainfall patterns and regimes. Interaction among the conformation of the coastline, the south-western trade winds, and the cold Benguela current produces a set of isoclines of diminishing rainfall parallel with the coast. From an average total of over 1400 millimeters, the rainfall diminishes to as low as 200 millimeters per annum in the interior.

The first to observe the ecological correlations with this rainfall pattern was the French botanist and ecologist Auguste Chevalier, who described a succession of forest and savanna zones (Figure 7-3). Subsequent workers have detailed how the human settlement pattern is intimately related to the ecosystem distribution with respect to these rainfall regimes.

The Moist Forest Zone

Until European colonial times in Africa the moist forest zone exhibited a pattern of swidden agriculture or shifting cultivation societies. Many of these remained in comparative isolation in the category of Early Agricultural groups. They must have possessed density-dependent population regulatory mechanisms; otherwise there would not have remained enough land to permit regeneration of the forest areas, which they cropped for 3 or 4 years before abandoning them for between 15 and 40 years. They cultivated forest-clearing crops such as oil palm and *Dioscorea* yam, which supposedly originated 5,000 or 6,000 years ago in this area.

Imposed on this swidden agriculture pattern in the forest were societies based on permanent settlements like Kumasi, Benin, and Enugu. They had become trading centers, the seats of tribal authority, and the focal points of political power for the Advanced Agricultural societies that dominated the forest region. These cities appear to have been in a phase of continuous population growth; surplus members emigrated to form colonies in many other West African towns. Probably in these cities whatever density-

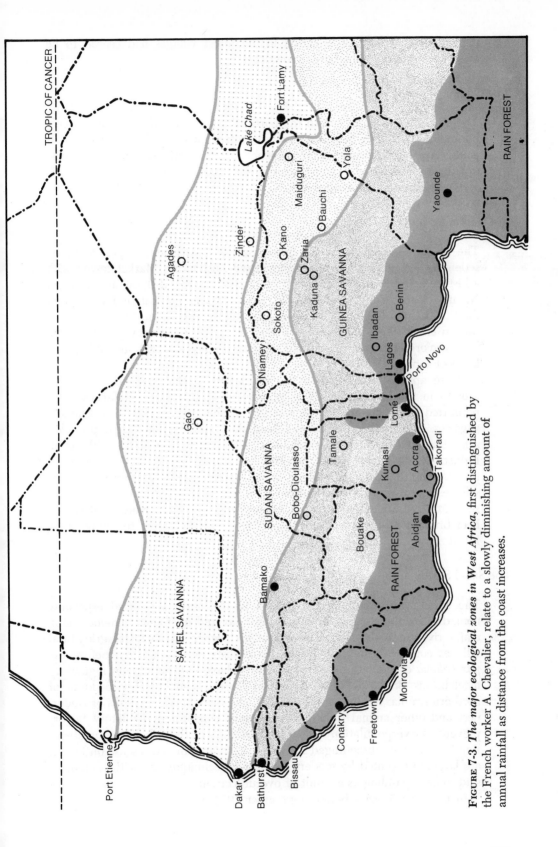

TROPIC OF CANCER

RAIN FOREST

● Fort Lamy

Lake Chad

○ Zinder

○ Maiduguri

○ Yola

○ Bauchi

● Yaounde

○ Agades

○ Kano

○ Zaria

GUINEA SAVANNA

○ Sokoto

○ Kaduna

○ Ibadan

○ Benin

○ Lagos

● Porto Novo

○ Niamey

● Lomé

○ Gao

Takoradi

○ Tamale

○ Kumasi

● Accra

SUDAN SAVANNA

● Abidjan

○ Bobo-Dioulasso

RAIN FOREST

○ Bouake

SAHEL SAVANNA

● Bamako

● Monrovia

○ Conakry

○ Freetown

Port Etienne

○ Bissau

● Bathurst

● Dakar

FIGURE 7-3. *The major ecological zones in West Africa*, first distinguished by the French worker A. Chevalier, relate to a slowly diminishing amount of annual rainfall as distance from the coast increases.

dependent controls that operated in the forest villages had ceased to be effective in limiting population growth.

The Guinea Savanna Zone

The adjoining savanna belt to the north, the Guinea savanna, was scarcely populated, and there is still no entirely satisfactory explanation for this. It could have been because of the presence of human disease such as malaria, or animal diseases such as trypanosomiasis, or both, a seasonal lack of domestic water, or any number of other causes.

The Sudan Savanna

The second savanna zone, the Sudan savanna, was heavily populated by sedentary agriculturalists practicing mixed husbandry. Market towns with a history going back at least 1,000 years are established there, among them Niamey, Kano, Zaria, and Bamako. Lines of communication ran through the Sudan zone linking these towns and across the Sahara to the Mediterranean cities. They were walled towns, developed originally perhaps from a pattern originating in the Tigris-Euphrates Valley, and with much the same kinds of activity.

There appear to have been no regulatory mechanisms on population growth in the Sudan zone. The permanent occupation of a farmed area permitted the accumulation of reserve supplies to buffer against climatic vagaries. The market towns reinforced this buffering effect and provided a holding area for surplus populations. Although not yet describable as an Industrializing society, this Advanced Agricultural society was already preadapting to a colonizing phase at the beginning of this century, and few villages or towns in the rest of West Africa lacked a "Hausa quarter." In such external areas the Hausa handled the meat trade, which distributed beef brought on the hoof from the Sudan and Sahel zones to the meatless forest zones.

The Sahel Savanna

The Sahel zone adjoining the Sahara to the north supported especially nomadic pastoralists, but included a few favored microenvironments along major rivers and the shores of Lake Chad, where permanent agricultural villages could be established. The pastoralists must originally, before the *pax colonia* of the late nineteenth century, have had population regulatory mechanisms similar to those described for the camel-owning Bedouin of northern Arabia. Raiding, local drought, and occasional outbreaks of rinderpest and other animal diseases would have constituted regulatory factors preventing overpopulation and overexploitation of the Sahelian ecosystems. Raiding has been discouraged for some years, and animal disease outbreaks are largely prevented by modern veterinary techniques. Now the Sahara is said to be expanding as a result of overutilization.

In the 4 or 5 years before September 1974, rainfall was so low in this

Sahel zone that insufficient grass grew to feed the animal herds, and agricultural crops failed. The flow of water in the two great rivers of the area, the Niger and the Senegal, greatly diminished. The rivers flowing into Lake Chad were similarly affected, so that the margins of this shallow lake retreated far from the villages located on its shores. Many pastoral groups lost all their animals, and because of the general crop failure were themselves reduced to starvation. As of the summer of 1973, an estimated 10–13 million people of the possible total of 40 million inhabiting the Sahel zone had to exist entirely on famine supplies brought in from the outside world. Such local disasters are always liable to overtake human populations that occupy marginal ecosystems, but because of severe overstocking of the Sahel with both man and beast, the suffering and loss are exacerbated. For the Sahel, climatic vagaries represent the one density-dependent factor regulating population growth that has not yet been removed.

Modern Limitations of Population Growth

This admittedly oversimplified categorization of West African communities has been included to illustrate the problems that now confront developing societies, even when their net primary productivity is sufficient to permit a successional advance to a further societal category.

In Nigeria, the Hausa peoples of the Sudan zones are in a colonizing phase, but already their political influence over the whole country is resented and resisted. A federal prime minister from this society (Abu Bakr) was assassinated several years ago.

Advanced Agricultural societies of moist forest areas are also lacking any population control. The Yoruba of western Nigeria have inserted colonies into virtually every forest settlement in West Africa. The Ibo of eastern Nigeria, who had done the same in Nigeria, have been confronted with an intertribal power struggle that precipitated a genocidal war of distressing dimensions and consequences.

Unless immediate and drastic curbs on population growth in such forest and Sudan zone societies can be introduced, genocidal conflicts of these tragic dimensions are unavoidable. Meanwhile the habitats of these two areas are being changed irreversibly, as is that of the Sahel zone.

Growth and Stress

The inevitability of genocidal strife of one form or another would appear, in the continuing absence of population control measures, to threaten all Advanced Agricultural societies, which still accommodate a considerable majority of the world's 3-billion-plus inhabitants. Honduras and San Salvador make war essentially for the reason that the growth of the agricultural population has exceeded the supply of land. Trouble breaks out between Kikuyu and Luo in Kenya, which is discovering that the end of colonial rule has not solved problems of population pressures. Malay and Chinese-origin inhabitants of Malaysia come into conflict; China and Russia stage a military confrontation over a relatively unproductive section of a disputed territory. Emigrants from the West Indies and Pakistan, fleeing from population pressures in their homeland, provoke color discrimination in Britain, a country with the longest traditions of liberalism of any extant society. A little nearer home, the central sections of all great U.S. cities become ghettos for uncontainable multitudes of citizens of particular ethnic groups, tongues, or religions, mostly immigrating there from burgeoning Advanced Agricultural societies. We return to such matters again several times in this and the succeeding two chapters.

Fundamental Considerations of Population Growth

Population ecologists have long been concerned with the dynamics of population growth, and all standard ecological texts describe this concept and related terms. As applied to human population dynamics, the concept is very simply illustrated in the following way.

Theoretical Population Increase

Assuming that we are dealing with one of the occasional human situations in which consanguineous marriages are permitted (as in the Egyptian royal dynasty to which Cleopatra belonged) and that each such marriage produced two boys and two girls who lived to marry and produce two boys and two girls of their own, we could express the population growth in tabular form as in Table 7-4.

If it can be assumed that there is an interval of 30 years between each generation and that each parent survives to age 60 as a grandparent but does not become a great-grandparent, then in ten generations and in a time span of three centuries, the original pair have generated no fewer than 3,584 descendants.

When plotted in graph form, this apparently modest ambition of raising

TABLE 7-4 *Theoretical Population Growth in Initially Consanguineous Marriages Yielding Two Male and Two Female Offspring*

Number of generations	Number of marriages	Number of children	Total population
1	1	4	6
2	2	8	14
3	4	16	28
4	8	32	56
5	16	64	112
6	32	128	224
7	64	256	448
8	128	512	896
9	256	1,024	1,792
10	512	2,048	3,584

two boys and two girls per family assumes a *geometric* or *logarithmic* rate of growth (see Figure 7-5 on page 266). In current idiom this form of growth is most commonly referred to as *exponential growth*. Applied to demographic data, all three terms, *geometric, logarithmic,* and *exponential,* refer to the same type of growth. In an economic sense, we are most familiar with this kind of growth in its guise as *compound interest*.

The compound interest formula is

$$A = P \left(1 + \frac{r}{n} \right)^{nt}$$

where A = the amount of money after t years
P = the principal, that is, the original sum of money invested at a rate of r per cent annual interest
n = the number of times interest is compounded during the year

If we were not to compound the interest, the rate of increase of our money would be a standard amount each year. This *arithmetic* rate could be expressed as a straight line, as in Figure 7-4A. Compounding the interest means that each interest increment is progressively larger. The rate of increase then has to be expressed by an ever-steepening curve, as in Figure 7-4B, and this is a typical exponential growth curve. If the basic interest rate on our money remains constant, this rate expressed on a logarithmic vertical scale for growth becomes a straight line as in Figure 7-4C.

When births in a population exactly equal deaths, we have in effect no interest. The original sum, that is, the original number of individuals, remains constant. Demographers call such a stable situation a *stationary population*. Should there be even a slight excess of births over deaths, we get into a compound interest situation and develop an exponential rate of population growth. If the excess remains the same, demographers call this a *stable population*. It is important to note that ecologists use the term

FIGURE 7-4. *Arithmetic and geometric increase.* In A the interest on an investment is *not* compounded. That is, the interest earned does not in turn, earn interest. The increase of the principal is by arithmetic progression. Compound interest (and populations) follow the curve shown in B because interest is able to generate interest, just as increments to population in their turn produce additional increments, leading to an *exponential* or *geometric* rate of growth. Should this growth be plotted using a logarithmic scale, the rate again appears as a straight line relationship, as shown in C.

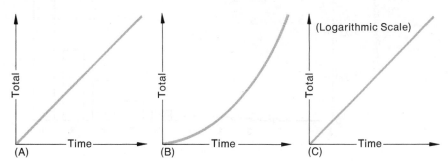

stable for the condition social demographers call *stationary*. How the type of population growth that characterizes a given population may be determined becomes clear when the parameters of birth, death, and natural increase rates are further considered. Planners often take a "guestimate" of population increase at 2 per cent per annum. If they fail to remember that this will be an exponential increase, that is, a compound rate, and plan the expansion of essential facilities at an *arithmetic* rate of 2 per cent, things soon begin to go awry.

Birth Rates

Theoretical Calculations

As defined for demographic purposes, the birth rate is calculated from the total number of births in a year divided by the total population as of the midpoint in that year, July 1, multiplied by 1,000.

For the final generation of the hypothetical example illustrated in Table 7-4 and Figure 7-5, assuming the number of births remains constant both through the year and through the 30-year generation interval, the following birth rate results:

Total number of births in 30th year = 68.3

Total population, midpoint of 30th year = 3,584 − 34 (births) +

$$8 \text{ (deaths)} \longrightarrow 3,558$$

$$\text{Birth rate} = \frac{68.3}{3,558} \times 1,000$$

which gives a birth rate per thousand of 19.8.

FIGURE 7-5. *Graph illustrating the rate of increase in the theoretical population presented in Table 7-4.* Even the medium rate of 4.0 is sufficient, as can be seen, to cause a population explosion. As explained in the text, this rate apparently is still a generally *desired* value; probably for much of the time and for a variety of reasons, it also was approximately the value of the *achieved* rate in our ancestral populations.

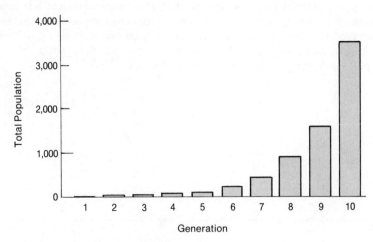

This figure for the birth rate is about half of what was considered even after World War II as the theoretical maximum. Demographers argued that per thousand population 500 would be women. Of these 500, 410 would be too young, too old, unmarried (or at least unmated), or infertile. Of the 90 remaining, 45 would have just completed parturition and would have insufficient time to conceive and complete another pregnancy during the year. This left 45 in various stages of pregnancy or about to conceive, who would provide a statistic of a maximum possible birth rate of 45 per thousand per annum.

Birth Rate Statistics

In actuality a number of countries have been able within the last 20 years to surpass easily this theoretical maximum. Niger holds the dubious distinction of heading this list with a 1973 figure of 52.

As already noted, comparisons of demographic statistics have to be made with caution. One reason underdeveloped countries feature so prominently in the upper portion of the contemporary selected national birth rates listed in Table 7-5 is that a high proportion of their female population is young and still nubile.

TABLE 7-5 *Birth Rates from a Range of Nations About the World*

These are given per thousand of population. Many countries still far exceed what was until comparatively recently believed to be a theoretical maximum of 45. Belgium, with the lowest recorded birth rate, is one of the very few countries with an almost stabilized population (Table 7-7). The relative infrequency of rates in the 30s and 20s is not fortuitous. Nations tend to adopt population control measures on an extensive scale, in which case their birth rates soon fall to 20 or below, or they continue with none, which leaves their rates up in the 40s and 50s. It seems that a given country will in general either totally ignore or totally adopt birth control practices.

Niger	52	Uganda	43
Dahomey	51	India	42
Pakistan	51	Venezuela	41
Algeria	50	Egypt	37
Sudan	49	Costa Rica	34
Upper Volta	49	Sri Lanka	30
Kenya	48	China (Peoples Republic)	30
Guinea	47	Chile	26
Tanzania	47	Singapore	23
Ghana	47	Australia	20.5
Malagasy Republic	46	Hong Kong	20
Ivory Coast	46	U.S.S.R.	17.8
Colombia	45	France	16.9
Zaire	44	Italy	16.8
Mexico	43	United States	15.6
Guatemala	43	United Kingdom	14.9
		Belgium	13.8

Although reliable figures as to birth rates of the past are lacking, it seems reasonable to suppose this has not fluctuated by more than perhaps 50 per cent during the 2 or 3 million years of *erectus-sapiens* history. When life expectancy has been lower, there has been a correspondingly higher proportion of nubile females in the population. This circumstance has tended to maintain the level of the birth rate.

Mortality Rates

Theoretical Calculations

In much the same kind of way as in birth rates, for demographic purposes the *mortality* or *death rate* is calculated from the total number of deaths during the year divided by the total population as of the midpoint of that year, multiplied by 1,000.

The hypothetical example of Table 7-4 gives

Total number of deaths in the 30th year = 17.1

Total populaton, midpoint of the 30th year = $3,584 - 34 + 9 = 3,558$

Mortality rate = $\dfrac{17.1}{3,558} \times 1,000$

which gives a mortality rate per thousand of 4.8.

Mortality Rate Statistics

There are several ways of showing mortality statistics, including age-specific mortality rates, life expectancy, infant mortality, and mortality rate as considered here, which is the total annual death rate per thousand persons (Table 7-6). With regard to this last figure, records extending over two centuries suggest that in the eighteenth century this statistic, as far as can be judged, had probably not varied a magnitude of more than one. It probably had varied by a magnitude of at most two over the whole period of *erectus-sapiens* history.

World War II focused attention on several discoveries that dramatically changed this situation on a world-wide basis. Sulfa drugs and antibiotics for the first time provided a simple but effective means of treating diseases such as dysentery, cholera, syphilis, yaws, pneumonia; DDT and other pesticides became available for widespread control of disease vectors. Their use is thought to have resulted directly or indirectly in the control of 70 diseases, including malaria. Immunization serums developed to supplement that for smallpox were individually effective against many bacterial and virus diseases such as typhoid and yellow fever.

Although these treatments were achieved in the modern industrial societies, they could be copied readily by all societies, and were. Dramatic reductions in mortality rates resulted, the most spectacular being in some Advanced Agricultural societies, where these rates fell from as high as 35 or 40 in the 1930s to as low as 7 or 8 by the 1950s, as shown in Table 7-6, although the qualification in the legend of this table must be noted.

This dramatic and universal reduction in mortality rates throughout the

TABLE 7-6 *Death Rates from a Range of Nations About the World*

These are given per thousand of population. The statistics are in part misleading; no conclusions should be drawn from them without simultaneously taking account of *population structure*. The crude death rates in West African countries are genuinely high, but age-specific mortality in Hong Kong, Singapore, and Costa Rica is not necessarily low. The probability is that these last countries will be found to have populations with a very high proportion of young persons, who suffer few mortality losses. Belgium, with a virtually stationary population, appears, because of this misleading feature of mortality rates, to have a medium-high crude death rate.

Upper Volta	29	Belgium	12
Guinea	25	United Kingdom	11.9
Malagasy Republic	25	Colombia	11
Ivory Coast	23	France	10.6
Zaire	23	Mexico	10
Tanzania	22	Italy	9.6
Kenya	18	United States	9.4
Ghana	18	Chile	9
Uganda	18	Australia	8.5
Pakistan	18	Venezuela	8
India	17	Sri Lanka	8
Guatemala	17	Costa Rica	7
Algeria	17	Singapore	5
Egypt	16	Hong Kong	5

world in the 1950s is the most significant individual causal factor of the present population explosion. The sociofacts, mentifacts, and artifacts relating to public health produced by Advanced Industrial societies were disseminated by cultural exchange at this time in the same manner as any other sociofacts, mentifacts, and artifacts. Cultural exchange in the modern world has, however, become almost instantaneous—as rapid as the electronic waves of radio or television—and the power of modern culture to mass-produce artifacts is immense. It is difficult to think of any other cultural adaptations that had such a profound effect on human populations, and certainly no other had such an immediate impact as these measures for death control. The only comparison is perhaps with the discovery of fire-making, and it seems to have taken some 300,000 years for that cultural advance to have spread to all human populations.

Although it is apparent that *erectus-sapiens* populations everywhere had gradually, over their long history, begun the processes of pollution, over-exploitation, irreversible modification of ecosystems, and annihilation of other species, the utter devastation we are now wreaking on our world will be dated by historians as effectively commencing in the decade from 1940 to 1950. This may be anticipating somewhat; first the effect of lowered mortality rates on population dynamics must be considered.

It must be emphasized that these quite recent reductions in mortality have had no significant effect on individual longevity. We may now have a better chance at birth of attaining the biblical three-score years and ten, but once that goal has been reached, actuarial estimates provide little encouragement as to enhanced possibilities of continuing long from there. Current interest in populations with apparently high numbers of centenarians has focused on remote areas of Peru, Pakistan, and the Georgian Caucasus, where modern medical facilities are not extensively available. In the United States there has been little increase in life expectancy since the decade of the 1920s. An adult of 65 can anticipate a mean of 15 more years, only 2 years longer now than then. A white female born in 1900 had a life expectancy of 52 years. The fact that this has now risen to 75 years has had little effect on population increase statistics, for death control for females has extended merely the postreproductive years. However, in underdeveloped areas, where death control during this same period has modified mortality within the fecund years, it has had profound economic significance. In the 1920s, for example, more than half of the females dying in India were under 24 years of age. Death control in this instance, as will be discussed later, has a very pronounced effect on demographic statistics.

Rate of Natural Increase

Theoretical Calculations

Subtracting for our hypothetical example the mortality rate of 4.8 from the birth rate of 19.8, 15.0 is the *rate of natural increase* for the hypothetical population in the final year considered.

The percentage rate of increase is therefore 1.5 per cent, the statistic most commonly quoted in demographic figures. When the birth rate exceeds the death rate, the natural increase is positive, the population is *expanding*. When the reverse holds, the increase is negative, and the population is *declining*. When birth rate equals the death rate, the natural increase is zero, the population growth rate is zero, and the population is *stationary*.

To provide some comparative figure analogous with the half-life of radioactive isotopes, demographers commonly convert the percentage rate of population increase to *population doubling time*. This calculation involves only the standard method of calculating the rate of compound interest from a bank rate. An approximation of the doubling time can be determined by rule of thumb simply by dividing a constant, 70, by the percentage crude rate of natural increase. In integers of increase, doubling times are as follows:

Per cent natural increase	Doubling time in years
1.0	70
2.0	35
3.0	23
4.0	17

For the hypothetical example in Table 7-4, with a rate of natural increase of 1.5 per cent, the doubling time is 47 years. Some current doubling time estimates are listed in Table 7-7.

Historical National Increase

With these theoretical calculations and qualifications in mind, it is possible to return to the 2 or 3 million years of *erectus-sapiens* history and understand how population growth remained so low for so long, and why it has suddenly moved into an explosive phase.

A nubile human female is capable of becoming pregnant and giving birth during the ages of approximately 13 to 45 years. As gestation takes 9 months and conception can occur within 2 or 3 months of parturition, there is a theoretical possibility of an individual female producing some 42 offspring.

In actuality the maximum recorded number of single live births for a female is 24. There are many factors accounting not only for the reduction of the theoretical expectation to this value but also for the usual occurrence of an even lower figure. For example, nursing an infant for a period of up to 3 years seems to be associated sometimes with a failure to resume ovulation following parturition. It appears that the drain on the nursing mother's metabolism prevents the attainment of the minimum threshold of tissue

TABLE 7-7 *Predicted Population Doubling Times*

Given is the year by which each country listed will have doubled its population, on the basis of present rates, starting from 1973. Costa Rica no longer has the unenviable distinction of leading the field for a number of years; several nations presently have an estimated doubling time of as little as 21 years. At the other end of the scale in 1973, Luxembourg is stationary, and East Germany has a slowly declining population. Only countries described in this text as Advanced Industrial societies have attained or are approaching stationary populations. Likewise only nations definable here as Advanced Agricultural societies have very short doubling times.

Philippines	1994	Zaire	2006
Venezuela	1994	China (Peoples Republic)	2014
Syria	1994	Russia	2043
Mexico	1994	United States	2060
Pakistan	1994	Italy	2072
Colombia	1994	France	2090
Honduras	1995	Sweden	2204
Libya	1996	United Kingdom	2204
Kenya	1996	Ireland	2212
Ghana	1997	Belgium	2320
Swaziland	1998	Austria	2673
Brazil	1998	Luxembourg	stationary
Costa Rica	2003	Germany	
India	2004	(Democratic Republic)	declining
Taiwan	2005		

fat content necessary before ovulation can proceed. This factor alone re-duces the reproductive potential from some 42 offspring to eight or nine. An average life expectancy of 30 years would halve this value, providing for a maximum number of issue per female of a mere four or five. Child mortality losses of as high as 50 per cent are not unusual, even in the con-temporary world, occurring especially immediately after weaning. This could bring the maximum number of children actually reaching adulthood per female down to two or three, and the birth rate would then barely suffice to maintain existing numbers.

Rate of Natural Increase Statistics

All these factors have operated in the past to reduce population growth. The general practice of prolonged nursing in many tropical African societies perhaps represents the persistence of a very ancient practice. Supposedly it accounts partly for the fact that number of live births per female never seems greatly to exceed nine in any contemporary society (Table 7-8). Similarly, estimates of life span at various periods of *erectus-sapiens* history suggest that until the twentieth century women generally died some years before they had completed the menopause and become barren; even in some contemporary societies women die before this stage.

Figures for infant mortality (Figure 7-6) show that despite the knowledge

TABLE 7-8 *Variations in the Number of Live Births per Female in Different Populations*

These fertility figures vary considerably, depending primarily on whether the population has yet undergone the "demographic transition," the cultural adjustment of birth rate to a lower mortality rate. The countries that are usually categorized by some such term as "underdeveloped" are inevitably the ones that have escaped from the cultural controls reducing fertility in the earliest forms of societies and that have failed so far to complete this further cultural transition.

Cultural category	Live births per female (to nearest integer)
Hunters (category 1)	
Eskimos	3
North American Indians	3–4
Hunter-gatherers (category 1)	
Australian Aborigines	5
Early Agriculturalists (category 2)	
Sumatra	4
Advanced Agriculturalists (category 3)	
Central Africa	circa 6
India	6–8
Advanced Industrial Societies (category 6)	
United States (mid-twentieth century)	3
United States (beginning of nineteenth century)	7

of modern hygiene and the techniques and materials available for use, some contemporary societies can still lose approximately one quarter of the children born alive before they have reached the age of 12 months.

Two of the three factors affecting the rate of natural increase, reduction in infant mortality and a longer life expectancy, are largely determined by the level of hygiene, which in turn represents the extent of health care available. This is most readily dispensed in an urban environment, which also encourages earlier weaning, because it offers both the baby food substitutes and the employment that will induce the mother to leave her child during the day.

It is not surprising, therefore, that the massive reductions in infant mortality, the lengthening of life expectancy to cover the nubile years, and the shortening of the nursing period all coincide with the industrial revolution and the greatly accelerated concentration of populations in urban centers. Nor is it surprising that both the colonizing phase of social succession and the beginnings of our present population crisis have their origin in this quantum cultural advance in public health. The graphs reproduced in Figure 7-7 illustrate this coincidence between the demographic incidence of population expansion and the cultural process of industrialization.

There is a twofold reason why western nations could undergo this industrialization and population explosion process a century ago without pro-

FIGURE 7-6. *Infant mortality* is expressed as the number of deaths of infants under 12 months of age per 1,000 live births. These figures from a selection of countries illustrate that although the various disease-control measures introduced following World War II reduced the rate in all countries, the effect was proportionately greatest in those countries with an already low mortality rate.

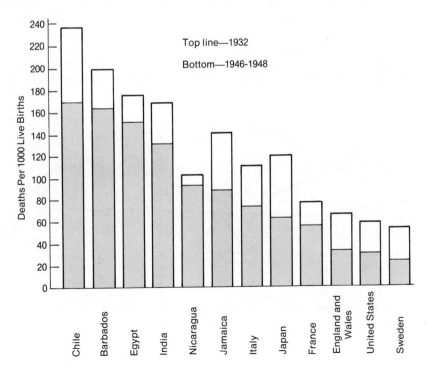

ducing effects comparable with what happened in a parallel process between 1940 and 1960. First, the frontiers of the world were still open, and the excess populations could be accommodated in sparsely settled lands. Second, the actual proportion of the total human population of the world that these western nations represented was low. Although their relative explosion in numbers was comparable with that which occurred a century later, in absolute terms their utilization of additional resources was not so extensive that the increased demands could not be met simply by overexploitation of existing supplies. We were dealing then with populations increasing by the millions, not, as now, by the billions.

One example from the many statistics that could be produced will serve to illustrate the dramatic effects of the post-World War II reduction in mortality on the rate of natural increase. In Table 7-9 are shown the main

FIGURE 7-7. *Industrialization of societies and natural increase. A.* A comparison of birth rates and death rates between an Advanced Industrial society (Sweden—solid line) and an Advanced Agricultural society (Ceylon—broken line). Hygienic innovations improving public health following industrialization have reduced the mortality rate in both societies, allowing for the small time lag necessary for cultural exchange (outstanding in Ceylon is the resulting control of malaria, associated with a fall in the death rate from 22 in 1945 to 9 in 1963). However, the desire for fewer children or the cultural knowledge of how to achieve this, which has also depresesd the birth rate in the industrialized society, has scarcely affected that of the agricultural society. In *B*, which compares an Advanced Industrial society (Scandinavia—solid line) with one in which succession to this stage occurred approximately half a century later (Japan—broken line), the birth rate and the death rate are in this instance simultaneously declining. In the year 1955, at the peak of the population decline that proceeded in Japan from 1947 until the early 1960s, some 1.2 million legal abortions were performed, halving the population growth. It appears that the stimulus of succession from an Advanced Agricultural to an Advanced Industrial stage is necessary to initiate a decline in the birth rate. (From Kingsley Davis, "Population," *Scientific American,* **209**[3]: 68, 70, 1963. Copyright © 1963 by Scientific American, Inc. All rights reserved.)

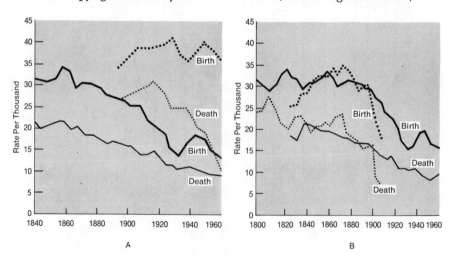

TABLE 7-9 *Population Growth in India*

During this century, the birth rate has remained constant, fluctuating about a mean, while the death rate has fallen almost two thirds. The population growth rate has increased some 2,500 per cent since the beginning of the century.

Census year	Total population in millions	Birth rate	Death rate	Rate of natural increase percentage
1901	236	46	44	0.1
1911	249	51	43	0.8
1921	248	49	48	0.1
1931	276	46	36	1.0
1941	313	45	31	.1.4
1951	357	39	27	1.3
1961	439	40	21	1.9
1965	490	41	17	2.4
1969	537	43	18	2.5

population statistics for India at intervals during this century. Remarkably, under the circumstances, the birth rate continues to rise. So therefore does the rate of natural increase, although the recent upward trend of the mortality rate, if real, will reduce this in future years.

The actual family size India has to attain if it is to achieve zero population growth at present death rates may be estimated from the theoretical figures in Table 7-10 at about 2.5 children per female. This statistic, known

TABLE 7-10 *Zero Population Growth in Terms of the Number of Offspring*

Assuming in a sample population of 1,000 that 200 are nubile females—and making several other smaller assumptions to simplify the calculation—the following levels of fecundity would be necessary to maintain zero population growth at the given mortality rates.

When death rate reaches this figure	Birth rate to maintain zero population growth must be	Fertility (number of offspring per female) to provide this birth rate is
45	45	6.7
35	35	5.2
25	25	3.7
15	15	2.2
10	10	1.5
5	5	0.7

A fertility ratio of a little over two is therefore compatible with zero population growth only when the death rate is at least 15 per thousand. As can be seen from Table 7-6 all Advanced Industrial societies are already below this figure; the others shortly will be. Very soon two definitely will *not* do for India and many other nations.

as the *total fertility rate,* is somewhat confusing on first sight. It might be considered that a fertility rate of precisely two would in an ideal world maintain a zero population growth; however, it will do so only in a population with a mature structure. Because of the dramatic reductions in death rates over the past 30 years, the population in many underdeveloped nations contains a disproportionate number of young people. A birth rate based on a figure-per-thousand population is not then offset as much as it should be by a given death rate, because, being younger, many do not die when they should according to the mortality statistics. Thus India, with a death rate down to 17 or 18, should not permit a fertility rate higher than 2.5. Hong Kong and Singapore, with death rates down to five (Table 7-6), will have to attain a fertility rate of 0.7 to stabilize their populations at zero growth. This is tantamount to having three girls in ten remaining single while the rest get married and have one child each, and it is understandable that no society presently could contemplate discouraging childbearing to this extent.

The rates of natural increase for a range of other countries, together with the birth and mortality rates, are shown in Table 7-11.

Population Structure

These crude rates as explained here need further sophistication before they can be used accurately to describe a given human population. One of the obvious deficiencies is that they take no account of population structure (Figure 7-8).

In the theoretical example discussed previously it is known that half the children produced are male, half female, and that all survive to an age of somewhere between 60 and 90 years. It is then possible to construct diagrams illustrating the number of males and females in three categories as follows:

1. Children (0–14 years)—male and female
2. Adults (15–44 years)—male and female (*nubile* females)
3. Old people (45–90 years)—male and female

A somewhat similar categorization of population structure is employed to estimate a primarily economic rather than demographic statistic, the *dependency load.* In this case the youngest age group (less than 15 years) is assumed not to be employed, adults 15–64 constitute the work force, and the elderly (65 years and over) are the retired or senior-citizen category. The dependency load is expressed as the ratio of the numbers in the work-force category to those in the other two groups. Although the size of the other two groups varies considerably between one population and another, the proportion of people in the work-force group, perhaps surprisingly, remains fairly constant.

The diagram in Figure 7-8 shows what a balanced population structure can look like. If, on the other hand, one third of the old folk were to live to between 90 and 120, and as a consequence there was a reduction of one

TABLE 7-11 *Rate of Natural Increase from a Range of Nations About the World Together with Their Birth and Death Rates*

Countries such as the Ivory Coast with very high birth rates have a medium natural increase because their death rate still is high. When this is drastically reduced, even a somewhat lowered birth rate still permits the same high rate of natural increase, as seen by contrasting the statistics for Ivory Coast and Hong Kong. Belgium has achieved a rate of natural increase that gives it virtually a stationary population; one very small country, Luxembourg, has actually reached zero population growth. Some statistics, like those for Kuwait (43:7:9.8), are considered so exceptional because of a high migration factor that it would be misleading to include them in this table.

	Birth rate	Death rate	Natural increase		Birth rate	Death rate	Natural increase
Colombia	45	11	3.4	Taiwan	27	5	2.2
Venezuela	41	8	3.4	Egypt	37	16	2.1
Rhodesia	48	14	3.4	Malagasy			
Morocco	50	16	3.4	Republic	46	25	2.1
Algeria	50	17	3.3	Upper Volta	49	29	2.0
Pakistan	51	18	3.3	Australia	20.5	8.5	1.9
Mexico	43	10	3.3	China			
Kenya	48	18	3.0	(Peoples			
Ghana	47	22	2.9	Republic)	30	13	1.7
Costa Rica	34	7	2.7	Chile	26	9	1.7
Guatemala	43	17	2.6	Japan	19	7	1.2
Uganda	43	18	2.6	United States	15.6	9.4	0.8
India	42	17	2.5	Italy	16.8	9.6	0.7
Hong Kong	20	5	2.4	France	16.9	10.6	0.6
Ivory Coast	46	23	2.4	United Kingdom	14.9	11.9	0.3
Guinea	47	25	2.0	Sweden	13.8	10.4	0.3
Singapore	23	5	2.2	Belgium	12.8	12.0	0.2
Sri Lanka	30	8	2.2	Luxembourg	11.8	11.9	0.0

third in the birth rate for a generation, the pyramid would have the form of a declining population (Figure 7-9B). Should all the adults die at 45, and a consequent increase occur of one third in the birth rate, the pyramid

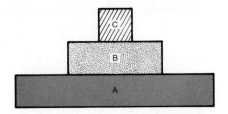

FIGURE 7-8. *Population structure* as developed from the theoretical example illustrated in Table 7-4 and Figure 7-5. The population is divided into three categories only: (*A*) infants and juveniles, (*B*) reproductive adults, (*C*) postreproductive adults. The numbers of the first class far exceed those of the second, so that unless a very high mortality rate occurs there, the population has to be unbalanced and experiencing a high rate of natural increase.

would have the form of a young and unstable population (Figure 7-9A).

It is customary to express such diagrams of population structure in intervals of 5 years, with males on the left and females on the right of the diagram. In Figure 7-9 two typical diagrams are shown, illustrating expanding and declining populations. The usefulness of this method of graphic display is illustrated in Figure 7-10, which analyzes further the demographic statistics for India contained in Table 7-9 and compares them with the statistics for England.

Computerized Printouts of Population Pyramids

The population pyramids that have been illustrated in Figure 7-10 were prepared by somewhat tedious hand methods. Hence many of these illustrations have been used over and over again in college texts. Using readily available demographic statistics as input data, computer programs are now available to calculate the group or *cohort* values. These values can then be transferred to a graphic program that prints out the population pyramid. Programs for the first operation have been prepared by

FIGURE 7-9. *The structure of actual populations:* that of Algeria in 1954 in *A*, Sweden in 1955 in *B*. At this time the first three age classes in the Algerian population considerably exceeded the size of the reproductive classes—this population was increasing rapidly. The Swedish population, by contrast, was declining. The first age class and also the 15–19, 20–24, and 25–29 classes were smaller than their predecessors, so that even without any mortality at all, the reproductive section of the population was being reduced in numbers.

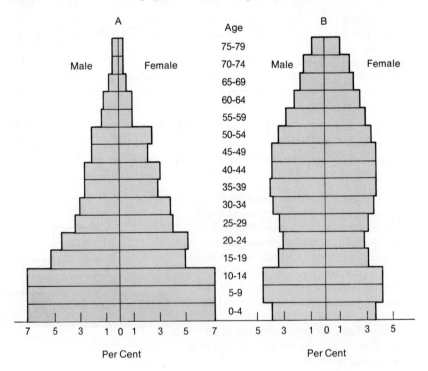

Nathan Keyfitz. The graphics program was written by James Pick. All the pyramids illustrated in subsequent pages here have been prepared in this way. For convenience they show also the calculated *stable* and *stationary* as well as the actual (observed) values for the same population. Although the *observed* population is a real statistic and the *stationary* figures calculated from observed values are real, the *stable* data represent an *estimate* of a hypothetical situation.

Concealed Variations in Demographic Statistics

With the calculated values of crude birth rates, crude death rates, rates of natural increase, and doubling times, it might be imagined that sufficient statistics had been assembled for a valid comparison to be made between

FIGURE 7-10. *Comparison of the population structure of an expanding population* (India, 1931) *with a stabilizing one* (England, 1945). Each 5-year age class in the Indian population, with one exception, so far exceeds its immediate predecessor that unless very considerable mortality occurs at each level, the whole population must have substantially larger numbers in 5 years' time. The English population, by contrast, shows the effects of the declining birth rate first associated with the financial depression of the early 1930s, which continued until World War II even after economic recovery had been achieved. The numbers in the younger age classes do not provide for the maintenance of the age-class numbers in the middle age classes, even when no mortality occurs in these intermediate classes.

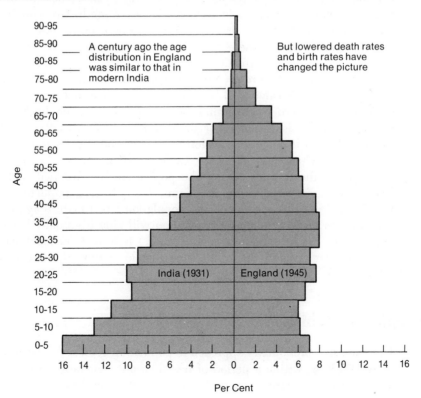

different populations, always assuming that immigration and emigration effects can be discounted. This is not the case.

For example, in the hypothetical case in Table 7-4, suppose that during the tenth generation all the mortality occurred not among the grandparents, as previously hypothesized, but entirely among infants. Without change in the number of children per nubile female, or in the number of deaths in the population, the birth rate then becomes 18.2 per thousand instead of 19.8, the death rate 5.1 per thousand instead of 4.8, and the rate of natural increase 13.1 instead of 15.0. Clearly, calculations of birth, death, and natural increase rates have validity for purposes of comparison only if they are made on populations with the same *structure* of age groups and in which mortality and natality are occurring *at the same age group levels.*

In practice, this is very far from being the case. Populations vary considerably in age structure, sex distribution, marriage age, life expectancy, and age-specific mortalities and natalities. Any comparisons and predictions based on demographic statistics must make appropriate allowances for such variation.

The most common refinement of demographic statistics is obtained by calculating the number of live births per annum per thousand females in the population aged 15–44, that is in the nubile category. Sometimes this fecund period is extended downward to age 10, or upward to age 49, or both. This figure is known as the *fertility rate;* it can vary from about 1.5 to 8.0. For some purposes—for example, as will be seen later for population projection—it is more accurate to use the *age-specific fertility rates.* These are the mean rates per thousand women in each of the six reproductive cohorts represented in the 15–44 group classes. Unfortunately, even in industrial nations like the United States, these data are not always routinely recorded. For underdeveloped nations they can only be obtained by specific surveys.

The *net reproduction rate* (NRR) is a measure of the fertility rate sometimes utilized. It is calculated by taking 1,000 female babies and following their fertility record to determine how many live female births occur to them. The NRR is this number divided by 1,000. Actually the NRR is calculated from the age-specific fertility rates modified by the age-specific death rates, which give the Life Table Survival Rates. When the NRR value is 1.0, the fertility rate of the population has attained *replacement level;* that is, in more colloquial terms, there is zero population growth (ZPG). At least this is so when the population has a mature age structure. When it has a young one, the population will only eventually become stationary. The gross reproduction rate (GRR) is sometimes mentioned. This is the number of live female births per female merely adjusted for the sex ratio at birth.

Population Stabilization

This extensive preliminary approach to population growth has been included to explain how, within a period of less than one generation, we have been presented with a population problem that we do not know how to contain immediately. Indeed, we did not even recognize it as a problem until it was persistently spelled out for us. For all the past millennia in human history each female had to give birth on average to five or six children; this has suddenly become far too many, and the rest of this chapter will explain why.

It is not simply a question of some nations, some segments of some societies, or some individuals maintaining too high a fertility rate. Fecundity is universally too high for it to be possible for nations, societies, or individuals to continue at their present fertility rates without trespassing onto some of the resources that rightly belong either to others or to everybody. Advanced Agricultural societies have to reimpose controls if they are to preserve the net primary production necessary for their progression to industrialized societies and are to avoid still further reduction of productivity potential by habitat degradation. Societies entering colonizing and industrializing phases presently have no unoccupied territory available to accommodate their surplus populations; they must impose severe restrictions on population growth if their net productivity is not to be drastically reduced on a per capita basis by overdispersion among too high a population. Advanced Industrial societies are either multiethnic or economically multistructured, or both. Differential growth rates that presently exist between one population component and another must all be equalized at zero population growth if city habitats as well as national resources are not to be irreversibly destroyed by a final overexploitation.

Global and national population growth figures do little more than give the dimensions of our world population crisis. There is a danger indeed that they can have a tranquilizing effect. It is easy to extrapolate from miracle rice and miracle wheat to how easy it will be to avoid the predicted famines for overcrowded populations of many underdeveloped areas. This tends to ignore the circumstance that such areas have sometimes been characterized by low productivity for so long that the people have been selected for, among other features, small stature. For 2 or 3 million years *erectus-sapiens* populations have been penetrating into all areas of the earth, and with a few special exceptions like Madagascar and Oceania, by about 20,000 years ago they had entered all ecosystems capable of supporting human life, unless we are now to start living in ice caves, subterranean caverns, or skyscraper configurations.

A few Advanced Industrial societies have managed to achieve a population growth rate of zero; that is, they have simultaneously stabilized their population number and structure. To further illustrate population growth, we will examine the contemporary world situation.

World Population Growth

Various references have been made in the first part of this chapter to global estimates of population size in human populations. The evidence on which these are based, which is always scanty, makes these estimates subject to considerable margins of error. Even with modern census techniques, it is impossible to obtain demographic information that is completely without error.

Forecasts of Population Size

Despite the reservations that must be made on account of such census deficiencies, it is possible to prepare tables of past, present, and future world populations. Nathan Keyfitz has estimated, by backward projection of current demographic statistics, the total number of individuals of our species who have ever lived. He arrives at a figure of 70 billion, of whom about one half survived to adulthood and of whom one twentieth are presently alive. One of the first estimates of the distribution over time of this total human population was published by H. F. Dorn, and many of the tables and figures illustrating this statistic are based on his classical work. A recent paper by J. Frejka has updated the total population estimates and projected them on into the next millennium. A compilation of these several estimates of total population distribution is shown in Table 7-12. It is possible, as Frejka maintains, that the total world population will level off at about 8 billion persons; it may go higher but almost certainly at a much reduced rate of growth. It is inconceivable that the rate of population growth that has been attained in the last three centuries can persist for many years of the next millennium. However, when the world population once more becomes stationary, it will have substantially different demographic and ethnic structures from those which characterize it today.

From Table 7-12 and Figure 7-11 it may be observed that the world will double its population between now and 2000 AD. This staggering addition of between *3 and 4 billion* people is beyond normal human comprehension and must be broken down into more local increases to become credible. For example, the total increase in the population of Latin America during the last 50 years of this century will equal the total increase of population in the

TABLE 7-12 *World Population Increase in the Christian Era*

Until the beginning of the industrial revolution in Europe, the doubling times for the total world population are estimated to have remained very modest; such increases would in any case be contained. The 1975 figure of 37 years is close to producing with each new generation an *additional* human population estimated to be greater in size than the total of all the people who have ever lived and died in the whole of our 4-million-year or so history as a genus.

Year (AD)	World population (billions)	Doubling time (years)
1	0.25 (?)	1650 (?)
1650	0.50	200
1850	1.1	80
1930	2.0	45
1975	4.0	37
2013	8.0	?

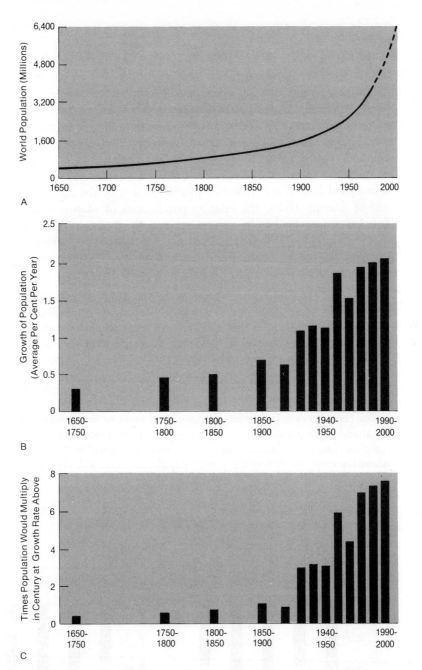

FIGURE 7-11. World population growth between 1650 and 1950. The first graph, *A*, represents the total persons alive at any given time. *B* shows the rate of population increase at various times, whereas *C* indicates the number of times the world population would double in a century, at the same time intervals. It is possible that the 1970 census figures when fully analyzed will show a slight reduction in the rate of increase and an extension of the doubling time, but unfortunately will inevitably have to record still further immense increases in the total world population.

world prior to European settlement in the western hemisphere. The increase in Asia for the same 50 years will approximately equal the total population of the world as recently as 1958. Following Keyfitz, we might estimate that during the last 50 years of this century, there will be an addition to our global population roughly equal to the number of people who have ever lived. Much of this population increase will be Asian. Along the same lines some idea of what Frejka means by imminent change in the nature of the world population can be obtained from Figure 7-12, which has been developed from a diagram prepared by P. Demeny. In this diagram major temperate areas of the world are matched against adjoining tropical and subtropical land masses. In two instances, that of the American continents and of Europe-Africa, the existing populations of each pair are of roughly equal size; i.e., the development continent, North America, has a population roughly equal to that of its undeveloped pair, South America. In the other three pairs, partly because the tropical regions in each instance greatly

FIGURE 7-12. *The geographical distribution of national populations in relation to development.* The industrialized nations are found essentially within the cooler and higher latitudes of the North and South Temperate Zones, underdeveloped ones at lower latitudes—that is, within the Equatorial Zone and the subtropical portions of the Temperate Zones. In this figure depicting 1974 population size (A) and rate of increase (B) on a regional basis, underdeveloped areas at these lower latitudes are below the zero line; they are matched geographically against developed areas in higher latitudes, above the line.

The current much higher rate of population increase in the underdeveloped and equatorial regions is still further increasing the imbalance of population between these matched temperate-developed lands and the tropical-underdeveloped regions, where the major portion of the world's population increase is now occurring.

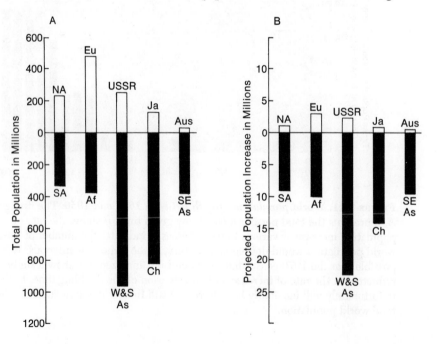

exceed the temperate pair member in area, the populations of the under-developed regions already greatly exceed their corresponding temperate areas. As natural increase is presently occurring at a fairly uniform rate of 2.3 to 2.4 in the underdeveloped areas everywhere, while that of the developed areas is usually below 1.2, the world's population by the next millennium will be concentrated in tropical and subtropical regions, which at present are typically also the underdeveloped ones.

Many of the population growth forecasts made in the past, and not a few even now, have been carried out by merely extending into the future the gross rate of population increase. As discussed earlier, a stable growing population will continue to increase at a constant rate. However, such projections do not allow for variations in the rate. Demographers sometimes include these by labeling the first rate the *median* expectation and also providing a spread from a *low* estimate and a *high* estimate to permit respectively a downward and upward change in rate.

More accurate short-term forecasts are provided by summing the known and anticipated behavior of individual cohorts. In a stable population this will provide a reasonably accurate forecast. However, fertility patterns have been so changeable, especially in this country, that demographers are loath to project much more than 15–20 years ahead.

Population Growth in the United States

It is appropriate at this point to look critically at the recent demography of the population of the United States. There are several reasons why this provides extremely valuable information. First, many individuals—particularly academics such as Garrett Hardin, Paul Ehrlich, and Georg Borgstrom—not only are publishing semipopular works dealing with the population crisis but also are vigorously engaged in lecture and seminar programs specifically focused on the necessity of achieving world population equilibrium. Moreover, virtually every science conference held in this country now passes resolutions urging population control in the United States as well as elsewhere. Second, and the papers about to be discussed provide evidence on this point, birth control devices are both more readily and more universally available in the United States than in any other population of comparable size. Finally, this country's population presently forms an incompletely integrated mosaic of ethnic, religious, and economic groups, illustrating the immense complexity of demographic studies and the difficulty of summarizing observable trends in population growth and behavior on a national basis.

Fertility in the United States

Every 8½ seconds a child is born in the United States; every minute a new immigrant is admitted. Every 17 seconds there is a death, and every 23 seconds an emigrant leaves. This is a somewhat flamboyant expression of the rate of population growth, which presently is very close to 1 per cent per year and provides for a doubling of the population every 70 years. These statistics, now requiring continuous updating, are shown graphically in Figure 7-13.

Recent levels of the fertility rate in the United States are illustrated in Figure 7-14. The lowest point previously recorded for the general fertility rate was in 1936, with figures of 73.3 for white and 95.9 for nonwhite women. After rising again to a peak of 117.7 and 163.0, respectively, in 1957, the rate fell to an estimated 77 and 115, respectively, in 1969. Combining both white and nonwhite, this would be a rate of 123 in 1957 and 82 in 1968, very close to the figure for 1936 in the depression years.

There was a slight rise in the total fertility rate in 1970, when it reached 87.6, only to fall again to 82.3 in 1971. By 1973 this falling trend continued and for the first time in American history the total fertility rate fell below 80. There is little possibility this decline will continue through the second half of the decade, and it may not be resumed until the early 1980s. The population structure of the U.S. is unbalanced because of the "baby boom" of the 1950s; now and for the next few years America will contain an undue proportion of women in their fecund years. The two cohorts 20–24 and 25–29, where most births occur in western societies, will increase by about one third during this time.

There is as yet no full explanation for these fluctuations, which contradicted all demographic forecasts. Advanced Industrial societies now appear to have developed an economic feedback mechanism for population control,

FIGURE 7-13. *Increase in population size in the United States* since the first census in 1790. Immigration has at times contributed significantly to the natural increase illustrated here from the 10-year census figures, more especially during the opening years of this century. If the United States were immediately to achieve zero population growth, the present maximum immigration rate would account for approximately 15 per cent of the annual additions to the population that otherwise would come from births.

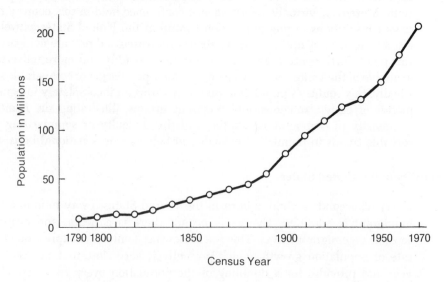

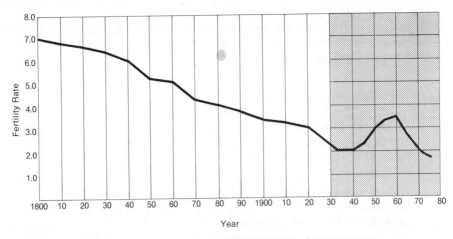

FIGURE 7-14. *The fertility rate in the United States.* The average number of live births occurring per female between the ages of 15 and 45 provides a more accurate statement of the level of reproduction occurring in the population than any of the statistics illustrated in Figure 7-12. The fertility rate in the United States is now at an unprecedented low value and may be expected to continue to fall in response to the various socioeconomic pressures now in force, which are discussed further in the text. Because of the young population age structure, ZPG in the U.S. will nevertheless not be attained for several decades.

which has been called the "Madison Avenue effect." This and other aspects of fluctuations in population and fertility are discussed in the next chapter.

The position of the United States in regard to population growth was well expressed in the President's message on population control to Congress on July 18, 1969, the main points of which are reiterated here as follows:

Rate of Population Growth in the United States

During the 1960s the rate of natural increase in the United States ran at about 1.0, a little high for an industrial nation, as the statistics in Table 7-11 indicate. In this current decade there has been some further decrease. In 1970 the crude birth rate was 18.2 and the crude death rate 9.4. Even at this lower rate of natural increase of 0.88, the total rate is over 1.0, because 300,000 immigrants annually have to be added, providing an additional 0.15 per cent. Short of immense unforeseen disasters, the U.S. is locked into a demographic pattern which has to result in a population totaling 300 million by the end of this millennium.

Although the 1970 rate of U.S. population growth of approximately 1 per cent is the result of a decline that began in the eighteenth century, and this rate of increase is certainly well below the world average, it still is a significant figure. Factors that tend to prevent further decline to zero population growth include the large proportion of women of childbearing age, the desire for a particular family size, and increased longevity.

It took 300 years for the population of the United States to exceed 100 million, in 1917, but the next 100 million was added in 50 years. Even at the present declining growth rate, the third 100 million will have been added by 2000 A.D. (Figure 1-13).

Location of Increased Population

Of the third 100 million Americans who will be located in the U.S. by the end of this millennium it is estimated that 75 million will live in urban areas. Reviewing this situation, the National Commission on Urban Growth has recommended the creation of 100 new towns of about 100,000 each and 10 new cities of at least 1 million each. However, this still leaves another 55 million to be located, about the present total population of the British Isles. It is also, as it happens, the figure by which the total population of India is presently increasing *each year*.

Commission on Population Growth and the American Future

To meet this kind of problem, a sizable enough one even in the U.S., Congress in 1970 at the request of the President established a Commission on Population Growth and the American Future that was charged with three main responsibilities. First, it was to project population growth, internal migration, and other demographic processes up to 2000 A.D. It was assisted in this work by the organization of census data by country in the decennial census, beginning with that of August 1970. Computer summaries of this census were first available in April 1971. The commission also was to review the resources in the public sector of the economy required to deal with the estimated population increase. These included needs related to education, domestic water, recreational areas, and highways. Third, it was to survey the effect of this population growth on federal, state, and local government.

The final report of the Commission on Population Growth and the American Future was presented in 1972. It urged that the U.S. "welcome and plan for a stabilized population . . . ." With this object in view, it recommended an array of educational, health, social, and economic programs considered essential for the early achievement of this stated objective.

Freedom of Choice

To quote from the President's 1969 statement, requesting the establishment of a population commission, none of the proposed measures would "be allowed to infringe upon the religious convictions or personal wishes and freedom of any individual . . . ." It is with this last issue that many scientists would dissent, and problems are raised that will be discussed in Chapter 9, dealing with population control. It will be very apparent that family planning, no more and no less than any other expression of socioeconomic planning, ultimately resolves into the kind of value-judgment issue about which a unanimous consensus of opinion is rare.

BIBLIOGRAPHY

References

Anonymous. "Future is famine for ten million south of Sahara," *Smithsonian,* 4(6): 72–78, 1973.

Birdsell, J. B. "Some environmental and cultural factors influencing the structuring of Australian aboriginal populations," *American Naturalist,* 87: 171–207, 1953.

Birdsell, J. B. "On population structure in generalized hunting and collecting populations," *Evolution,* 12: 189–205, 1958.

Borgstrom, G. *Too Many,* New York: Macmillan, 1969.

Braidwood, R. J., and Reed, C. A. "The achievement and early consequences of food production," *Cold Spring Harbor Symp. Quant. Biol.,* 22: 19–29, 1957.

Chevalier, A. "Le territoire geobotanique de l'Afrique tropicale Nordoccidentale et ses divisions," *Bull. Soc. Bot de France,* 80: 4–26, 1933.

Davis, K. "Population," *Scientific American,* 209(3): 62–71, 1963.

Demeny, P. "The populations of the underdeveloped countries," *Scientific American,* 231(3): 148–59, 1974.

Dorn, H. F. "World population growth: an international dilemma," *Science,* 135: 283–90, 1962.

Durand, J. D. "The modern expansion of world population," *Amer. Philosoph. Soc. Proc.,* 3(3): 136–45. 1967.

Ehrlich, P. R. *The Population Bomb,* San Francisco: Ballantine, 1968.

Frejka, T. "The prospects for a stationary world population," *Scientific American,* 228(6): 22–29, 1973.

Hardin, G. "The tragedy of the commons," *Science,* 162: 1243–46, 1968.

Harrison-Church, R. J. *West Africa,* 4th ed., New York: Wiley, 1963.

Hauser, P. M. "The census of 1970," *Scientific American,* 225(1): 17–25, 1971.

Haynes, C. V. "Elephant hunting in North America," *Scientific American,* 214(6): 104–12, 1966.

Idyll, C. P. "The anchovy crisis," *Scientific American,* 228(6): 22–29, 1973.

Keyfitz, N. "How many people have ever lived on earth?" *Demography,* 3: 581–82, 1966.

Keyfitz, N. "On the momentum of population growth," *Demography,* 8(1): 71–80, 1971.

Lee, R. B., and DeVore, I. (eds.) *Man the Hunter,* Chicago: Aldine, 1966.

Orians, G. H. *The Study of Life,* Boston: Allyn and Bacon, 1969.

Rappaport, R. A. *Pigs for the Ancestors,* New Haven: Yale University Press, 1967.

Ryder, N. B. "The family in developed countries," *Scientific American,* 231(3): 108–20, 1974.

Sahlins, M. D. "The origin of society," *Scientific American,* 203(3): 76–87, 1960.

Taeuber, I. B. *The Population of Japan,* Princeton, N.J.: Princeton University Press, 1968.

U.S. Commission on Population Growth and the American Future. *Population and the American Future,* Report of the National Commission, New York: Signet Books, 1972.

Westoff, C. F. "The populations of the developed countries," *Scientific American,* **231**(3): 108–20, 1974.

Further Readings

Bernard, B., Cox, P. R., and Peel, J. (eds.) *Resources and Population,* New York: Academic Press, 1973.

Davis, K. "The amazing decline of mortality in underdeveloped areas," *Amer. Economic Rev.,* **46**: 305–18, 1956.

Desmond, A. "How many people have ever lived on earth?" *Population Bull.,* **18**: 1–18, 1962.

Dubos, R. *Man Adapting,* New Haven: Yale University Press, 1965.

Ehrlich, P. R., and Ehrlich, A. H. *Population Resources Environment,* 2nd. ed., San Francisco: Freeman, 1972.

Garlick, J. P., and Keats, A. W. J. (eds.) *Human Ecology in the Tropics,* Oxford, England: Pergamon, 1970.

Keyfitz, N. *Introduction to the Mathematics of Population,* Reading, Mass.; Addison-Wesley, 1968.

Kliser, C. W., Grabill, W. D., and Campbell, A. A. *Trends and Variations in Fertility in the United States,* Cambridge, Mass.: Harvard University Press, 1968.

Mangin, W. "Squatter settlements," *Scientific American,* **217**(4): 21–29, 1971.

Price, D. O. (ed.) *The 99th Hour—The Population Crisis in the United States,* Chapel Hill: University of North Carolina Press, 1967.

Rainwater, L. *Family Design, Marital Sexuality, Family Size and Contraception,* Chicago: Aldine, 1965.

Revelle, R. *Rapid Population Growth; Consequences and Policy Implications,* Baltimore: Johns Hopkins Press, 1971.

Stockwell, E. G. *Population and People,* Chicago: Quadrangle Books, 1968.

Thompson, W. S., and Lewis, D. T. *Population Problems,* 5th. ed., New York: McGraw-Hill, 1965.

von Foerster, H., Mora, P. M., and Amoit, L. W. "Doomsday Friday 13 November, A.D. 2026," *Science,* **132**: 1291–1905, 1960.

Population
Fluctuation

The previous chapter considered the basic features of human population dynamics, especially the current population explosion that, through the resulting resource overexploitation, threatens to destroy most of the ecosystems we occupy and to degrade in some measure all others. Even though the early prognostications of this environmental catastrophe went largely unheeded, the population explosion that was the root cause of the potential for disaster was not exactly unforeseen. The early warnings were not formulated in the precise terms used today, but many writers from Greek and Roman scholars onward discussed population growth in a manner indicating they appreciated the significance of mortality and natality rates in regard to population growth. Terms like *teeming population* and *burdensome numbers* feature in these classical writings. During the middle ages the subject was apparently largely ignored, perhaps because developing urban civilizations suffered periodic checks to their population growth that prevented undue natural increase. Population problems lay more in the difficulty of establishing sufficient numbers than in controlling increasing ones. Not until the close of the eighteenth century was the question of the possibility of uncontrolled growth again raised, in a now classical work by Thomas Malthus.

Malthusian Theory

Thomas Robert Malthus was an English economist generally acknowledged as the founder of demography, the study of human population growth. He prepared a text entitled *An Essay on the Principle of Population as It Affects the Future Improvement of Mankind with Remarks on the Speculations of Mr. Godwin, M. Condorcet and Other Writers,* which was published anonymously in London in 1798. In this work Malthus presented the inaugural essay on population growth. This recognized that all biological populations have a *potential* for increase that is larger than the *actual* rate of increase and that the resources required for the support of increase are limited. Malthus supposed that the difference between the potential and the actual population increase resulted from this limitation, which continuously exerted pressures on the population, restricting realization of its potential growth. Stated in modern ecological terms, increase is density dependent; the optimum population is the maximum population.

Publication of this work by Malthus provoked violent criticism from contemporary authorities, particularly because the theory seemed to contradict the possibility of obtaining a utopia for all, and especially for the poor, which had begun to appear as one of the aims of contemporary eighteenth-century society. Malthus subsequently issued a number of further publications under his own name. He modified his theory sufficiently to introduce the element of moral restraint in the procreation of children, which could perhaps act as a check to population growth.

In his *Essay on the Principle of Population,* he presents the following propositions:

1. Population is necessarily limited by the means of subsistence
2. Population invariably increases where the means of subsistence increases, unless prevented by some very powerful and obvious checks
3. These checks and those that repress the superior power of population and keep its effects on a level with the means of subsistence are all resolvable into moral restraint, vice, and misery

Some other parts of this theory have been extensively reproduced, for example, the proposition that unchecked population increases geometrically, whereas food production can only increase arithmetically.

Much of Malthus' writing, although a little quaint in its style to twentieth-century readers, contains extraordinarily acute observations. For example, in discussing population growth he supposes that "the passion between the sexes is necessary and will remain in its present state." Later he states: "I do not know that any writer has supposed that on this earth man will ultimately be able to live without food. But Mr. Godwin has conjectured that the passion between the sexes may in time be extinguished. . . . But toward the extinction between the sexes, no progress whatever has hitherto been made."

Malthus also forecasts quite accurately the present rate of population growth of the United States. "In the United States of America, where the means of subsistence have been more ample, the manners of the people more pure, and consequently dejects to early marriages fewer than any modern states of Europe, the population has been found to double itself in 25 years." He goes on to predict the world population in the following terms: "the human species would increase as the numbers 1, 2, 4, 8, 16, 32, 64, 128, 256, 512, etc. and subsistence as 1, 2, 3, 4, 5, 6, 7, 8, 9, 10, etc. In two centuries and a quarter the population would be to the means of subsistence as 256 to 9; in three centuries as 4096 to 13, and in two thousand years the difference would be almost incalculable, the produce in that time would have increased to an immense extent."

As to checks on population, Malthus notes that "a country in pasture cannot support so many inhabitants as a country in tillage, but what renders nations of shepherds so formidable is the power which they possess of moving altogether and in the necessity they feel in exerting this power in search of fresh pasture for their herds."

He notes that "from all the accounts that we have on nations of shepherds . . . the actual population kept equal to the means of subsistence by misery and vice," and qualifies it by adding that the "commission of war is vice, and the effect of it misery, and none can doubt the misery of want of food."

Mortality Rates

The dire proposition of Malthus in his various works that poverty, disease, and war would be the only factors restricting the undesirable realization of the full potential growth in human population has been extensively examined during the nineteenth and twentieth centuries. A paper by Marston Bates on the role of war, famine, and disease in controlling population is representative of such definitive writings. After reviewing the considerable evidence for universal cannibalism among hominids as established from their remains, Bates notes that the cultural evolution of *Homo sapiens* has also been characterized by intraspecific strife, which has been an important limiting factor on human population growth. He adds, however, that the period from 1650 to 1950, one in which the world population growth was quite spectacular, could hardly be called a period of peace, but rather was characterized by a succession of regional wars.

The Effect of Famine

Bates proceeds to examine the effect of famines on population growth and concludes that Advanced Agricultural societies are particularly susceptible to famine and malnutrition. He quotes figures for China suggesting that between 108 BC and 1911 AD there were 1,828 famines, nearly one a year. The worst of these in modern times were caused by severe drought, which occurred in the years 1876 and 1897. The area affected was about the size of New England, the Middle Atlantic states, Ohio, Indiana, and Illinois, and 9 to 13 million people are estimated to have died from one cause or another. In 1920–21 half a million people are believed to have died in China following a further famine.

In the western world Bates notes that 201 famines were recorded for the British Isles between 10 AD and 1846, the last eight representing the Irish famine episode during which it is estimated that 1 million people died and 1 million people emigrated from that country. The population of the island still is only about one half that cited in the 1841 census, which estimated a population of over 8 million.

The Effect of Disease

Discussing death from bubonic plague or "black death" in medieval Europe, Bates notes that this disease appeared suddenly in central Europe in 1348 and that a quarter of the population then died from it. Further epidemics occurred in 1361, 1371, and 1382, but after that the plague persisted only locally, although it broke out again in 1663 and 1668. During 1664, in the first of these later outbreaks, one quarter of London's population is believed to have perished. Bates observes that both plague and leprosy, which seems also to have been prevalent in Europe in medieval times, have disappeared unaccountably. He notes that the most dramatic effects on human populations are caused by the contagious diseases, of which smallpox is the most notorious. Apparently introduced by Spaniards to the New World, this disease is believed to have been as much responsible for wiping out Amerindian populations as their actual slaughter by the invaders. In return, the New World is thought to have contributed a more virulent form of venereal disease that at various times has been equally devastating in the Old World.

The Effect of War

Statistics generally seem to bear out Bates' ideas; in World War I, for example, an estimated 9 million men were killed. Despite this removal of men from society at the height of their reproductive life, the population growth of western Europe and North America did not show any significant downward trend until a decade later. Then the incidence of severe economic stress in the western world following the Wall Street crash in 1929 appears to have had a determining effect on population growth. In World War II even more men died, but again population growth throughout the western world showed an upward rather than a downward trend following the end of hostilities in 1945. Even global wars as they were waged in World War I and World War II would therefore appear to be without significant depressing effects on world population growth.

This has not always been the case, when more local effects of war are considered. The work of Rappaport in New Guinea on the warfare feedback mechanism already has been discussed. S. F. Cook has produced evidence for supposing that in pre-Columbian Central America warfare, and the human sacrifice with which it was associated, effectively checked population increase and prevented overexploitation of the habitat. There is less well-documented evidence that the same ecological factors were holding West African populations in a steady state before the period in the last two centuries when captured individuals were disposed of by export as slaves instead of being sacrificed.

Such local examples of past population control do not negate the assertion that modern warfare is ineffective in leading to regulation of population increase. If a tragic example were needed, it is provided in the apparent ability of a small country like North Vietnam to sustain indefinitely war casualties estimated as approximately ten times those that even a large nation like the United States found painful.

Effects of Atomic Warfare

The same ineffectiveness could not be predicted for international wars utilizing atomic as opposed to conventional weapons. It is to be fervently hoped that the utterly disastrous effects of such an outbreak can be avoided, because there is every indication that the first one would be sufficiently devastating to destroy all world populations. This prospect is considered again in Chapter 10, which deals with air pollution.

Effects of Disease on Population Growth

There is ample evidence that in this century the far higher frequency of personal contacts in our essentially urban populations has considerably increased the possibility of death from infectious and contagious diseases. The greater concentration of peoples in cities has presented more massive exposure to bronchial infections, which are effectively spread in crowded buses, trains, stores, offices, and restaurants.

Influenza

The outbreak of influenza in 1919 following the end of World War I spread rapidly through all the cities of the world. It is very difficult to estimate the number of deaths from this particular epidemic, but it has been conservatively placed at from 10 to 15 million, thus considerably exceeding the number of men killed in the actual fighting. In more recent years a new form of influenza known as Asian flu spread through the whole world in 1967–68, and again caused extensive mortalities, estimated to be in the region of 9 million.

The molecular structure of the influenza virus now is known in detail. Each virus particle is made up of an internal ribonucleoprotein antigen, surrounded by an envelope composed of RNA coils. This envelope includes two glycoprotein antigens, hemagglutinin and neuraminidase. The hemagglutinin enables the virus particles to adhere to a target cell. Adherence is prevented by antibodies specific to the hemagglutinin. This is the basis for laboratory tests for particular strains of the virus, which determine the extent of adherence of virus particles to chicken erythrocytes, causing them to agglutinate. Neuraminidase forms a glycosidic bond in the cell membrane of the host cell, releasing the new virus particle. Inhibition of this glycoprotein will thus prevent the spread of new virus particles to other host cells.

Using the agglutination test, types A, B, and C of the virus have been distinguished. Type A is responsible for widespread and B for local epidemics; C seems not to cause influenza outbreaks at all. Strains A0, A1, and A2, which result from changes in the hemagglutinin or neuraminidase antigens, also have been detected. A0 appears presently to be dormant, A1 has caused most of the recent influenza outbreaks, and A2 produces Asian or "Hong Kong" flu. Once every year or so *point mutations* (substitution of one amino acid for another at one or more sites on the polypeptide chain) may occur. These produce the kinds of variation exhibited among strains of the A series. With a frequency of some 10–12 years, mutations involving more extensive genetic modification can appear. These may involve considerable alteration of the amino acid composition of one or both antigens. It is possible that such mutations result from *recombination* between two virus strains coexisting in some nonhuman animal host; apparently such coexistence does not occur in humans.

The antibodies active against one strain of influenza have little effect on another, so that the amount of infection developing in a community will depend on its past history of exposure. Pandemics of particular strains appear for which there are neither natural antibodies in the population nor antisera available. It seems that such new strains are most likely to appear after the virus has passed through other animals, especially horses or chickens. Populations living in close proximity to their chickens, as in the Far East, are therefore those in which influenza epidemics associated with new strains are most likely to start. Although it is still difficult to produce quickly the antiserum necessary for immunization against new strains arising from recombination, modern research is shortening the time needed to produce antisera against new antigens arising by point mutation.

Effects of Increased Communication

Increased communication is beginning to render agricultural populations as susceptible to infectious diseases as urban concentrations of people. During this century the territorial patterns that have persisted for hundreds of thousands of years, and have probably prevented extensive contact between peoples except in marketplaces, have begun to break down. The spread of smallpox in tropical areas was much more rapid at the beginning of this century because of the *pax Europea*. The occupation of tropical colonial areas by European powers had brought about a cessation of hostilities between the peoples of adjoining village territories and heralded an era of travel on a previously unprecedented scale. Such local travel is now a characteristic of many agricultural societies. Sleeping sickness, or trypanosomiasis, is another good example of a disease whose incidence has spread with this breakdown of territoriality. Infected individuals from centers of endemic infection are liable to move out and infect new areas.

More rapid communication also has increased greatly the chances of introducing epidemic diseases such as yellow fever from one area into another presently free of the disease, but with certain conditions such as the presence of a suitable vector predisposing toward its establishment.

Vast new agricultural schemes are likely to increase the incidence of water-borne diseases such as schitosomiasis, or bilharzia, which has spread widely through the continent of Africa during this century, although it previously had apparently been restricted to Egypt.

Looking again at Malthus' three principles of war, disease, and famine, it would appear that disease will not now have any significant or immediate effect on population fluctuation, although a highly virulent form of bronchial disease would prove uncontrollable initially. War, if of a nuclear type, would undoubtedly annihilate all human populations in one generation, as well as all other higher forms of life on this planet. Famine, and its alternative manifestation as malnutrition, remains therefore as the most urgent factor to be considered as a potential cause of major population fluctuation, as first enunciated by Malthus.

Famine

An address was published several years ago by I. L. Bennett, chairman of a panel of the President's Science Advisory Committee that was instructed to study the world food problem. This competently surveys the contemporary situation relative to world food supply. From this authoritative review it is possible to assess the modern impact of Malthus' last prediction in regard to population fluctuations resulting from famine and starvation.

Bennett groups his principal conclusions under five headings:

1. *Unless the Situation Changes Markedly Food Shortage and Famine Are Inevitable.*

Bennett notes that estimates of the precise date when famine will reach disaster proportions ranged from the mid-1970s (Paddock and Paddock) through 1980 to 1984 (U.S. Department of Agriculture); to this can be added 1985 (McElroy). These global famine disasters will occur when food transfers from "have" nations no longer compensate for the shortfalls in domestic food production to the "have nots."

One of the earlier unequivocal statements of the lag between grain production and consumption was prepared by Lester R. Brown. Figure 8-1, taken from his publication, with the addition of estimates which update his data, illustrates graphically how this lag is increasing annually.

It can be seen from the figure that from 1967 the improved total grain production in the world was steadily reducing the drain on the world's surplus stocks characterizing the previous five-year period. Despite this, the

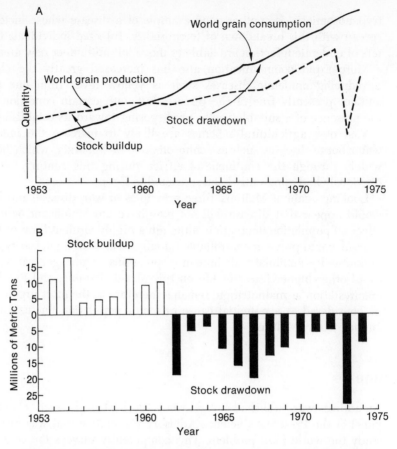

FIGURE 8-1. *Deficiency in world grain production as compared with world consumption.* The first graph, *A*, shows (not to scale) how increasing consumption of food by 1961 had converted a world overproduction to a world underproduction, and created a deficiency that has subsequently steadily increased each year. Graph *B* is another way of representing the same basic figures, showing how world grain stocks were rapidly being depleted at the close of the last decade, and are no longer available to meet grain production shortfalls.

underdeveloped countries were forced to nearly triple their grain imports. Revelle estimates that during the period from 1949 to 1972, underdeveloped countries increased their gross imports of cereals from 12 million tons to 36 million tons. Moreover, 1972 saw a sudden catastrophic deterioration in the world grain situation, as Figure 8-1 indicates, due to low rainfall. The shortfall between total grain production and demand in 1972–73 Revelle estimates at about 60 million tons. As a consequence there was extensive trading on the international cereal grain market, and the world cereal reserves were reduced to an estimated 30-day supply in the spring of 1974. By the summer of 1974 India was reported to be shopping for 10 million tons of grain. This represents only about 1 per cent of the total annual world production (say, 10 per cent of total U.S. production), but world grain stocks were so low it was apparent that India would not be able to purchase even this quantity on the international market. It was an inescapable deduction from these circumstances that in 1975 many thousands of Indians would die of starvation, just as in 1974 an estimated minimum of 100,000 people died in Ethiopia and the Sahel savanna region of West Africa.

Figures such as these have prompted the current predictions of inevitable famine conditions on a catastrophic scale. Although it is true that some societies do not depend on grain for their main food source, these exceptions are minor in relation to the total world population.

Additional proof to substantiate further the self-evident conclusions from such statistics can be obtained from food import figures for individual countries such as those provided in Table 8-1 for India. That country is far from moving toward independence in food production; its demands for food imports now are increasing annually and soon will attain gigantic proportions. Reference to Table 7-9, which shows the natural increase of India's population, will confirm this.

Some idea of increased food demands on a regional basis may be obtained from Table 8-2. At first sight it might appear that local improvements in yield, such as that shown for Mexico, and production of export surpluses in countries such as the United States and Canada might cover this deficiency. However, a consideration of the figures in Tables 8-1 and 8-2 will show this is impossible. The increased food production in Mexico in the decade to 1955 actually totaled less than 1 million metric tons annually. This was less than *one fourteenth* of India's shortfall in 1966. Meanwhile

TABLE 8-1 *Indian Import of Cereal Grains*

Some variation occurs dependent on the nature of the annual Indian crop, but there is a definite upward trend that overrides this variation from year to year. In the years 1967–69 the Indian grain crop increased with the introduction of "miracle" seeds, and imports fell to an average estimated 63 million metric tons. Continuing extension of the new "miracle" seed area enabled India temporarily to reach self-sufficiency by 1972. Partial crop failures in 1972–73 saw imports rise again, reaching a need in 1974 probably in the region of at least 10 million metric tons. That is, India was virtually back to the 1966 position; population increase had wiped out the advantage gained by the introduction of "miracle" seeds.

Year	Amount in million tons
1957	3.6
1958	3.2
1959	3.9
1960	5.1
1961	3.5
1962	3.6
1963	4.6
1964	6.0
1965	7.5
1966	14.0

TABLE 8-2 *Additional Food Requirements Estimated on a Regional Basis*

The first table (*A*) shows overall additional needs for 1975, based on the projected population growth. The second table (*B*) shows that for some countries, such as Mexico, conditions may permit the realization of the required increase in food. (See text for major qualifications of this statement.)

A Region	Estimated regional population at current growth rate in millions		Additional food production needed 1965–75 in per cent
	1965	1975	
East Asia	876	1,040	20
South Asia	875	1,250	28
Africa	311	404	30
Latin America	248	335	35
Total	2,400	3,000	26

B Year	Cultivated area (hectares)	Yield (kilograms/hectare)	Production (metric tons)
1945	500,000	750	330,000
1955	790,000	1,100	850,000
Increase:	48 per cent	48 per cent	158 per cent

India has added something like another 40 million to its population. Mexico, like all Latin American countries, has also added to its population and has little good land remaining uncultivated.

These overwhelming figures can be reinforced by some crude generalizations. The total annual wheat crop of the United States is nearly 40 million metric tons. In 10 years' time the population of India will increase by about 200 million. These extra mouths will require the *total* of the current U.S. wheat production to sustain them, even if by superhuman efforts India has by then so increased its agricultural production as to be self-supporting with respect to the *present* populations. According to most authorities inside and outside India, this seems unlikely.

Another way of presenting the trading method of avoiding world famine has been followed by T. Kristensen, who projects food demand and production in terms of *costs* (Table 8-3). He assumes that by the year 2000 AD agricultural technology in developing countries would be capable of expanding to produce annual food surpluses valued at 35 *billion* dollars (23 per cent over their needs), but it is inconceivable that the less-developed nations could at that time find such an enormous annual sum of money to purchase these surpluses. The sum indeed will represent about one twentieth of the gross national product of the United States, or *twice* what the 93 less-developed nations presently spend on public health and public education. Even if these less-developed nations gave up *entirely* their expenditure

TABLE 8-3 *Value of Projected Food Shortage in Less-Developed Countries*

This table assumes this shortage could be made good by trading for surpluses that could be produced in developed countries. In 1960 trade balances were about to complete the change from the situation that had long prevailed, in which less-developed nations had traded food surpluses to meet food deficiencies in developed nations.

| | *Total Value in Billions of Dollars* | | | | | |
| | *Developed countries* | | | *Less-developed countries* | | |
	1960	*1980*	*2000*	*1960*	*1980*	*2000*
Demand	80	113	151	47	89	179
Production	78	125	186	48	77	135

on military defense, this presently represents only one half of the 35 billion dollars needed to buy food.

2. *Present Efforts Will Not Do the Job.*

The scale of internationally organized assistance intended to solve food shortages is entirely inadequate, and the situation continues to deteriorate.

It is sometimes stated that introduction of new high-yielding crop varieties, increased use of fertilizers, and extension of agricultural acreages will suffice to restore the deficiency between food requirements and actual production. The statistics on which this argument is based are illustrated in Table 8-2. The average regional shortfalls are shown by such figures to vary between 20 and 35 per cent, an increase easily attained in the past, as shown by the statistics for Mexico. It must be recognized, however, that Mexico is a special case. First, as shown in Table 8-2, it was able to increase its acreage of cultivated land by 48 percent in the decade 1945–55. Second, it had in 1944 a cadre of trained and experienced agricultural scientists to call on in a sophisticated program originally extensively supported by money from the United States and skill from the Rockefeller Foundation. India has no such reserve of potential agricultural land; contact between village farming practice and the findings of the few sophisticated agricultural research centers India presently operates offers great logistic problems. As Paddock and Paddock explained a few years ago, combined use of fertilizer, irrigation, pesticides, improved varieties, and double or triple cropping could result in fantastic increases in production of grain crops in India—as indeed they did. However, there is not sufficient fertilizer presently produced in the whole world to permit the universal use of it in India at the same rate as in, for example, Japan. In terms of economics, such improvements have to be bought, and already in 1967, according to the Paddocks, half the rupees circulating in India were *American* rupees. India even more so today totally lacks the foreign reserves necessary to purchase the necessary supplies and equipment.

The development of "miracle rice" has been described by R. F. Chandler.

Whereas average unimproved rice yields in tropical Asia are about 1,500 kilograms per hectare, and more efficient production in Japan gives some 5,000 kg/hectare, the first new rice variety, IR8, produced by the International Rice Research Institute in the Philippines, can yield over 9,000 kg/hectare. Chandler states that even better varieties will become available from the Institute. However, fertilizer applications are necessary to obtain the highest yields with the new varieties. Indeed, it is difficult for plant breeders to improve long-established varieties of any major crop that have been selected empirically for performance under local conditions. The mixed genotypes of these established varieties contain sufficient diversity to ensure that whatever the conditions of a given season, they will be optimal for some genetical component of the plant population. Replacing these all-purpose, local varieties with a general improved one normally requires a greater degree of cultural care to ensure optimum growing conditions to meet the more exacting requirements of the new variety.

3. Increased Food Must Come from Farming.

Although this point is reiterated by numerous agricultural authorities, the issue appears debatable. N. W. Pirie, for example, has developed unconventional protein foods to the point where they have become a practicable and acceptable proposition. His work has concentrated on unconventional sources of *protein* because he maintains that shortage of protein is currently the main dietary deficiency. After noting that ruminant animals should be utilized as a protein source only when they are maintained on nonarable (noncropable) land, Pirie states that not only legumes such as soybeans but also some varieties of cereal crops contain significant amounts of protein, e.g., sorghum (3–4 per cent), wheat (3.8 per cent), and oats (4.8 per cent). There is also a new corn variety with increased lysine (protein) content. When leaf proteins, even of trees, are taken seriously as a food source, many crops not now thought of as food plants will come into use.

4. There Is Still Hope.

But this belief is essentially based on point 3. The huge success of "miracle rice" as described by R. F. Chandler and similar improvements in wheat suitable for underdeveloped areas appear to emphasize that this feeling of optimism is not unfounded. This question is discussed further in a later chapter.

5. Population Control Alone Is Not a Solution.

The current estimate is that food needs of "have-not" areas will double in the next 20 years. Even if present efforts at population control are entirely successful, this will reduce by only about one fifth the increased demand for food.

Bennett illustrates the reason for this by referring to India, where maintaining even present nutritional levels *if no more children were born in* the period 1965–75 would require 20 per cent more food. This is because more than half the population is less than 15 years old, and food requirements continue to increase approximately up to the age of 19 years. Under the same circumstances a 30 per cent increase in food supplies would be necessary to raise the level of nutrition to that recommended by the Food and Agricultural Organization of the United Nations. This increase in 10 years, Bennett notes, about equals that currently being achieved by improved agricultural practices. It is also more than twice the level of the present U.S. shipments to India, and it *allows for no reproduction whatsoever*. Unless radical action is taken, widespread famine disasters in India are inescapable *in the decade of the 1970s*. Unfortunately the example of India is far from unique; it is cited constantly because of the immensity of its problems. The same circumstance will be found to apply to any tropical underdeveloped area. India represents only a quarter to a fifth of the Third World of underdeveloped nations, depending on how these are defined.

Shortcomings in Present Approaches

The inevitable inadequacy of present steps to avoid world famine, which becomes apparent with even a summary review such as this, is not primarily the result of callousness or neglect of the problem. Literally billions of dollars in foreign aid are expended from both governmental and private sources in this country alone, and funded through an imposing array of international agencies—FAO, WHO, UNESCO, UNICEF, and IBRD. Again Bennett has laid out the reasons for the comparative ineffectiveness of these programs. These are as follows:

1. THE COMPLEXITY OF THE WORLD PROBLEM. Hunger and malnutrition, as every agricultural scientist is aware, basically are not technological agricultural problems, but the result of inadequate economic development. The complex intricacies of the economic aspects make the appropriate solution elusive.

2. THE REAL COMPLEXITY IS NOT APPARENT. Solutions to the problems of hunger and malnutrition are thus almost inevitably oversimplified. Unforeseen consequences have an unfortunate habit of arising on the eve of spectacular successes. A premium thereby is placed on short-term solutions, whereas a carefully planned long-term systems approach is essential. Georg Borgstrom describes some of the pitfalls in ignoring such a need.

3. THE PROBLEM HAS BEEN STATED ONLY IN GENERALITIES. The desire to evoke humanitarian response has necessitated brief generalizations, and some kind of improvement has all too frequently lulled sympathizers into a false sense that all is now well.

4. FOOD SHORTAGES AND POPULATION GROWTH ARE INSEPARABLE. Solutions of the one problem will not at all automatically solve the other; both problems must be solved simultaneously.

5. SOLUTIONS TO BOTH FOOD SHORTAGE AND POPULATION GROWTH CAN ONLY BE ACHIEVED WITH INDIVIDUAL COOPERATION. No amount of governmental

legislation will increase agricultural production and reduce population reproduction unless individuals cooperate willingly.

It is general economic experience that the supply of foodstuffs made available by farmers is closely related to the amount of attractive goods they can purchase with the cash received in return. In subsistence farming, enough is produced to meet family needs; surpluses commonly are converted immediately into alcoholic beverages and consumed in one prolonged and glorious bout of intoxication. Production beyond this subsistence amount has to be coaxed by other attractive and economic incentives.

As will be discussed in the next chapter, these economic incentives appear the most practical for obtaining the individual cooperation necessary to achieve population control.

The Triage System

The Paddock brothers, who were among the first writers to present the inevitable world famine crises in a well-documented and convincing form, have suggested the triage system to cope with the problem of deciding what action developed nations can take when the famine they foresee begins to occur on a massive scale in 1975. Perhaps a little cynically, they comment that of the only four countries—Argentina, Canada, Australia, and the United States—that will have any exportable food to supply, the first three will *sell* on the open market. For them there will be no problem, but for the United States, which may be expected to continue somehow to *give* surplus food away, there will be an acute problem of selecting the recipient nations.

Triage is a well-established military practice for the handling of battle casualties at saturation levels. When it becomes necessary to decide whom to treat and whom to ignore, the wounded are divided into three groups. Members of one are dying and cannot be saved; they will receive no treatment other than pain suppressants. A second group, those who can survive without immediate treatment, are held off temporarily. The third group are those likely to survive if treated right away. The limited medical resources are concentrated on these, and experience has shown that this action saves a maximum number of lives under these conditions.

Triage is not a pleasant method to apply, but it would identify the regions to receive famine assistance on the certainty that this will be in limited supply. Just how limited began to be apparent in 1973, when virtually the total surplus U.S. wheat stocks were *sold* to Russia. Apart from immediately obvious detrimental effects on the U.S. economy, this action preempted options for action that might otherwise have been available to help deal with famines that subsequently developed that year in West Africa, Sri Lanka (Ceylon), India, and Pakistan.

Nations selected to receive aid under the triage system will be those in temporary imbalance, where the rate of natural increase is decreasing and the food production rising, but too poor to purchase enough imports to meet the temporary deficiency. Some more fortunate nations may have almost closed the gap and can find the foreign exchange necessary to import food until they become self-supporting. These can make it on their own. Others, in the third category, will not yet have begun to reduce their population increase or to move in the right direction toward self-sufficiency in food production. They will have to be abandoned until malthusian phenomena so reduce their population size and rate of natural increase that they can become self-supporting, or nearly enough so to qualify for aid.

The opportunity for political mischief-making when such a triage system has to be used may be too tempting for some nations to miss. The "Age of Famines" will be a most difficult time for all nations, and the greatest amount of tolerance and forbearance will be needed if the world is to pull through what will be the greatest crisis it has ever faced. William Paddock, who is still professionally involved in food production matters, like most other food production experts at the time when this book was written, can be described as having pessimistic expectations; he is reported to consider there is no valid reason for amending his original prognostication that the "Age of Famines" should commence in 1975.

The FAO Plan

For 6 years the Food and Agricultural Organization of the United Nations has been considering means of averting disaster in the underdeveloped world as a result of the population explosion. This international organization has ratified a plan entitled "Indicative World Plan for Agricultural Development" (IWP). This plan, reviewed in papers by G. Chedd and M. Allaby, analyzes the major crises facing the world in the next decade or so and makes recommendations not only as to how to solve the imminent threat of world famine but also as to the achievement of economic and social well-being for the whole of mankind.

All projections in the plan take 1962 as the base year, with the world divided into three categories of nations. Zone A contains developed market economies like the United States and most western countries. Zone B has "certainly planned," i.e., communist, economies, and zone C "developing" economies, the underdeveloped nations.

If agriculture in a zone C country is to provide sufficient food by the end of the plan period (1985), it has to attain an annual increase that has risen from 2.8 per cent in 1962 to between 3.2 and 3.8 per cent. The difficulty of reaching this seemingly unambitious goal is illustrated by the fact that recently agricultural production in some zone C countries has not increased at all.

The following five objectives are identified in the IWP plan:

1. Producing sufficient food to cope with 2.5–3 per cent natural increases in population, especially by emphasizing cereal production
2. Producing more of the protein that rising standards and increasing urbanization demand, especially by improving the production of animal protein

3. Strengthening the foreign exchange situation in zone C countries
4. Increasing the number of jobs in "agro-allied" industries
5. Intensifying agricultural production

The plan acknowledges that zone C countries, which in 1965 totaled a population of 1.5 *billion,* will by 1985 have a combined population of 2.5 billion, requiring an 80 per cent increase in food production. IWP allows for some of this to be met from the importation of food to the value of 26 *billion* dollars (total 2 billion in 1962). The problems become more acute as they are successively considered. Provision of increased amounts of protein is difficult, and developed nations, it is suggested, can best help by providing processed milk on a massive scale. The issue of unemployment has driven even the basically optimistic sponsors of IWP almost to despair. Some suggestions for its cure seem reminiscent of New Deal days in the United States, which is not surprising, for the problems are very similar.

We tend to clutch at any straw in the hope of having not to believe what we do not want to believe. A plan like IWP will not work if we use it merely to reassure ourselves. Only frenzied activity and massive sacrifice by the citizens of all zone A and many zone B countries could make it work.

It would be possible to continue almost indefinitely with opinions as to the nature of the world food crisis. This all too brief review may best be concluded by references to two papers. The first, by H. F. Robinson, presents the findings of a Presidential committee panel on this issue, and its conclusions are worth quoting here.

1. The scale, superiority, and duration of the world food problem is so great that a massive, long-range innovative effort unprecedented in human history will be required to master it.

2. The solution of the problem, which will exist after about 1985, demands that programs of population control be initiated now. For the immediate future, the food supply is critical.

3. Food supply is directly related to agricultural development and, in turn, agricultural development and over-all economic development are critically interdependent in hungry countries.

4. A strategy for attacking the world food problem will, of necessity, encompass the entire foreign economic assistance of the United States in concert with other developed countries, voluntary institutions, and international organizations.

The second paper is by H. D. Thurston, who reviews the prospects for improvements in tropical agriculture. He concludes that the most realistic short-range solution to the world food crisis is an increase in tropical food production, necessitating a fundamental change from a previously cash-crop and plantation-type agricultural economy. One of the major contributions industrialized temperate countries can make is by intensifying and extending the training of personnel from these tropical areas in the latest methods of tropical agricultural technology. The arguments raised in these two papers, and many other issues regarding food production, are discussed exhaustively and clearly in a comprehensive survey of the world food situation published by an internationally recognized authority on this topic, Georg Borgstrom, in 1973.

Nontraditional Methods

Of the nontraditional methods of increasing agricultural production, perhaps the closest approach to realization has been achieved by various techniques for desalinating salty water for irrigation or hydroponics (soilless culture). The economics of this process has been reviewed recently and it has been concluded that the costs of desalting operations, inasmuch as it is possible to foresee these over the next 20 years, are a whole order of magnitude greater than the agricultural value of the resulting irrigation water. This discouraging conclusion relates not only to projects for desalination of sea-water but also to desalting of drainage or underground water in such areas as the San Joaquin Valley of California, where salt accumulation may render such water resources unacceptable for agricultural purposes.

The same kind of economic difficulties intervene whenever any other such nontraditional way of food cropping is explored, whether it is planktonic harvesting of the sea, raising fungi on petroleum or molasses, or leaf harvesting as developed by Pirie. Time is quite inadequate for the research and

TABLE 8-4 *Disparities in Protein Consumption*

Mean per capita daily amount of meat, milk, and fish protein in grams in the diet in various selected countries. Advanced Industrial societies generally have at least ten times the meat consumption of Advanced Agricultural societies. In a few of these latter, protein consumption is high, just as in a few of the former (Portugal, Japan) it is low, although somewhat offset by a high fish consumption. It is tempting to conclude from such figures that a high animal protein diet is a luxury placing an unreasonable strain on ecosystem productivity. It would be less disruptive if human populations in countries in the upper section of this list functioned as primary consumers and adopted vegetarian diets.

Country	Meat	Milk	Fish
Australia	317	561	13
Argentina	300	394	10
New Zealand	288	724	19
United States	254	750	13
Canada	212	758	18
France	195	563	30
United Kingdom	194	591	28
Netherlands	121	776	20
Brazil	81	158	6
Mexico	65	258	13
Taiwan	50	17	56
Portugal	46	97	85
United Arab Republic	37	114	14
Turkey	36	301	7
Japan	15	50	62
Pakistan	11	114	9
Ceylon	8	32	18
India	4	130	6

development work required on all such methods. For this reason in particular nontraditional cropping may find its most immediate practical application in food enrichment programs, in remedying the nutritional imbalance of much of the world's present diet.

Nutrition

The important subject of nutrition cannot even be dismissed adequately in one short paragraph. There are sufficient statistics to demonstrate that a high proportion of the population of underdeveloped nations is suffering from malnutrition. Even in an affluent society like that of the United States approximately one half of the 24 million persons judged on economic grounds to be living below the poverty line were believed to be suffering from malnutrition. Such severe food shortage contributes to a higher mortality rate but apparently has little direct effect on the fertility rate. It restricts full physical and mental development not only in "hungry" nations but also in impoverished pockets of developed nations such as the United States. Malnutrition is partly the result of ignorance, but mostly it is a question of economics and therefore has an economic as well as a dietary solution.

Two examples of statistics suffice to illustrate this relationship between economics and diet. Table 8-4 shows the mean amount of animal protein calculated on a national basis for selected countries at different economic levels. Table 8-5 illustrates the proportional amounts of animal protein in the human diet on a regional basis. As regards the world picture, 1 billion people in the developed countries have half again as many calories and

TABLE 8-5 *Composition of Regional Diets*

This expresses the major food elements as a percentage of the total diet and can be compared with the information in Table 8-4, but regional means obscure national differences within these statistics. However, the same conclusion is valid: a diminished exploitation of ecosystems in North America and perhaps some other areas could be attained by reducing the proportion of animal protein in the diet.

| | *Carbohydrates* | | | *Proteins* | | | *Fats and Oils* |
	Cereal	*Other plant carbo-hydrates*	*Plant proteins*	*Meat*	*Milk*	*Fish*	*Plant oils*
Far East	69	14	8	2	2	1	4
Near East	64	14	6	3	5	1	7
Africa	50	28	7	4	3	1	7
Latin America	40	28	7	9	6	1	9
Europe	44	22	2	8	11	1	12
North America	21	25	2	23	13	1	15
Oceania	27	23	2	24	11	1	12

five times the animal protein to eat as the 2 billion of the less-developed areas (Table 8-6). One inescapable conclusion from such statistics is that an Advanced Industrial society can function without the animal protein-rich diets that presently pertain in some such nations. Another is the ecological conclusion that changing from an animal protein-rich diet to place more dependence on plant carbohydrate foods would either permit a reduced human exploitation of local ecosystems or provide a human food surplus. A new technique in plant breeding, inducing by irradiation mutant forms of crops that have a much higher protein content, offers another seemingly feasible method of preventing malnutrition.

Disease

Fluctuations in population size due to disease have been touched on already, and some factors discussed that tend to increase any such effects. Two kinds of selection processes have tended to reduce or minimize the effects of disease on human population increase—selection of physical or biochemical characteristics that reduced the effects of the disease, and selection for cultural behavior that avoided it or limited its incidence and spread.

Disease Resistance

Using a science fiction writer's license, H. G. Wells ended his *War of the Worlds* by having the Martian invaders succumb to bacterial infections against which human populations are protected by antibodies in their blood. In employing this idea, Wells was making use of immunological knowledge regarding the coevolution of *Homo sapiens* populations and various parasitic diseases.

Immunological processes in the human body will produce antibodies as a defense against invasion by a wide range of microbial and viral infections. This reaction was first appreciated in the last century by Pasteur, who produced sera that would induce reactions providing immunity against infection for rabies and smallpox. Such *clinical immunity* can be obtained for a wide range of diseases. A *natural immunity*, which will develop phenotypi-

TABLE 8-6 *Comparison of the Daily per Capita Total Calorie and Animal Protein Diet in Developed and Less-Developed Countries*

	Developed countries	Less-developed countries
Daily calorie consumption per person	2,941	2,033
Total daily amount of protein per person in grams	84.0	52.4
Total daily amount of animal protein per person in grams	38.8	7.2
Population in millions	1,089	1,923

cally after contraction of many of these diseases, has been fixed genetically in a number of human populations. Some of this genetic resistance appears to be correlated with the possession of particular major blood group characters. For example, individuals with A type blood appear more resistant to bronchial infections than individuals with other blood types. The converse is true in the case of A type as regards resistance to smallpox.

Genetic Resistance

In all temperate zone populations, there seems to be some degree of resistance to bronchial infection, whether caused by influenza or pneumonia. Eskimo populations, certainly isolated for 15,000 years, and perhaps longer, from temperate zone populations in which these diseases were endemic, apparently have not developed any resistance to them. Individual Eskimos who migrate into temperate zone communities not infrequently die from such bronchial infections.

Rather in the same way, Early Mongoloid stocks apparently did not develop any resistance to smallpox and were decimated by this disease when it was introduced by Europeans to the New World. Europeans, in turn, have developed little or no resistance to infection by some of the tropical intestinal parasites, which apparently now cause tropical populations no inconvenience.

T. Dobzhansky has suggested that resistance to tuberculosis infection has been developed in some temperate zone populations (Figure 8-2). He predicts that the widespread use of antibiotic drugs to control this disease may halt the further adaptation of these and any other human populations to tuberculosis.

Cultural Avoidance

Many cultural practices through empirical selection have enabled human populations to avoid the fluctuations in number that result from disease outbreaks of epizootic proportions. J. M. May provided examples of a number of these. He cites the case of a small Chinese village in which one half of the inhabitants were suffering from a very heavy hookworm infection, while the other half were largely free of this disease. The infected individuals were found to be mostly rice farmers, from whom the healthy ones bought rice, with whose cultivation they were not concerned. The healthy group were silkworm farmers, who had no need to venture into paddy fields. The rice farmers, by contrast, spent all their day in the water of the rice paddies, which had been manured with "night soil," and consequently were exposed to infection through their skin by hookworms whose eggs were introduced to the paddy in the human fertilizer.

Another example of the cultural avoidance of disease is the Islamic injunction in the Koran forbidding the consumption of pork. Around subtropical villages the pig is the universal scavenger, feeding even on human feces. Consequently its flesh contains a number of parasites that have evolved with man as an alternate host.

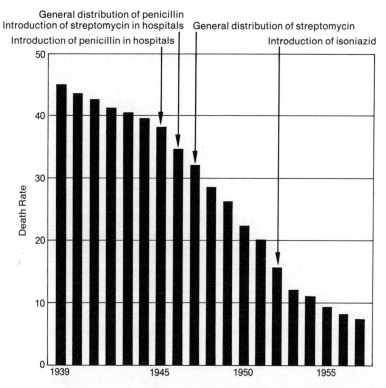

FIGURE 8-2. *Selection for tuberculosis resistance in a contemporary population.*
The tuberculosis death rate per 100,000 persons declined dramatically with the
introduction of antibiotics about the time of World War II. There is now a
reduced threat to tuberculosis-susceptible individuals, and the selection pressure
operating against them is greatly reduced. (From Theodosius Dobzhansky, "The
present evolution of man," *Scientific American,* **203:** 208, 1960. Copyright ©
1960 by Scientific American, Inc. All rights reserved.)

Recent Fluctuations

The causes of fluctuations in population size and rate of increase considered
so far are essentially malthusian; they arise from famine, war, or pestilence.
It has been noted that although in the past fluctuations in numbers from
these causes or combinations of them have been rather local, there now is
a potential for huge variations in numbers to be associated with these
factors. World famine is estimated to be certain, there remains only the
question of when its intensity will reach disaster proportions, in 5, 10, 15,
20, or 30 years. Nuclear war, it is very apparent, would destroy all human
populations. We risk pestilence on an ever-expanding scale because of the
continually more crowded dimensions of our increasingly urban life, and
our loss of genetically acquired resistance to infectious diseases, particularly
those of the respiratory and digestive systems, which are the areas of our
most intimate contact with the environment.

Also contributing to population fluctuation are losses due to emigration and gains due to immigration; throughout history both have significantly affected population growth. Emigration, as already noted, enabled the nations of Western Europe to undergo a population explosion as they moved into the Advanced Industrial category of society without devastating their metropolitan ecosystems. It permitted the Japanese to do likewise nearly a century later without drastically curtailing the rate of natural increase within their home islands.

Immigration over 20,000 years ago first populated the virgin continents of the Americas and Australia, and later the equally virgin islands of Oceania. Subsequent waves of immigrants in historical time caused tremendous upsurges in the populations of these and other continents. These various effects are perhaps best exemplified by examining the greatest immigration movement of all, that into the United States, which took place during the first part of this century, and continues although at a diminished rate.

Immigration into the United States

In 1965 Congress revised the Immigration Act of 1924 to permit a total annual immigration into the United States of 290,000 persons. In 1969 demographic figures this statistic, which has to be included as part of the natural increase of the population, represented 8.4 per cent of the annual United States gain in population. If this country is soon to achieve a zero population growth, this quota of immigrants will represent some 15 per cent of the annual population gain. This appears to be a higher immigration rate than any other major developed country is now prepared to sustain. It is a figure of sufficient significance for demographic studies to have to make the necessary allowance in the statistics.

The 1924 act had established an immigration quota of a maximum of 2 per cent per year of any national group as represented in the 1890 census. This set an annual maximum of 162,000 immigrants for the United States, very different from the over 1 million annual total that had been reached for several years at the beginning of this century prior to 1913.

In 1860 foreign-born Americans represented 13.1 per cent of the population; in 1900 they accounted for 13.6 per cent. At zero population growth, the present immigration quota, if fully taken up, will result in a population that continues to have about the same level of foreign-born citizens as registered in the 1860 and 1900 statistics, a remarkable level of great biological as well as demographic significance, as is discussed in its various aspects several times in this text. In 1972, 384,685 immigrants were actually admitted to the U.S. By 2000 AD legal immigrants and their descendants will number 15 million, almost 25 per cent of the projected U.S. population growth to this time. The ZPG organization proposes the numbers of immigrants be reduced to equal the number of emigrants.

Immigration and Outmigration

The extent of immigration into the United States is atypical of the world situation. Most national frontiers are used as barriers to minimize the flow of new permanent residents into a given country. Population mobility in the modern world is not so much a question of movement *between* countries as *within* countries. Sociologists generally use the terms *inmigration* and *outmigration* for this kind of movement rather than *immigration* and *emigration,* which are reserved for movements across national boundaries. The term *mobility* may also refer, not to a physical transference of residence, but to an upward or downward movement from one social class of society to another.

All nations now have a more or less acute population problem as a result of a universal tendency toward inmigration to urban centers from rural districts.

Inmigration and Urban Crises

This urban inmigration movement has already been discussed in Chapter 5, where it was noted that approximately 80 per cent of the U.S. population will soon be urbanized. As far as this country is concerned, the urban/rural movement is virtually complete. Existing cities will not show any major increase in size if ZPG can be achieved. This is far from true of other countries, particularly of underdeveloped ones, where ZPG is far from being realized, and urban inmigration may only just have commenced.

From Figure 5-12 it can be seen that in the United States in 1790, when the first census was taken, about 6 per cent of the population were city dwellers. Thereafter, the proportion rose exponentially, 15 per cent in 1850, 40 per cent in 1900, nearly 75 per cent in the last census, 1970. The proportion of urban residents in the population of India is estimated as about 11 per cent in 1900, in 1970 about 20 per cent. India's urban revolution thus appears to be about a century behind that of America. Other underdeveloped nations are in much the same position, some rather more delayed even than this. In any case their existing cities are simply exploding with people. The population of Lagos, Nigeria, is growing at 14 per cent per annum. Lusaka, capital of Zambia, has a rate of 14 per cent, Accra, Ghana, 8 per cent, Nairobi, Kenya, 7 per cent.

Kingsley Davis provided forward projections of such inmigration figures, which forecast that by George Orwell's date of 1984 half the world's population will be urbanites. On these simple projections, by 2020 everyone will be living in cities, the largest of which will contain 1–4 *billion* people. Of course, city organizations could not long absorb such a rate of urbanization. Already in Bombay and Calcutta, for example, hundreds of thousands of unassimilated inmigrants live and sleep in the streets. Mexico City sprouts shanty towns on its boundaries, Rio de Janeiro has its notorious *favelas.* Even industrialized societies encounter difficulties in absorbing city inmigrants.

Regional Inmigration

The factors that promote this universal city inmigration were discussed in Chapter 5. Even where, as in the U.S., the progress is almost complete, this is not the end of population mobility. Regional differences in climate, economic and educational opportunities, political attitudes, and many other factors combine to promote long-term regional population movements. Figure 8-3, taken from a paper presented several years ago by Arthur Boughey, James Pick, and Gordon Schick, illustrates this regional in- and outmigration in America. The inmigration is especially into the Atlantic and Pacific seaboards and the Rocky Mountain states. However, the latter have also a high rate of outmigration; the population is highly mobile. In the southern Atlantic region, particularly Florida, many of the inmigrants are retirees, and they stay. The same tendency occurs in California, but many

FIGURE 8-3. *Interstate in- and outmigration in the Pacific region of the United States.* The lines labeled A show the progressive net biological increase in the residential population of the Pacific region (line represents total population without migration). The lines labeled B show this value plus total inmigration; lines C show value A minus total outmigration. The last line, D, shows the net population, or the total population after the natural increase and the net migration data are added. The model predicts these values at 5-year intervals commencing from 1967. Of the approximately 8 million population increase over 15 years from 1967, about 4 million will probably be inmigrants proceeding to Orange and San Diego counties.

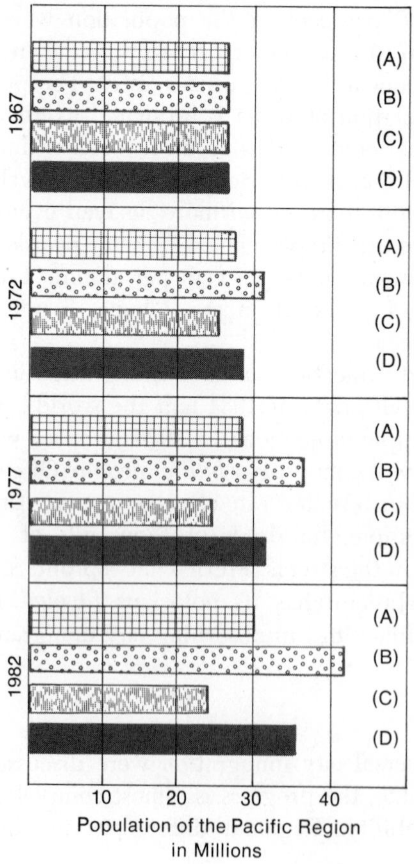

	Inmigration	Outmigration	Total Population
1967	0	0	26,522,631
1972	3,775,259	2,175,951	28,934,971
1977	7,771,193	4,489,767	31,917,569
1982	11,868,740	6,876,654	34,969,005

10 20 30 40
Population of the Pacific Region
in Millions

inmigrants are also in early adult age groups. Consequently, California in this decade with 20 million people has become the most populous state and is still growing mostly by net inmigration. Its population is already much larger than that of a majority of European and many African and Asian nations.

Mobility Ages

This kind of population mobility in an industrialized nation is distinctly age specific; probably it is also age specific, although to a lesser extent, in a nonindustrial one. In Figure 8-4 the frequency of occurrence of in- and outmigration in the American population is plotted on a graph. There are two peaks, one at about 22 years of age, the other among infants. These are not entirely unexpected. Students graduate from college at about age 22; they often wish to move or have to move elsewhere for employment. Young marrieds in this age cohort take their infants with them, thus attaching to the infants a susceptibility to an in- or outmigration label.

Age-Specific Urban Mobility

The same age-specific mobility can be observed in population pyramid diagrams of city population structure. In Cincinnati, for example, it is apparent that many white individuals, both males and females, in their 30s

FIGURE 8-4. *Age pattern of mobility.* In the United States individuals in the 20–24 age cohort, and to a lesser extent the 25–29 one, are the most likely to move from one area to another. This mobility may be local (within the district) or interstate (migration). It is reflected in the first cohort of children up to 4 years of age, who get moved with their parents.

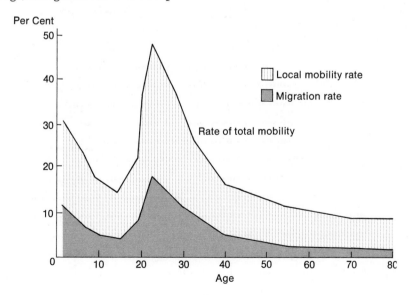

and early 40s become outmigrants (Figure 8-5). In Pittsburgh this same pattern is especially true of black males (Figure 8-6). It is tempting to suggest the explanation is lack of employment opportunity in a city where the population still is undergoing exponential growth. Further proof would be needed before such an assertion could be substantiated, but the outmigration is indisputable. In Norfolk, Virginia, inmigration of military personnel causes an entirely disproportionate representation of white males in the 20–24 cohort (Figure 8-7). We could go on more or less indefinitely with such examples of the effect of age-specific population mobility in the United States and elsewhere.

One last effect of inmigration must be mentioned, that it can convert what had once become a *stationary* population back again to a stable and exploding one. This has happened, for example, in Orange County, California (Figure 8-8). This county, by virtue of massive inmigration of individuals

FIGURE 8-5. *A large Midwestern city with appreciable outmigration.* As this pyramid compiled from 1970 census data shows, the white population of Cincinnati has a calculated stable population approximately coinciding with the calculated zero population increase parameter. This is despite the increased fertility of the last two decades, which is reflected in the increased size of the corresponding cohorts toward the base of the pyramid. A very considerable reduction in the size of the youngest cohort is partly the consequence of outmigration of young parents. It also reflects the current trend toward a reduction of fertility rates. The outmigration in the 30–44 cohorts would appear to be a contributory factor toward reducing the rate of natural increase in the Cincinnati white population.

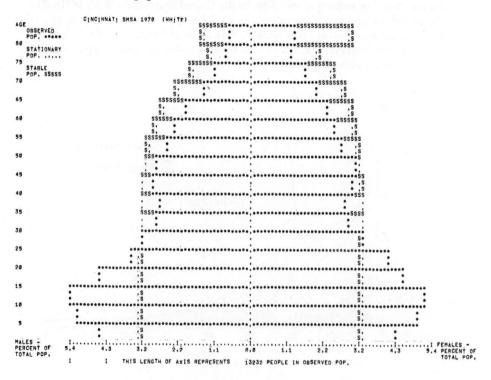

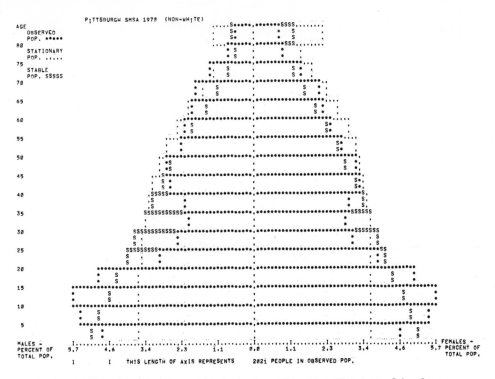

FIGURE 8-6. *An expanding city population.* This pyramid prepared for the nonwhite (in this instance, essentially the black) population of Pittsburgh from 1970 census data shows a population still set in a moderately high rate of exponential growth. As is the case with the black segment of the population in many U.S. cities of the Midwest and the East, there is heavy outmigration, especially of males in the 20–45 age groups. Outmigrating females, although not so numerous, take infants with them, which is reflected in the relatively smaller sizes of the two youngest age cohorts. It may be surmised that were it not for these outmigrating females, the rate of natural increase of nonwhites in Pittsburgh would be even higher.

in the two most mobility-susceptible classes, now has all the economic and social problems that beset an underdeveloped nation with a similar exploding population pyramid. It also has a higher than average representation of elderly widows, who because of their greater life expectancy have survived their mates to unbalance this section of the population pyramid.

Economic Influences on Population Fluctuation

There are other causes of population fluctuation, both negative and positive, than those considered so far. These are as yet imperfectly understood and they have sometimes in the past made demographers appear very bad prophets. In this century, the cause of neither the downward trend of population in most western nations in the 1930s nor the upward surge in many of these during the 1950s has been explained satisfactorily. Within the last decade all nations have shown a decline in their crude birth rate

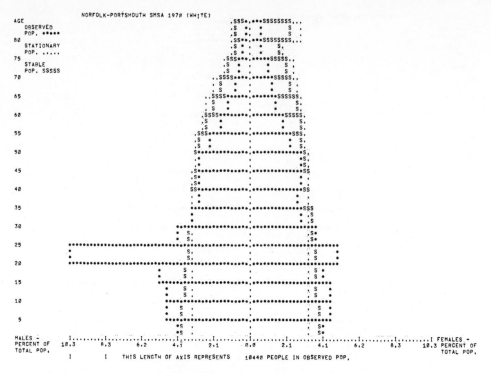

NORFOLK-PORTSMOUTH SMSA 1970 (WHITE)

AGE
OBSERVED POP. •••••
STATIONARY POP. ,,,,,
STABLE POP. SSSSS

MALES – PERCENT OF TOTAL POP.

FEMALES – PERCENT OF TOTAL POP.

THIS LENGTH OF AXIS REPRESENTS   10440 PEOPLE IN OBSERVED POP.

Figure 8-7. *An urban area with zero population growth.* The pyramid for the white population of the Norfolk-Portsmouth area at the 1970 census shows that the observed and calculated stable and stationary populations are beginning to approximate closely. The exaggerated size of the male 20–24 age cohort and the slight increase in the corresponding female cohort reflect inmigration due to employment in the military establishment of this area. Despite this anomaly, the presence of these potentially fecund individuals does not apparently cause any observable departure from the stationary population structure.

Figure 8-8. *Changes in population structure in Orange County, California.* All nations and many lesser geographical units normally move from a young population structure to a mature one as they proceed through the demographic transition. Large-scale inmigration caused Orange County to do the reverse, as shown from these two pyramids constructed from the 1960 (*A*) and the 1970 (*B*) census data. During the first part of this century, up to the time of World War II, Orange County population approached stationary in structure. Then very high inmigration of persons in the 30–49 age cohorts, together with their children in the up-to-14 age cohorts, converted the 1960 structure (*A*) to that of a young population. The large increase in the proportional representation of the younger age cohorts created many pressures, for example on educational institutions, job markets, recreational facilities, and highways.

By 1970 (*B*) the peak of inmigration had passed and the population was beginning again to approach the stationary structure it had in 1940.

The effects of the inmigrant children (now in the 10–24 age cohorts) and of those children the inmigrants had soon after arrival (5–9 age cohort) will cause overcrowding of various facilities for several generations to come. As of 1974 they were beginning to increase local college enrollments and will continue to do so for about 10 years.

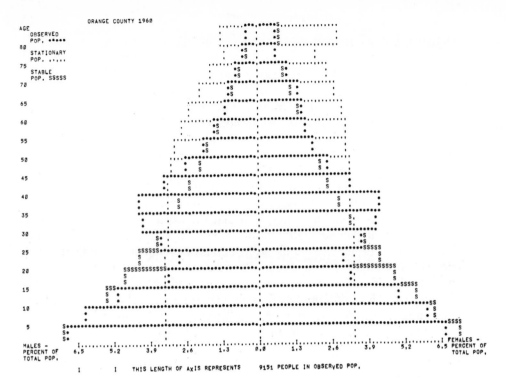

A

B

statistics, but no general explanation for this has so far been accepted. As regards industrialized nations, factors that propel populations into and through the demographic transition have long been considered primarily economic ones. The issue of economic incentive is the basic operative factor in models of the *demographic transition* process described by H. Frederiksen.

This process is modeled in the second scheme (B) in Figure 8-9; it essentially represents the process of ecological succession from an Industrializing, through a Colonial, to an Advanced Industrial society as described in Chapter 5. The first model (A) in Figure 8-9 is the neo-malthusian one, which holds the society in an Advanced Agricultural category without any possibility of further succession occurring—productivity remains too low. A number of workers have suggested that the existence of feedback mechanisms providing population control in western Europe in the eighteenth and nineteenth centuries permitted its nations to complete these final transitions of societal succession. Frederiksen did not of course intend that these models should represent economic factors tending to reduce individual fertility rates, for this would make a circular argument. The actual economic pressures can rather be lumped together under what may be termed a "Madison Avenue effect" (Figure 8-10).

The Madison Avenue Effect

The Madison Avenue effect operates through interaction between the acquisitive desires of individuals, as subtly exploited by agencies employing pressure advertising techniques through the mass media, and the complete separation between sex and reproduction that has been achieved in Advanced Industrial societies. Methods of birth control that have been available since at least biblical times have during the twentieth century been the subject of such intensified research that the new generation of nubile females is able to have complete control of reproduction without suffering any sexual deprivation. The time at which a pair bond, whether sanctified by legal matrimony or not, embarks on the procreation of children can therefore be postponed, or permanently delayed, in favor of a new home, a second home, a second car, a yacht, a world cruise, or whatever glamorous alternative the advertising industry is currently promoting.

This effect has been described several times by different workers, although in different terms. H. F. K. and K. Organski state: "in an industrial society the family's standing depends heavily upon the occupation of the head of the household and upon the standard of living of the whole family as evidenced by the material goods it can display. A family with eight children cannot live in as good a neighborhood as a family with two; it cannot dress as well, drive as new a car, have as many conveniences and luxuries, or travel as much. The children will not receive as expensive an education, and this is particularly important in determining their social status when they become adults."

The nation in which this Madison Avenue effect had the most dramatic results is Japan. From being a Colonial society in the first half of this century, Japan not only has moved into the final category of an Advanced

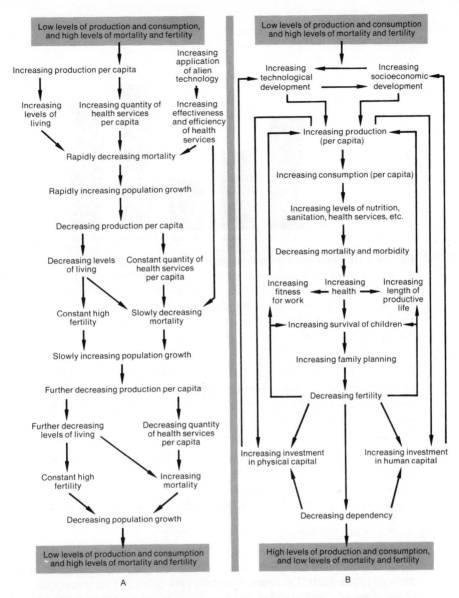

FIGURE 8-9. *Models of economic and demographic transition.* A. The neo-malthusian model, which assumes increasing productivity is *not* accompanied by reduced fertility. B. A model that assumes the opposite, so that a low increasing productivity is not absorbed by increasing population growth.

Industrial society but also has nearly stabilized its population structure and growth. A combination of legalized and readily obtained abortions, linked with extended use of contraceptive practices, has enabled young couples to respond materially to the pressure of the Madison Avenue factor. Japan, however, was not the first major national group to respond in this way; most of the industrialized nations of the New World and western Europe had shown response, but not recognized it, following the 1929 collapse of the

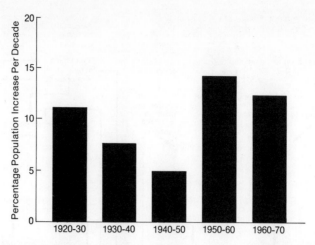

FIGURE 8-10. *Population growth in some western industrialized societies* during the present century. Despite various financial inducements intended to restore the falling birth rates and a reduction in death rates, the rate of natural increase in industrialized societies continued to fall through this century until the early 1940s. The trend was then reversed and the rate of natural increase took an upward swing until about the middle 1950s.

Wall Street stock market and the acute economic depression of the early 1930s. In these countries this was a period when a Madison Avenue factor operated under reinforcement from an economic crisis. Although contemporary birth control practices still involved some inconvenience and considerable restraint, in order to respond to some of the Madison Avenue pressure despite economic restriction, family size had to be cut drastically. Population pyramids of all western nations show this fluctuation (Figure 8-11); those of nonindustrialized nations do not (Figure 8-12). Nor does Japan, whose economy in the decade of the 1930s was not closely linked with world

FIGURE 8-11. *Final phases of the demographic transition in the United States.* These two pyramids are for the total population of the U.S. in 1920 (A) and in 1970 (B). The 1920 U.S. census showed a population still set in an exponential phase of growth, with the observed population in the younger age cohorts considerably exceeding the calculated theoretical stable position. The 1920 pyramid shows no observable departure from the rate of exponential growth prevailing since 1835. In the pyramid prepared from the 1970 census data (B), the first cohort, that is, the five-year group up to the day before the fifth birthday, approximates to the calculated stable population size. This is nearing the stationary position. The effect of the population explosion that peaked in the later part of the 1950s is very obvious in the size of the 10–14 age cohort. The great drop in fertility in the 1930s, a result of the Wall Street crash and the industrial world trade recession, shows very conspicuously in the 30–34 and 35–40 age cohorts. As of 1970, despite the lowering fertility rate, the U.S. population was still set in an exponential pattern. The early age cohorts must approximate to the stationary position indicated in B before the nation can achieve zero population growth.

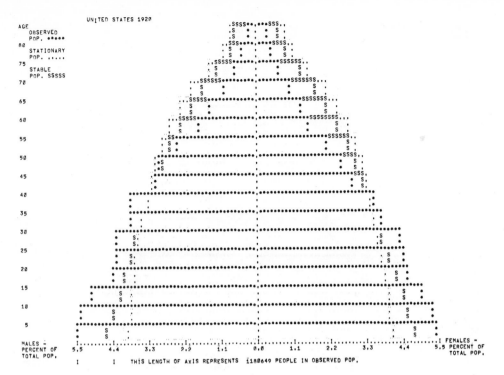

A

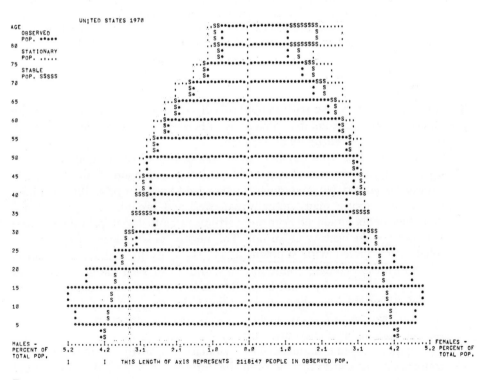

B

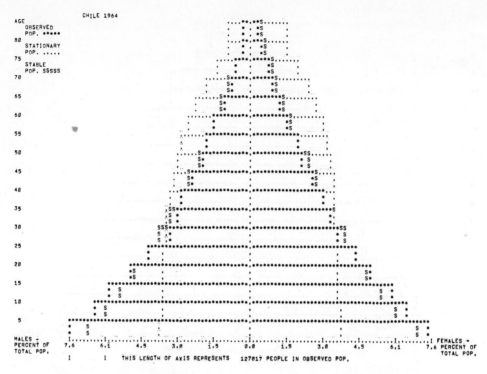

FIGURE 8-12. *The population of Chile.* Nonindustrialized nations have not yet experienced the demographic transition. This pyramid constructed from the 1964 figures for Chile shows a population set in an exponential mode of ever-increasing steepness. Fertility in underdeveloped nations such as Chile is not controlled by the same economic factors as that in industrialized nations. Accordingly, the effect of the depression of the 1930s, so obvious in Figure 8-11, cannot be detected here.

trade. As was the case with nonindustrialized nations in the decade of the 1930s, Japan could not be influenced either by a Madison Avenue or by any general economic factor at this time.

The early 1930s were the first time in the history of the United States and other western nations that a general voluntary reduction of the birth rate ever had occurred. No governments recognized the great prize that was then within their grasp; many actually panicked and brought in various kinds of inducements aimed at *restoring* fertility rates to levels prevalent at the beginning of the century in Europe. This mostly took the form of income tax relief, often linked with supplementary payments for child maintenance. Many of these practices continue today. Indeed in 1966 the government of Roumania, alarmed by a fall in fertility in that country to a rate of 1.9, reversed its permissive abortion law; the following year the fertility rate doubled, reaching 3.7. Subsequently, illegal abortion and contraceptive practices have again begun to reduce fertility.

Despite economic inducements, population growth in these western countries continued to decline until the outbreak of World War II (Figure 8-10). From then on, international and regional war has coincided with a continuing arrest and reversal of the trend toward a reduced population growth in these areas. The most immediate explanation for this reversal would appear to be that surviving veterans long for a return to stability, which they imagine to be a permanent and legalized pair bond with its associated home, family, and material provisions. (Without such incentives, and lacking any stimulation of a group territorial behavioral reaction, it is unlikely that men would have been willing to fight in the first place.)

Demographers believe that this post-War baby boom resulted only partially from a change in the fertility rates; there was also a modification of the age of childbearing, consequent partly on a reduction of the age of marriage. The effect on the birth rate was therefore temporary, and the steady decline in fertility which has characterized all Advanced Industrial nations in the middle of this century has now been resumed. As of 1974 most of these countries have a fertility rate at or about replacement value. The U.S. fell below this, to 1.9, in 1973; West Germany, with a current fertility

TABLE 8-7 *Falling Death Rates in the Twentieth Century*

Figures per thousand persons on a regional basis (A) and selected national basis (B). The reductions obtained in individual countries of particular regions, compared with other national figures for the same regions, show how estimates of further extensive falls in regional death rates may well be realized. Figures in the last column of A are projections.

A Region	1935	1950	1965	1980
World	25	19	16	13
Latin America	22	19	16	8
Asia	33	23	20	13
Africa	33	27	23	18

B Region	Selected countries	Current death rate
Latin America	Cuba	8
	Mexico	10
	Chile	11
	El Salvador	13
	Guatemala	16
Asia	Hong Kong	5
	Taiwan	6
	Singapore	6
	Sri Lanka	8
	South Korea	11
	India	17
Africa	Senegal	22
	Ivory Coast	25
	Upper Volta	25

rate of 1.5, has declined to the lowest national fertility rate in the world. Significantly perhaps, West Germany presently has the strongest currency in Europe.

A general relationship between birth rates and economic factors has been established since over 30 years ago. V. Galbraith and D. S. Thomas showed business cycles in the United States from 1919 to 1937 affected both birth and marriage rates; D. Kirk demonstrated the same correlations for Germany about the same time. J. J. Spengler more recently developed the concept that family increase was dependent on (1) the preference system, (2) the price system, and (3) income. The last two items are self-explanatory and relate to the cost of additional children relative to earnings or savings available to cover it. The preference system refers to the choice the pair-bond couple makes as to whether to use this money for family increase or the purchase of material possessions. Spengler's analysis provides an approach to fertility essentially similar to that of the Madison Avenue effect.

Summarizing this chapter on population fluctuation, it appears reasonable to deduce that our present culture has incidentally evolved an economic feedback mechanism in Advanced Industrial societies that has limited and may well continue to limit population fluctuation and to contain natural increase at or below zero population growth.

For underdeveloped nations in other successional categories, and for underprivileged sections of Advanced Industrial societies, there presently is no such feedback mechanism operative. Among these populations, falling death rates (Table 8-7) will insure a continuation of the high rate of natural increase that already prevails. No prospect faces such populations other than the certainty that within the present generation they will have to endure the malthusian disasters of famine, pestilence, and at least internal war. The figures for population density provided in Table 8-8 illustrate that even if there were no significant use of space other than for food cropping none of the Advanced Agricultural societies has a sufficiently productive agricultural

TABLE 8-8 *Per Capita Acreage of Cultivable Land*

For various countries. Japan, with the lowest acreage on this list, by enormous expenditure of individual effort contrived to become self-supporting in food. The United Kingdom has never been so in this century, nor has India, partly because of low productivity. The Soviet Union has also at times been a food importer, despite its favorable position in regard to cultivable land.

Country	Per capita acreage
United States	2.68
Soviet Union	2.59
India	0.82
China	0.50
Germany	0.48
United Kingdom	0.42
Japan	0.17

technology combined with enough land to render it self-sufficient in food production. The next chapter considers what measures can be taken to re-introduce human population control feedback mechanisms into these eco-systems, which are imminently in danger of total collapse through the absence of any such cybernetic interaction.

BIBLIOGRAPHY

References

Allaby, M. "One jump ahead of Malthus," *The Ecologist,* (1): 24–28, 1970.

Anonymous. *World Population Prospects, as Assessed in 1963,* United Nations Population Studies No. 41, New York: United Nations, 1966.

Armelagos, G. J., and Dewey, J. R. "Evolutionary response to human infectious disease," *Bioscience,* 20: 271–75, 1970.

Attah, E. R. "Racial aspects of zero population growth," *Science,* 180: 1143–51, 1973.

Bates, M. *The Prevalence of People,* New York: Scribner, 1955.

Bennett, I. L. *Problems of World Food Supply,* in XIth International Botanical Congress, Seattle, Wash., All-Congress Symposium, World Food Supply, Aug. 28, 1969.

Borgstrom, G. *World Food Resources,* New York: Intext, 1973.

Boyko, H. "Salt-water agriculture," *Scientific American,* 216(3): 89–96, 1967.

Brown, L. R. "The world outlook for conventional agriculture," *Science,* 158: 604–11, 1967.

Chandler, R. F. *New Horizons for an Ancient Crop,* in XIth International Botanical Congress, Seattle, Wash., All-Congress Symposium, World Food Supply, Aug. 28, 1969.

Chedd, G. "Mutations v. malnutrition," *New Scientist,* 45: 450–53, 1970.

Clawson, M., Landsberg, H. H., and Alexander, L. T. "Desalted seawater for agriculture: is it economic?" *Science,* 164: 1141–48, 1969.

Cook, S. F. "Human sacrifice and warfare as factors in the demography of pre-colonial Mexico," *Human Biol.,* 18: 81–101, 1946.

Dobzhansky, T. "The present evolution of man," *Scientific American,* 203(3): 206–17, 1960.

Frederiksen, H. "Feedbacks in economic and demographic transition," *Science,* 166: 837–47, 1969.

Frejka, T. "The prospects of a stationary world population," *Scientific American,* 228(3): 15–23, 1973.

Galbraith, V., and Thomas, D. S. "Birth rates and the inter-war business cycles," *J. Amer. Statist. Assn.,* 36: 465–76, 1941.

Harrar, J. G., and Wortman, S. "Expanding food production in hungry nations: the promise, the problems," in C. M. Hardin (ed.), *Overcoming World Hunger,* Englewood Cliffs, N.J.: Prentice-Hall, 1969, pp. 89–135.

Idyll, C. P. *The Sea Against Hunger,* New York: Crowell, 1970.

Kirk, D. "The relation of employment levels to births in Germany," *Milbank Mem. Fund Quart.,* 28: 126–38, 1962.

Kristensen, T. "The approaches and findings of economists," *Internat. J. Agrarian Affairs,* **5:** 139, 1967.

McElroy, W. D. "Biomedical aspects of population control," *Bioscience,* **19:** 19–23, 1969.

Malthus, T. R. *Essay on the Principle of Population,* 7th ed., London: J. M. Dent and Sons Ltd., 1816.

Maugh, T. H., II. "Influenza (II): a persistent disease may yield to new vaccines," *Science,* **180:** 1159–61, 1215, 1973.

May, J. M. "The ecology of human disease," *Ann. N.Y. Acad. Sci.,* **84:** 789–94, 1960.

Paddock, W., and Paddock P. *Famine 1975,* Boston: Little, Brown, 1967.

Pirie, N. W. "Food from forests, *New Scientist,* **40:** 420, 1968.

Pirie, N. W. *Plants as Sources of Unconventional Protein Foods,* in XIth International Botanical Congress, Seattle, Wash., All-Congress Symposium, World Food Supply, Aug. 28, 1969.

Population Reference Bureau. "The future population of the United States," *Population Bull.,* **27**(1): 4–31, 1971.

Rappaport, R. A. *Pigs for the Ancestors,* New Haven: Yale University Press, 1967.

Revelle, R. "Food and population," *Scientific American,* **231**(3): 160–170, 1974.

Robinson, H. F. "Dimensions of the world food crisis," *Bioscience,* **19:** 24–28, 1969.

Spengler, J. J. "Demographic factors and early modern economic development," *Daedalus,* **97:** 433–46, 1968.

Thurston, H. D. "Tropical agriculture: a key to the world food crises," *Bioscience,* **19:** 29–34, 1969.

Young, G. "Dry lands and desalted water," *Science,* **167:** 339–43, 1970.

Further Readings

Abelson, P. H. "Malnutrition, learning, and behavior" (editorial), *Science,* **164:** 17, 1969.

Adams, E. S. "Unwanted births and poverty in the United States," *The Conference Board Record,* **6**(4), 10–17, 1969.

Brown, L. R. *Seeds of Change,* New York: Praeger, 1970.

Coale, A. J. *The Growth and Structure of Human Populations,* Princeton, N.J.: Princeton University Press, 1970.

Connell, K. H. *The Population of Ireland 1750–1845,* Oxford, England: Clarendon Press, 1950.

Dalrymple, D. C. "The Soviet famine of 1932–34," *Soviet Studies,* **14:** 250–84, 1964.

Davis, K. "The population impact on children in the world's agrarian countries," *Population Rev.,* **9:** 17–31, 1965.

Davis, K. "The urbanization of the human population," *Scientific American,* **123**(3): 41–53, 1965.

Day, L. H., and Day, A. T. *Too Many Americans,* Boston: Houghton Mifflin, 1964.

Dumont, R., and Rosier, B. *The Hungry Future,* New York: Praeger, 1969.

Feiss, J. W. "Minerals," *Scientific American,* 209(3): 128–36, 1963.

Fiennes, R. *Man, Nature and Disease,* London: Weidenfeld and Nicolson, 1964.

Gulland, J. A. (ed.) *The Fish Resources of the Ocean,* Surrey, England: Fishing News Ltd., 1971.

Hance, W. A. *Population Migration and Urbanization in Africa,* New York: McGraw-Hill, 1970.

Hauser, P. M. *The Population Dilemma,* Englewood Cliffs, N.J.: Prentice-Hall, 1969.

Holden, C. "Water commission: no more free rides for water users," *Science,* 180: 165, 167–68, 1973.

Keyfitz, N., and Fliefer, W. *Population: Facts and Methods of Demography,* San Francisco: Freeman, 1971.

Langer, W. L. "The black death," *Scientific American,* 210(2): 114–21, 1964.

Leaf, A. "Getting old," *Scientific American,* 229(3): 44–52, 1972.

Lessing, L. "Power from the earth's own heat," *Fortune,* 79(7): Part 2, 138–41, 1969.

Lovering, T. S. "New fuel mineral resources in the next century," *Texas Quart.,* 11: 127–47, 1968.

Markert, C. L. "Biological limits on population growth," *Bioscience,* 16: 859–62, 1966.

Maugh, T. H., II. "ERTS(II): a new way of viewing the earth," *Science,* 180: 171–73, 1973.

Pichat-Bourgeois, J., and Si-Ahmed, T. "Un taux d'accroissement nul pour les pays en voie de developpement en l'an 2000: reve ou realite?" *Population,* 5: 957–74, 1970.

Pirie, N. W. "Orthodox and unorthodox methods of meeting world food needs," *Scientific American,* 216(2): 27–35, 1967.

Population Reference Bureau. "The story of Mauritius from the dodo to the stork," *Population Bull.* 18: No. 5, 1962.

Postgate, J. "Bat's chance in hell," *New Scientist,* 58: 12–16, 1973.

Potts, M. "Make the possible available," *New Scientist,* 58: 84–85, 1973.

President's Science Advisory Committee Panel on the World Food Supply. *The World Food Problem,* 3 vols., Washington, D. C.: Government Printing Office, 1967.

Revelle, R. "Water," *Scientific American,* 209(3): 92–108, 1963.

Rhyther, J. H. "Photosynthesis and fish production in the sea," *Science,* 166: 72–76, 1969.

Schurr, S. H. "Energy," *Scientific American,* 209(3): 110–26, 1963.

Scrimshaw, N. S. "Food," *Scientific American,* 209(3): 72–80, 1963.

Simpson, D. "The dimensions of the world food crisis," *Bioscience,* 19: 24–29, 1969.

Stamp, J. D. *The Geography of Life and Death,* New York: Collins Fontana Press, 1964.

Stearn, E. W., and Stearn, A. E. *The Effect of Smallpox on the Destiny of the Amerindian,* Boston: Humphries, 1945.

United Nations. *World Population Situation,* Economic and Social Council, Population Commission Report (E/CN9/231).

Waterbolk, H. T. "Food production in prehistoric Europe," *Science,* **162:** 1093–1102, 1968.

Williams, R., et al. "Trophic value food versus maintenance diets," *Proc. Nat. Acad. Sci.,* **70:** 710–13, 1973.

Wrigley, E. A. *Population and History,* New York: McGraw-Hill, 1969.

Chapter

9

Population Control

As was described in Chapter 7, human populations have the potential to achieve a geometric, that is, an exponential, rate of increase. It has already been noted several times elsewhere that throughout our long history we have had long periods when our population size has been more or less stationary, albeit interspersed with brief episodes of exponential expansion. In this respect human populations are no different from the vast majority of plant, animal, and microbial populations inhabiting the earth. The fact that populations of *Homo sapiens* have not until the past hundred years ever long sustained the *continuous* realization of their potential rate of increase is because, as in all other natural populations, regulatory mechanisms of one sort or another have previously prevented its other than *intermittent* occurrence.

Reference has already been made to the nature of some of these regulatory mechanisms, and the reasons human populations have recently tended to evade such controls were discussed in the last chapter. Stated very briefly, this ability of our species to escape from environmental limitations on natural increase is attributable especially to our unique characteristic that permits adaptation to different ecological conditions by *cultural* rather than *physical* evolution.

Early Methods of Population Control

The reason why exponential rates of population growth have previously been of only intermittent occurrence and relatively short duration has been considered by various workers.

Some ecologists, notably V. C. Wynne-Edwards, have developed the idea first presented by A. M. Carr-Saunders that even the earliest human communities possessed some form of non-malthusian population control, either voluntarily or involuntarily practiced. Anthropologists also have considered this possibility, and Table 9-1 summarizes one such review of the extent of control through abortion and infanticide exercised by selected early groups.

In a previous chapter it was noted that hunter-gatherer societies in many instances must have been regulated by density-dependent factors, especially by the *seasonal* limitations of food and water supply imposed on such nomadic groups. It also has been observed that the Early Agricultural groups representing the next successional category of societal evolution were occasionally regulated by density-dependent feedback mechanisms. These mechanisms and the responses they triggered represented a cultural response to selection for stability in the relations between the human population and the environment. The reasons such feedback mechanisms and responses were apparently lost during the successional progress to Advanced Agricultural societies have been discussed already, and it has been considered that in some populations

TABLE 9-1 *Population Control Measures Practiced by the Earliest Categories of Human Societies*

Numerous observances and taboos in such groups may be responses to feedback from density-dependent situations. Evolving cultural techniques permit later categories of societies to override these situations by achieving a greater productivity, inevitably to the detriment of ecosystem stability and diversity.

Category	Intercourse avoidance	Abortion	Infanticide
1. HUNTER-GATHERER GROUPS			
Bushmen (southern Africa)	No	No	Yes
Eskimos	No	Yes	Yes
Australian Aborigines	No	Yes	Yes
Tasmanians	No	Yes	Yes
Amerindians	Yes	Yes	Yes
2. EARLY AGRICULTURAL GROUPS			
African	Yes	Yes	Yes
Amerindians	Yes	Yes	Yes
Oceania	Yes	Yes	Yes

of the earliest successional societies in less favorable environments, either density-dependent malthusian factors or the voluntary practices of abortion and infanticide continued to regulate population numbers.

The writings of Malthus epitomize the extent of knowledge of human population fluctuation and control at the beginning of this century. Since that time the science of *historical demography* has become established. This provides more accurate and factual evidence as to the nature of control practices, some of which have persisted into modern industrial societies.

Historical Demography

Absolute figures concerning the demography of populations were not generally available until periodic census-taking began in the western world during the nineteenth and early twentieth centuries. Before this, information had to be gleaned from such sources as parish registers and tombstones. This precensus demographic research constitutes the new science of historical demography, whose aims are reviewed very briefly by Roger Revelle in introducing a volume of the periodical *Daedalus* devoted entirely to historical population studies.

In this introduction Revelle remarks on historical variations in fertility in various regions of Europe at different times. He notes that even before modern population control methods become extensively utilized in this area during the present century, social customs consciously or unconsciously produced fertility control. Notably these were postponement of the time of marriage and long-term nursing of children as opposed to early weaning. Revelle also observes that the marriage pattern of the time included within large households a considerable reservoir of unmarried individuals, principally servants and unmarried relatives. These categories of single persons did not necessarily remain either continent or infertile. It was not unusual at this time for unwanted newborn infants simply to be abandoned. Many died, others were rescued by orphanages, then in their most active period. Nevertheless, the number of children surviving from unsanctified unions was small relative to those produced in wedlock. The European marriage pattern of this period can thus be regarded as a fertility control device that could respond quickly to a change in economic circumstances. A disaster such as plague could wipe out as much as one half of the population in a large city. One immediate response of the population to the need for replacing such losses was an increase in the marriage rate among the survivors of this considerable reservoir of previously single individuals.

The demographic patterns of western Europe were studied in some detail by J. J. Spengler, who describes the unique marriage pattern that appeared in that area during the fifteenth century and seems to have become universal by the seventeenth century. In western Europe during this period marriage took place at a much later age than that prevailing elsewhere, later, even than in eastern Europe (Figure 9-1). Associated with the family as Revelle noted, were many unmarried relatives and servants, and the prevalence of various religious and other orders requiring celibacy. This marriage pattern resulted in a comparatively low fertility rate even before the spread of effective birth control measures. Spengler estimates that the gross reproduction rate, the average number of daughters born to a woman, very rarely exceeded 3.8. This low value for the gross reproduction rate resulted in a high level of per capita productivity and capital accumulation, and to some extent prevented the overexploitation of land and resources. He estimates that the resulting rises in average incomes facilitated the launching of the industrial revolution in western Europe.

Spengler points out that the age structure of modern Advanced Agricultural societies is far less favorable to productivity than that existing in European countries at a comparable stage of development. Quoting 1965

FIGURE 9-1. *Marriage age and fertility* in western Europe. The lateness of marriage that characterized western Europe in the eighteenth and nineteenth centuries was associated with a comparatively low fertility. The introduction into this area of more effective contraceptive methods permitted increasingly earlier marriage (upper graph) with simultaneously a decreasing fertility (lower graph).

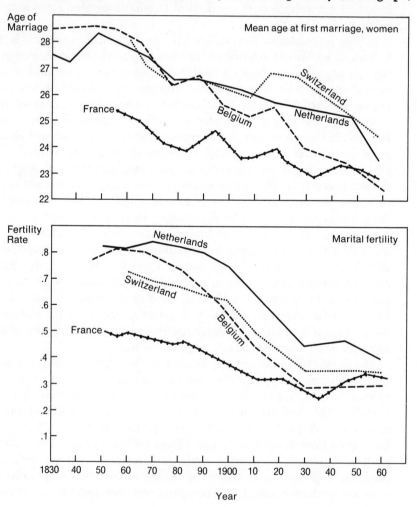

figures, he notes that the fraction of the population represented by persons aged 15–64 years (now used to calculate the dependency load) was as follows: east Asia (excepting Japan), 51.7 per cent; south Asia, 54.7 per cent; Africa, 54.2 per cent, ranging from 51.7 per cent in West Africa to 56.1 per cent in East Africa; Central America, 56 per cent; Philippines, 53 per cent. The high dependency loads are associated with annual percentage natural increase rates ranging between 2.5 and 3.5. These figures represent life expectancy at birth of less than 40 years in Africa, 40–50 years in Asia, and 50–55 in Latin America. Such high rates of population growth, according to Spengler, will absorb most of the additional capital resources that are generated and will prevent a parallel increase in per capita income.

D. M. Heer has discussed fertility reduction in the population of the western world and illustrated this with figures for the white population of the United States from 1869 to 1965 (Figure 7-13). He observes that the agricultural societies of eighteenth-century America, which had an unlimited supply of land, viewed children as productive assets that could be used in working additional parcels of agricultural fields. Compulsory education laws and a social aversion to industrial child labor prevented a similar attitude from being generally taken toward factory work during the industrial revolution.

He also notes, as we have observed already, that from 1790 through 1960 the urban population of the United States increased from 5 to 70 per cent of the total population; without these changing attitudes toward labor, the industrial revolution would have had an even greater economic impact. Heer summarizes the factors associated with the industrial revolution that have resulted in a general downward trend in fertility rates, despite some minor reversals of a temporary nature. These factors are institution of the social security system, suppression of child labor, compulsory education for children, changes in relative cost of living space, and increasing cost of child care, all economic features of a type discussed in the previous chapter under the heading of the "Madison Avenue effect."

It is apparent from this rapid review of early practices and factors associated with population control that there is a marked tendency in human populations, during historical times at least, for the *need* for population control to produce a *method* of control, voluntary or involuntary and however effective. Fortunately, individual methods of control have become both effective and extensively available, at least in all industrialized countries.

Pregnancy Avoidance

All modern methods of population control in human societies basically involve either the avoidance of reproduction in the individual nubile or fecund female, or the elimination of offspring at some stage. Leaving aside for the moment the latter aspect, many methods of pregnancy avoidance of varying degrees of efficiency have been known and practiced in human societies for a considerable period of time. Practices of earlier times now seem as amusing to us as they must have proved ineffective in our forebears.

These range from pessaries of crocodile dung, honeyed tampons, potions of mule kidneys, to various superstitious acts that at least had the merit of being more sanitary.

It is easy to become involved in semantic argument as to precisely which of the many procedures since adopted can be regarded as *contraceptive,* preventing conception, and which are *abortifacient,* producing abortion. The issue for some is still a matter of conscience, religion, or philosophy. For the purposes of this work all such methods are categorized as *pregnancy control,* for pregnancy still generally takes at least days to diagnose. After this diagnosis, the artificial or natural termination of pregnancy short of parturition can be termed *abortion.* It is equally permissible to call pregnancy avoidance "fertility control," but to some this may suggest simply a population statistic, which indeed it is, rather than the personal choice that "pregnancy avoidance" connotes.

Coitus Interruptus

Probably the oldest of the partially effective methods of pregnancy control is coitus interruptus, or external deposition of seminal fluid after initial intromission. Dating at least to biblical times, it is still in extensive use. In western Europe, for example, it seems to have been one of the principal techniques used to reduce fertility in the early 1930s, when fertility rates as low as 1.5 were recorded (Czechoslovakia).

Rhythm Method

Depending on a "safe period" and the avoidance of coitus when ovulation is about to take place, the rhythm method may also have been practiced for a considerable time, and at least until recently was extensively recommended among some groups. It demands an accurate identification of the time of ovulation, usually the fourteenth day of the menstrual cycle, and abstinence for an appropriate time before and after this, usually the eleventh to eighteenth days. The rhythm method was the first technique involving a basic scientific understanding of ovulation. Requiring a sophisticated approach, it is unreliable; its most frequent scientific use now is in reverse. That is, subfertile couples wishing to reproduce are advised to concentrate on achieving coitus during the identified *nonsafe* period.

Condoms

The technique of using condoms is believed to have been devised in the sixteenth century by an Italian anatomist, Fallopius. The opening up of the New World had added an additional hazard to sexual intercourse in the form of syphilis, and Fallopius apparently recommended his linen sheath as a protection against venereal disease rather than pregnancy.

The invention of rubber processing by vulcanization in the mid-nineteenth century provided an eminently more satisfactory material for the manufacture of such sheaths, or condoms as they are technically described. Although no nation was at first readily willing to admit their extensive use by its own citizens (the French referred to *capots anglais* and the English to "French letters"), the use of condoms by the beginning of this century had become extensive. Although there is no documentary evidence, it nevertheless seems reasonable to suppose that the appropriate use of condoms greatly contributed to the unprecedented reduction in the birth rate of western nations in the 1930s. It is estimated that presently 700–800 million condoms are manufactured each year in the United States. A few simple calculations show that their use is presently of greater significance in relation to the birth rate than use of oral contraceptives.

Spermatocides and Diaphragms

Spermatocides and diaphragms, used separately or in conjunction with each other, also became popular in the twentieth century. The vaginal diaphragm had the advantage of a greater aesthetic acceptability, the spermatocidal jelly or foam of being obtainable without medical consultation. Spermatocides are sometimes also used *after* coitus as douches.

Intrauterine Devices

All the techniques mentioned so far, and several other minor ones, were based on preventing sperm from reaching an ovum by imposing a physical or chemical barrier, or both. The intrauterine device (IUD) does not prevent conception in this manner, but in some way still incompletely understood prevents the fertilized egg from developing into an implanted embryo in the uterus.

Various types and shapes of plastic devices that can be inserted into the uterus are now available (Figure 9-2), and they are provided with an indicator, in the form of an attached thread, of whether "fallout" has occurred. One of the early difficulties experienced with use of IUDs was detecting whether the device was still where it should be. Besides this expulsion problem, two other difficulties have been experienced with IUDs, pregnancy and pain and bleeding necessitating surgical removal. As seen from Table 9-2, these three principal side-effects vary considerably according to the structural design of the device. The most successful device to date, the Dalkon Shield, has short projections along its sides that increase the prospects of retention but do not cause it to become so deeply embedded in uterine tissues that it becomes partially ineffective or causes medical problems. As the contraceptive effectiveness of an IUD is known to be proportional to its surface area in contact with the lining of the uterus, the Dalkon Shield is also the most effective device in protecting against pregnancy.

The current understanding of this action is that an IUD stimulates a "foreign body response" in the uterus, so that larger devices cover the intrauterine area more completely. However, they have a higher frequency of rejection unless designed to counteract this tendency. The "foreign body response" in the uterus is basically similar in its effect on preventing implantation of a fertilized ovum to the rejection of surgically implanted organs such as hearts or kidneys from another individual. This effect

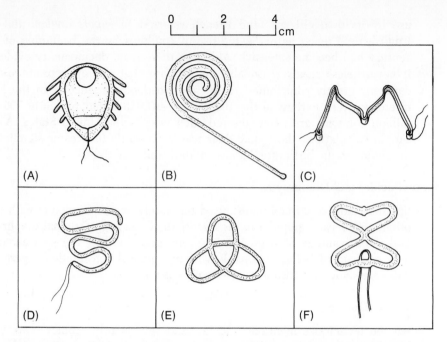

FIGURE 9-2. *Forms of intrauterine devices.* As it is now believed that the "foreign body response" plays an important role in the contraceptive effect of IUDs, their surface areas has assumed a greater significance. Although materials used would likewise be expected to have an important effect, the commoner models continue to be made of plastic. The latest models, such as the Dalkon Shield, are provided with short projections that increase retention and lessen the risk of embedding, thus reducing the dangers of loss of effectiveness or the creation of a need for medical removal. The types illustrated are (*A*) Dalkon Shield, (*B*) Gynekoil, (*C*) M Device, (*D*) Lippes Loop, (*E*) Shamrock, (*F*) Birnberg Bow. Statistics on their effectiveness are provided in Table 9-2.

TABLE 9-2 *Performance of Intrauterine Devices*

The comparative figures provided in Table 9-3 indicate that the IUD is among the three most effective readily reversible methods of pregnancy control generally available. However, as shown from the statistics in this table, which are extracted from a recent presentation by E. L. Parr, there is some variation among types of IUD. Successful designs are about three times as effective as the least successful and induce fewer gynecological complications.

	Events per 100 Woman Years of Use		
	Pregnancy	*Expulsion*	*Medical removal*
Dalkon Shield	1.1	2.3	2.0
Gynekoil	1.3	25.8	22.1
M Device	1.7	2.7	13.5
Lippes Loop	2.8	10.4	14.0
Shamrock	4.0	13.0	17.0
Birnberg Bow	4.7	2.6	14.3

may possibly be increased when the IUD is made of a metal such as copper instead of plastic. Experiments on this metal effect are in progress.

"The Pill"

Like the IUD, oral contraceptives, which first became available in the United States in an FDA-approved commercial form in 1961, employ a different method of pregnancy control than the earlier techniques referred to here. They function by regulating the balance of reproductive hormones so as to prevent ovulation.

Some half million eggs are present initially in the ovaries of human females. At ovulation each month in a mature woman one egg usually is released. The mechanism that ensures this release and suppresses other eggs is modified by chemicals in the pill so that *all* egg release is inhibited. The various stages in this process, the chemical substances involved, and the endocrine and glandular reactions have been described succinctly by R. W. Kistner and others. Table 9-3 shows the comparative effectiveness of the pill as compared with the other methods mentioned here in preventing pregnancy.

The first pill to be mass-tested on human females, marketed as Enovid, contained the hormones norethynodrel and mestranol. Testing has been in progress since 1956 in Puerto Rico. There have been no reports from this source of serious consequences of its use during the considerable interval that has elapsed since the testing began. However, elsewhere there have been confirmed instances of blood clotting associated with the taking of some forms of oral contraceptive.

TABLE 9-3   *Comparison of Various Methods of Pregnancy Avoidance*

The pregnancy rate is a percentage figure expressing the number of women out of 100 who became pregnant in 1 year using the method stated. The more successful tests with IUDs produce a rate of 1.5–3.0. The relatively high rate with sequential pills here is explained by "patient failure"—forgetting to take the pills for 1 or 2 days. These figures emphasize the need for a "backup" abortion facility. They also stress that the rhythm method is partially effective as well as free of cost and direct psychological objections.

Method	Pregnancy rate
None	61
Douche	31
Rhythm	21
Jelly only	20
Coitus interruptus	18
Condom	14
Diaphragm	12
IUD	2.6
Sequential pills	2.0
Combination pills	0.1

Enovid is no longer the sole prescription for purposes of pregnancy control. It has been superseded by other brands, of which there are presently 30. This multiplicity of brands containing varying amounts of the hormones progestin and estrogen permits a physician to prescribe a pill that will avoid any side-effects such as sore breasts, weight increase, bleeding, nausea, and diminished awareness that might arise in individual instances.

In 1969 it was estimated some 8.5 million women in the United States were using some brand of pill. Since then, this method has gained in popularity, being now generally favored over the IUD. Testing of Enovid (about which there is most information because it is the oldest brand marketed) shows that in 14,840 women and 116,000 cycles, only three pregnancies resulted, a pregnancy rate of 0.028 per cent.

Vasectomy and Tubal Ligation

Pregnancy avoidance by surgical operation has long been practiced by various human populations. Whereas surgical procedures for the removal of the testes or ovaries have during this century virtually been restricted to medically recommended operations, castration of males was at one time a common practice to produce individuals who could not interfere sexually with groups of females entrusted to their care. The word *eunuch* indeed has not yet disappeared from the modern English vocabulary.

Vasectomy and tubal ligation are not to be confused with such surgical operations for removal of the sex glands. They have no biological effect on sexual activity, merely preventing the release of sperm in the one case or of ova for fertilization in the other.

Tubal ligation of females is achieved by tying off or severing the two fallopian tubes. Ova released from the ovaries cannot then pass down the tubes to be fertilized. The operation is potentially reversible, but involves surgery. Unfortunately the operation of tubal ligation is not yet fully perfected; too tight tying off may eventually sever the tubes, too light tying may permit them to open up again.

Vasectomy of males is a much simpler operation. It can be performed under a local anesthetic in a physician's surgery, in an operation taking only a few minutes. The two vas deferens tubes are severed or tied off, preventing the release of sperm into the ejaculate, the liquid portion of which is secreted by the prostate gland, which of course remains intact. Because the vascular supply to the testes also remains intact, there is no interference either with the flow of hormones between the testes or with the vascular system that distributes them about the body. A recent development of the vasectomy operation is to form a plug in the vas deferens with silicone rubber. This may increase the possibility of reversal of the operation, which is not high when the vas deferens tubes are severed.

The number of vasectomies performed in the United States is reported to have risen from over 200,000 in 1969 to about 650,000 in 1972. If this rate of use of the method were to continue, and there were no overlap of contraceptive procedures, vasectomy could reach an equal significance in fertility control with that of the pill before the end of this decade. It could have an even greater significance in population control in underdeveloped nations if two difficulties could be overcome. These difficulties are, first, persuading a mostly illiterate population that there are no harmful sexual consequences of the operation and second, obtaining sufficient medical assistance to handle many millions of adult males. In India, for example, these and other difficulties seem momentarily at least to have caused an abandonment of a government-sponsored vasectomy program. Currently vasectomy is the most favored method of pregnancy avoidance in the U.S. among couples in which the woman is aged 30–40 years.

In the U.S., fears of the irreversibility of vasectomy have promoted the foundation of commercial frozen sperm banks, presumably by males who wish to take out insurance against a change of mind. Knowledge of the effectiveness of present methods of storing human sperm is not certain beyond a period of about three years.

Other Methods

Pills to control male fertility so far have not received much research attention, perhaps because of cultural factors, but doubtless also because of the simplicity of the vasectomy operation. Many types of substances that interrupt the female reproductive processes are presently under intensive investigation. One of the most promising new chemicals is the steroid *progestin*. Applied in low dosages this avoids many of the deleterious side-effects of present contraceptive pills. One of the new ways of employing low-dosage progestin is as a time-capsule, implanted under the skin and lasting up to 25 or 30 years. Another is as a cervical ring, needing monthly replacement.

Many such new techniques are presently undergoing laboratory, clinical, or field testing and will be released for general use, if successful, before the end of this present millennium. Being cheap, simple, and effective, they will obviously be of greatest service to population control in underdeveloped countries. For more sophisticated societies a "morning-after" technique is now available, comprising large doses of estrogens. This drastic treatment is said, however, usually to produce unpleasant temporary symptoms.

Control at the Population Level

It is apparent from the review presented here and in the previous chapter that attempts at population control presently are largely but not entirely uncoordinated and essentially on an individual rather than a population basis. They are inconsistent and still place considerable emphasis on the freedom of choice. If any really effective methods of population control are to be established in time to prevent the completely disastrous onset of conditions described in the previous two chapters, it is clear that a more coordinated and consistent approach to the problem must be made on a world-wide as well as a regional basis.

Various organizations have been working on such schemes for a number of years; a now classical review by B. Berelson presents a detailed survey of the various approaches currently considered possible.

Methods of Population Control

The survey presented here of the various methods of population control applicable to our present stage of development largely follows the outline of Berelson's paper. He begins his account by emphasizing the four suppositions rephrased below:

1. The population problem is among the most pressing world problems requiring immediate attention
2. This problem is most urgent where population increase is hampering required social and economic development, that is, in developing countries
3. Solution to the problem is vital; any postponement of action only makes an ultimate solution more difficult to achieve
4. Whereas there is general agreement that population increase should be reduced, there is no consensus as to how this should be achieved

As Berelson remarks, family-planning programs that have been operating in most advanced societies for some years are for many reasons not applicable to the urgencies of the present population situation. The primary reason for this seems to be that these programs are concerned with the "spacing" of the interval between children rather than with reducing the number of children. A recent report by T. Eisner and some colleagues suggests this is true even for the most sophisticated individuals. Figure 9-3, showing statistics regarding American women on the one hand and Taiwanese on the other, indicates that many women in the population at large still desire to have an average of four children. As programs to reduce births below this figure must be developed, the various procedures which are available can, following Berelson, be categorized on the following basis:

1. Intensification of Voluntary Fertility Control

a. EXTENSION AND LEGALIZATION OF FREE ABORTION SERVICES TO THE COMMUNITY AS A WHOLE. Statistics from countries where abortion practices have not yet been legalized usually reveal a vast extent of illegal abortion. In a country such as Japan, where abortion has been legalized and made readily available, the results in population control have been demonstrably effective. During the period of very rapid decline in natural increase in Japan following World War II, the annual number of abortions performed was 1.5 million. Failure to terminate these pregnancies would almost have doubled the Japanese birth rate. In the United States, current unofficial estimates put the number of illegal abortions performed at an annual figure of over a million. This is more than one quarter of the actual annual number

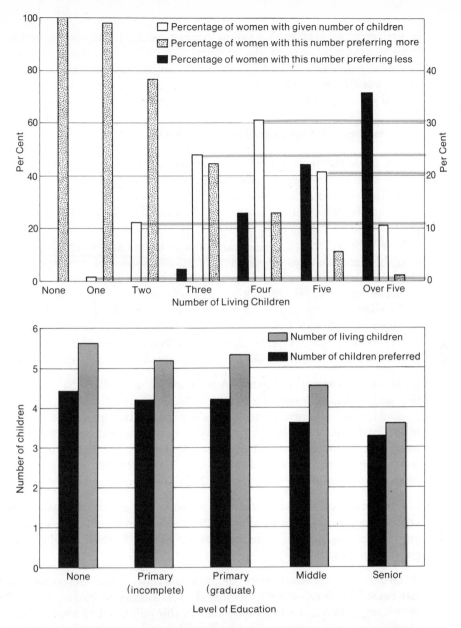

Figure 9-3. *Realized as compared with desired fertility.* Upper: An area of Taiwan, representing an underdeveloped nation. Lower: The United States. In both cases it seems that the majority of women presently prefer a family of four children, and all want some children; very few women sampled in these two societies like having over five children.

of live births, and suggests that freely available legalized abortion might assist considerably in stabilizing American population growth. Now that abortion has been legalized in New York State, it is expected 100,000 operations will be carried out there annually. Table 9-3 indicates the advisability of having a backup system of legalized abortion.

b. PROVISION OF MATERNAL CARE IN VILLAGE AREAS. In developing countries three quarters or more of the population is still concentrated in villages. Maternal care there should be placed on an institutionalized basis, which makes possible the provision of both education and services in population control simultaneously with the provision of prenatal care.

2. Introduction of Compulsory Fertility Control

a. USE OF A FERTILITY CONTROL AGENT ON A REGIONAL BASIS. Government action in this direction is advocated by a number of writers, including P. R. Ehrlich. The introduction of such group treatment would be designed to reduce fertility in the area treated to from 5 to 75 per cent below the present rate.

It is believed that a considerable range of suitable substances will be available in from 5 to 15 years as a result of present research. Such a chemosterilant substance could readily be introduced into the water supply of urban areas, where often it is especially needed. A parallel suggestion is the addition of similar temporary sterilants to staple foods such as bread or sugar, which might be more practical alternatives in some instances. The Paddocks suggest that the United Nations consider instituting a "no birth year," which would best be obtained with such mass sterilants.

b. ISSUING "CHILDREN LICENSES." A number of authorities have suggested that each woman be issued a license to have whatever number of children is established as necessary to maintain a population growth rate of zero in her particular community. In many areas this would be arranged so as to achieve an average of 2.2 children per female, but the actual number would vary from 7 to 0.7 (Table 7-10). These children certificates could be exchanged as gifts or sold, and would be completely negotiable.

c. TEMPORARY STERILIZATION OF GIRLS BY MEANS OF REVERSIBLE TIME-CAPSULE CONTRACEPTIVES. The reversibility of this control device should be permitted only when the popular vote decides to permit further development of population growth.

d. COMPULSORY STERILIZATION OF MEN HAVING THREE OR MORE LIVING CHILDREN. This is extensively advocated as an alternative to (c). A corollary of this suggestion is not to sterilize men at this point, but to require induced abortion for all pregnancies resulting beyond a particular number of children.

The difficulties of enforcing such a method are quite apparent.

3. Intensified Educational Campaigns

a. DISSEMINATION OF EDUCATIONAL MATERIALS ON POPULATION CONTROL AT ALL PRIMARY AND SECONDARY LEVELS. Opposition in various regions of the United States to sex education in schools illustrates the kind of difficulty that may be anticipated in a universal implementation of this suggestion, even in sophisticated societies. There appears some reason for the belief that

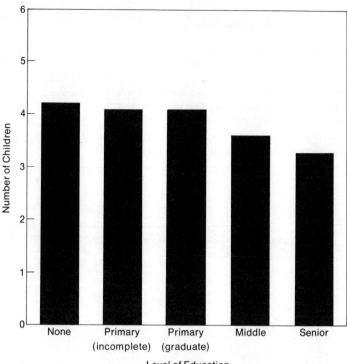

Figure 9-4. *Education and family size.* The effect of educational experience on actual fertility among women 35–39 years old. (The use of contraception most extensively by the most educated is correlated with an approximation on average to desired family size in this group.) Progressively less educated women wanted more children, but progressively had more even then than they wished, correlated with a decreasing contraceptive practice.

education in general, rather than mere instruction in birth control techniques, exercises a strong controlling influence on population (Figure 9-4 and Table 9-4).

b. USE OF SATELLITES FOR EXTENSIVE TELEVISION PROGRAMS DISSEMINATING INFORMATION ON POPULATION CONTROL, FAMILY PLANNING, AND ATTITUDES TOWARD POPULATION GROWTH. This is a very practical temporary measure to cover the situation where local technical facilities may be still in an early stage of development. It would need to be supplemented with a scheme for communal TV sets and could encounter the same objections as in (a) but on an international rather than a regional basis.

4. Incentive Programs

a. PAYMENTS MADE DIRECTLY TO MARRIED COUPLES WHO REFRAIN FROM PRODUCING CHILDREN DURING PARTICULAR TIME INTERVALS. These would include payment for the acceptance of sterilization or for the effective employment of contraception.

TABLE 9-4 *Education and Fertility*

The total average number of children born to individual U.S. white women expressed in terms of their educational exposure: figures from the 1940 census. College graduates in 1940 were beginning to approach the fertility value necessary to achieve zero population growth; all lesser educational levels progressively exceeded this figure. The same analysis carried out on this contemporary generation might well indicate that college graduate fertility has reached the essential low figure, but would certainly reveal that in other educational classes fertility was still far too high.

Schooling completed (years)	Children per married woman
None	4.97
Grade school	
1–4	4.54
5–6	3.97
7–8	3.04
High school	
1–3	2.61
4	2.03
College	
1–3	2.07
4 or more	1.83

b. BONUSES FOR SPACING CHILDREN OR FOR ACHIEVING "NONPREGNANCY." Rewards might take the form of savings certificates for each 12-month period in which no child is born, a prize for each 5 years of childless marriage, or special lottery tickets for the childless, and so forth. If such incentive schemes designed to *increase* fertility have been successful, the reverse should apply. They also should work successfully when negative, that is, when they are the reward for nonreproduction.

5. Tax and Welfare Benefits and Penalties

Such measures would represent a reversal of the present social service principles, which have largely developed as a result of the previous population decline in the 1930s. These favor the production of children in that they give cash payments in support of children, tax relief with respect to child dependents, or both. Rather than encouraging childbearing in this way, tax and welfare benefits would be reversed so as to discourage it by the following procedures:

a. WITHDRAWAL OF MATERNITY BENEFITS AFTER THE BIRTH OF A GIVEN NUMBER OF CHILDREN.

b. IMPOSITION OF A TAX ON BIRTHS AFTER A DETERMINED NUMBER. Neither of these measures would be particularly helpful in industrial societies, for among them it is only the poorest element that needs to reduce its natural increase (see later in this chapter), and it is too poor to pay any further tax or sustain withdrawal of any child benefits. The same applies to (c) below.

c. LIMITATION OF GOVERNMENTALLY PROVIDED MEDICAL TREATMENT, HOUSING AND FINANCIAL AID FOR SCHOOLING TO FAMILIES WITH FEWER THAN A SPECIFIC NUMBER OF CHILDREN.

d. COMPLETE REVERSAL OF TAX BENEFITS. Instead of strongly penalizing single persons as at present, tax measures should favor the unmarried and the parents of fewer rather than more children. This is an entirely practical measure that can immediately be implemented in a society. It will considerably assist in the creation of the "uncles" and "aunts" for whom Garrett Hardin pleads.

e. PROVISION BY THE STATE OF A PARTICULAR ALLOCATION OF FREE SCHOOLING TO EACH FAMILY. This free schooling could then be distributed over a family of any size as desired, giving, for example, a university education for two or a grade school education for eight. Such a measure would work only in an underdeveloped country where educational facilities were still at a premium.

f. PROVISION OF PENSION FOR IMPOVERISHED PARENTS WHO HAVE PRODUCED FEWER THAN A GIVEN NUMBER OF CHILDREN. This would be a form of social security for their retirement years for low-income couples who decided not to have children, a measure applicable to any economically structured society.

6. Shifts in Social-Economic Institutions

This involves fundamental changes in institutional arrangements with the intention of lowering fertility. It would include the following:

a. BY LEGISLATION OR THROUGH IMPOSITION OF EFFECTIVE FEES, RAISING THE MINIMUM AGE FOR MARRIAGE. This might be achieved on a voluntary basis by payment of marriage benefits only to brides and grooms over a particular age limit, or through issuing governmental loans for wedding ceremonies when the bride is over a particular age.

This is a measure with potentially disturbing social overtones, for the creation of a large group of unattached but very vigorous females could have disruptive effects on previously stable social institutions. Perhaps the measure should only be applied where it can be combined with a longer period of education for the women so released from early marriage.

b. ACTION TO REQUIRE PARTICIPATION OF WOMEN IN THE LABOR MARKET OR TO PROMOTE THEIR VOLUNTARY PARTICIPATION. This might provide other roles for women as an alternate or supplement to marriage. In effect this is what the Madison Avenue effect referred to in earlier chapters achieves. Married couples prefer the higher standard of living which the dual income provides to the self-elected hardships of family costs.

c. FUNDAMENTAL RESTRUCTURING OF SOCIETY TO MINIMIZE THE SOCIAL POSITION OF THE FAMILY IN COMMUNITY LIFE. This proposition needs rephrasing perhaps, e.g., "so as to emphasize alternatives without denigrating the central social importance of the family."

d. INSTITUTION OF TWO TYPES OF MARRIAGE. One kind would be designed to be of a more temporary nature and not to result in the production of children. The other would be licensed to be more permanent and to be the legitimate vehicle for the production of children. In effect this is almost what we have now in some countries in the "common law marriage," but it has to be disassociated from the procreation of children.

e. SOCIAL IMPROVEMENTS OF AN ECONOMIC KIND. These could be achieved by the provision of new productive industries to encourage the employment of women and to raise their status in society, and in general to refocus attention on social contributions other than marriage. Perhaps this also could be better phrased. There is no need to destroy pair bonding as such, but merely to demonstrate that children are not an essential feature of this association.

f. STRENUOUS ATTEMPTS TO LOWER INFANT AND CHILD MORTALITY RATES. These are considered necessary in order to obviate the feeling of the necessity for large families in order to ensure the survival of some descendants, a point well taken for many underdeveloped societies with recent histories of high child mortalities.

7. Political Channels and Organizations

a. INSISTENCE ON POPULATION CONTROL PRIOR TO DELIVERY OF FAMINE RELIEF SUPPLIES. This might be regarded as political blackmail, for it requires exertion of political pressure on governments before they can receive foreign aid favors. This, at any rate, is not so coldly inhumane as the Paddock brothers' "triage" system, which may well have to be implemented within the next decade.

b. MASSIVE REORGANIZATION OF NATIONAL AND INTERNATIONAL AGENCIES TO DEAL MORE EFFECTIVELY WITH POPULATION PROBLEMS. In the United States this would require institution of a federal department of population and environment, empowered to legislate for the establishment of a calculated and desired population size. In underdeveloped areas it supposes the creation of ministries for population control and, on an international scale, the organization of a specialized agency parallel to but larger than the World Health Organization to effectively promote family limitation techniques throughout the world, particularly in the underdeveloped areas.

c. PROMOTION OF THE CONCEPT OF ZERO POPULATION GROWTH. This must be considered as the aim of all human societies. Steps to achieve acceptance of this idea as of now could be taken in order to provide the necessary urge to accept even further lowered fertility measures in the present critical situation. Although private enterprise and initiative already have established such organizations in this country and elsewhere, governmental attitudes have not yet been modified to any significant extent (see the end of this chapter).

8. Augmented Research Efforts

a. MORE SOCIOLOGICAL RESEARCH ON ACCEPTANCE OF NECESSARY FERTILITY OBJECTIVES.

b. RESEARCH ON PRACTICAL METHODS OF SEX DETERMINATION.

c. INCREASED RESEARCH EVEN BEYOND THAT PRESENTLY IN HAND. Still mainly of a private enterprise nature, these efforts are directed toward the improvement of contraceptive technology rather than such problems as social acceptance of zero population growth.

Implementation of Population Control Measures

For reasons that already have been mentioned, and notably the difficulty in obtaining accurate census figures, it is exceedingly difficult to assess the effect of the introduction of particular population control measures, even when details of the latter are available. D. Kirk has nevertheless attempted a review of the prospects for reducing birth rates in underdeveloped nations. He concluded in 1967 that within a decade, birth rates of 20 to 25 would be attained in the most progressive parts of east and southeast Asia and that the same levels may be reached in India and mainland China in two decades. He prognosticated that by this time the whole underdeveloped world will have begun to move toward this figure, with Moslem communities bringing up the rear. Kirk also foresees an inevitable doubling, for example in the populations of most Latin American countries. As can be seen from the latest available figures, which are entered in Table 7-5, these prognostications appear to have been fairly accurate.

Kirk's optimistic report might be more encouraging were it not for several other considerations. First, Table 7-11 shows that with a death rate in most nations already fallen below 15, birth rates of 20 to 25 are not going to achieve anything approaching zero population growth. Second, and perhaps more important, powerful interests always have resisted, or are beginning again to resist, the introduction of population measures. In Japan, for instance, industrial support for the fertility control program has reportedly been withdrawn, because of the mounting expense of Japanese labor compared with that of surrounding countries.

Power Politics and Population Control

The influences of sectional and national aspirations on population growth have not always been apparent, but they have never ceased to be extremely powerful. The most openly debated example has been that of the effect of the Roman Catholic Church, which long has been accused of seeking to maintain the numbers of its national and international congregations by promoting natural increase among Catholics. It is true that many countries with high birth rates are Catholic, but as shown in Table 7-5 Moslem countries can also have high birth rates, and so can Buddhist or animistic societies.

Less obvious perhaps are minority group situations in many countries. When, as in the United States, employment of minority group personnel is based on a system of counting the number and color of noses, it would be unreasonable to expect that factional leaders would wholeheartedly support programs whose primary object if realized would remove the very source of their power. As shown in Figure 9-7B minority groups can have much higher birth rates than those prevailing in the majority populations, but Figure 9-8A indicates that this is not invariably the case.

The same, what can be termed "politics of natural increase," is believed to apply on a national scale. The figures in Table 9-5 indicate the relationship which is often believed to hold between national populations and national power. France and Italy decline in the pecking order of nations as their proportional populations decline, while the United States, United Kingdom, and Germany improve their relative positions.

The lack of direct control and direct causal relationships in such interactions is stressed by Kingsley Davis, who emphasizes that one of the demographic factors *weakening* a nation's power is a high birth rate. Before World War II, both Mussolini and Hitler were endeavoring to raise their countries' birth rates. Had they fully succeeded, their population structures would have been loaded with noncontributing members of society; that is, the dependency load would have been high. Women and material resources would have had to be withdrawn from the war effort

TABLE 9-5 *Correlation Between National Population Size and National Standing*

At the beginning of the nineteenth century France was a leading world power, perhaps *the* world power. She joined with Russia to resist Germany, and largely ignored Italy, the United Kingdom, and the United States. A century later France, while still allied to Russia, had to form closer ties with the United Kingdom and Italy in an endeavor, despite her weakening condition, to contain a more powerful Germany. This last country hoped, in the event of war, to keep the now powerful but isolated United States neutral.

Country	Population in 1800 (millions)	Country	Population in 1900 (millions)
Russia	37	Russia	140
France	27	United States	92
Germany	25	Germany	65
Italy	18	United Kingdom	45
United Kingdom	16	France	40
United States	5	Italy	35

to care for these dependents, and the chances of winning World War II would have been still further reduced.

The Basic Problem of Population Control

Of all the characteristics we have inherited from our ancestral populations, whether by biological or cultural procedures, the demographic pattern can be regarded as having become the most embarassing. It will be apparent from ideas presented in these last three chapters that during 4 million years or so as hunter-gatherer populations we and our ancestors adapted by natural selection to a pattern that provided a stable population size and structure with a crude birth rate in the region of 35 to 40 and an average fertility level of 5 to 6 births per female.

As various technological advances during our long cultural evolution proceeded to reduce the mortality rate, so in some circumstances were they associated with a reduction of the birth rate. Such societies remained stable and in a steady state within their ecosystems. Unfortunately the great majority of evolving populations were able to adopt the technological improvements that provided for lower mortality rates without parallel adjustments in birth rates. Many even succeeded in simultaneously increasing their birth rates. Whether this failure suitably to lower birth rates was incidental or deliberate, the effect was always the same, a population explosion. Many societies are already having to pay an appalling price for this simple demographic omission.

It would seem that we ought to be able to resolve this most basic problem in population control by combining the findings from three different approaches. First, we have both paleoanthropological and historical records of family size: a critical analysis of the available data should permit us to relate rates of natural increase both in time and in space to the several factors that control this rate, and to determine whether these factors are innate in our own genomes, or are environmentally determined features of the ecosystems we occupy. Second, as sociologists we can observe present populations, and their individual aspirations in terms of children and posterity in relation to their more immediate economic aims. Third, we can attempt biological prognostications of the future in terms of actual realized fertility, desired family size, and genetically or culturally determined behavioral patterns.

Biologists, as they plunge deeper into the social behavior of nonhuman animals, encounter more and more examples of breeding systems in which the amount of reproduction in a population is environmentally regulated. Inseparable from such environmental regulatory feedback mechanisms is usually a basic division of the population into breeding and nonbreeding individuals. Is such a dichotomy inherent also in our human societies? Do we need to maintain, as we have already noted we once did at several of our societal levels, a reserve of nonbreeding individuals? Is Garrett Hardin thereby justified in proposing that we build up once more to substantial proportions this nonbreeding sector of our population? Should we then maintain family size within the breeding sector as it has always apparently been, that is, with four or five surviving children per pair-bonded couple, and direct the individuals of the nonbreeding sector into a hedonistic world of unisex?

Sex is the oldest and the only universally practiced recreational activity of our human species. We have now finally succeeded in fashioning cultural procedures that make it possible to separate entirely this recreational activity and the act of procreation. It is as comparable a situation as would occur, say, if we were to have all the fun of eating without any metabolic consequences. Some of us have lived through two severe depressions in the rate of natural increase apparently facilitated by this separation of sex and procreation, and seemingly provoked by economic stresses. Yet we still have no clear understanding of the reason for these depressed fertility statistics for the early 1930s and the continually decreasing figures for the decade of the 1960s and the early 1970s. It may be we are now beginning to appreciate our own sexual requirements, but until we achieve a more complete understanding of our procreational behavioral needs, it will be impossible to fashion totally acceptable procedures for the population regulation measures that we must now so patently and so urgently introduce.

United Nations Declaration

In order to reinforce the urgency and seriousness with which the various proposals presented in this chapter must be considered, and wherever possible implemented, the following passages are abstracted from a declaration of the United Nations on Human Rights Day, December 1967. The italics have been inserted here for emphasis.

The peace of the world is of paramount importance. . . . But another great problem threatens the world—a problem less visible but no less immediate. That is the problem of *unplanned population growth*.

It took mankind all of recorded time until the middle of the last century to achieve a population of one billion. Yet it took less than a hundred years to add the second billion, and only thirty years to add the third. At today's rate of increase there will be four billion people by 1975 and nearly seven billion by the year 2000. *This unprecedented increase presents us with a situation unique in human affairs and a problem that grows more urgent with each passing day.*

The numbers themselves are striking, but their implications are of far greater significance. Too rapid population growth seriously hampers efforts to raise living standards, to further education, to improve health and sanitation, to provide better housing and transportation, to forward cultural and recreational opportunities, and even in some countries to assure sufficient food. In short, the human aspiration, common to men everywhere, to live a better life is being frustrated and jeopardized.

As heads of government actively concerned with the population problem, we share these convictions:

We believe that the population problem must be recognized as a principal element in long-range national planning if governments are to achieve their economic goals and fulfill the aspirations of their people.

We believe that the great majority of parents desire to have the knowledge and the means to plan their families; that the opportunity to decide the number and spacing of children is a basic human right.

We believe that the objective of family planning is the enrichment of human life, not its restriction; that family planning, by assuring greater opportunity to each person, frees man to attain his individual dignity and reach his full potential.

Recognizing that family planning is in the vital interest of both the nation and the family, we, the undersigned, earnestly hope that leaders around the world will share our views and join with us in this great challenge for the well-being and happiness of people everywhere.

This declaration was signed by the UN representatives of the following countries:

Australia	Finland	The Philippines
Barbados	Ghana	Singapore
Colombia	India	Sweden
Denmark	Indonesia	Thailand
Dominican Republic	Iran	Trinidad
Nepal	Japan	Tunisia
The Netherlands	Jordan	United Arab Republic
New Zealand	Korea	United Kingdom
Norway	Malaysia	United States
Pakistan	Morocco	Yugoslavia

No more critical and serious a declaration ever has been issued by such an august body in the history of this earth. Significantly, by 1970 almost all these countries had a governmental population-planning policy. It would be a very fitting note on which to close this chapter on population control but for the fact that it hardly goes far enough to meet the extreme urgency of the situation.

It will be apparent that there is a considerable distinction between the attitude displayed in the UN declaration and that ascribed earlier to Berelson's population council insofar as its survey of possible control measures is concerned.

The UN declaration emphasizes an individual freedom of choice in the procreation of children. Faced with the population statistics presented in the last three chapters, many will consider it totally unrealistic to suggest that there is still time to exercise such a freedom. There can be no question but that retention of the individual's right to decide how many children to strive for severely restricts the selection of population control measures that can quickly be brought into operation.

Nor is there any question that we have accepted other restrictions on our individual freedoms in order to achieve a betterment of social conditions for all. For example, in the vast majority of countries it is *compulsory* to pay some form of income tax. In this country we have in addition lost our right to cross the road where we please, to maintain a private army, to mint our own money, to have two or more legal wives or husbands, to send our children to work rather than to school, to visit any country we wish, to attend a state university.

It may be possible to establish the principle that to be allowed to procreate children is a *privilege,* not a right, that it is just as necessary to demonstrate a suitability for parenthood as it is to prove scholastic ability in order to enter a university.

Since social life always has entailed some obligation, it seems unrealistic to insist at this late hour that we cannot contemplate one more restriction on our liberty in such a critical cause as the survival of our species. At any rate the disastrous effects of continuing to insist on our individual rights in this matter will be made apparent to those of us living now even in our own lifetimes, and our declining years could be blighted by the folly of our own inaction. The future may weigh heavily on the representatives of those ninety nations who refused to respond to even this relatively innocuous UN declaration on population control.

The Bucharest Conference

The year 1974 was declared by the UN to be "World Population Year." One event scheduled for this year was the UN population conference in Bucharest to consider a "world plan of action" for population control. Among the national groups present at the Bucharest conference, the indifference with which some had regarded the 1967 UN declaration appeared to have hardened in 1974 into an adversary position. Although, with the exception of the Vatican, national representatives did not appear to dissent from the broad view that the global human population must eventually be brought into equilibrium, the major attention of the conference was directed toward establishing the opinion that preoccupation with population increase should not divert attention from the critical issue of world development. Population growth, it was considered, was not inevitably an obstacle to development, although under some circumstances it could be so.

The 136 countries attending the 1974 Bucharest conference did succeed in passing by acclamation a World Population Plan of Action, outlining strategies for dealing with world population problems. The plan has yet to be adopted by the UN General Assembly. It embraces not only fertility reduction but also the elimination of poverty, the need to raise the status of women, and the reduction of mortality and of nonrenewable resource depletion. Recognizing that "All couples and individuals have the basic right to decide freely and responsibly the number and spacing of their children," the Plan recommends that "in the exercise of this right (couples and individuals) take into account the needs of their living and future children, and their responsibilities towards the community." The recognition that we are moving from an elitist to an egalitarian world is embodied in the resolution which reads, "Recognizing that per capita use of the world resources is much higher in the more developed than in the developing countries, the developed countries are urged to adopt appropriate policies in population, consumption, and investment, bearing in mind the need for fundamental improvement in international equity."

In view of this attitude in the underdeveloped nations it seems likely that pregnancy avoidance will be left essentially on an individual basis, as it basically has been during the demographic transition in the 31 nations generally recognized as industrialized. Population issues have now become embroiled in wider political issues, and it is unlikely any government can afford to take unilateral action to limit its population on a national basis without negotiating some prior compensatory economic trade-off. Nevertheless with or without government pressure, individual determiners appear to be having some sporadic effects. For example crude birth rates in Barbados, Chile, Cuba, Hong Kong. Jamaica, Mauritius. Puerto Rico, Singapore, South Korea, Taiwan, and Trinidad have fallen by 25 to 50 per cent in the last few

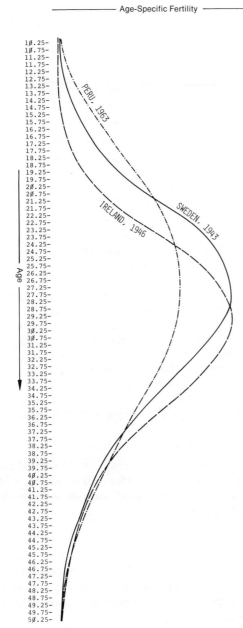

FIGURE 9-5. *The effects of population control on age-specific fertility.* The three curves here show age-specific fertility of the female populations of Ireland 1946, of Sweden 1943, and of Peru 1963. In Peru no modern methods of birth control were believed to be practiced at that time. In Ireland fertility was limited by postponement of marriage associated at least theoretically with celibacy. In Sweden even in 1943 birth control was widely practiced. Without any form of population control, the highest age-specific fertility is either in the 20–24 or in the 25–29 cohorts. With postponement of marriage it can be transferred to later cohorts. Extensively practiced birth control in some societies concentrates high fertility in the 20–24 cohort, in other societies in the 25–29 cohort.

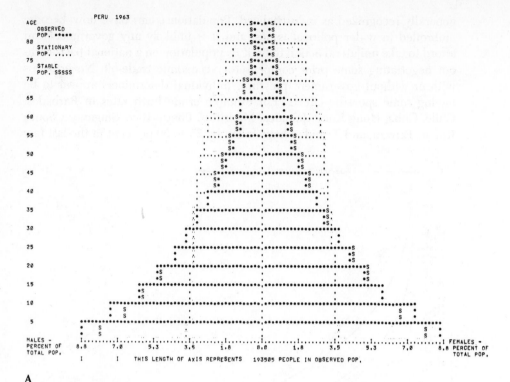

A

FIGURE 9-6. *The population of Peru.* The 1963 population data for Peru, the latest figures available for this analysis, show the country set in a pattern of extreme exponential growth. In the pyramid (*A*), the youngest age cohort greatly exceeds even the numbers that would hold the population in its present stable exponential pattern. As can be seen from the scale at the base of the pyramid, the first age cohort is approximately two and a half times that necessary to hold the population in a stationary position. The age-specific fertility graph (*B*) illustrates the reason for this extreme exponential rate of growth. The high fertility rate is achieved early and maintained through the most fertile years. On the basis of this fertility curve, it would be surmised that no methods of social population control were practiced in modern Peru up to 1963 and that the total fertility rate approached the maximum potential fecundity of Peruvian women under the conditions of the time.

years, and probably that of China is falling appreciably too. In critical areas vigorous government action may even now be under consideration, as for example in Bangladesh, where the possibility of sterilizing all couples when they have two children was recently explored.

Population Stabilization

The various population control measures described in this chapter are generally, to a varying degree, all in operation. If we look again at the scheme of societal succession, it is possible to recognize a pattern of human behavior in respect to these control measures (Figure 9-5).

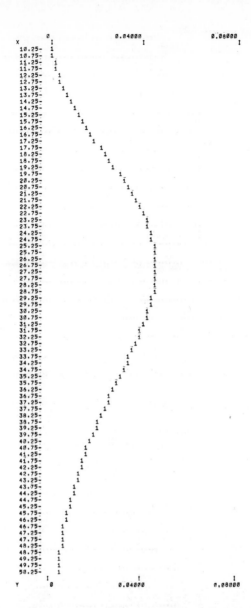

Hunter-gatherer societies and Early Agriculturalists have become so rare that they can be ignored in this survey. Advanced Agriculturalists, by contrast, still constitute at least one half of the world's population, probably nearer two thirds. The extent to which they utilize population control measures is thus a critical issue.

Population Control Measures in Advanced Agricultural Societies

The population structure of a typical Advanced Agricultural society, Peru, is shown in Figure 9-6A. This is a stable population structure that shows no sign of any general population control measure. The age-specific fertility rate (Figure 9-6B) runs close to the maximum for all cohorts, and

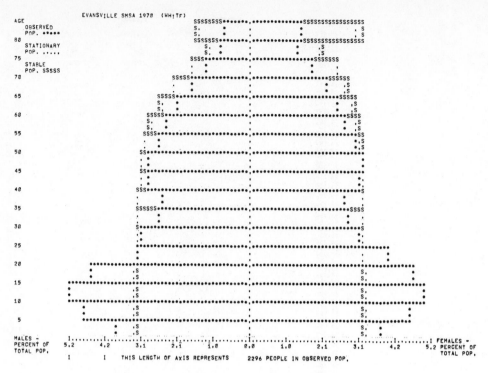

A

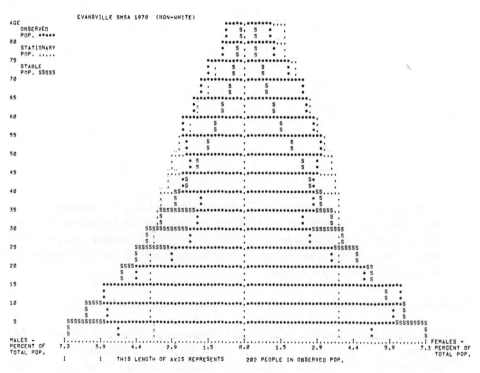

B

FIGURE 9-7. *Contrasting demographic features within the same urban ecosystem.*
The population pyramids for whites (A) and blacks (B) in the city of Evansville,
Indiana, illustrate the wide variation still encountered even within the same urban
areas. The pyramid of the whites shows an estimated stable population (SSSS)
very close to the estimated replacement value line (. . . .). In the case of the
blacks the stable population (SSSS) is very far from the replacement value
(. . . .). Both sectors of the population lose by outmigration, the whites especially
in the two cohorts from age 30, the blacks in the two from age 20. In both
instances infants in the less-than-5-years-old cohort accompany the outmigrants,
thus reinforcing any lowering of age-specific fertility rates that reduces the size
of this first cohort.

there is no sign even of containment of fecundity by deliberate or incidental
postponement of marriage age.

This pattern is very persistent as long as this way of life persists. In
Figure 9-7, the structure of the white and black populations of Evansville, Indiana, is shown. The blacks, presumably because they are still held
in rural activities, still have a stable population pyramid in an exploding
mode characteristic of an Advanced Agricultural society. The white pyramid also is approaching a stable structure, but near replacement value,
and is characteristic of an Advanced Industrial society.

In South Africa (Figure 9-8), the whites of the cities are mostly in
what may be termed an Industrializing societal stage. There is nevertheless
a large white rural element in the population, so it is not surprising to find
that it has not yet generally adopted any population control measures to
take it through the demographic transition. Nor, as would be imagined, has
the black portion of the South African population, essentially an Advanced
Agricultural society.

Population Control Methods in Advanced Industrial Societies

Some Advanced Industrial nations have nearly achieved zero population
growth. This is the case with, for example, Sweden (Figure 9-9), which has
had a population approaching stationary values through most of this century.
Likewise some elements of national populations are also beginning to operate
at replacement values. This applies, for example, to the white population of
Beaumont-Port Arthur. It does not apply to the black population, which from
the point of view of the successional scheme presented here could be considered to be in an Industrializing societal phase. In Figure 9-10 it can be
seen how white fertility rates are held very low through the 15–19 and
20–24 age cohorts, the highest fertility rate is permitted in the 25–29
cohort, and the rate is promptly reduced in all subsequent cohorts. This is a
characteristic curve for a population under effective population control
restrictions. The curve for blacks, by contrast, shows little application of
population control measures in the 15–19 and 20–24 cohorts, and the high
fertility of the 25–29 cohort is allowed to continue until it falls away for
reasons other than imposition of population control measures.

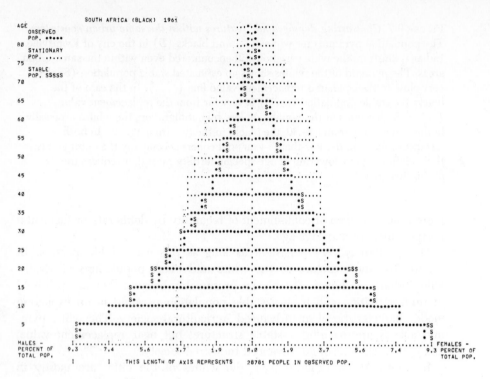

A

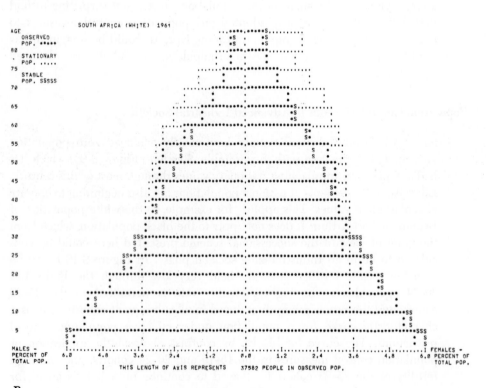

B

FIGURE 9-8. *Ethnic components of national populations.* In some cases at least, national population parameters appear more responsive to environmental conditions than to ethnic or religious affiliations. In these two pyramids for the population of South Africa in 1961 showing the white population (A) and the black population (B), both appear at that time to be set in a pattern of increasing exponential growth. The white segment of the population is derived especially from Dutch and British settlers; both home countries are now approaching zero population growth. Religious affiliations of the whites are mostly with the Dutch Reform Church or with other Protestant groups. The pyramid for the black population (B), which is mostly of Bantu origins and largely animalistic in religious affiliation, shows it set in an expanding mode with an even higher rate of increase. The demographic situation in South Africa would thus appear to support the viewpoint that environmental influences other than cultural behavior and religious traditions have the strongest effects in determining demographic behavior.

Population Stabilization in the United States

The message of Hardin, Ehrlich, Borgstrom, and now many others—insofar as this country is concerned—is that we must take *immediate* steps to stabilize at replacement level this nation's population by limiting to a maximum of two the number of children an individual may have. Hardin in particular stresses equally the need to adjust behavioral rituals to an acceptance of childless bachelors and spinsters as *normal* members of society.

An authoritative paper by Judith Blake several years ago stated this last need very succinctly:

Individuals who—by temperament, health, or constitution—do not fit the ideal sex-role pattern are nonetheless coerced into attempting to achieve it, at least to the extent of having demographic impact by becoming parents. . . . The rigid structuring of the wife-mother position builds into the entire motivational pattern of women's lives a tendency to want at least a moderate-size family. . . . The desired number of children relates not simply to the wish for a family of a particular size but relates as well to a need for more than one or two children if one is going to enjoy "family life" over a significant portion of one's lifetime. . . . The notion that most women will "see the error of their ways" and decide to have two-child families is naive . . . .

Blake addresses a considerable part of her argument to countering the popular proposition that approximately 15 per cent of the U.S. population classified as "poor" have more children per family than higher-income groups, not because they desire it, but through lack of information or resources to prevent it. She presents evidence (Table 9-6) that suggests that "poor" families actually desire more children and favor birth control less (Table 9-7), and she concludes that the majority of lower-income couples already use birth control when they choose to do so.

Blake's 1969 paper immediately evoked a strong response from O. Harkavy and some of his associates. Their statements are extremely pertinent, not only in the context of Dr. Blake's paper, but also as they concern the efforts epitomized by Hardin, Ehrlich, and Borgstrom. They

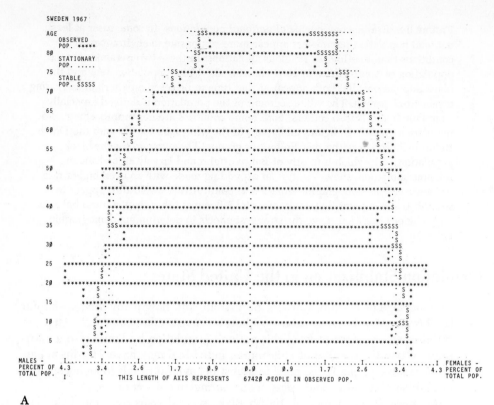

A

FIGURE 9-9. *Structure of a stationary population.* The 1967 data for Sweden provide a pyramid of a population which has been close to a stationary one for some time (*A*). The stable (SSSS), stationary (....), and observed (* * * *) lines almost coincide. This is a typical pyramid for an Advanced Industrial society. The higher fertility following the ends of World War I and of World War II can be observed. This last in particular is responsible for the slight increase in the first cohort. The larger cohort of postwar females aged 20–24 is producing a slightly larger first cohort of infants, although age-specific fertility rates remain constant.

From the age-specific fertility curve (*B*) it can be observed that birth control practices rather tightly restrict births to the two cohorts covering ages 20–30, with a steep peak at ages 23 to 26. This contrasts very sharply with the graph for Peru, Figure 9-6B.

note: "There has never been official policy regarding the virtue or necessity of reducing the U.S. population growth, much less achieving population stability," and reproduce various official pronouncements to support their contention that federal policies in this country are directed at improvements in health and economic prosperity, *not* population control. Included in the very interesting data presented in their paper is Table 9-8.

The data presented here indicate that "poor" and "near-poor" families, which represent 40 per cent of American families, contain no fewer than 58 per cent of those with six or more children, but only 14 per cent of

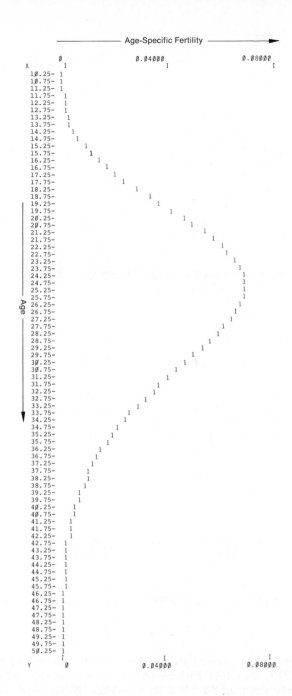

Age-Specific Fertility ⟶

B

those with only one child. An even more narrowly defined group of "poor" families, constituting 26 per cent of the United States total, included 42 per cent of the United States total with six or more children, but only 9 per cent of those with a single child.

Faced with pressures such as are being exerted by Hardin, Ehrlich, and Borgstrom, with wide divergences in specialist opinion such as exists be-

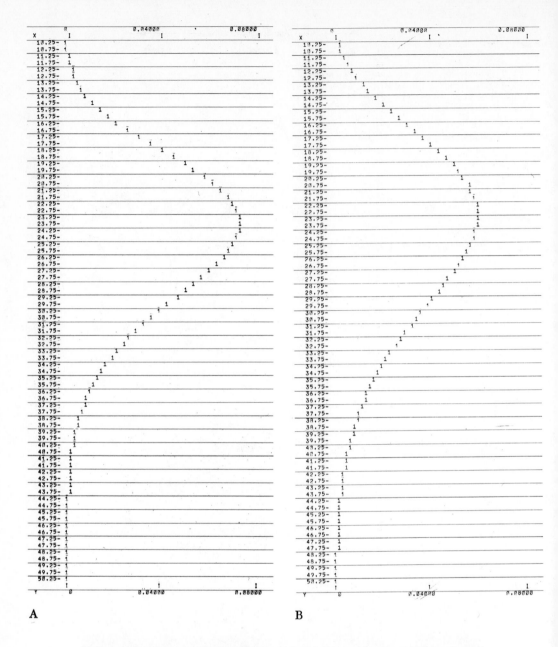

A   B

FIGURE 9-10. *Age-specific fertility rates in the Beaumont–Port Arthur district, Texas, 1970.* The rates for whites (A) show a maximum fertility rate attained only briefly about age 23. Those for nonwhites (B) show a maximum fertility rate reached almost two years earlier and moreover mantained at a high level right through the 20–24 age cohort. Among nonwhites the fertility rate also increases more rapidly earlier and falls more slowly later. These statistics suggest that birth control is much more extensively utilized for family planning among whites in the Beaumont–Port Arthur district than among nonwhites.

TABLE 9-6 *Number of Children Considered Ideal by Non-Catholic Women in the United States*

Participants were arranged into four levels of economic status ranging from "high," 1, to "low," 4. The 1943 figures reflect the low fertility levels of the previous decade. From considering fewer children more desirable than did more wealthy categories, the lowest-income group upgraded their optimum fertility almost 50 per cent, while those in the highest category hardly varied in their opinions between 1943 and 1968.

Year	Age range	Economic status 1	2	3	4
1943	20–34	2.9	2.7	2.7	2.5
1952	21+	–3.3–		3.3	3.3
1955	18–19	3.2	3.1	3.2	3.5
1957	21+	–3.3–		3.2	3.5
1959	21+	–3.3–		3.5	3.6
1960	18–39	3.2	3.3	3.5	3.4
1963	21+	3.3	3.3	3.5	3.5
1966	21+	3.2	3.2	3.4	3.7
1967	21+	3.3	3.2	3.1	3.4
1968	21+	3.2	3.0	3.4	3.6

tween the authors of the two papers just discussed, and with a general dearth of information on possible economic interactions such as the Madison Avenue effect, it is hardly surprising that professional politicians still tread warily and appear slow to respond with positive edicts on United States population policy. Nor are decisions any easier for the individual citizen.

The kind of population analysis procedures presented in the previous two sections of this chapter could be used to help with the resolution of some

TABLE 9-7 *Attitude Toward the Use of Birth Control*

These data are for attitudes expressed in different economic levels in the United States in polls taken between 1959 and 1966. Persons participating were aged 21 to 44 and divided into four categories of economic status, progressing from the most, 1, to the least prosperous, 4. Figures represent percentages of individuals who expressed approval of birth control.

Economic status	Approval Men	Women
1	89	87
2	84	82
3	83	80
4	74	78

TABLE 9-8 *Relation of Poverty to Size of Family*

Figures represent the percentage of the total of all U.S. "poor" and "near-poor" families with the number of children indicated.

Number of children	"Poor" families	"Poor" and "near-poor" families
1	9.3	14.1
2	10.2	15.6
3	12.8	21.3
4	18.6	30.9
5	27.7	42.5
6 or more	42.1	58.1

population control arguments. These graphic display techniques facilitate a critical examination of the U.S. population, as they would of any national or regional unit, and identify the fertility high-spots where population control measures are most urgently needed. They would also permit the close monitoring of the effectiveness of any control measures that are introduced. The method of display of the data is new, but the statistics are already available. The trouble with such demographic statistics is that even for the specialist their meaning is difficult to perceive. For the uninitiated the sheer mass of apparently boring data obscures the meaning of even the most startling statistics. From computerized analysis and resulting presentations such as are illustrated in this and the previous two chapters, a more immediate comprehension of population trends may be perceived even by a tyro.

There is no other way to conclude this chapter except on a somewhat ominous note; in fact, we have no real choice. The Club of Rome resource models are wrong only insofar as they run to a total and universal disaster. For several reasons population growth will fall away before reaching this point. It has already begun to do so on a global scale, although whether ZPG will be achieved before the end of the first century of the next millennium is still a contentious matter. The whole point of organizing in societies is that we thereby can arrange to minimize distress and hardship and maximize well-being. If we were to make no societal rules, there would be no point in society. The question now is not whether we shall reach ZPG and when, but *by what means* will we elect to stabilize our individual national communities, including the populations of the United States, and proceed *immediately* to initiate the measures necessary to control any further increase in the size of the human global population.

BIBLIOGRAPHY

References

Adams, E. S. "Unwanted births and poverty in the United States," *The Conference Board Record,* **6**(4): 10–17, 1969.

Anonymous. *Recreation in the Nation's Cities: Problems and Approaches,* National League of Cities, Dept. of Uurban Studies, Dept. of the Interior, 1968.

Ayala, F. J., and Falk, C. T. "Sex of children and family size," *J. Heredity,* **62**(1): 57–59, 1971.

Berelson, B. "Beyond family planning," *Science,* **163**: 533–43, 1969.

Blake, J. "Abortion and public opinion; the 1960–1970 decade," *Science,* **171**: 540–49, 1971.

Blake, J. "Reproductive motivation and population policy," *Bioscience,* **21**: 215–24, 1971.

Calderone, M. S. *Manual of Contraceptive Practice,* Baltimore: Williams and Wilkins, 1964.

Carr-Saunders, A. M. *World Population,* Oxford, England: Clarendon Press, 1936.

Coale, A. J. "The history of the human population," *Scientific American,* **231** (3): 40–51, 1974.

Darity, W. A., Turner, C. B., and Thiebaux, H. J. "Race consciousness and fears of black genocide as barriers to family planning," *Perspectives from the Black Community, PRB Selection No. 37,* June 1971.

Davis, J. B. "Population policy for Americans: is the government being misled?" *Science,* **164**: 522–29, 1969.

Davis, K. "The demographic foundations of national power," in M. Berger, T. Abel, and C. H. Page (eds.), *Freedom and Control in Modern Society,* New York: Van Nostrand, 1954.

Davis, K. "Population policy: will current programs succeed?" *Science,* **158**: 730–39, 1967.

Day, A. T. "Population control and personal freedom: are they compatible?" *The Humanist,* Nov.–Dec. 1968, 7–10.

Dickson, E. M. "Model for zero population growth," *Bioscience,* **20**: 1245–46, 1970.

Djerassi, C. "Prognosis for the development of new birth control agents," *Science,* **166**: 486–73, 1969.

Djerassi, C. "Birth control after 1984," *Science,* **169**: 941–51, 1970.

Ehrlich, P. R. *The Population Bomb,* New York: Ballantine, 1968.

Eisner, T., van Tienhoren, A., and Rosenblatt, F. "Population control, sterilization, and ignorance," *Science,* **167**: No. 3917, leader, 1970.

Enke, S. "The economics of having children," *Policy Sciences,* **1**: 6–30, 1970.

Frejka, T. "Reflections on the demographic conditions needed to establish a U.S. stationary population growth," *Population Studies,* **22**: 379–97, 1968.

Hardin, G. "Abortion—or compulsory pregnancy?" *Marriage and the Family,* 30: 246–51, 1968.

Hardin, G. "Not peace, but ecology," in "Diversity and Stability in Ecological Systems," *Brookhaven Symposia in Biology,* No. 22, 1969, pp. 151–61.

Harkavy, O., Jaffe, F. S., and Wishik, S. M. "Family planning and public policy: who is misleading whom?" *Science,* 165: 367–73, 1969.

Heer, D. M. "Economic development and the fertility transition," *Daedalus,* 97: 447–62, 1968.

Heisel, D. F. "The emergence of population policies in sub-Saharan Africa," *Concerned Demography,* 2(4): 30–35, 1971.

Jaffe, F. S. "Public policy on fertility control," *Scientific American,* 229(1): 17–23, 1973.

Kangas, L. W. "Integrated incentives for fertility control," *Science,* 169: 1278–83, 1970.

Kirk, D. "Prospects for reducing natality in the underdeveloped world," *Ann. Amer. Acad. Pol. Soc. Sci.,* 369: 48–60, 1967.

Kistner, R. W. *The Pill,* New York: Delacorte Press, 1968.

Kourides, I. A. "Freedom of birth," *Medical Science,* 18(8): 25–31, 1967.

Paddock, W., and Paddock, P. *Famine—1975,* Boston: Little, Brown, 1967.

Parr, E. L. "Contraception with intrauterine devices," *Bioscience,* 23: 281–84, 1973.

Pincus, G. "Control of conception by hormonal steroids," *Science,* 153: 493–500.

Revelle, R. "Introduction to 'Historical Populations Studies,'" *Daedalus,* 97: 353–62, 1968.

Segal, S. J. "The physiology of human reproduction," *Scientific American,* 231(3): 52–62, 1974.

Spengler, J. J. "Demographic factors and early modern economic development," *Daedalus,* 97: 433–46, 1968.

Taylor, H. C., and Berelson, B. "Maternity care and family planning as a world problem," *Amer. J. Obstet. Gynecol.,* 100: 885, 1968.

Tietze, C., and Lewit, S. "Abortion," *Scientific American,* 220(1): 21–27, 1969.

van de Walle, E., "Marriage and marital fertility," *Daedalus,* 97: 486–501, 1968.

Wynne-Edwards, V. C. "Self-regulating systems in populations of animals," *Science,* 147: 1543–48, 1965.

Further Readings

Bajema, C. J. "The genetic implications of population control," *Bioscience,* 21: 70–75, 1971.

Barnett, L. D. "Population policy: payments for fertility limitation in the U.S.," *Social Biol.,* 16(4): 239–48, 1969.

Behrman, S. J., Corsa, L., Jr., and Freedman, R. (eds.) *Fertility and Family Planning; A World View,* Ann Arbor: University of Michigan Press, 1969.

Bumpass, L., and Westhoff, C. F. "The 'perfect contraceptive' population," *Science,* **169**: 1177–82, 1970.

Clarkson, F. E., Vogel, S. R., Broverman, I. K., Broverman D. M., and Rosenkrantz, P. S. "Family size and sex-role stereotypes," *Science,* **167**: 390–92, 1970.

Cook, R. C. "California after 19 million what?" *Population Bull.,* **22**: 29–57, 1966.

Freedman, R., and Tokeshita, J. Y. *Family Planning in Taiwan, An Experiment in Social Change,* Princeton, N.J.: Princeton University Press, 1969.

Friedlander, M. V., and Klarmov, J. "How many children?" *Environment,* **11**(10): 3–8, 1969.

Hardin, G. *Population, Evolution and Birth Control; A Collage of Controversial Readings,* 2nd ed., San Francisco: Freeman, 1969.

Hardin, G. *Birth Control,* New York: Pegasus, 1970.

Johnston, S. *Life Without Birth,* Boston: Little, Brown, 1970.

Kalman, S. M. "Effects of oral contraceptives," *Ann. Rev. Pharmacology,* **9**: 363–78, 1969.

McElroy, W. D. "Biomedical aspects of population control," *Bioscience,* **19**: 9–23, 1969.

Meier, R. L. "The social impact of a nuplex," *Bull. Atomic Scientists,* March 1969, pp. 16–21.

Pradervand, P. "International aspects of population control," *Concerned Demography,* **2**(2): 1–16, 1970.

Ransil, B. J. *Abortion,* New York: Paulist Press Deus Books, 1969.

Stycos, J. M. "Family planning: reform and revolution," *Family Planning Perspectives,* **3**(1): 49–50, 1971.

Sweezy, A. "The economic explanation of fertility changes in the U.S.," *Population Studies,* **25**(2): 255–67, 1971.

Tietze, C., Poliakoff, S. R., and Rock, J. "The clinical effectiveness of the rhythm method of contraception," *Fertility and Sterility,* **2**: No. 5, 1951.

Westhoff, L. A., and Westhoff, C. F. *From New to Zero: Fertility, Contraception and Abortion in America,* Boston: Little, Brown, 1971.

Willing, M. K. *Beyond Conception: Our Children's Children,* Boston: Gambit, 1971.

World Health Organization. "Biology of fertility control by periodic abstinence," *Technical Report No. 360,* 1967.

10

Air Pollution

Throughout the previous chapters of this book it has been apparent that for some 4 or 5 million years, from the earliest hominids to contemporary man, one human characteristic has remained unchanged and invariable— we have always produced garbage and litter. Even on our latest adventure into space, we have sent our debris ahead of us, and we have vented it on the surface and into the atmosphere of the new world where we have landed. Indeed, without such waste we should have little record of our prehistoric past, for it is from kitchen middens, discarded tools, unscattered remains of meals, feces, and similar wastes that we have been able to piece together something of our earlier and otherwise unrecorded cultural life. Unfortunately our global population, as noted in Chapter 7, now numbers about one twentieth of all the humans who have ever lived. Even assuming that we were to produce no more wastes per capita than we have ever done, this presents us with the problem here and now of disposing of one twentieth of all the discarded materials we and nature previously had 4–5 million years to recycle or reuse.

With global human populations totaling a million individuals, all in a hunter-gatherer phase, waste discards accumulated very slowly and quite locally. With a total population now estimated at 3 billion plus, the global pile of excrement alone increases at the rate of about 1½ *million* tons per day. Yet this is among the least of our problems. More and more of us now live in cities, and it has been calculated that the average city dweller produces 1,500 pounds of garbage per year—about 5 pounds per day. A garbage collector's strike brings home to us the dimensional realities of this rate of waste discard.

We need no strike, however, to make us appreciate that not all this garbage is carted away for disposal and that some instead rises into the air in the form of solid pollutants. A global total of 140 million tons of solid pollutants is now emitted into the air annually. Some of this coats our buildings and soils our hands and clothes; some of it is inhaled and deposited on the tissues of our lungs. These solid pollutants, together with gaseous and other material released into the atmosphere, present us with the problem, or rather the series of allied problems, now broadly described by the term *air pollution*.

The often spectacular examples of release of wastes into our common environment illustrate the kinds of problems that have arisen through overloading of the biogeochemical pathways of urban ecosystems until they cease to operate effectively. The waste materials at various trophic levels accumulate in the form of trash and garbage, instead of being broken down by decomposers and their elements recycled. In considering such ecosystem failures, it is convenient to call all such accumulations from overloaded pathways pollution and subdivide this further both according to the nature of the environmental media polluted, that is, air and water, and according to the type of pollution, for example, noise and thermal. In this text, the term *air pollution* is widened to embrace major disturbances in the proportions in which gases such as carbon dioxide, oxygen, ozone, and carbon monoxide occur in the air. A common way of classifying air pollutants is to categorize them into combustion products, radionuclides, and pesticides. The first is considered in this chapter, the second especially in Chapter 11, and the last in Chapter 12.

Air Pollution

Waste accumulation in this century is, as already noted, unprecedented only in regard to the total quantities produced. The release of particulate matter into the air and the consequent pollution of the atmosphere are far from new. The walled cities of medieval Europe 700 years ago already had such problems with the wood smoke that belched from the chimneys of every city house. The introduction of coal as a domestic and industrial fuel in the early thirteenth century seems, however, to have really ushered in the smog era. By the middle of that century London and other British cities were said to have become almost uninhabitable. The death penalty was introduced to reinforce laws prohibiting the burning of coal in London while Parliament was in session and is believed to have been carried out on one unfortunate citizen.

Smog and Particulate Matter

As the centuries passed, solid pollutant problems were reported from other cities. Edinburgh, Scotland, became affectionately known as "Auld Reekie"; the British industrial belt in Staffordshire and Warwickshire was called the "Black Country." London, however, remained notoriously pre-eminent. This unenviable distinction was reaffirmed in modern times during four critical days in December 1952, when in the worst smog attack of its recorded history one out of every 2,000 Londoners died, an estimated total of 4,000. This does not include uncounted thousands who were made seriously ill but recovered sufficiently to live on for a considerable time—although probably not to their original life expectancy.

Although the term *smog* applied originally to London's special brand of "pea-souper" combining smoke and fog, it became a generic term embracing all earlier names for the condition existing when solid pollutants are visible in the air. It is also now applied to situations where *invisible* particulate matter is present that produces some visible reaction. Particulates with a cross-section of less than 0.1 micron are invisible optically, but may be a major source of air pollution. An idling automobile emits from its exhaust pipe 1×10^{11} such particles per second in a normally invisible plume. Although invisible, they have, as will be described, very significant effects.

In this text, air pollution has been given a wider definition than is sometimes employed. It is taken here to include any atmospheric disturbance resulting from human activity that has a modifying effect on the role air plays as an abiotic element of natural ecosystems.

Smog and Toxic Effects

There have been other manifestations of air pollution not so immediately obvious as particulate matter. Some 5 per cent of U.S. soldiers stationed in the greater Tokyo area in the late 1940s developed a disorder known as "Yokohama asthma"—a breathing difficulty from which they recovered on transfer from the area. The City Health Department of New York calculated during 15 days of heavy smog in 1963, the normal death toll was increased by 650. In statistics gathered by a Senate committee, it is estimated that by 1960 every American city with over 1 million population had a major air pollution problem, as did to a lesser degree 64 smaller ones. One quarter of the U.S. population live in cities with a major, 30 million in areas with a moderate, and 3½ million in areas with a minor problem, the report stated.

The city of Los Angeles has come to be synonymous with smog. Medical advisors are reported to have told at least 10,000 people to leave the city for their health's sake. School children are officially advised to minimize outdoor activity during smog alerts. In 1973 federal employees in the city were requested to stay home and not come in to work during one prolonged smog alert period. This action was taken to reduce the number of autos entering the city. It is no longer a question confined to problems with inversions in the Los Angeles Basin; ponderosa pines are dying in the San Gabriel and San Jacinto Mountains at altitudes upward of 2,500 meters. Various species of desert plants are disappearing from as far afield as the Coachella Valley. The smog even reaches out over Palm Springs, where the affluent have fled in an attempt to preserve their lungs, or at any rate their vocal cords.

Many other cities about the world are in no better shape. Tokyo during the last three decades has gone from bad to worse. Affluence in Teheran has brought automobiles and smog. Mexico City, as urban industry expands, is subjected to progressively more severe smog incidents. No urban area, whatever its topographical or geographical location, remains totally unaffected.

In order to understand these smog incidents and the distribution and dispersal of the *aerosols* formed in the atmosphere by suspended particulate matter, the physical reactions that they promote, and the several chemical reactions that take place following release of various liquid and gaseous wastes, it is necessary to examine some basic climatic phenomena.

Atmospheric Circulation

Air masses, like water masses, are not dispersed evenly over the earth's land surfaces, but may be segmented into *air sheds* that correspond to some extent with watersheds (Figure 10-1). The composition of the air masses in these air sheds—the proportional mixture of gases, their temperature, their movement, and the solid, liquid, and gaseous pollutants they contain—depends very much on the nature of the ecosystems included within each air shed. Thus, whereas each human being processes perhaps 30 pounds of air every 24 hours, the average car will need about 1 *ton* during a 200–300 mile run, or in using one tank of gasoline. Burning 1 pound of coal requires 14 pounds of air; 1 pound of natural gas, 18 pounds of air; 1 gallon of fuel oil, 19 pounds of air.

Air Shed Utilization

The air shed of a city of some 3 million inhabitants such as Los Angeles is obviously therefore not so much affected by the 90,000 tons of air its citizens process a day as by the half million tons that would pass through its automobiles if every citizen daily used half a tank of gasoline. The effect of the presence of the human population is to remove some of the oxygen and to add some carbon dioxide. Even within the city's air shed, the producer populations of trees, grasses, and other plants of the urban ecosystem could take care of this perturbation and maintain the original atmospheric balances.

A   Airshed in Coastal Southern
and Central California

FIGURE 10-1. *Air sheds in coastal
southern and central California.*
The air sheds of the southern
Pacific Coast commonly recognized
are over the Central Valley and
Bay Area, the Greater Los Angeles
Basin, and the San Jose Valley.
This last can drain over the Outer
Coast Range into the Central
Valley Basin, but it is then shut in
by the uninterrupted mass of the
Sierra Nevada. The Los Angeles
Basin drains through the Santa
Ana River Gap and the San
Gorgonio Pass.

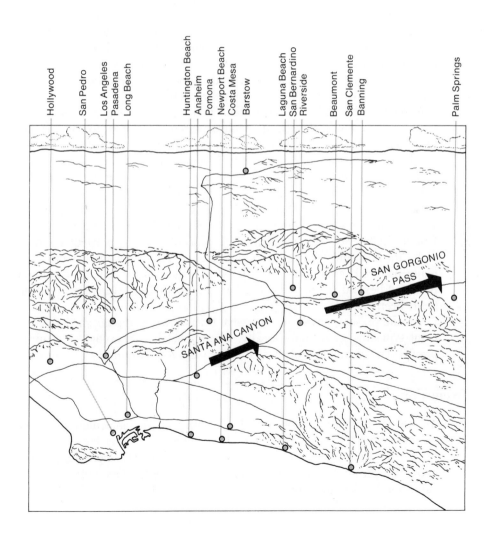

At some value between the amount of expiration by the human population and the total of vented exhaust gases from the automobiles, the non-linear dose-response threshold of the atmosphere is exceeded. As the vented effluents from automobile exhausts are more than any naturally occurring ecosystem can handle, only the redistribution of air from the city's air shed, and its replacement by fresh air, will remove the automobile emissions or air pollutants from the atmosphere of the city. Because of special features of meteorology and topography in Los Angeles and some other cities, this air shed exchange is not always possible. When it is not, a smog attack develops.

Air Flow

Owing to the general circulation of air masses in the northern hemisphere (Figure 10-2) these pass from one air shed to another in an easterly direction across the United States from the Pacific to the Atlantic Coast, then across the North Atlantic to Europe, across Eurasia, and over the Pacific. During the ocean passages in particular it is assumed that the oxygen to carbon dioxide balance is restored through the photosynthetic activity of marine phytoplankton. Whatever solid pollutants have been added settle out in the oceans, if fallout has not already deposited them on a land mass. Rain and snowfall assist considerably in the process of removing particulate matter from the circulating air masses.

Although very little is known about this replenishment of air sheds with clean air as a result of such atmospheric circulation, it is not unreasonable to suppose that the settling-out and gas-ratio homeostatic mechanisms of the air shed ecosystems could eventually become saturated, and that each time around the globe just a little less oxygen would be added, just a little more carbon dioxide and particulate matter left in.

Recent work by V. J. Schaefer and others suggests that the process of self-cleansing may be beginning to break down over many parts of the United States. Schaefer measured the concentration of particulates during eight transcontinental flights, which included passage over most of the major polluted centers, and considered that the general fallout of pollutants persisted beyond the limits of the local air sheds into which they had originally been vented.

For the present, however, a massive breakdown in the air shed replenishment typically occurs only locally, when special meteorological conditions prevail. The commonest of these is known as a *temperature inversion*.

Inversions

As any jet traveler can observe by listening to the information supplied en route over the cabin intercom, air temperature in the lowest region of the earth's atmosphere, the *troposphere*, decreases with increasing distance above the ground. This temperature decrease is maintained at a steady value until a height above the surface of 8 to 18 kilometers (the *stratosphere*) is reached.

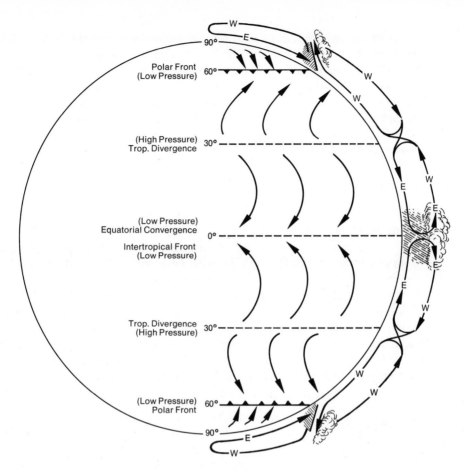

Figure 10-2. *The pattern of planetary winds* that produces prevailing westerly winds in the temperate regions of North America. Air sheds in this area therefore tend to drain toward the east. The smog generated on the Pacific Coast is drawn over the Sonoran, Mohave, and Great Basin deserts; that of the Eastern seaboard is carried out over the North Atlantic Ocean.

The actual *rate* of temperature loss is known as the *lapse* rate. If the lapse rate becomes negative, *temperature inversion* occurs; instead of decreasing with increasing height, the air temperature actually *increases*. Above 8 to 18 kilometers, where the *troposphere* is replaced by the *stratosphere*, there is a permanent inversion. There may be a number of inversions at various heights in the troposphere, and meteorologists classify them by their various modes of origin and the several effects they produce such as haze, fog, turbulence, fronts, and cyclones.

Characteristics of Inversions

Two conditions are especially associated with inversions. The first is the presence of a slowly moving high-pressure system in the troposphere. The mass of cooler dense air in such a system prevents circulation of the surface layers of the atmosphere beyond the inversion layer, and the lack of movement produces only light winds insufficient to dissipate the air mass. Second, the topography of a particular region may tend to enclose an inversion so that when adiabatic cooling of the air included in it occurs, a balance is eventually achieved at the level of the inversion layer, where further air circulation in the air shed beyond the inversion layer is impossible. The loss of heat that occurs adiabatically as the air rises and expands causes it to cool. It then becomes denser and tends to descend again.

Topographic Contributory Factors

In a number of areas in the United States these two conditions of high atmospheric pressure and local topography are commonly met. The city that has gained most notoriety in this respect is Los Angeles. The San Gabriel and San Bernadino ranges, with peaks rising over 10,000 feet, form the boundaries of an air shed over the Los Angeles Basin that in summer frequently contains cold air rolling in from the sea. As the lower layers of this cool coastal air become heated by contact with the ground, they rise, losing heat as they expand until they form a low inversion layer over the basin anywhere from several hundred to several thousand feet above sea level. High winds are needed to break up this layer, because the mountain barriers tend to stabilize the air shed and prevent its dispersal inland.

Smog Composition

The visible material that becomes trapped in the air below an inversion layer is known as *particulate matter*. As noted earlier, particulate matter is defined as suspended solid-waste particles with a diameter over 0.1 micron (μ). Particles larger than 0.1μ but less than 5.0μ are usually called *smoke* or *fume;* over 10μ they constitute *dust* or *fly ash*, which quickly settles. Below 0.1μ in diameter suspended matter is invisible and is generally known as an *aerosol*. This term is sometimes applied to larger particles that do not readily settle. Particulate matter is not only readily visible, it is also the easiest of the smog elements to measure and was at one time known as *soot*. The U.S. Public Health Service has sampled the air sheds of various major cities in the United States to determine the amount of particulate matter present at any given time per square mile up to 100 feet. The figures are reproduced in Table 10-1.

Primary Pollutants

Particulate matter is finely divided carbon particles; it is associated with a number of other substances that are less visible but constant components of all smog. A partial list of these substances includes carbon monoxide, sulfur oxides, and nitrogen oxides (Table 10-2). The sulfur is mostly in the form of sulfur dioxide, forming sulfuric acid near the source of emission, which may in its turn become converted into sulfates. The nitrogen occurs

TABLE 10-1 *Amount of Particulate Matter per Square Mile Over Various U.S. Cities*

Amounts measured over the first 100 feet of the troposphere in 1961. Some of these cities, like Pittsburgh, have subsequently introduced clean air control measures that have substantially reduced the quantities shown here.

City	Tonnage	City	Tonnage
Detroit	153	Washington	58
Chicago	124	Houston	57
Los Angeles	118	San Francisco	46
New York	108	Pittsburgh	45
Philadelphia	83	Salt Lake City	24
Atlanta	61		

generally as nitrous oxide. Air pollutants of this kind are known as *primary pollutants;* gasoline is among them. In Los Angeles the amount of gasoline released into city air was estimated in 1967 to be 8 per cent of the quantity actually sold there.

When an attempt is made to prepare an emissions balance sheet for a given air shed by calculating the emissions released and comparing these figures with the amounts actually measured, less of the first and more of the second are sometimes found for a given pollutant (Table 10-3). The

TABLE 10-2 *Sources of Major Pollutant Emissions in the United States During 1965*

Amounts expressed in millions of tons per year. Motor vehicles operating along thoroughfares and thus representing a line source of emission are responsible for approximately two thirds of the total air pollution. In Europe, before electrification of most of the rail systems, coal-fired railway engines constituted a major line source of pollution. Industry and power plants in the U.S. are the major generators of sulfur oxides and particulate matter.

	Particulate matter	Carbon oxides	Sulfur oxides	Hydro-carbons	Nitrogen oxides	Per cent of total
POINT SOURCE						
Industry	6	2	9	4	2	17
Power plants	3	1	12	1	3	14
LINE SOURCE						
Motor vehicles	1	66	1	12	6	60
AREA SOURCE						
Space heating	1	2	3	1	1	6
Refuse disposal	1	1	1	1	1	3

explanation advanced for this imbalance is that some of the primary pollutants have been converted to *secondary pollutants*.

The smog observed in European and eastern cities in the United States, sometimes called London smog, is predominantly confined to primary pollutants and is said to be of a *reducing* type. In this type the particulate matter is largely soot, and the irritating element is sulfur dioxide. In the Los Angeles type of smog there is an overriding amount of oxidizing activity. Although there is also some soot, it is not so heavy as in the reducing type of smog, and most irritation of the human body comes from the products of the interactions promoted by the oxidizing activity.

Secondary Pollutants

From the pollutant balance sheet for Los Angeles-type or oxidizing smog, it appears that gasoline emissions persist in smog only in about one half the expected amounts, while certain acids and aldehydes appear for which there is apparently no traceable source of direct emission. Such oxidized compounds apparently have been produced from the reactive hydrocarbons in the gasoline emissions. Another feature of Los Angeles-type smog is a very high ozone level, something like thirty times that of country air. It is now considered that the London type of reducing smog, characterized by its large amounts of soot and sulfur dioxide, also contains the same types of secondary pollutants as the Los Angeles type, but their presence is masked.

Photochemical Reactions

The knowledge of what reactions lead to the development of ozone and other secondary pollutants is based mainly on the work of A. J. Haagen-Smit, who studied Los Angeles smog from the California Institute of

TABLE 10-3   *An Emissions Balance Sheet for Major Pollutants*

By calculating the amount of each pollutant released at its source, then determining the amount measurable in smog, it is possible to draw up an emissions balance sheet such as that supplied below. Primary pollutants observed to have values *below* those calculated are assumed to have been converted by photochemical reactions into secondary pollutants.

	Parts per Million by Volume	
Pollutant	*Measured*	*Calculated*
Carbon monoxide	3.5	3.5
Oxides of nitrogen	0.08	0.10
Sulfur dioxide	0.05	0.08
Total hydrocarbons	0.20	0.40
Aldehydes	0.07	0.02
Organic acids	0.07	0.03

Technology. This institution is situated in one of the worst of that city's smog belts. Haagen-Smit designed a laboratory smog simulator and explored the generation of ozone by photochemical reactions involving organic material and many types of hydrocarbons, oxides of nitrogen, alcohols, aldehydes, ketones, and acids. The reactions involve the formation of peroxide radicals as intermediate products and may be represented by the following equations:

(1) $NO_2 \xrightarrow{\text{light}} NO + O$ (formation of atomic oxygen)
(2) $O + HR \longrightarrow R + HO$ (free radical formation)
(3) $R + O_2 \longrightarrow ROO$ (peroxyl radical formation)
(4) $ROO + O_2 \longrightarrow O_3 + HO$ (ozone formation)
(5) $ROO + NO_2 \longrightarrow ROONO_2$ (peracyl nitrate formation)

These reactions are represented schematically in Figure 10-3. It should be noted that they are limited to very low concentrations of reactants, which is one of the unique features of air pollution. These atmospheric reactions occur in concentrations on the order of *one millionth* of normal fast laboratory reactions. This explains why ozone and reducing agents can coexist and even free radicals can survive.

The majority of organic compounds are not readily oxidized when quite pure, but in the presence of peroxides they are subject to autoxidation initiated by the removal of hydrogen from the carbon chain. Photolysis, either acting directly or on oxygen, will likewise remove hydrogen. Although most hydrocarbons lack absorption bands in the wavelengths covered by sunlight, other substances may react photochemically and serve as oxygen donors. Nitrogen dioxides act in this way as oxidation catalysts.

Sources of Pollution

The major sources of air pollution in the United States have been described first by Haagen-Smit, then by a number of later workers. From Table 10-4 it can be seen that dust and fumes from heavy industry, refineries, home

FIGURE 10-3. *The photochemical reactions occurring in Los Angeles-type or oxidizing smog.* These convert primary pollutants such as nitrogen oxides into secondary ones such as ozone and PAN (peracyl nitrate), accounting for the differences appearing in emissions balance sheets as illustrated in Table 10-3.

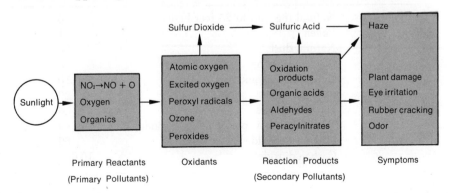

incinerators, and public dumps contribute to smog, but they are not the major source. Automobile exhaust emissions have been held accountable for most of the organic vapors (hydrocarbons), carbon monoxide, nitrogen oxides, and carbon dioxide in smog as shown in this table. Estimated figures of the amount of car emissions are currently 12 million tons of organic vapors, 66 million tons of carbon monoxide, 6 million tons of nitrogen oxides (see Table 10-4).

The general effect of this level of smog persistence and intensity is summarized in an extract from the Clean Air Act of 1963 (88th Congress), which states:

growth in the amount and complexity of air pollution brought about by urbanization, industrial development, and the increasing use of motor vehicles, has resulted in mounting dangers to the public health and welfare, including injury to agricultural crops and livestock, damage to and the deterioration of property, and hazards to air and ground transportation.

London Smog

The first type of smog encountered in cities about the world was a point-source or area type, commonly both. London smog started as an area type with the particulate matter released from the burning of coal for domestic heating and cooking. It worsened with the expansion of coal-fired industrial plants and the development of many point-source emissions. This London-type smog characterized also the earlier smog development in cities of the industrialized U.S. Northeast like Pittsburgh and Cleveland.

The main primary pollutant in this so-called London-type is sulfur dioxide, which is toxic to plants and animals. Sulfur occurs in coal as an impurity and oxidizes on combustion to yield sulfur dioxide. In the United States, where oil early began to replace coal in industrial plants, the sulfur

TABLE 10-4 *Major Air Contaminants over the United States*

Amounts expressed in millions of tons per annum. Some of these air contaminants, such as fog or pollen, might not usually be regarded as air pollutants, but as fluctuations in their amounts can result from human activity, a pollutant budget must include them.

Natural fog (up to 25 feet high)	15
Pollen	1
Natural dust	30
Smoke (carbon)	5
Industrial dust and ash	10
Sulfur oxides	20
Nitrogen oxides	6
Miscellaneous vapors (mostly organic)	40
Carbon monoxide	66
Carbon dioxide	10,000

dioxide emissions continued, for sulfur is also a contaminant of petroleum, or at least of North American petroleum.

Near the source of the emission, sulfur dioxide commonly forms sulfuric acid in the atmosphere; hydrogen sulfide, sulfurous acid, and various sulfates may also be released by the burning of coal or oil. The estimated quantities of total sulfur emissions are enormous, nearly 6 million tons a year for Great Britain, 26 million tons for the U.S., 80 million tons for the whole world. As the present total world production of sulfur by mining is only 30 million tons, this represents a classical case where nonrenewable resources could be conserved by devising an internalized sulfur extraction technology, with considerable economic and environmental gain. Rainfall and snow remove sulfur dioxide and other sulfur pollutants from the atmosphere so that their residence time cannot much exceed 6 weeks.

There is now a growing suspicion that the element in this type of smog that is most damaging to living tissues is the *sulfate*. During the 1952 London smog episode, the sulfur dioxide concentration at no time reached alert levels, which suggests some other pollutant was more actively lethal. Sulfates add to the particulate load and are presently under extensive investigation as possible toxic agents.

Los Angeles Smog

The special features of Los Angeles smog arise from the topographical features that limit movement in the air shed, the meteorological conditions that produce a temperature inversion associated with strong sunlight, and the excessive concentration of motor vehicles—a conservative estimate of 3.9 million in the greater Los Angeles area. These vehicles burn at the lowest estimate about 7 million gallons of gasoline a day (21,500 tons), emitting about 1,800 tons of unburned hydrocarbons, 500 tons of oxides of nitrogen, 9,000 tons of carbon monoxide. In importance automobiles far outweigh any other source of pollutants.

Los Angeles is essentially a commuters' city, working fairly strictly on a 5-day week. Through Saturday and Sunday the haze above the city slowly clears until by early Monday morning the 10,000-foot peak of the San Gabriel Mountains and the nearly 6,000-foot peak of the Santa Ana Mountains are momentarily visible, only to disappear between 8 and 9 AM when the smog haze of the inversion layer has been reestablished. A similar but daily cycle of pollutants may also be observed (Figure 10-4).

One of the effects of smog in Los Angeles and elsewhere results from accumulation of carbon dioxide concentrations in the inversion layer in quantities much greater than that in an unpolluted air shed. This produces what is known as the *greenhouse effect*.

The Greenhouse Effect

The so-called *fossil fuels*—coal, petroleum, natural gas, and lignite—are believed to have been formed during geological periods when the productivity of the producer trophic level of the totality of world ecosystems

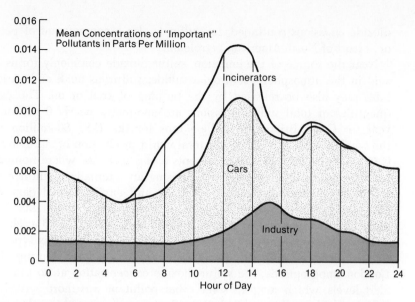

FIGURE 10-4. *The daily cycle of primary pollutants in Los Angeles* showing the emissions from the three principal sources in parts per million over a period of 24 hours in 1955. Although pollution from incinerators has been drastically reduced, both the number of cars and the industrial activity greatly increased, producing a considerably higher air pollution peak in the middle of the day.

was not entirely taken up by consumers and decomposers. The amounts of atmospheric oxygen released during photosynthesis were not therefore balanced by the oxygen-utilizing processes of consumer or decomposer organisms. One consequence of this would be that the amount of carbon dioxide returned to the atmosphere during these oxidative processes would be lower, and its proportionate representation in the troposphere would fall. However, when the fossil fuels are mined and extensively burned, this reduction in the amount of carbon dioxide is abruptly reversed. The new figure can overshoot the original atmospheric concentration. It is estimated that the burning of fossil fuels has increased the transfer rate of carbon dioxide to 3.5 grams per square meter per year, thus releasing 9×10^9 tons $(2 \times 10^{13}$ lb) a year into the atmosphere. This is now adding 0.23 per cent per year to the carbon dioxide content of the atmosphere; the content of this gas in air has risen from 290 to 330 parts per million (ppm) in the present century. As measured by a station situated on the summit of the mountain Mauna Loa in Hawaii, far from polluting industrial activity, the concentration of atmospheric carbon dioxide continues to rise at a rate of 0.72 parts per million (ppm) a year. Although plants absorb carbon dioxide, it has been determined that a new forest greater than the total area of the United States would be needed to absorb the carbon dioxide resulting from the combustion of fossil fuels. If all the remaining reserves were now to be mined and burned, the carbon dioxide content of the atmosphere would increase an estimated seventeen times.

Carbon dioxide gas differentially absorbs the longer-wavelength portion of solar radiation. Light reflected from the earth's surface, which has been converted during reflectance from shorter-wavelength ultraviolet to longer-wavelength infrared, is therefore absorbed instead of being reflected off into the stratosphere again after striking the earth. The accretions of heat energy from this longer-wavelength radiation tend to raise the temperature of the lowest atmospheric layers (Figure 10-5).

It has been calculated that a rise in the amount of carbon dioxide in air to 600 ppm would raise the earth's temperature by 1.5°C, although global air circulation and cloud effects might modify this figure. As the Pleistocene ice ages involved decreases of temperature in the region of 7–9°C, such an increase would be expected to have significant effects on all the world's ecosystems. The build-up of additional heat could be so great that the polar icecaps would begin to accelerate their rate of melting. The International Geophysical Year, 1957 to 1959, produced observations which demonstrated that, with one exception, all the glaciers of the world were retreating. If the polar icecaps also are melting, soon there will be evidence of an extensive shift in sea levels. However, countereffects have appeared as is described in the next few paragraphs.

The Earth's Albedo

It is suspected that, in addition to the greenhouse effect, air pollution has other influences on weather, more especially arising from the presence of particulate matter. Some of these may indeed modify the greenhouse effect. They do so by modifying the albedo of the earth, that is, the effect of reflectance or re-radiation of the solar radiation received at its surface.

E. Aynsley reports a number of accounts evidencing build-up of turbidity in the atmosphere, that is, an increase in atmospheric dustiness. Over Washington, D.C., this has amounted to 57 per cent, over Switzerland 88 per cent. Continuously increasing turbidity is also observed from Mauna Loa, as is the case with carbon dioxide. Some regional estimates provide evidence of increases in particulate matter of about ten times over the last few years. This turbidity not only reduces the amount of solar radiation reaching the earth's surface, but it may also initiate cloud condensation. Aynsley quotes figures from La Porte, Indiana, some 30 miles downwind from the industrial centers of Gary and South Chicago, which indicate that since 1965, La Porte had 31 per cent more rain, 38 per cent more thunderstorms, and 245 per cent more days with hail than adjoining communities not directly downwind of a major industrial area.

In the same way, cloud cover over the North Atlantic is held to have increased by 5–10 per cent because of the particulate matter ejected in high-flying jet aircraft vapor trails; once supersonic aircraft come into operation, an even larger effect may be anticipated. Sometimes under these circumstances "overseeding" may occur. The ice crystals that form about each nucleus may not coalesce but drift away to fall as precipitation elsewhere, and so *reduce* rainfall in the affected area.

V. J. Schaefer, who cites automobile emissions as the most common source of nuclei for ice-crystal formation, finds that present antismog devices

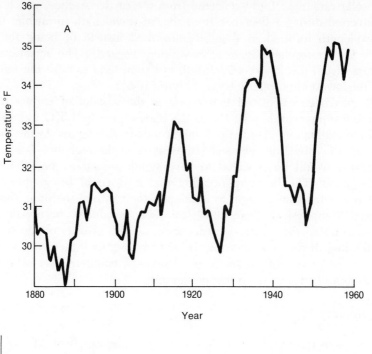

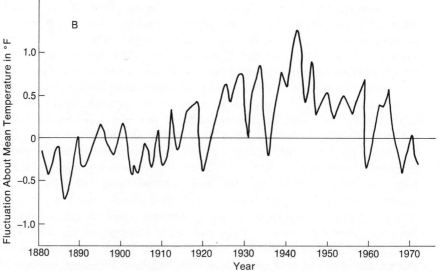

FIGURE 10-5. *Air temperature fluctuations over the last century.* The chart A shows the mean January temperatures of New York City, projected 1871–1958. The upward trend of the general temperature cycle during this period is commonly ascribed to the greenhouse effect. However, a contrary cooling effect may have been introduced by an increase in particular pollutants. This cooling trend can be observed in B, which is a summarized and smoothed expression of the average annual temperature in the Northern Hemisphere over the same time period, in terms of fluctuation about the mean value. It is possible the present cooling trend is the result of a long term climatic oscillation, which effect may or may not be reinforced by increasing atmospheric pollution.

for cars have no effect in reducing this source. The effect of massive release of such nuclei is to condense water vapor in such small droplets that stabilized clouds are formed. He claims that such effects are already changing the weather patterns over hundreds of thousands of square miles.

Total cloud cover over the world is estimated to average about 31 per cent. An increase in only 5 per cent in this cloud cover would so drop temperatures that the earth would be plunged into a new ice age. For the skeptics, a dramatic demonstration of the chilling effect of cutting off even temporarily a portion of the solar radiation can be obtained from experiencing an eclipse of the sun. In contrast with the findings of the International Geophysical Year up to 1960, the subsequent decade is believed to have witnessed a slight *decrease* in the earth's temperature. In 1968 ice coverage in the North Atlantic was the highest recorded so far for this century. When the general temperature trend over the last hundred years is charted by plotting the annual mean temperature for the Northern Hemisphere (Figure 10-5), some of these broad climatic trends can be observed. Until 1940 there was a half-century of general temperature amelioration. Since then, there has been a measurable cooling trend. Climatologists fear that this cooling trend may continue and perhaps increase the number and extent of perturbations in the mean annual rainfall belts of the world such as have already recently occurred.

Episodic Conditions

Over most portions of the earth's land surface a high-pressure weather system will produce a temperature inversion on warm sunny days. Even with the topographically unfavorable conditions described here for Los Angeles, this still permits a fair degree of air circulation. The critical air pollution episodes that sporadically occur are the result of the simultaneous occurrence of both a temperature inversion and a *radiation inversion*. This latter occurs on still nights when clear skies permit the emission into space of long-wave radiation. The resultant loss of heat from the land surface cools also the lowest air layers and creates a low inversion layer in which emitted pollutants are trapped. If the air is moist, as it is on the coast, a *radiation inversion* also develops a fog blanket. This can be dense enough to prevent the sun's rays from penetrating to the ground and breaking up this low inversion layer as the day proceeds.

All major air pollution episodes have resulted from the persistence of this condition where a shallow inversion layer has formed under a stable inverted stagnant air mass. Such episodes can occur almost anywhere over the flatter portions of a great continent like North America, but coastal cities backed by mountain ranges, as is Los Angeles, are particularly exposed to this kind of episodic occurrence. Maps have been prepared for the U.S. indicating in terms of high pollution potential days per annum the expected frequency of occurrence of these stagnating inverted high-pressure conditions. These maps show two systems, one centered on southern California, the other on Appalachia. Elsewhere in the world such pollution potential maps have been extended into an actual smog forecast service linked with enforceable smog relief measures. In the U.S. no such standing measures are routinely enforced, only a broadcast forecast suggestion that children and elderly adults stay indoors and minimize activity while the smog conditions persist.

Effects of Smog on Public Health

Although the greenhouse effect and other meteorological influences on the ecosphere are extremely significant in relation to stability in the earth's climatic ecosystems, certain direct effects of pollutants of more immediate concern to us have been observed in human populations. These range from inconvenient to incapacitating illnesses, and in all too many instances to fatal diseases. Effects on morbidity are most conspicuous during episodic shortperiod high concentrations of pollutants; probably carbon monoxide and ozone produce their more serious effects on public health in this manner. By contrast, particulate matter and certain substances like lead and asbestos are probably more closely related in their health effects to the total amount received over a relatively long period of time. Indeed it may take 40 years for the effects of exposure to asbestos minifibers to be observable.

The particulate matter or soot component of smog is known, from extensive experimental work, to induce skin and subcutaneous cancers when painted on or injected into mice. Such cause-effect relationships demonstrated from animal laboratory experiments are based on use of pollutant concentrations far higher than those encountered under actual field conditions. For this reason, environmental health workers prefer the epidemiological approach, where pollutants effects on a human population are studied over a wide range of conditions for a considerable period of time.

Lung Cancer

In England, there is a correlation between the occupation of chimney sweep and the incidence of lung and scrotal cancer. Follow-up investigations extend this observation to all British city dwellers. Among English immigrants over 30 in New Zealand, deaths from lung cancer were 75 per cent greater than in native New Zealanders. Similar observations were made on immigrants to South Africa and the United States.

No less than 32 polynuclear aromatic hydrocarbons have been identified in automobile emissions, including *benzopyrene*. Table 10-5 indicates the concentrations of benzopyrene in 22 heavily polluted U.S. cities contrasted with nonurban areas. Benzopyrene is believed to be the primary carcinogenic agent in inhaled cigarette smoke responsible for inducing lung cancer. This respiratory disease is estimated to kill 47,000 Americans a year. Its present high incidence is attributable to the habit of cigarette smoking, and correlations between death from lung cancer and the amount of cigarette smoking practiced are undeniably established.

TABLE 10-5 *Benzopyrene Concentration in American Cities*

The highest concentration found in any of the 28 nonurban areas sampled was 51 micrograms of benzopyrene per gram of particulate matter. Of the 94 urban areas sampled some were less than this; the 22 with figures greater than 100 are listed below. Amounts are expressed in micrograms per gram of suspended particulate matter.

City	Micrograms benzopyrene per gram suspended particulate matter
1. Richmond, Virginia	410
2. Montgomery, Alabama	340
3. Charlotte, North Carolina	290
4. Hammond, Indiana	280
5. Altoona, Pennsylvania	280
6. Knoxville, Tennessee	210
7. St. Louis, Missouri	200
8. Youngstown, Ohio	190
9. Raleigh, North Carolina	180
10. Portland, Maine	180
11. Roanoke, Virginia	160
12. Des Moines, Iowa	160
13. Wheeling, West Virginia	140
14. Tampa, Florida	140
15. Flint, Michigan	140
16. Indianapolis, Indiana	120
17. Columbia, South Carolina	120
18. Chattanooga, Tennessee	120
19. Orlando, Florida	110
20. Dearborn, Michigan	110
21. Duluth, Minnesota	110
22. Cleveland, Ohio	110

Although this correlation is indeed indisputable, the Public Health Service has published figures from a survey suggesting that the urban resident of an average-sized American city inhales daily the same quantity of benzopyrene as would be obtained from smoking a third of a pack of cigarettes. In a city where air pollution is heavy, this amount must be raised to the equivalent of a full pack. It has recently been stated that breathing the air of Mexico City is equivalent, in terms of the ensuing lung damage, to smoking two packs a day. The death rate from lung cancer in California is rising (Figure 10-6), as it is in the rest of the United States. For Britain the death rate during this century from lung cancer has increased by a factor of 80; the larger the urban center, the greater the incidence. In Los Angeles, a positive correlation has been established between residence in the vicinity of heavy industry and the incidence of deaths from lung cancer (Figure 10-7).

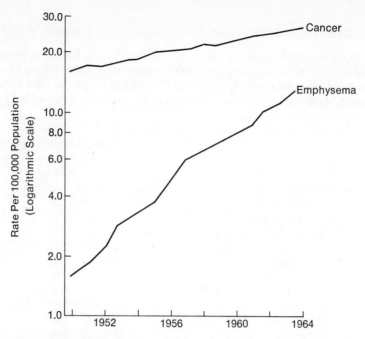

FIGURE 10-6. *Death rate from respiratory diseases.* A recent increase in the death rate from respiratory diseases is illustrated by the incidence of two such ailments, lung cancer and emphysema, in California. In California also, a greater tendency to develop lung cancer was found among people residing near heavy industry (**FIG. 10-7.**) The same holds true for Great Britain (**FIG. 10-8**).

Emphysema

Much controversy still surrounds the precise allocation of causal factors of lung cancer, but there seem to be unequivocal correlations established for environmental effects just as there have been for cigarette smoking. To illustrate with one typical correlation, Haagen-Smit—in noting that emphysema is the fastest growing cause of death in the United States—observes that between 1950 and 1959 deaths among males rose from 1.5 per 100,000 to 8 per 100,000.

Emphysema is a bronchial condition in which the ultimate fine air sacs of the lungs, the alveoli, become inflated and inflexible. The lungs are then unable to *inspire* and *expire* air. The afflicted patient is said to be "short of breath" and is commonly unable even to blow out a candle flame. Progressive deterioration of the cardiovascular system is usually associated with this lung condition, which soon proves terminal. The rapid increase in deaths from emphysema in California, one of the most heavily afflicted regions in the United States, is illustrated in Figure 10-6.

Other Respiratory Diseases

Emphysema is, however, only one recognizable syndrome in a complex of chronic constrictive respiratory conditions that may result from breathing polluted air; there are numerous other examples. Chronic bronchitis is

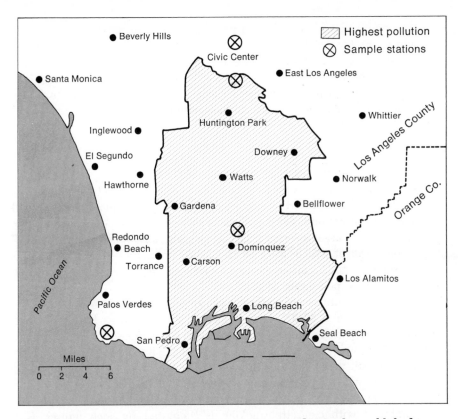

FIGURE 10-7. *Death rate from lung cancer in Los Angeles.* Studies published in 1974 showed that the half-million white males living in the districts indicated on this map had a 40 per cent higher incidence of lung cancer in the last decade than those living elsewhere in Los Angeles. The area is characterized by a higher benzopyrene concentration in air and soil than adjoining areas and by its large number of heavy industries, including oil refineries.

said to represent 10 per cent of the British Medical Health scheme's patient load. That it is not exclusively an English disease was ascertained when chronic bronchitis was diagnosed in 21 per cent of a group of men 40 to 59 years old in the United States. The added strain placed on the heart by difficulties in breathing may be fatal, and the cause of death is then recorded as cardiac failure.

In the Japanese city of Ube, which has heavy air pollution, the death rate from diphtheria was observed to be 7.8 times higher than in a less polluted city, from lung cancer 2.4 times higher, from heart disease 2.2 times higher. In 1959 deaths recorded as resulting from pneumonia in New York City were 50.6 per 100,000 as compared with 38.6 in upstate cities and 24.2 in upstate rural areas.

Possible Synergistic Effects

One major question still unanswered regarding the effect of polluted air on human health is whether there are *synergistic* effects, that is, whether the combined effects of several pollutants are multiplied rather than simply added. As the result of animal experiments, it is suspected that pollutants act synergistically, that inhaling a mixture of two pollutants will have five or six times the effect of inhaling either one singly. One of the most interesting possibilities regarding a synergistic effect has just begun to be appreciated—tobacco smoke as a secondary pollutant.

Cigarette Smoking

The controversy over toxic effects of cigarette smoking on the individual smoker has tended to obscure possible population effects in which tobacco smoke can be regarded as another form of air pollution. Cigarette smoke is known to contain inorganic pollutants such as carbon monoxide and nitrogen dioxide and organic ones such as phenols, aldehydes, benzopyrene, and acrolein, all of which are well-known air pollutants. It also contains some substances—for example, hydrogen cyanide—that never have been reported as general air pollutants.

The concentration of carbon monoxide in cigarette smoke has been reported as high as 42,000 ppm. Concentrations as low as 100 ppm can cause headaches and dizziness. The smoker survives supposedly because he blows away his own smoke and does not reinhale it continuously. In the same way he can avoid the full impact of the 250 ppm of nitrogen dioxide in smoke, of which 5 ppm is believed to represent a dangerous level.

The possible existence of synergistic effects when heavy air pollution is combined with cigarette smoking is suggested by such figures as those shown in Figure 10-8, where the chances of contracting lung cancer or chronic bronchitis are apparently correlated with residence in industrial cities in England and Wales. The microecosystems of large buildings must be considerably modified by the addition to their internalized and partially closed-circuit ventilation systems of considerable quantities of cigarette smoke. The smoker's action therefore goes beyond the question of individual choice and may have significant repercussions on a larger population.

V. J. Schaefer has suggested one further synergistic effect of cigarette smoking. He notes that in addition to ventilating the lungs with a smoke particle concentration 10–100 times greater than that of badly polluted air, the smoker draws air through a burning zone in the cigarette of considerable intensity. The external air frequently contains 10,000–100,000 pollution particles per cubic centimeter. Some of these particles will be vaporized during passage through the burning zone and drawn into the lungs in highly reactive condition. Many additional chemical interactions may therefore occur in the lungs.

Some workers have proposed that the synergistic effects of pollutants are responsible for reversal of the infant mortality decline in this country, observable since 1957. Currently the United States, despite its massive hospital facilities, stands eleventh lowest in the world among the 15 Advanced Industrial societies usually selected for comparison of infant mortality statistics.

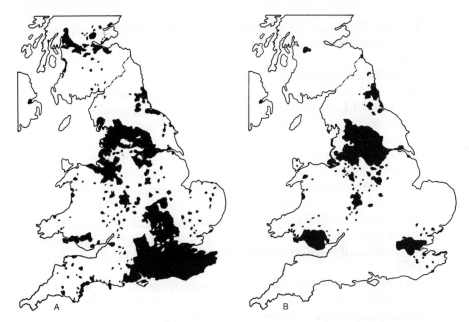

Figure 10-8. *The distribution of lung cancer and chronic bronchitis in England.*
Areas in which the incidence of these diseases is above the national average are
blocked in black. In the case of lung cancer (A) not only is the incidence of
this disease above average in the major industrial centers, but many minor urban
centers, also appear to be affected. Chronic bronchitis (B), on the other hand,
appears to be associated more especially with the large industrial complexes in
South Wales, the Midlands, the North, and the London area.

In considering infant mortality rates it is impossible to ignore another
form of air pollution which has not yet been mentioned, that arising from
radioactive fallout. Hotly debated in the 1950s, this subject has again
begun to arouse considerable controversy.

Radioactive Fallout

A meeting of the Health Physics Society in 1969 considered new information, presented by E. J. Sternglass, indicating that at least one of three
children who died before their first birthday during the 1960s may have
succumbed to the effect of strontium-90 contained in the fallout from the
peaceful nuclear testing carried out in America during that decade.

This contention was vigorously denied by the U.S. nuclear establishment, and by a large majority of scientists. An independent refutation also
was quickly published. Sternglass' argument is therefore not yet considered
fully acceptable, but the evidence on which it is based is summarized here.

Effects on Infant Mortality

Sternglass first considers an increase in the incidence of infant mortality
along the path of the fallout cloud from the first atomic test in New Mexico

in 1945. This was obtained from a detailed correlation of state-by-state infant mortality excesses with the early changes in strontium-90 levels in milk.

By 1963, changes in infant mortality approached one additional death in the United States per 100 live births from the release to that date of 200 megatons of fission energy. Extrapolating, he suggests the release of 20,000 megatons anywhere in the world could result in eliminating any surviving infants. This actually is the amount of fission energy needed in offensive warheads for an effective first strike of defensive ABM warheads.

From 1935 to 1950 infant mortality rates in the country showed a steady decline. After 1950 this drop, according to Sternglass, continued everywhere except in the states downwind of Alamogordo, site of the first series of atomic tests. By 1950 the infant mortality rates in Texas, Arkansas, Louisiana, Mississippi, Alabama, Georgia, and both the Carolinas showed deviations from the mathematical model projecting the declining rates from 1935 to 1950.

More than 1,000 to 1,500 miles away from the test site, in Arkansas, Louisiana, and Alabama, the mortality rates of between 3 and 4.5 per hundred live births increased by 20 to 30 per cent.

Thus the Alamogordo blast appears to have resulted in infant deaths of 1 per cent of children in the area downwind. No effect was observed in Florida, which is south of the fallout cloud path, or in any of the states to the north.

Radioactivity in Milk

During the early 1950s it was discovered that radioactive strontium becomes concentrated in cows' milk and is transmitted along with calcium to the bones of the developing human fetus. Such effects, however, are slow acting, and from studies on young women working in such jobs as painting luminous watch dials, it was found that relatively large amounts of such radiation over long periods were necessary before bone cancer or leukemia could result.

It has been estimated that during 1963 the average dosage of strontium-90 in newly formed bone in England was about 20 millirads, less than one fifth of the radiation received from the natural background. Extrapolating to the end of this century, the accumulating dosage in bone would be about 260 millirads, the natural background amount received in about 32 months (Tables 10-6 and 10-7). Moreover, survivors of the two atomic bomb attacks at Hiroshima and Nagasaki showed no serious long-term effects from radiation. Atomic nuclear weapons were therefore tested in Nevada until 1958, and Pacific tests continued until 1963 (Figure 10-9). In that year the U.S.S.R. and U.S. signed the treaty banning atmospheric nuclear tests. France and the People's Republic of China, neither of whom were signatories, have continued atmospheric testing. French tests in the South Pacific during 1973 aroused vigorous international protest.

TABLE 10-6 *Natural Radiation*

Estimates of the mean dose-rates the world population receives from natural sources, in millirads per year. These amounts should be compared with the figures supplied in Table 10-7 showing comparable fallout statistics.

	Gonads	Cells lining bone surfaces	Blood-forming cells
EXTERNAL			
Cosmic rays	50	50	50
Terrestrial radiation	50	50	50
INTERNAL			
Potassium-40	20	15	15
Uranium and thorium series:			
Radium and decay products	1.3	14.0	1.6
Lead and polonium	0.3	3.6	0.4
Radon	3	3	3
Carbon-14	0.7	1.6	1.6
TOTAL	125	137	122

TABLE 10-7 *Comparisons Between Radiation Received from Natural Radiation and Weapons Testing Fallout*

The estimated doses from 1954 to 2000 AD to which the world population has been exposed as a result of the 1962 tests. Figures express the doses of beta and gamma radiation received in millirads.

	Gonads	Cells lining bone surfaces	Bone marrow
EXTERNAL			
From short-lived nuclides	21	21	21
From cesium-137	29	29	29
INTERNAL			
From strontium-90	—	174	87
From strontium-89	—	0.3	0.15
From cesium-137	13	13	13
From carbon-14	13	13	13
TOTAL	76	257	163
Time during which similar dose is received from natural background (months)	9	32	20

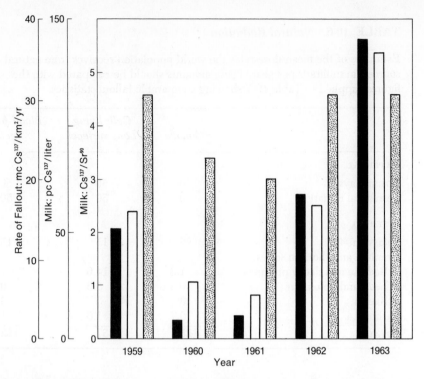

FIGURE 10-9. *Relationship between the amount of strontium in cows' milk and atomic test fallout.* The rate of fallout (solid ground) is determined by recording the amount of cesium-137 in soil. The amount of strontium-90 in milk is estimated by determining the amount of cesium-137 (plain ground) and the cesium-137 to strontium-90 ratio in milk (stippled ground). Although this ratio showed little variation from 1959 through 1963, the amount of cesium and therefore of strontium showed a strong correlation with the amount of fallout each year.

Prenatal Effects

Meanwhile, observations on pregnant women who had been exposed to X-rays indicated the possibility that ova and embryos could be from twenty to fifty times more sensitive to the possible development of leukemia than the mature adults. In support of this contention Sternglass quotes from a 1958 publication by A. Stewart of Oxford University, England, showing that women who had received a series of three to five abdominal X-rays produced children nearly twice as likely to die of leukemia, or other forms of cancer, than children born to mothers who had not been X-rayed during pregnancy. This was somewhat surprising, considering that a single diagnostic exposure only provided a radiation dose of the same magnitude as that received from natural radiation such as cosmic rays (Table 10-6).

Stewart's findings, according to Sternglass, were confirmed by B. Mac-Mahon at the Harvard School of Public Health, who examined the records of 800,000 children born in the larger New England hospitals. He found a 40 per cent increase in the cancer rate among children born to mothers who received diagnostic X-rays during pregnancy.

Leukemia and other cancers are second only to accidents as causes of death among children aged 5–14 years. Examination of the leukemia rate for the state of New York showed a rise and fall in this most susceptible age group correlated with the sequence of individual test series in Nevada between 1951 and 1958. Moreover, the stillbirth rate showed an upward trend beginning within a year after testing began in Nevada in 1951. Instead of steadily declining as it had done from 1935 to 1950, from 1957 to 1963 this stillbirth rate leveled off to about 23 per thousand live births. In 1964 it rose to 27.3 per thousand only to decline again in 1965 and 1966 as the amount of fallout in milk and foods was reduced throughout the United States following the cessation of aerial tests.

By contrast, in California, which is upwind of the Nevada test site, the stillbirth rate continued a steady decline, although this decrease was less up to three years following the onset of hydrogen bomb tests in the Pacific in 1954.

The effects of fallout on the stillbirth rate, however, are more serious than the figures for childhood leukemia, because for every case reported an estimated five or six are not, and those which are recorded are more than ten times the deaths from childhood leukemia.

Correlations with Fallout in Soil

The amount of strontium-90 deposited in soil is recorded readily. Sternglass obtained the relevant statistic where this had been measured and plotted it against the excess of fetal mortality compared with expected mortality, based on the 1935–50 decline.

Birth curves in New York showed a correlated decrease in rate of climb coinciding with the temporary halt of nuclear testing in 1958–61, and both curves showed a sharp rise beginning with the large-scale Soviet test series in 1961. However, after the test ban agreement in 1963, both the fetal death rate and the amount of strontium-90 in soil began to decline.

Infant Mortality

Figures for infant mortality are known more accurately than those for childhood deaths from leukemia or for the number of stillbirths. Like these, according to Sternglass, infant mortality had shown a steady decline in the period from 1935 to 1950. From the onset of the Nevada test in 1951 until the test ban in 1963, however, this decline in the rate leveled off in the United States. It did not do so in other Advanced Industrial societies such as Sweden, Holland, and Norway or southern hemisphere countries like Chile and New Zealand until late in the 1950s. At this time hydrogen bomb testing in the South Pacific and Siberia began to produce world-wide fallout on a much increased scale. The United States infant mortality figures did not continue the 1935–50 decline again until 1965; the observed infant mortality rate per thousand live births during the testing period was 5.4 instead of the estimated 2.7. During the tests, the rate in Sweden still declined to 2.6 per thousand.

Differential Infant Mortality Rates in the United States

Moreover, this halting of the decline in infant mortality was not universally distributed in this country. Within 2 years of the onset of atomic testing in Nevada in 1951, resistance to further decline in infant mortality was most marked in eastern, midwestern, and southern states. This coincided with the known pattern of accumulated radioactive strontium on the ground and in the diet, as the amount carried down from the atmosphere is partly correlated with the amount of rainfall.

Clinical Effects of Radioactivity

K. G. Luning and his coworkers published some 10 years ago the results of experiments demonstrating that small amounts of strontium-90 injected into male mice 3 or 4 weeks before mating produced an increase in fetal deaths among their offspring. The same results were not obtained with radioactive cesium-137.

Sternglass states that evidence produced in May 1969 to an international symposium on radiation biology of the fetal and juvenile mammal, demonstrated severe chromosome damage, fetal death, and congenital malformations in the offspring of female mice injected with strontium-90 before and during pregnancy. Similar effects are reported for very small quantities of tritium, which is produced by both A bombs and H bombs. In this connection Sternglass notes that, following the Hiroshima and Nagasaki bombs, the rate of death from cancer among Japanese children up to 14 years old increased by more than 200 per cent between 1949 and 1951.

To indicate the magnitude of the effects of excess infant mortality, which Sternglass associates with this air pollution by radioactivity fallout, it should be noted that in the 1950s about 3.5 to 3.0 infants of every hundred born in the United States died before attaining the age of 1 year. As about 4 million children were born each year at this time, this means that approximately 40,000 infants of up to 1 year old died as a result of the atomic testing—a total of some 375,000 by the mid-1960s.

As the effects of strontium-90 appear to lie in such factors as lowered birth rate and reduced ability to resist infection, children who have received adequate medical care are more likely to survive than those who do not.

Population Effects of Fallout

Considering the probable population effects of a large nuclear war, the detonation of a single small tactical-sized nuclear weapon on the ground in the western United States appears to have led, Sternglass maintains, to one out of every 100 children born subsequently dying before reaching the age of 1 year. This nation has 8,000 such tactical nuclear weapons to protect western Europe. Obviously this protection, if used, would jeopardize all future human generations in the world.

Although admittedly controversial and disputed by several sources, this presentation by Sternglass of the projected effects of atmospheric pollution by radioactive fallout clearly demands the most serious consideration. It revives a controversy that may erroneously have been considered in these last few years less significant than other aspects of air pollution surveyed in this chapter. The Atomic Energy Commission in 1974 attempted to allay growing fears regarding the possibility of damage from accidental nuclear contamination in the U.S. by publication of a report finding that the chance of sustaining lethal accidental damage could be assessed at $\frac{1}{300}$ million. This contrasted with the $\frac{1}{4,000}$ chance of being killed in a car accident in the U.S.

Other Effects of Air Pollution

The reducing type of air pollution, especially the sulfur dioxide portion of it, is reported to cause extensive damage to agriculture and forestry. Many years ago it was necessary to remove the coniferous plantings maintained by the Royal Botanic Gardens in Kew to a country area outside London. More recent reports of damage have been summarized by Haagen-Smit. Smelters in British Columbia, emitting 600 tons of sulfur daily, almost completely destroyed firs and pines for dozens of miles around. Orange trees are reported damaged in Florida and California, spinach growing severely curtailed in California, tobacco leaves damaged in Maryland, and ponderosa pine trees killed on the San Bernardino Mountains above Los Angeles. Commercial orchid growing is now impossible in any American city, and total agricultural losses from smog damage are put at 500 million dollars annually. Quite recently, G. E. Likens, F. H. Borman, and others, citing new studies, have re-emphasized such general types of smog damage.

Because much of the early damage of this nature was attributed to sulfur dioxide, the change from solid fuels such as coal to liquid types such as diesel oil seems to have reduced these particular effects of air pollution. The smog constituents now held responsible for most agricultural damage in the United States are ozone in the east and peroxyacetyl nitrate (peracyl nitrate, or PAN) in the west. Under smog attack metals, paper, rubber, plastics, and fabrics deteriorate, and buildings must be cleaned and restored frequently. Expensive smog-filtering devices have to be fitted to the air intakes of factories engaged in the manufacture of critical electronic components. Curiously, perhaps, such care does not seem to be taken in the preparation of oxygen cylinders. Oxygen used for the critically ill in hospitals consequently often has a higher proportion of pollutants than the air from which it was prepared.

Some air pollution effects are very complex—for example, the possible reduction in the depth of the ozone-containing layer that girdles the earth in the stratosphere. This ozone layer so effectively reduces the further penetration of ultraviolet radiation from the sun that at the surface of the earth human populations only rarely sustain lethal damage from such radiation. Atmospheric nuclear device testing, which produces nitric oxide, may have initiated chemical reactions that destroy some of the ozone in this layer. Supersonic aircraft flying in the stratosphere and releasing nitrogen oxides in their exhaust plumes may do likewise. It is even suggested that *freon,* the medium extensively used in a variety of domestic aerosol spray cans, is now reaching such levels that it may penetrate the stratosphere and destroy some of the ozone layer. One of the earliest symptoms of reduction in the depth of the ozone layer that could be anticipated would be an increase in human skin cancer, especially in populations provided genetically with a minimum of skin pigmentation.

Control of Air Pollution

Action taken to restore air quality varies from none in many instances to a complete ban on the sale of internal combustion engines (as proposed in 1969 by the California State Assembly but not adopted). In between these extremes, state and federal agencies are enforcing various types of control that reduce at least partially the incidence of one pollutant or another.

"Smokeless" fuel zones in cities reduce the amount of particulate matter and often of sulfur dioxide released. The establishment of such a zone in London following the 1952 disaster is reported to have prevented the recurrence of any of its infamous "pea-souper" fogs. In most large cities there are generally regulations concerning domestic and larger-scale burning of garbage. Natural gas and diesel oil are in any case the most common forms of fuel now used in the United States.

From 1968 all new cars sold in this country had to be fitted with some kind of exhaust emission control device. The first device was concerned with crankcase hydrocarbon emissions; the PVC valve, as this type of device was called, was intended to reduce crankcase emissions 100 per cent.

Further regulations for control of vehicular emissions were implemented in 1971 and again in 1975. The devices on the 1971 models controlled evaporative emissions of hydrocarbons from the carburetor and gas tank. Figure 10-10 shows the calculated combined results of fitting the 1968 and 1971 devices. The 1975 standards were set to control particulate emissions. As it seemed impossible within the time available to produce a device handling lead particulates, it was prescribed that these 1975 models be operated on lead-free gasoline. It is estimated that these several requirements could reduce emissions of hydrocarbons, carbon monoxide, and oxides of nitrogen in California to a minimum by about 1985. Moreover, these figures might prove to be lower than at present, where there has been a reduction in 1971 and later models of carbon monoxide to a third and hydrocarbons to 23 per cent, even allowing for an appropriate increase in the number of motor vehicles (Figure 10-11).

Starting with 1975 models, all new cars sold in California must be fitted with a catalytic converter and operated with lead-free gasoline. Additionally all 1966–1970 used cars registered in the counties sharing the Los Angeles air shed must before mid-1975 be fitted with an additional smog control device. This device will reduce nitrogen oxide emissions, which were not

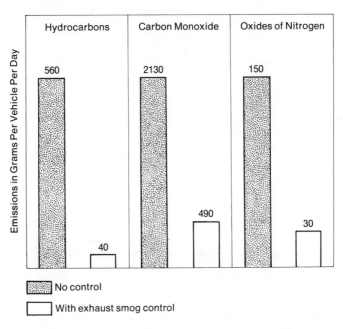

FIGURE 10-10 *Comparison of exhaust emissions from uncontrolled and controlled sources.* Amounts represent grams per vehicle per day as projected for 1972. The three graphs are on a different scale to reduce them to comparable size. The emission controls are those imposed by the new California emissions standards; the three major primary pollutants in smog represented here are very considerably reduced, and the secondary pollutants will likewise be affected.

affected by the control devices fitted to new models until 1971. However, taking into account the imperfect function of many control devices and increases in the atmospheric load of pollutants from world-wide sources, it is not certain that the automotive controls just described will result in a significant reduction of urban smog in Los Angeles or in the country as a whole. The chances of some reduction occurring have increased with recent trends toward the use of smaller cars, lower driving speeds, better mileage performances, and increased availability of bus and other rapid transit systems.

A detailed study of the literature, or a period of residence in one of the more heavily polluted cities, may persuade many people that the need for air pollution control is far more urgent than this and that to wait patiently for some reduction to reach its maximum effect in 1985 is not nearly good enough. Unfortunately, conclusions from scientific experiments or analyses of the effects of pollution are not proved; and where so many vested interests are vitally affected, it is difficult to introduce and enforce legislation on the basis of a *probable* reduction in air pollution.

Environmental Control Agencies

Environmental concern reached a climax in the U.S. in 1970, with vigorous representations made by students, scientists, and other citizens, governmental

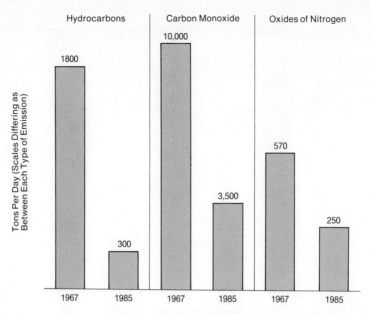

FIGURE 10-11. *The estimated effects of control regulations on vehicular emissions.*
These are estimated figures over the 1967–85 period for vehicles registered in the
Los Angeles metropolitan area, together with the extent of control established
and the amount of estimated emissions.

and industrial groups, and the press, all with a common interest in halting
further environmental degradation. The response at the federal level was
to create the Council on Environmental Quality, which recommended the
amalgamation of various scattered federal environmental responsibilities
into a single body, the Environmental Protection Agency.

The Environmental Protection Agency, better known by its acronym
EPA, was established in 1971 and charged with the formulation and enforcement of laws for the protection of the environment. One of its first
responsibilities was implementation of the Clean Air Act of 1970, which
called for control of emission sources, and especially stringent curtailment
of automobile emissions. By employing an 1899 Refuse Act which had
lately been little used, the EPA was also able to move vigorously to discourage pollution on navigable waterways.

During the early years of this decade numerous countries established
agencies similar to the EPA. Moreover the increasing gravity of global
environmental problems promoted the organization in 1972 of the United
Nations conference on the Human Environment in Stockholm, Sweden. The
Stockholm Conference produced three major concepts, a declaration of
broad environmental principles, an environmental management plan, and
a global environmental agency. The last, known as the Governing Council
for Environmental Programs, has been located in Nairobi, Kenya. It remains at this time in its establishment phase; little progress has been made
as yet in the implementation of the international plan of action for environmental management agreed upon in Stockholm.

Noise Pollution

Because sound waves are transmitted by air, and because the internal combustion engine is simultaneously one of the main sources of both pollutants and noise, the phenomenon described as *noise pollution* is associated increasingly with urban life.

The centers of the world's great cities have in any case long been subjected to excessive noise. London, Paris, and New York most likely sell great quantities of sedatives to permanent residents who cannot otherwise sleep.

The cacophony of sound in the modern world does not, however, emanate solely from such urban sources. The universality of public address systems, linked with the ubiquitous transistor radio, has provided a background of noise against which people everywhere must strive to communicate or concentrate. There is evidence that this causes loss of efficiency. There also is much evidence that this exposure produces irreversible deterioration of the auditory apparatus and that many city dwellers have suffered a serious loss of auditory sensitivity by middle age, as compared with those in Advanced Agricultural societies, where little deterioration occurs up to or past the age of 65. Recent studies reveal that a partial hearing loss has occurred among 85 per cent of the adult Eskimo population of Baffin Island, following the introduction of snowmobiles and hunting rifles. Standards developed by the U.S. Air Force recommend the use of ear "defenders" if noise levels exceed 85 decibels. Unfortunately, "ordinary" noise too frequently exceeds this figure.

The effects of noise pollution on "aggression," using this term in its broadest sense, have not been extensively investigated, although in the United States there was recently a case of a man seizing an axe and dashing into the street to demolish the persistent chimes of a mobile ice cream vendor. In this country also, an appreciable percentage of neighbor dispute incidents to which the police are called are classifiable as noise pollution problems—too loud a party, excessive noise from hi-fi equipment, hammering in the late or early hours, dogs barking, and cocks crowing. Perhaps for the suburbs the most widespread source of irritation is low-flying aircraft, both commercial and private planes. There is a growing body of public opinion favoring insistence on stricter regulations regarding jet engine noise and abandoning of all plans to introduce commercial supersonic planes, although Britain, France, and Russia currently have tested models in commercial production. This latter type of aircraft would benefit very few while subjecting many to shattering and frequent sonic booms.

Problems of noise pollution are more likely to be debilitating than disastrous; the other immediately critical environmental problem that closely parallels air pollution is *water pollution*. Just as it is difficult to define air pollution precisely, so water pollution lacks any concise definition. It is best considered in relation to the function of water in the world's ecosystems, just as air pollution can be treated as a disturbance of the vital role that air plays as a major abiotic element of natural ecosystems.

BIBLIOGRAPHY

References

Anonymous. "Air pollution control in California," *1968 Annual Report, Air Resources Board,* State of California, 1969.

Anthrop, D. F. "Environment noise pollution: a new threat to sanity," *Bull. Atomic Scientists,* **25**(5): 11–16, 1969.

Aynsley, E. "How air pollution alters weather," *New Scientist,* **44**: 66–67, 1969.

Bach, W. *Atmospheric Pollution,* New York: McGraw-Hill, 1972.

Boffey, P. M. "Ernest J. Sternglass: controversial prophet of doom," *Science,* **166**: 195–200, 1969.

Boggs, D. H., and Simon, J. R. "Differential effect of noise on tasks of varying complexity," *J. Appl. Psychol.,* **52**: 148–53, 1968.

Bregman, J. L., and Lenormand, S. *The Pollution Paradox,* New York and Washington, D.C.: Spartan Books, 1969.

Breidenbach, A. W., and Eldredge, R. W. "Research and development for better solid waste management," *Bioscience,* **19**: 984–88, 1969.

Brodine, V. "Episode 104," *Environment,* **13**(1): 2–27, 1971.

Broecker, W. S. "Man's oxygen reserves," *Science,* **168**: 1537–38, 1970.

Brown, R. "Assessing damage from sonic booms," *New Scientist,* **41**: 116–17, 1969.

Cadle, R. D., and Allen, E. R. "Atmospheric photochemistry," *Science,* **167**: 243–49, 1970.

Carter, L. J. "SST: commercial race or technology experiment?" *Science,* **169**: 352–55, 1970.

Changnon, S. A., Jr. "The La Porte weather anomaly—fact or fiction? *Bull. Amer. Meteorol. Soc.,* **49**(1): 4–11, 1968.

Changnon, S. A., Jr. "Recent studies of urban effects on precipitation in the United States," *Bull. Amer. Meteorol. Soc.,* **50**: 411–21, 1969.

Charlson, R. J., and Pilat, M. J. "Climate: the influence of aerosols," *J. Appl. Meteorol.* **8**: 1001–2, 1969.

Cohen, A. "Effect of noise on psychological state," *Amer. Soc. Safety Eng. J.,* **14**: 11–15, 1969.

Commoner, B. "Review: the closing circle," *Environment,* **14**(3): 25, 40–52, 1972.

de Nevers, N. "Enforcing the Clean Air Act of 1970," *Scientific American,* **226**(6): 14–21, 1973.

Ehrlich, P. R., and Holdren, J. P. "Review: the closing circle," *Environment,* **14**(3): 24, 26, 31–39, 1972.

Gentilli, J. *A Geography of Climate,* Perth, Western Australia: University of Western Australia Text Books Board, 1952.

Golden, J., and Morgan, T. R. "Sulfur dioxide emissions from power plants: their effect on air quality," *Science,* **171**: 381–83, 1971.

Goldsmith, J. R., and Landaw, S. A. "Carbon monoxide and human health," *Science,* **162**: 1352–59, 1968.

Haagen-Smit, A. J. "Reactions in the atmosphere," in A. C. Stern (ed.), *Air Pollution,* rev. ed., Vol. 1, New York: Academic Press, 1968.

Haagen-Smit, A. J. "Air conservation," *Scientia,* 163: 359–67, 1969.

Hexter, A. C., and Goldsmith, J. R. "Carbon monoxide: association of community air pollution with mortality," *Science,* 172: 265–67, 1971.

Hinkley, E. D., and Kelly, P. L. "Detection of air pollutants with tunable diode lasers," *Science,* 171: 635–39, 1971.

Holland, G. J., Benson, D., Bush, A., Rich, G. Q., and Holland, R. P. "Air pollution simulation and human performance," *Amer. J. Publ. Health,* 58(9): 1684–91, 1968.

Kryter, K. D. "Sonic booms from supersonic transport," *Science,* 163: 359–67, 1969.

Lave, L. B., and Seskin, E. P. "Air pollution and human health," *Science,* 169: 723–33, 1970.

Likens, G. E., Bormann, F. H., and Johnson, N. M. "Acid rain," *Environment,* 14(2): 33–40, 1972.

Likens, G. E., and Bormann, F. H. "Acid rain: a serious regional environmental problem," *Science,* 184: 1176–78, 1974.

Loutit, J. F., and Scott Russell, R. "Criteria for radiation protection," in R. Scott Russell (ed.), *Radioactivity and Human Diet,* Oxford, England: Pergamon Press, 1966, p. 40.

Luning, K. G., Frolen, H., Nelson, A., and Ronnback, C. "Genetic effects of strontium-90 injected into male mice," *Nature,* 197: 304–5, 1963.

Machta, L., and Hughes, E. "Atmospheric oxygen in 1967 to 1970," *Science,* 168: 1582–84, 1970.

Middleton, J. T., and Haagen-Smit, A. J. "The occurrence, distribution, and significance of photochemical air pollution in the United States, Canada and Mexico," *J. Air Pollution Control Assn.,* 11: 129–34, 1961.

Parmeter, J. B., Bega, R. V., and Neff, T. "A chlorotic decline of ponderosa pine in southern California," *U.S. Department of Agriculture Plant Disease Report,* 46: 269–73, 1962.

Rankin, R. E., "Air pollution control and public apathy," *J. Air Pollution Control Assn.,* 19(8): 565–69, 1969.

Schaefer, V. J. "The nuclei from auto exhaust and organic vapors," *J. Appl. Meteorol.,* 7: 113, 1968.

Schaefer, V. J. "The inadvertent modification of the atmosphere by air pollution," *Bull. Amer. Meteorol. Soc.,* 50: 199, 1969.

Schaefer, V. J. "Some effects of air pollution on our environment," *Bioscience,* 19: 896–97, 1969.

Scott, J. A. "The London fog disaster," *Proc. 20th Ann. Clean Air Conf. Glasgow,* 1953, pp. 25–27.

Shy, C. M., Creason, J. P., Pearlman, M. E., McClain, K. E., and Benson, F. B. "The Chattanooga school children study: effects of community exposure to nitrogen dioxide," *Air Pollution Control Assn.,* 20: 582–88, 1970.

Spar, J. "Temperature trends in New York City," *Weatherwise,* 7(6): 149–51, 1954.

Stein, J. "Coal is cheap, hated, abundant, filthy, needed," *Smithsonian,* 3(11): 18–27, 1973.

Sternglass, E. J. "Has nuclear testing caused infant deaths?" *New Scientist,* 43: 178–81, 1969.

Stewart, A. "The pitfalls of extrapolation," *New Scientist,* 43: 181, 1969.

Thompson, D. N. *The Economics of Environmental Protection,* Cambridge, Mass.: Winthrop, 1973.

Westberg, K., Cohen, N., and Wilson, K. W. "Carbon monoxide: its role in photochemical smog formation," *Science,* 171: 1013–15, 1971.

Wise, W. *Killer Smog,* New York: Rand McNally, 1968.

Further Readings

Bertine, K. K., and Goldberg, E. D. "Fossil fuel combustion and the major sedimentary cycle," *Science,* 173: 233–35, 1971.

Brodine, V. *Air Pollution,* New York: Harcourt Brace Jovanovich, 1973.

Eighty-eighth Congress. An Act to Improve, Strengthen, and Accelerate Programs for the Prevention and Abatement of Air Pollution, Public Law 88-206, 88 Congress, H.R. 6518, 1963.

Environmental Protection Agency. *The Clean Air Act,* Washington, D.C.: Government Printing Office, 1970.

Ferguson, F. A., Semran, K. T., and Monti, D. R. "SO₂ from smelters: by-product markets a powerful lure," *Environ. Sci. Technol.,* 4(7) 562–68, 1970.

Freeman, A. M., Haveman, R. H., and Kneese, A. V. *The Economics of Environmental Policy,* New York: Wiley, 1973.

Haagen-Smit, A. J. "Carbon monoxide levels in city driving," *Arch. Environ. Health,* 12: 548–50, 1966.

Hamburg, F. C., and Cross, F. L., Jr. "A training exercise on cost-effectiveness evaluation of air pollution control strategies," *J. Air Pollution Control Assn.,* 21(2): 66–70, 1971.

Heller, A. *The California Tomorrow Plan,* Los Altos, Calif.: Kaufmann, 1972.

Hepting, G. H. "Damage to forests from air pollution," *J. Forestry,* 62: 630–34, 1964.

Hohonemser, K. "Onward and upward," *Environment,* 12(4): 22–27, 1970.

Landsberg, H. E. "Man-made climatic changes," *Science,* 170: 1265–74, 1970.

Miller, M. E., and Holzworth, G. C. "An atmospheric diffusion model for metropolitan areas," *J. Air Pollution Control Assn. Amer.,* 17: 46–50, 1967.

Miller, P. R., Parmeter, J. R., Taylor, O. C., and Cardiff, E. A. "Ozone injury to the foliage of *Pinus ponderosa,*" *Phytopathology,* 53: 1072–76, 1963.

Miller, R. W. "Delayed radiation effects in atomic-bomb survivors," *Science,* 166: 569–74, 1969.

Munn, R. E. "Air pollution meteorology," *Occup. Health Rev.,* 20(3–4): 1–8, 1968–69.

Newell, R. E. "Modification of stratospheric properties by trace constituent change," *Nature*, **227**: 697–99, 1970.

Panofsky, H. A. "Air pollution meteorology," *American Scientist*, **57**(2): 269–85, 1969.

Scott Russell, R. (ed.) *Radioactivity and Human Diet*, Oxford, England: Pergamon Press, 1966.

Scorer, R. S. *Pollution in the Air*, London: Routledge and Kegan Paul, 1973.

Sterling, T. D., et al. "Measuring the effect of air pollution on urban morbidity," *Arch. Environ. Health*, **18**: 485–94, 1969.

Stern, A. C. (ed.) *Air Pollution*, 2nd ed., New York: Academic Press, 1968.

Sternglass, E. J. "Infant mortality," *Environment*, **11**(10): 9–13, 1969.

U.S. Public Health Service. *Proceedings of the National Conference on Air Pollution of 1962*, 1963.

Wayne, W. S., et al. "Oxidant air pollution and athletic performance," *J. Am. Med. Assn.*, **199**(912): 151–54, 1967.

Weinstock, B. "Carbon monoxide: residue time in the atmosphere," *Science*, **169**: 224–25, 1969.

Willie, A., and Weyermuller, G. H. "Pollution control system saves plant $60,000," *Chemical Processing*, **32**(1): 15–20, 1969.

Woodwell, G. M. "Radioactivity and fallout: the model pollution," *Bioscience*, **19**: 884–87, 1969.

11

Water Pollution

The last chapter reviewed the pollution problems that confront us because industrial societies presently release into air sheds a heavier load of waste products than the reducing and recycling processes of their ecosystems can handle. It is entirely arbitrary and purely for convenience that the process of *air pollution* is separated in this text from *water pollution*. Radioactive fallout in the air sooner or later contaminates all natural waters. Disturbance of gas ratios in the atmosphere affects their concentrations in the water of lakes, rivers, and oceans. Toxic particulate matter released into the atmosphere must eventually settle or be washed down onto water as well as land surfaces. Sources of air pollution may equally well also be sources of water pollution.

There are, however, certain differences between air and water pollution that provide real substance as well as some convenience to the distinction. Not the least of these is the circumstance that air and water pollution are usually investigated by somewhat different techniques. Thus separate research groups conventionally handle these two interrelated aspects of environmental wastes.

Recycling Processes

In any ecosystem, water is essentially the vehicle through which recycling of nutrients is achieved. Water is in any case necessary in some form or another for the continued existence of most producer, consumer, and reducer populations, but it is especially with respect to this recycling process that the water transport of essential substances is most critical for the continued functioning of a given ecosystem.

The two water-soluble substances most critical in these recycling processes are *nitrates* and *phosphates,* both of which are limiting to the productivity of most natural ecosystems. Any disturbance of the process of recycling dissolved nitrates and phosphates through an ecosystem will therefore generally have most critical effects on it. Decomposer populations in aquatic ecosystems usually release the more soluble nitrates quicker than the less soluble phosphates. This is readily observable in lake ecosystems.

Lake Ecosystems

Most lakes are comparatively short-lived, at least in a geological sense. During the Pleistocene there were extensive lake systems extending over much of North America, but these are represented now by only a few reduced water surfaces, together with a vast and awe-inspiring system of fossil lake beds. The most familiar of these is the Great Salt Lake in Utah, the

FIGURE 11-1. *The natural process of ecological evolution of a lake, or eutrophication.* This leads through a series of successional stages, from oligotrophic through mesotrophic to eutrophic, in which the amount of nutrients recycled and the biomass that accumulates in bottom sediments gradually increase, eventually filling in the lake.

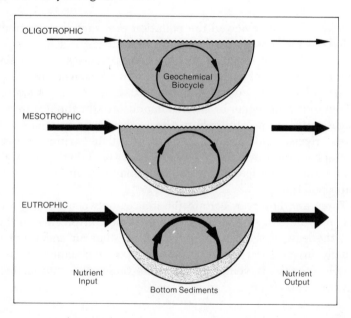

shrunken remnant of the once immense Lake Bonneville. About the world the majority of lakes are of Holocene, or Recent, date; they are shallow cyclic bodies of water that appear, mature, silt up, and disappear in a matter of centuries rather than millennia.

If the drainage basin of any of these various lakes is infertile and supplies few nutrients, it is said to be *oligotrophic.* On the other hand, when it is located in an area of readily weathered rocks, which release considerable amounts of nutrients, the lake ecosystem productivity is high and it is said to be of a *eutrophic* type.

Oligotrophy and eutrophy are really the extremes of a continuous range. Because available nutrients in a lake accumulate with time, only a small percentage being lost in the outflow waters, even an oligotrophic lake eventually becomes eutrophic; that is, it undergoes a successional ecological process of *eutrophication* (Figure 11-1). It does so because the biomass produced at various trophic levels, if not consumed by organisms of higher trophic level, sinks to the bottom of the lake where decomposers reduce it, eventually releasing inorganic nutrients. The more soluble nitrates are carried away rapidly in any outflow; the phosphates tend to be lost more slowly, as already noted.

At the bottom of the lake the decomposer organisms may be limited in their activity by one or more ecological factors, such as low oxygen pressure or low temperature. The organic matter then tends to accumulate instead of being broken down immediately. This, as was observed in an earlier chapter, is the main source of fossil fuels, if bottom deposits in lakes are included with those in marine basins and river systems.

Disposal of Waste

Until human populations reached the size and densities achieved following the Industrial Revolution, the amount of supplemental organic matter added to the larger bodies of water reached only local significance. Human excrement indeed was very often simply thrown into any drainage system existing in the streets. The invention of the indoor toilet in the early part of the nineteenth century changed this situation relatively little, for sewage was then flushed untreated into the nearest watercourse where it had eventually ended up anyway. It is estimated that 20 per cent of U.S. sewage still is handled in this manner. In addition, many waterways and marine basins lack laws or enforcement of laws forbidding the use of open heads, that is, directly flushed toilets, on boats.

Adding untreated sewage to an aquatic ecosystem seriously disturbs its process of biogeochemical circulation by vastly increasing the accumulation of organic matter and the demand for oxygen. Sewage treatment does not prevent this, for whereas *primary* treatment consists of the removal of solid matter, *secondary* treatment is essentially a fermentation process in which nitrates and phosphates are released in large quantity. The resultant oxidation considerably reduces the "biochemical oxygen demand." In either case, the maturation from an oligotrophic to a eutrophic stage is accelerated greatly, and the process of *eutrophication* proceeds so fast that the normal successional phases are shortened and bypassed.

Waste production by animals in the United States has been estimated as approximately ten times that of the human population, attaining a figure in the region of 2 billion tons annually. Large increases in the number of domestic animals coupled with a modern tendency toward "factory" production result in such dense concentrations that their waste products also cause serious eutrophication problems.

Eutrophication

The addition of organic matter in the form of untreated human or animal excrement or outflow from a treated sewage plant rich in phosphates and nitrates thus results in accelerated eutrophication of the body of water into which the effluents drain. This is commonly described as *cultural eutrophication*, to distinguish it from the natural successional process. On a smaller

TABLE 11-1 *Estimated Amounts of Nitrogen and Phosphorus Reaching Wisconsin Surface Waters from Various Urban and Rural Sources*

Amounts are expressed in units of 1000 pounds per annum. These figures illustrate a number of points that occur repeatedly—e.g., the high phosphate content of sewage plant effluents (from detergents) and the high nitrogen content leached to ground water tables—but this comparatively rural state is below average in some features, such as those for industrial wastes.

	Nitrogen		Phosphate	
Source	Actual amount	Percentage figure	Actual amount	Percentage figure
URBAN				
Municipal sewage treatment installations	20,000	24.5	7,000	55.7
Private sewage systems	4,800	5.9	280	2.2
Industrial wastes	1,500	1.8	100	0.8
Urban runoff	4,450	5.5	1,250	10.0
Subtotal	30,750	37.7	8,630	68.7
RURAL				
Manured lands	8,110	9.9	2,700	21.5
Other cropland	576	0.7	384	3.1
Forest land	435	0.5	0.43	0.3
Pasture, woodlot, and other lands	540	0.7	360	2.9
Ground water	34,300	42.0	285	2.3
Precipitation on surface of water areas	6,950	8.5	155	1.2
Subtotal	50,911	62.3	3,927	31.3
Grand Total	81,661		12,557	

scale, individual septic tanks produce a similar effect, for the nitrates and phosphates they release find their way first to the water table, then to stream and river drainage systems.

Another cause of accelerated or cultural eutrophication is the large amount of soluble fertilizers now applied to most agricultural and horticultural crops in an attempt to boost yields. When organic manure was the sole fertilizer used by farmers, release of inorganic nutrients was relatively slow, and they were mostly taken up by the soil and plants rather than leached into the water table or carried off in flood waters. The liberal use of such compounds as ammonium sulfate and potassium phosphate now results in as much as one half of the added nutrients being lost in drainage water (Table 11-1). The nitrate and phosphate pollution of natural waters by such agricultural drainage may entirely disturb the balance of populations in an aquatic ecosystem (Figure 11-2). Although these usually would evolve eventually to a more eutrophic state, this accelerated process is another example of cultural eutrophication. And whereas the effluent from sewage plants may be treated further to remove particularly the nitrates and phosphates, it is more difficult and even more expensive to control water pollution from agricultural sources, and little attempt presently is made to do so.

FIGURE 11-2. *Production of carbon by algal photosynthesis in four Swiss alpine lakes. A.* Millstatter. *B.* Klopeiner. *C.* Worther. *D.* Constance (lower part). These lakes show progressively higher levels of eutrophication, as indicated by their increased production in milligrams of carbon per cubic millimeter per day. In *D,* however, the increased algal biomass restricts light penetration in depth into the lake, so that algal growth as measured by mg C/m³ per day production is virtually restricted to the surface 3 meters.

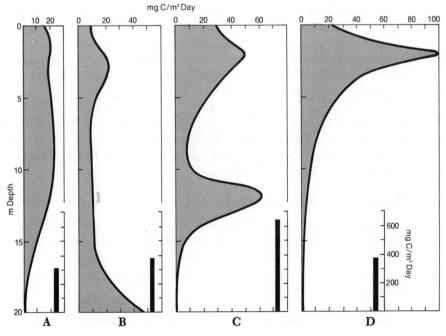

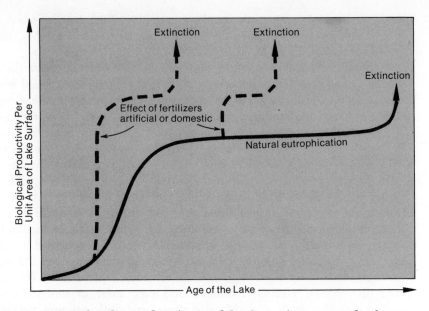

FIGURE 11-3. *Cultural eutrophication in a lake.* Curve A represents the slow process of ecological succession from an oligotrophic to a eutrophic condition illustrated in Figure 11-1. Curve B shows the effect when the increase in net primary productivity is advanced by the addition of nitrates and phosphates into the biochemical cycling processes of the natural ecosystem. Curve C represents even heavier dosages of such additions.

Rate of Eutrophication

Although all natural waters age and undergo a process of succession from an oligotrophic to a eutrophic state, a great acceleration of the rate by cultural eutrophication can be very disturbing (Figure 11-3). It is likely to result in a number of highly undesirable primary consequences such as depletion of oxygen, increased turbidity of the water, accumulation of organic matter, and blooms of toxic algae, together with a wide range of secondary consequences such as elimination of many fish species.

Cultural eutrophication has been observed in various lakes throughout the world. The first to be investigated was Lake Zürich in Switzerland. Fure Lake in Denmark and Lake Washington in the Pacific Northwest are two other early examples. The most extensive, in terms of surface area at least, are the American Great Lakes.

Detection of Cultural Eutrophication

As cultural eutrophication is basically an accelerated rate of increase in net primary productivity, it may be detected by any method that measures the rate of net primary productivity. Sometimes the change of rates is so obvious that it can be seen or smelled; it is expressed in the following features:

1. *Increased amounts of standing crop*—In particular the amount of algal growth is increased greatly (Figure 11-4); sometimes toxic species be-

come dominant. A thick algal mat frequently fills stream beds in areas where the septic tanks of vacation homes drain down from a watershed area.

2. *Diminished transparency*—as measured with a standard Secchi disc. One of the early signs that the accelerated eutrophication of Lake Washington was being slowed was the realization that the beer cans on the bottom of the lake had become visible again. The increasing turbidity localizes algal growth in the surface layers of water (Figure 11-2), and utilization of food materials becomes restricted to near-surface water.

3. *Oxygen production*—measured scientifically by a dark and light bottle technique. Lack of oxygen may become obvious by the smell when anaerobic processes in the deoxygenated lake result in the formation of odoriferous gases (Figure 11-5). Hydrogen sulfide in particular tends to accumulate in toxic concentrations when biochemical oxygen demand from untreated biological wastes deoxygenates lake waters significantly.

4. *Nutrient levels*—particularly of nitrates and phosphates, which become higher as eutrophication proceeds.

TABLE 11-2 *Physicochemical Characteristics of the St. Lawrence Great Lakes*

Lakes Erie and Ontario, into which the St. Lawrence drains, have a low concentration of oxygen in the hypolimnion, a characteristic of accelerated eutrophication. These two lakes also have high amounts of dissolved solids, which similarly characterize this condition.

Lake	Mean depth (m)	Transparency (Average Secchi disc depth, m)	Total dissolved solids (ppm)	Specific conductance (μmhos at 18° C)	Dissolved oxygen
Oligotrophic	>20	High	Low: around 100 ppm or less	<200	High, all depths all year
Superior	148.4	10	60	78.7	Saturated, all depths
Huron	59.4	9.5	110	168.3	Saturated, all depths
Michigan	84.1	6	150	225.5	Near saturation, all depths
Eutrophic	<20	Low	High: >100	>200	Depletion in hypolimnion <70% saturation
Ontario	86.3	5.5	185	272.3	50–60% saturation in deep water in winter
Erie, average for lake	17.7	4.5	180	241.8	
Central basin	18.5	5.0	—	—	<10% saturation, hypolimnion
Eastern basin	24.4	5.7	—	—	40–50% saturation, hypolimnion

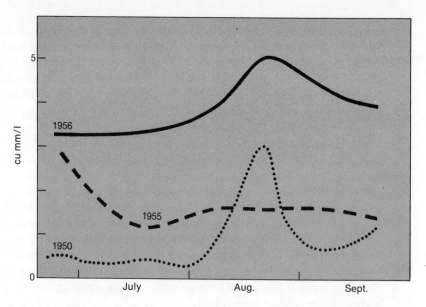

FIGURE 11-4. *The amount of algae as an indicator of cultural eutrophication.* The seasonal growth of algae, typically with a spring bloom, which characterizes even oligotrophic lakes, occurred in Lake Washington (near Seattle) in 1950 and began to persist throughout the year as eutrophication developed, until the volume of algae in 1956 persisted at a level higher than the seasonal level of the unpolluted lake.

The Great Lakes

In considering the rate of eutrophication, limnologists are by no means in agreement as to recognizable stages. It is convenient, however, to distinguish between *physicochemical* and *biological* conditions. The physicochemical conditions of a lake include water transparency, amounts of dissolved solids, electrical conductivity, and quantity of dissolved oxygen.

Physicochemical Conditions

Table 11-2 lists measurements of physicochemical values for the Great Lakes. It can be seen from this table that Lakes Erie and Ontario have a low concentration of oxygen in the *hypolimnions* (the water below the *thermocline*), which is characteristic of eutrophic conditions. The high total of dissolved solids and conductance values for these two lakes also indicate eutrophic conditions.

Biological Characteristics

The biological features of the Great Lakes, summarized in Table 11-3, indicate that all except Lake Erie are in an oligotrophic state as far as biological characteristics are concerned. This is evidenced by the dominance of salmonids—salmon, trout, and char—in the fish populations. Although

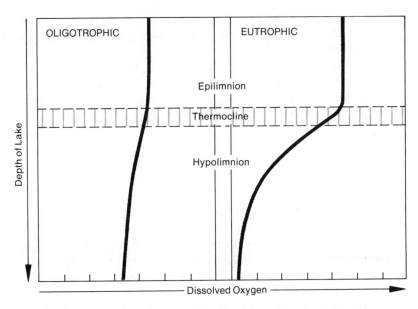

FIGURE 11-5. *Summer stagnation in a lake through oxygen depletion.* The thermocline that forms in summer, separating water circulation in the warmer upper section of the lake from that in the cooler deeper portion, acts as a barrier to rapid gaseous exchange. The oxygen content of the water below the thermocline becomes very low because of the activity of decomposer organisms that are reducing the increased amounts of organic matter settling out from the higher productivity of a eutrophic condition in the lake. In winter the surface layers of water are cooled and become more dense, overturn occurs, the thermocline disappears, a single water circulation develops in the lake, and the resulting enhanced gaseous exchange restores normal oxygen levels throughout the water depth.

species of these salmonid fish groups may still be present in the eutrophic conditions of Lake Erie, they have been extensively replaced by perch, bass, smelt, and dorem.

Degree of Eutrophication

Combining this physicochemical and biological information, A. M. Beeton considers that Lakes Huron and Superior are definitely oligotrophic, Lakes Michigan and Ontario mesotrophic. Because the water of the clearly eutrophic Lake Erie flows on into Lake Ontario via the Niagara Falls, it is not surprising that the dissolved solid content of this last lake also is typical of eutrophic conditions.

Supporting evidence for these conclusions regarding the St. Lawrence Great Lakes is sometimes confused and is drawn mostly from studies on the behavior of much smaller bodies of water. In regard to the chemical characteristics that Beeton summarizes, Lakes Erie and Ontario alone show significant increases in calcium and chloride whereas sulfates have increased in all except Lake Superior (Figure 11-6). Chlorides and sulfates are major components of domestic and industrial waste.

Some workers identify blooms of the blue-green planktonic alga *Oscilla-toria rubescens* with later stages of eutrophication. Such indicator blooms have occurred in various European lakes, but planktonic studies on the Great Lakes were at first too inadequate to provide any useful evidence as to the occurrence of this alga. Subsequent research, however, suggests that certain planktonic changes are taking place uniquely in Lake Erie.

Lake Erie

Since cultural eutrophication appears to have proceeded furthest in Lake Erie of all the Great Lakes, attention has been focused there on the causes for it. As in all five lakes, the change seems to have occurred during the past half century. During this time the population of the northeast central states grew from approximately 16 million in 1900 to some 36 million in 1960. Twenty-five million people are now estimated to be living in the communities around the shores of the Great Lakes; approximately 11 million of them—about one twentieth of the population of the United States—are concentrated on Lake Erie. The changes that have occurred in Lake Erie have gone much beyond mere eutrophication, owing to excessive release into it of phosphates and nitrates. The dumping of wastes and garbage of all kinds, combined with the discharge of domestic and industrial effluents, has given the lake the unenviable distinction of being the most polluted large body of water in the world.

The Dying Lake

To many, what has befallen Lake Erie is the first (and perhaps last) large-scale warning that we are now destroying by industrial exploitation the habitability of this earth for humans. There can be no doubt that it is a massive commentary on our loss of environmental quality.

The western end of the lake is dead, with no sign of life above or below the water. Patches of oil, trash, and sewage float on the surface; foul-smelling scum mixed with the bodies of waterfowl, fish, and assorted jetsam, among which scuttles an occasional rat, now line the shore. The Cuyahoga and Buffalo rivers have been declared *fire hazards,* and the former, which flows into the lake at Cleveland, has actually caught fire.

The somewhat spectacular estimate has been made that the lake now contains a billion tons of algae—enough to fill a freight train 40,000 miles long. The western quarter of the lake is becoming an algae marsh.

Fishing and Hunting

Commercial fishing, once a thriving industry on Lake Erie, has declined and almost disappeared with the loss of such species as whitefish, pickerel, sauger, and sturgeon (Figure 11-7). One fishing group on the Vermillion River, which once employed over 200 men, now has only nine or ten. An estimated 12,000 ducks were once reported killed on the Detroit River by oily pollutants, and in many areas the only animals left to hunt are rats.

TABLE 11-3 *Biological Characteristics of the St. Lawrence Great Lakes*

All except Lake Erie on this biological basis are still in an oligotrophic condition.

Lake	Bottom fauna and dominant midges	Dominant fishes	Plankton abundance	Dominant phytoplankton
Oligotrophic	Orthocladius-Tanytarsus type (Hydrobaenus-Calopsectra)	Salmonids	Low	Asterionella formosa Melosira islandica Tabellaria fenestrata Tabelleria flocculosa Dinobryon divergens Fragilaria capucina
Superior	Pontoporeia affinis Mysis relicta Hydrobaenus	Salmonids	Very low	Asterionella formosa Dinobryon Synedra acus Cyclotella Tabellaria fenestrata Melosira granulata Fragilaria crotonensis
Huron	Pontoporeia affinis Mysis relicta Hydrobaenus Calopsectra	Salmonids	Low	Tabellaria fenestrata Fragilaria construens Fragilaria pinnata Cyclotella kutzingiana Fragilaria capucina
Michigan	Pontoporeia affinis Mysis relicta Hydrobaenus	Salmonids	Low	Fragilaria crotenensis Melosira islandica Tabellaria fenestrata Asterionella formosa Fragilaria capucina
Ontario	Pontoporeia affinis Mysis relicta	Salmonids, ictalurids, percids	—	—
Eutrophic	Tendipes plumosus type	Yellow perch, pike, black bass	High	Microcystis aeruginosa Aphanizomenon Anabaena
Erie Central basin	Tendipes plumosus	Yellow perch, smelt, freshwater drum	High	Melosira binderana Stephanodiscus Cyclotella Fragilaria crotonensis
Eastern basin	Pontoporeia affinis Few Calopsectra	Yellow perch, smelt, few salmonids	High	Microcystis Aphenizomenon

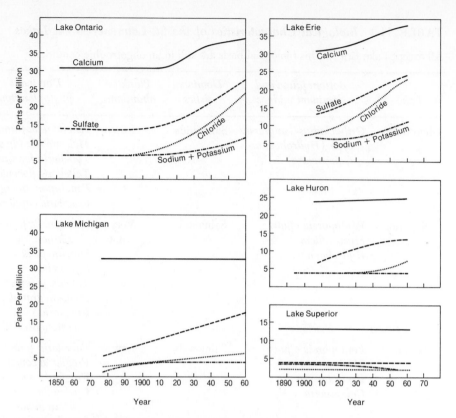

FIGURE 11-6. *Changes in chemical characteristics of the St. Lawrence Great Lakes.* All except Lake Superior show chemical changes indicative of cultural eutrophication. In Erie, and Ontario into which it drains, this affects the levels of all nutrients recorded, in Michigan it affects all except calcium levels, in Huron all except calcium and sodium + potassium.

The colonies of snowy egrets nesting on the islands in the Detroit River diminish each year.

Fall in Water Level

Associated with this pollution is a lowering of the water level of this shallow lake, which is mostly only about 15 meters deep, little over 50 meters at its deepest point. The Army Corps of Engineers dredged 3 million cubic yards of spoil, at a cost of 1 million dollars, and dumped this pollution-laden material back into Lake Erie. This is the traditional way of handling materials dredged to keep open the waterways to the big lakeshore cities.

Increased runoff from the surrounding rivers, bringing silt, pesticides, and fertilizers, has accelerated the rate at which the shipping channels fill.

Causes of Pollution of Lake Erie

Approximately 2 per cent of the sewage disposal plants in the country are considered adequate to deal with the living and household wastes they

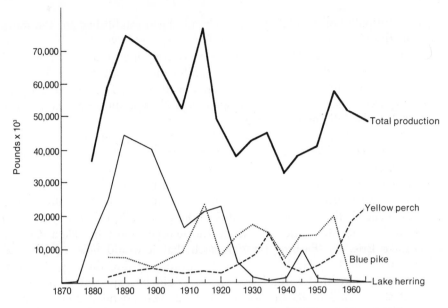

FIGURE 11-7. *Changes during the past century in the fish population of Lake Erie.* The catch of the two predators, blue pike and lake herring, has fallen to nil, whereas in their absence the numbers of the herbivorous yellow perch, and therefore the catch, have risen considerably.

should handle. The great majority, and those around Lake Erie are no exception, are at most 50 per cent efficient. Their effluents contain not only the phosphates and nitrates released from living matter but also the very high levels of phosphates that modern detergents release. From measurements made at the end of the last decade it has been estimated that the states of Pennsylvania, New York, Indiana, Michigan, and Ohio discharge an estimated 1.5 million gallons of sewage into Lake Erie every day. The lakeshore communities themselves pour in an estimated 18,000 tons daily of sewage, sediments, and fertilizers, including detergents. During rainstorms some 39 principal sewage overflow pipes release an estimated 50 million gallons of raw sewage into the lake.

The nearly 400 major industrial plants contribute an additional 10 million gallons of waste materials. Detroit provides wastes from its automotive, steel, chemical, paper, and petroleum industries. Toledo dumps the waste products from automotive, steel, petroleum, and glass factories; Cleveland from oil, cyanide, steel, automotive, chemical, phenol, and paper industries; and Buffalo from steel, chemical, and cement industries and from flour mills.

Control of Pollution in Lake Erie

Problems created by technology (as in the case of water pollution), whether in Lake Erie or elsewhere, have technological solutions. An issue that has already been raised in this text several times, it will appear again. Briefly, it is a matter of *internalizing* costs instead of *externalizing* them from a particular industry to the community as a whole.

Antipollution regulations have already been established for the industries in the Great Lakes region; some few factories have shut down or moved elsewhere because of the added cost of conforming to the required measures. It is unlikely that more than one quarter of the industries and one half of the cities will be able to accommodate to newly introduced antipollution requirements, estimated to require a capital expenditure of one-third billion dollars for full implementation.

Against this expenditure should be set for comparison the value of the estimated 17 *billion* dollars' worth of cars, steel, chemicals, rubber, paper, and petroleum products that the Great Lakes region is estimated to produce annually. It is projected that this will increase up to five times in the next half century. So will the pollution; what price are we willing to pay to prevent it? There is no doubt about who eventually has to pay, so it should not be difficult to identify where the decision must be taken. Concerning Lake Erie, its "flushing time" is such that it would take an estimated 20 years to restore the lake to something approaching the unpolluted, meso-trophic state that natural succession would be expected to have attained at this period of its history, even if all recommended controls were implemented immediately.

Lake Washington

To provide some contrast to this appalling story of Lake Erie, the recent history of Lake Washington illustrates what can be done to reverse changes in eutrophication and pollution processes if they are vigorously pursued.

The case of Lake Washington is somewhat simpler than that of the Great Lakes, for three reasons. First, industrial pollution was never a significant compounding factor. Second, surrounding activities in the Pacific Northwest are essentially of a forestry rather than an agricultural nature: runoff of pesticides and fertilizers is consequently minimal. Third, a convenient emptying ground, Puget Sound, is available to take treated sewage effluent.

Nevertheless the operations necessary to clear up Lake Washington and to reverse the cultural eutrophication there cost in excess of $120 million. This money seems to be regarded by the people of the Pacific Northwest, who provided most of it, as having been well spent.

An account of the earlier part of the Lake Washington story has been published by W. T. Edmondson and completed by him with an address given in September 1969 at the 11th International Botanical Congress in Seattle. The lake has had two episodes of pollution by sewage. In the first, which reached a maximum just before the depression of the 1930s, pollution by raw sewage became so extensive that in 1926 this was diverted instead into Puget Sound. This diversion ensured that by the mid-1930s the lake was back in good condition.

However, further human population growth in the area led to the erection of sewage plants releasing *treated* sewage effluent into the lake. By 1956 there were ten such plants handling 20 million gallons of raw sewage daily. Cultural eutrophication from the added phosphates and nitrates showed up in the usual way with depleted oxygen content in the deep water and summer phytoplankton populations persisting at values as high as previous spring blooms (Figure 11-4). In 1955 *Oscillatoria rubescens,* which was associated with the eutrophication occurring in European waters such as Lake Zürich, appeared also in Lake Washington.

As a result of public alarm at this increasing pollution of Lake Washington, legislation was introduced in the State of Washington (instigated by the mayor of Seattle and a committee he formed to deal with waste disposal) that permitted a concerted effort to be made to halt this cultural eutrophication. In 1963 work was begun on two main sewage lines, completed in 1968, which pick up all wastes from the area and deliver them to a main sewage treatment plant on Puget Sound. The treated effluent from this plant, which also handles the raw sewage previously dumped into the Sound, is released some distance offshore in an area of strong tidal flow at a depth of about 75 meters.

This is not just an externalization of the problem into a marine instead of a freshwater ecosystem, for because of an upwelling situation Puget Sound is a nutrient-rich habitat. The natural nutrient levels where the treated effluents are released are about the same for nitrates and phosphates as Lake Washington had at the point of maximum pollution.

The nutrient status of Lake Washington has now almost returned to the 1933 level, at least with regard to phosphates. Nitrates are reacting somewhat more slowly. The water is clear again, and the bottom has once more become visible near the shore. By 1971 the lake was completely restored to its nutrient status of 1933. Similar action to lower biochemical oxygen demand, phosphate content, and coliform bacteria contamination will probably save another famous beauty spot, Lake Tahoe. Agricultural use of treated effluents is also possible.

Pollution of Coastal Waters

The discharge of sewage effluent into coastal waters is not always accompanied by so little ecological disturbance as has resulted from the example just described for Puget Sound. Farther south along the Pacific Coast, in southern California, the explosive reproduction of sea urchins around sewage outfalls has been described by W. S. North and others. Here sea urchins can utilize the high concentrations of organic materials because of their apparent ability to absorb amino acids directly through their spines and tube feet. Their numbers increase enormously as a result, and they destroy kelp beds through their habit of gnawing away the holdfast, thus releasing the massive kelp fronds. Following destruction of the kelp beds, the urchins turn to destroying other marine invertebrates in the area of the sewage outfall.

North has pioneered a method of controlling such sea urchin outbreaks with the use of quicklime, which selectively kills the urchins, after which the kelp beds and their invertebrate populations become reestablished.

It has been estimated that in Pacific coastal waters in 1967, some 12 million fish were killed by pollution; moreover, reports of abnormalities in species such as flatfish and bass are received constantly. As many as 56 per cent abnormal individuals in a marine fish population off the California coast have been reported.

In an endeavor to facilitate a systems approach to the study of such marine pollution effects, D. L. Mayer has presented a "waste impact index." This is derived by the following equation:

$$\text{Waste impact index} = \triangle P_p + \triangle B_p + \triangle D$$

where $\triangle P_p$ = primary productivity of water (determined by the chlorophyll method)

$\triangle B_p$ = O_2 consumption by surface sediments (an estimate of benthic productivity)

$\triangle D$ = index of diversity of Foraminifera populations

Such quantitative approaches are clearly necessary both for pursuing fundamental studies of marine pollution and for any necessary enforcement of antipollution legislation.

Other Forms of Water Pollution

Apart from sewage and fertilizer pollution, modern industry has succeeded in dumping virtually every imaginable waste product into natural waters. Indeed the economic terms *externalization* and *internalization* have now been extensively applied by ecologists to this dumping process. Instead of going to the trouble and expense of removing all harmful substances from an industrial effluent by internalized processes, industry dumps its effluent, externalizing the problem on society at large, and saving itself much cost. This is happening in Lake Erie.

These externalized waste substances sometimes are grouped for convenience of discussion. One particularly dangerous group is the radioactive isotopes. As an example the amount of radioactivity in the water discharged from a nuclear power station is listed in Table 11-4. The question of absorption of radioactive isotopes already has been examined in the previous chapter.

Radioactive Isotopes

Along with two other groups of pollutants, actual poisons and pesticides, the most serious aspect of radioactive isotope discharge is that once introduced into an aquatic ecosystem, they may be absorbed selectively. In other words populations at one trophic level or another may concentrate them in a particular portion of the body at levels far greater than occur in the surrounding water. Although continual monitoring of the concentration of the isotope, poison, or pesticide in the *water* may indicate that this remains well below tolerance levels, selective absorption may be passing on from one trophic level to other quite toxic concentrations of the substance. There are many examples of the occurrence of this selective concentration, and the subject will be considered again in the next chapter when problems

TABLE 11-4 *Radioactivity Discharge in Waste Water from a British Nuclear Power Plant*

The figures indicate the mean monthly discharge in curies of beta emitters in 1962. Besides these elements, a further 34 radionuclides have been detected in the outflow from nuclear power stations in amounts varying from 200 curies monthly to traces only.

A tightening of the U.S. regulations regarding waste discharge of radioisotopes apparently was responsible for reducing the number of new nuclear power stations ordered from 17 in 1968 to three in 1969. Up to the end of 1968, 88 nuclear power stations had been ordered for the United States of which 12 were operative.

Ruthenium-106	1,916
Cerium-144	200
Ruthenium-103	153
Yttrium	125
Cesium-137	92
Strontium-90	85
Zirconium-95	78
Strontium-89	42

of pesticide and poison absorption are examined further. Some comment, however, must be made at this stage.

Radioactive effluents from nuclear power plants are, together with fallout from atomic tests, one of the main sources of artificially produced radioactive contamination of marine waters. The nuclear reactor at Hanford in the Pacific Northwest discharges effluents that have been measured as reaching 1,000 millicuries a day in the Columbia River estuary 350 miles downstream (Figure 11-8). Some of the isotopes present are of little significance because of short half-lives or poor absorption characteristics. Others, however, such as chromium-51 and zinc-65 have been detected in pelagic organisms some distance out in the Pacific Ocean.

An indication of the general contamination of marine waters on a worldwide basis resulting from atomic testing may be obtained from Table 11-5. The extent to which radioactive contamination of seawater from this, as from any other source, may be concentrated selectively in marine ecosystems is indicated in Table 11-6.

Mercury

In addition to general forms of water pollution such as eutrophication and contamination with radionuclides and other substances that are selectively absorbed as they pass through ecosystems, serious concern is being expressed regarding the dispersal and distribution of specific elements, one of which is mercury. Some of the most striking evidence as to occurrence of this element in various ecosystems has been provided from Sweden. Since the 1930s mercury compounds have been used in that country for dusting

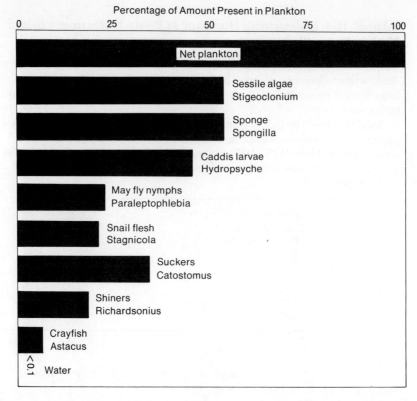

**FIGURE 11-8. *Amounts of radioactivity present in various organisms at different tropic levels in the Columbia River.* The relative amounts of beta-emitting radioisotopes have never been recorded as reaching hazardous proportions in any organism. Fish collected downstream of the reactors have, however, 100 times the dose of control fish, and bottom animals, especially herbivorous insect larvae, may be even more radioactive.

the seeds of many agricultural crops in order to suppress seed-borne diseases, and in some instances to provide protection against fungal attack during seed germination.

During World War II Swedish farmers extended this practice and used liquid preparations of mercury in a commercial form known as Panogen for such crops as wheat, oats, and barley. The actual form of mercury in this commercial preparation is methyl mercury dicyandiamide. Extremely toxic to all vertebrates, such methyl mercury compounds are effectively absorbed through the skin or by inhalation. They accumulate in the body because they are excreted only slowly.

The worst incident to date of mercury poisoning, a result of the human consumption of cereal grain seed treated with a fungicidal dressing, occurred in Iraq in 1972. Out of a total of 6,530 hospitalizations in the first part of that year attributed to mercury poisoning, there were 459 deaths. Seed grain of improved wheat and barley varieties imported by Iraq from Mexico had been treated with several methyl mercury fungicidal seed-dressing compounds. Extensive warnings of the poisonous nature of these

TABLE 11-5 *Concentration in Surface Layers of Seawater of Some Radioactive Isotopes*

Figures expressed in microcuries per gram of seawater $(\mu c/l)_s$ representing the effects of estimated world-wide fallout into the oceans. The only isotope occurring in significant concentrations in marine waters is potassium-40, which represents over 90 per cent of marine radioactivity. Contamination of seawater in general with artificially produced radioisotopes appears slight, but see Table 11-6 for the effects of selective concentration in marine ecosystems.

NATURAL (COSMIC-RAY PRODUCED)	
Carbon-14	−0.1
Tritium (radioactive hydrogen)	−4.0
Uranium-238	−1.0
Rubidium-87	−3.0
Potassium-40	−277.0
WEAPON-TESTING FALLOUT	
Cesium-137	−0.1
Strontium-90	−1.0
Carbon-14	−0.001
Tritium	−1.0 to 10.0

TABLE 11-6 *Selective Concentration of Radioactive Isotopes in Marine Ecosystems*

The concentration factor at various trophic levels: producer (algae), primary consumer (mollusk), secondary consumer (crustacean), and tertiary consumer (fish). For some isotopes the insignificant concentrations in seawater (Table 11-4) are concentrated by a magnitude of four and can have significant effects.

	Producers	*Consumers*		
	Edible	*Primary*	*Secondary*	*Tertiary*
Isotope	*red algae*	*mollusks*	*crustaceans*	*fish*
FISSION PRODUCTS				
Strontium-90	10^{-1}–1	10^{-1}–1	10^{-1}–1	10^{-1}–1
Cesium-37	1–10	10–10^2	10–10^2	10–10^2
Cerium-144	10^2	10^2	10^2	10–10^2
Zirconium-95	10^2–10^3	10–10^2	10^2	1–10
Niobium-95	10^2–10^3	10^2	10^2	1–10
Ruthenium-106	10^3	1–10^3	1–10^3	1–10
INDUCED ACTIVITIES				
Zinc-65	10^2	10^3–10^5	10–10^4	10^3–10^4
Iron-55	10^3–10^4	10^2–10^4	10^2–10^4	10^2–10^4
Cobalt-60	10^2	10–10^3	10–10^3	10–10^2
Manganese-54	10^3	10^3–10^4	10^2–10^4	10^2–10^3
Chromium-51	10–10^3	10^3	10^3	10^2–10^3

treated seed grains were disseminated. These warnings were not sufficient to prevent some treated grain from being used in baking bread for human consumption. Because of the relatively long delay before the onset of mercury poisoning symptoms, cautious peasants who first tested the treated grain on their chickens would not immediately observe any ill effects. Symptoms did not appear in humans for several weeks, or even months. Moreover, washing in water would remove the dye used to mark the treated grain; largely illiterate peasants would understandably suppose that such washing had also removed the poison, which it had not done. Incidents such as this outline the difficulties involved in trying entirely to prevent accidental contamination of human food with poisonous metals, even during such philanthropic enterprises as the extension of the Green Revolution.

The first concern over methyl mercury poisoning in Sweden was with possible effects on wildlife, but attention soon became focused on agricultural products. According to G. Löfroth and M. E. Duffy, in 1964 Swedish hens' eggs averaged 0.029 ppm of mercury, whereas those from six other European countries averaged 0.007 ppm. An examination of other farm products provided the results shown in Table 11-7.

These findings in Scandinavia were confirmed by investigations on the use of mercurial fungicides in Japan. However, further research there, and subsequently in Sweden, revealed that the main source of the introduction of methyl mercury into agricultural ecosystems was industrial discharge rather than its agricultural use as a fungicide. Pulp factories use a mercury compound (phenyl mercury acetate) to discourage fungal growth on their machinery. Chlorine factories using mercury electrodes release mercury into both air and water, as do some sections of the electrical industry. Certain fossil fuels when burned release small quantities of mercury compounds; even the sludge from sewage plants may contain sufficient mercury to make it unsuitable for use as a soil fertilizer.

In the United States, although the pattern of use of mercury compounds in agriculture and industry is similar to what has been described for Sweden

TABLE 11-7 *Amounts of Mercury in Scandinavian Meat Products*

Amounts are given in parts per million. An analysis carried out in 1965 of the following agricultural products in Sweden and Denmark revealed serious contamination with methyl mercury.

Raw Product	Mercury (ppm)	
	Swedish	Danish
Pork chops	0.030	0.003
Beef	0.012	0.003
Bacon	0.018	0.004
Pig's liver	0.060	0.009
Ox liver	0.016	0.005

and Japan, comparatively little is known about the circulation of mercury compounds in American ecosystems. A large additional use in this country is in paints, to prevent the development of mildew, and especially in anti-fouling paints used on boats and in other damp environments. Acrylic paints, according to S. Novick, may contain up to 1 per cent mercury fungicide. Mercury compounds also are used in making paper, and Novick states that it is difficult to obtain figures on the extent of this use in the paint industry.

Control of mercury pollution in industrial waste water is difficult; the selective absorption of mercury compounds and their passage through the trophic levels of natural and agricultural ecosystems provide for their persistence from 10 to 100 years in toxic concentrations affecting particularly the top carnivores in these systems. The most that can be done presently within the range of practicality is to hold levels of mercury contamination in our local and global ecosystems as low as possible and to monitor carefully the results of long-term exposure to specific low levels of mercury contamination.

Lead

Very similar problems are presented by lead, which, because of its growing use as a gasoline additive, is being introduced increasingly not only into the air sheds but also into the watersheds of the total global system. At one time lead entered human ecosystems as a water pollutant because of the extensive use of lead pipes for water transport. Like mercury, it is excreted only slowly from the human body; even small amounts of contamination from this source could therefore reach toxic concentrations eventually. Water pollution by lead seems likely, however, to have rather different but even more serious ecological consequences. Concentrations of lead in seawater are believed to be increasing until they are now approaching levels that may possibly have toxic effects on phytoplankton, particularly on diatoms, which represent a significant proportion of the phytoplankton of marine ecosystems in the open sea. As the phytoplankton constitute by far the largest element of the producer trophic level in the world's ecosystem, any reduction in their activity through water pollution by lead compounds could result in reduced carbon dioxide absorption and oxygen restoration in the atmosphere. The proportions of oxygen and carbon dioxide in the earth's atmosphere are well buffered against change, but the ultimate result of the lead poisoning of a major group of marine producer populations would be disastrous for all aerobic organisms on this globe.

Thermal Pollution

Pollutants affect aquatic ecosystems through modification of chemical processes that control productivity. Another form of water pollution becoming more prominent is *thermal* pollution, a physical modification in the form of increased temperatures. Thermal pollution is perhaps somewhat analogous to noise pollution and its significance in relation to air pollution, but with quite different results.

Water, extensively used as a *coolant* in machines ranging from internal combustion engines to thermal power plants, is the standard medium for the dispersal of unutilized heat in nuclear power stations. Temperature rises in the effluent water of such installations range generally from a low of 2°C to 10°C.

The last figure is known from van't Hoff's law to double the rate of any chemical reaction involved and obviously will cause profound disturbances to an aquatic ecosystem. Even a variation as low as 2°C may be critical and could raise the temperature of seawater, for example, beyond the tolerance limits for establishment of giant kelp beds. A possible means of utilizing the heat load has been discussed by L. J. Carter, who describes experiments on warm-water irrigation of agricultural crops.

Air pollution is forcing attention on a need to change from thermal to nuclear power stations for the production of electricity. At the same time, the large volume of cooling water required for a nuclear power station may generate new problems. One of them is thermal pollution, which may threaten the survival of aquatic ecosystems long regarded as part of a regional heritage.

Biological Pollution

All forms of pollution represent some kind of disturbance to the continuing function of a naturally occurring ecosystem. Air and water pollution operate through physical, chemical, or biological disturbance of productivity, either directly or indirectly. *Biological pollution* causes disturbance to an ecosystem by influencing the relationship between component species populations directly.

The classical case resulted from the opening of the Welland Canal during construction of the St. Lawrence Seaway. This permitted the penetration of sea lampreys from Lake Ontario, where the parasitic lampreys existed with their hosts in a balanced relationship, to Lakes Erie, Huron, and Michigan, where trout—one of two predatory fishes in the lake ecosystem—had no such evolutionary relationship with the lampreys.

The resultant decrease in numbers of trout in the three lakes had effects over and beyond a disastrous decline in commercial fishing. The reduction in population numbers of a top predator reduced the cropping of primary consumers such as alewives. Numbers of these primary consumers rose very rapidly, and occasionally great quantities of these commercially undesirable fish are cast up to rot on the lakeshores.

Construction of the Suez Canal might have resulted in similar ecosystem disturbance through mixing of the populations of the Mediterranean and Red Sea ecosystems, had the Bitter Lakes not continued to act as a barrier. The Bitter Lakes are a series of shallow lakes part way along the Canal, where water evaporation results in a salinity appreciably higher than that

PLATE 8. *The Kariba Dam* constructed on the lower reaches of the Zambezi River on the frontier between Zambia and Rhodesia, and first operative in 1958. The next year biological pollution by a small introduced water fern, *Salvinia molesta* (formerly *auriculata*), was reported, and within a year 10 per cent of the water surface was covered by this weed.

of seawater; it is so high that few marine organisms can survive long enough to move through them from one region to another. The prolonged closure of the canal is believed to be modifying the effectiveness of this barrier.

Proposals to blast a new Panama Canal operating entirely at sea level now threaten to mix Pacific and Atlantic marine ecosystems. If the canal is built, it appears certain that almost half of the combined marine species populations will disappear, because extensive competition will arise between ecological equivalents in the two oceans.

The disturbance to the tropical marine ecosystems of these two major oceans, which have evolved in complete independence for many millennia, would have extensive and harsh effects on their productivity for many years. It could also lead to extinction of numerous interesting and valuable species populations. The insertion of a freshwater lake somewhere along the new canal would effectively prevent this mixing of the two marine ecosystems.

Examples could be cited almost indefinitely. L. G. Holm recently surveyed major outbreaks of aquatic weeds, including water hyacinth (*Eichornia crassipes*), which now has entered virtually every major tropical and subtropical aquatic ecosystem in the world. A previously insignificant water

PLATE 9. *Biological pollution.* A bay of the developing young Lake Kariba in 1961 covered from shore to shore with a floating *Salvinia* mat on which both indigenous and introduced plant species have become established to form impenetrable floating colonies of "sudd." *Salvinia molesta* is a South and Central American species apparently introduced into the Zambesi system during the 1950 or 1940 decade. Mats of this floating weed disrupt aquatic ecosystems in the lake and impede navigation; no satisfactory control has yet been discovered.

fern (*Salvinia molesta*) assumed disaster proportions when allowed to pollute the new aquatic environment of Lake Kariba. The effects of a number of other such typical instances of biological pollution are cited in the discussion of conservation in Chapter 13.

One last aspect of biological pollution must be mentioned: as in the case of an air or a water pollutant, a biological pollutant does not need to be introduced from one ecosystem into another. It can develop by a disturbance to the steady state reached within one particular ecosystem. A dramatic example of this is provided by the recent explosive development of a giant starfish that is a predator of coral. This giant venomous starfish, *Acanthaster planci*, occurs in all tropical waters where coral is found. Although adult starfish are prey in their turn of the giant triton snail *Charonia tritenis*, a more significant predation is believed to be that of the larval stages of the starfish by coral polyps.

The most plausible hypothesis yet presented for the starfish "explosion," which has been noticed since 1963, is that destruction of large areas of coral by blasting and other means during harbor development created for the first time considerable areas of dead coral surfaces. In the shelter of these, starfish larvae could develop without being regulated by coral predation. The resulting mass of adults would then attack areas of coral adjoining the dead surfaces and thereby extend the dead areas in a continuing vicious circle. As one adult giant starfish is believed to be able to kill 1 square meter of coral per month, it can be imagined that it would not take long to destroy a coral reef completely.

During the last few years, there has been a tendency to consider many of the earlier reports of reef destruction by *Acanthaster* as grossly exaggerated. *Acanthaster* is now more frequently regarded as a *normal* component of coral reef communities. It may kill *local* areas of coral but is no longer believed capable of producing an unstoppable plaque. There is no doubt that it can reduce or even reverse the rate of coral reef growth and markedly influence the composition of coral communities. However, in so doing it may contribute to the *diversity* of reef habitats and so exhibit some redeeming features.

In 1970 it was estimated that 30 to 40 per cent of the Pacific coral reefs had already succumbed to starfish attack. Fish quickly desert such dead reefs, so that the many small coral islands of Oceania were threatened with loss of their major source of protein food. Moreover, if the dead reefs break up under wave erosion, the small islands previously sheltered by them eventually will go, too. This would provide one more unfortunate illustration of the dangers inherent in human disturbance of regulatory feedback mechanisms in natural ecosystems.

Waste Disposal

Aside from the pollutant effects of pesticides and accidental pollution—like the Santa Barbara channel oil spillage or the several recent oil tanker disasters—air and water pollution are essentially externalization and internalization problems whose control may be achieved by suitable techniques of environmental engineering. Given the appropriate economic stimulus, there are no obvious technological difficulties involved in solving any pollution problem. The release of hydrocarbons from internal combustion engines is preventable if electrically powered cars are used, provided the electricity generating stations are not themselves sources of pollution. Eutrophication is avoidable if nitrates and phosphates are first removed from sewage effluents and then applied to crop plants in such a manner that they are not drained off immediately.

Industrial wastes almost always contain recoverable substances. If an economically feasible recovery method can be applied technologically, this will be developed with alacrity. Even garbage can, in principle, be sorted and recycled.

It is all a question of regarding human byproducts as an inseparable element of one ecosystem or another and of ensuring that recycling of materials continues and the essential balance of particular ecosystems is not destroyed. For some time to come the energy that drives all our ecosystems will continue to originate as solar energy. The environments in which these ecosystems operate are the air, soil, and water of this planet. Over geological time, feedback mechanisms have developed in all natural ecosystems that permit effective exploitation of the available energy. Pollution of one form or another—air, water, thermal, noise, biological—is threatening to impose sufficient change on activities of these ecosystems to effectively destroy many and disrupt the rest. It will also cause the rapid extinction of numerous species that have slowly evolved over the last 3½–4 billion years.

Food Additives

Human populations must always have sought means to arrest the biogeochemical processes that, through the activities of microbial populations, cycle elements through ecosystems. Various attempts have been made to preserve food for days, weeks, or even years. With the development of urban ecosystems, this need for food preservation became even more urgent, and many substances were employed in an attempt to delay the inevitable reduction of food substances to their basic nutrient elements by reducer organisms.

Food pollutants, however, include an even wider range of substances than those added to improve taste, color, appearance, and consistency. The total of such alien substances is immense (there may well be nearly half a million), and it is almost impossible to investigate their possible deleterious effects. Although it was relatively simple to legislate, for example, against the use of formalin for the preservation of food, a demonstration that monosodium glutamate is ultimately harmful to the human system is exceedingly difficult. The outlawing of cyclamates as food and drink sweeteners illustrates how minimal experimental work on animals has to be extrapolated into possible human effects and action taken even before any of these effects can be directly and positively demonstrated. It is possible that many food additives of a preservative or of an improving kind, now widely employed, are having both long- and short-term effects to increase the risk of abnormal metabolism, abnormal offspring, or abnormal mortality rates. As many of these substances are utilized because of the disruptive effects they have on organisms involved in ecological recycling, it can be anticipated that their effects on these microbial reducer organisms may also be duplicated to some extent on organisms in other trophic levels.

BIBLIOGRAPHY

References

Aaronson, T. "Mercury in the environment," *Environment*, 13(4): 16–23, 1971.
Alberg, B., and Hungate, F. P. (eds.) *Radiological Concentration Processes*, Oxford, England: Pergamon Press, 1966.

Anonymous. "Radioactivity in airborne particulates and precipitation," *Radiol. Health Data Report,* 11: 85–93, 1970.

Bakir, F., et al. "Methylmercury poisoning in Iraq," *Science,* 181: 230–41, 1973.

Baldwin, N. S., and Saalfield, R. W. "Commercial fish production in the Great Lakes 1867–1960," *Great Lakes Fish. Comm. Techn. Rept. No. 3,* 1962.

Beeton, A. M. "Eutrophication of the St. Lawrence Great Lakes," *Limnology and Oceanography,* 10: 240–54, 1965.

Boughey, A. S. "The explosive development of a floating weed vegetation on Lake Kariba," *Adansonia,* 3(1): 49–61, 1963.

Bowen, D. H. M. "The great phosphorous controversy," *Environ. Sci. Technol.,* 4(9): 725–26, 1970.

Branham, J. M. "The crown of thorns on coral reefs," *Bioscience,* 23: 219–26, 1973.

Cairns, J. "Coping with heated waste water discharges from steam-electric power plants," *Bioscience,* 22: 411–19, 423, 1972.

Carter, L. J. "Warm-water irrigation: an answer to thermal pollution," *Science,* 165: 478–80, 1969.

Caswell, C. A. "Underground waste disposal: concepts and misconceptions," *Environ. Sci. Technol.,* 4(8): 642–47, 1970.

Chesher, R. H. "Destruction of Pacific corals by the sea star *Acanthaster planci,*" *Science,* 165: 280–83, 1969.

Coan, G. "Oil pollution," *Sierra Club Bull.,* March 1971, pp. 13–16.

Culp, R. "Water reclamation at South Tahoe," *Water and Wastes Engineering,* April 1969, pp. 36–39.

Davis, J. I., and Foster, R. F. "Bioaccumulation of radioisotopes through aquatic food chains," *Ecology,* 39: 530–35, 1958.

Dyer, A. J., and Hicks, B. B. "Radioactive fallout from the French 1966 Pacific tests," *Australian J. Sci.,* 30(6): 168–70, 1967.

Edmondson, W. T. "Water-quality management and lake eutrophication: the Lake Washington case," in T. H. Campbell and R. O. Sylvester (eds.), *Water Resource Management and Public Policy,* Seattle: University of Washington Press, 1968, pp. 139–78.

Eliassen, R., and Tohabanoglous, G. "Removal of nitrogen and phosphorous from waste water," *Environ. Sci. Technol.,* 3: 536–41, 1969.

Environment Staff Report. "A new river," *Environment,* 12(1): 36–41, 1970.

Gatz, D. F., and Dingle, A. N. "Air cleansing by corrective storms," in A. W. Klement (ed.), *Radioactive Fallout from Nuclear Weapon Tests,* Oak Ridge, Tenn.: Div. Tech. Information, 1965, pp. 566–81.

Grant, N. "Mercury in man," *Environment,* 13(4): 3–15, 1971.

Hammond, A. L. "Mercury in the environment: natural and human factors," *Science,* 171: 788–89, 1971.

Hasler, A. D. "Cultural eutrophication is reversible," *Bioscience,* 19: 425–31, 1969.

Holcomb, R. W. "Oil in the ecosystem," *Science,* 166: 204–6, 1969.

Holm, L. G., Weldon, L. W., and Blackburn, R. D. "Aquatic weeds," *Science*, 166: 699–709, 1969.

Jukes, T. H. "Antibiotics in animal feeds and animal production," *Bioscience*, 22: 526–34, 1972.

Kardos, L. T. "A new prospect," *Environment*, 12(2): 10–21, 27, 1970.

Kermode, G. O. "Food additives," *Scientific American*, 226(3): 15–21, 1972.

Klein, D. H., and Golberg, E. D. "Mercury in the marine environment," *Environ. Sci. Technol.*, 4: 765–68, 1970.

Lewis, R. "Last word in the radiation debate," *New Scientist*, 57: 610–11, 1973.

Löfroth, G., and Duffy, M. E. "Birds give warning," *Environment*, 11(4): 10–17, 1969.

McCaull, J. "The black tide," *Environment*, 11(9): 2–16, 1969.

Molte, H. L., et al. "Lead in soils and plants: its relationship to traffic volume and proximity to highways," *Environ. Sci. Technol.*, 4(3): 231–51, 1970.

Nisbet, A. "Crown-of-thorns in the Red Sea," *New Scientist*, 58: 74–78, 1973.

Novick, S. "A new pollution problem," *Environment*, 11(4): 2–9, 1969.

Odum, W. E. "Insidious alteration of the estaurine environment," *Amer. Fish Soc. Trans.*, 99: 836–47, 1970.

Peakall, D. B., and Lovett, R. J. "Mercury: its occurrence and effects in the ecosystem," *Bioscience*, 22: 20–25, 1972.

Rawson, D. S. "Algal indicators of trophic lake types," *Limnology and Oceanography*, 1: 18–25, 1956.

Regier, H. A., and Hartman, W. L. "Lake Erie's fish community: 150 years of cultural stresses," *Science*, 180: 1248–55, 1873.

Rivera-Cordova, A. "The nuclear industry and air pollution," *Environ. Sci. Technol.*, 4: 302–97, 1970.

Ryther, J. H., and Dunstan, W. M. "Nitrogen, phosphorus, and eutrophication in the coastal marine environment," *Science*, 171: 1008–13, 1971.

Ryther, J. H., and Dunstan, W. M. "Controlled eutrophication—increasing food production from the sea by recycling human wastes," *Bioscience*, 22: 144–52, 1972.

Sawbridge, D. F., and Bell, M. A. M. "Pacific shores," *Science*, 164: 1089, 1969.

Sawyer, C. N. "Basic concepts of eutrophication," *J. Water Pollution Control Federation*, 38: 737–44, 1966.

Selikoff, I. J. "Asbestos," *Environment*, 11(2): 2–7, 1969.

Topp, R. W. "Interoceanic sea-level canal: effects on the fish faunas," *Science*, 165: 1324–27, 1969.

Valentine, D. W., and Bridges, K. W. "High incidence of deformities in the serranid fish *Paralabrax nebulifer* from southern California," *Copeia*, 3: 637–38, 1969.

Wheeler, A. "Fish return to the Thames," *Science J.*, November 1970, pp. 28–32.

Wilson, B. R. (ed.) *Environmental Problems: Pesticides, Thermal Pollution, and Environmental Synergisms,* Philadelphia and Toronto: Lippincott, 1968.

Woodwell, G. M. "Radioactivity and fallout: the model pollution," *Bioscience,* **19:** 884–87, 1969.

Further Readings

Abelson, P. H. "Methyl mercury," *Science,* **169:** 237, 1970.

Anon. "Mercury in the air" (staff report), *Environment,* **13**(4): 24, 29–33, 1971.

Cain, S. A. "Ecology: its place in water management," *Water Spectrum,* **1**(1): 10–14, 1969.

Charnell, R. L., Zorich, T. M., and Holly, D. E. "Hydrologic redistribution of radionuclides around a nuclear-excavated sea-level canal," *Bioscience,* **19:** 799–803, 1969.

Clark, J. R. "Thermal pollution and aquatic life," *Scientific American,* **220**(3): 18–27, 1969.

Cole, L. C. "Thermal pollution," *Bioscience,* **19:** 989–92, 1969.

Curley, A., Sedlak, V. A., Girling, E. F., Hawk, R. E., Barthel, W. F., Pierce, P. E., and Likosky, W. H. "Organic mercury identified as the cause of poisoning in humans and hogs," *Science,* **172:** 65–67, 1971.

Eisenbud, M. *Environmental Radioactivity: Ecological and Public Health Aspects,* 2nd ed., New York: Academic Press, 1973.

Fonselius, S. H. "Stagnant sea," *Environment,* **12**(6): 2–11, 40–48, 1970.

Gofman, J. W., and Tamplin, A. R. "Radiation: the invisible casualties," *Environment,* **12**(3): 12–19, 1970.

Goldman, M. I. (ed.) *Controlling Pollution: The Economics of a Cleaner America,* Englewood Cliffs, N.J.: Prentice-Hall, 1967.

Grant, N. "The legacy of the Mad Hatter," *Environment,* **11**(4): 18–23, 43–44, 1969.

Hamilton, D. H., Flemer, D. A., Keefe, C. W., and Mihursky, J. A. "Power plants: effects of chlorination on estuarine primary production," *Science,* **169:** 197–98, 1970.

Hedgepeth, J. W. "The oceans, world sump," *Environment,* **12**(3): 40–47, 1970.

Hennigan, R. D. "Water pollution," *Bioscience,* **19:** 976–78, 1969.

Holcomb, R. W. "Waste-water treatment: the tide is turning," *Science,* **169:** 457–59, 1970.

Howells, G. P., Kniepe, T. J., and Eisenbud, M. "Water quality in industrial areas: profile of a river," *Environ. Sci. Technol.,* **4**(1): 26–35, 1970.

Joensuu, O. I. "Fossil fuels as a source of mercury pollution," *Science,* **172:** 1027–28, 1971.

Lagerwerff, J. V., and Specht, A. W. "Contamination of roadside soil and vegetation with cadmium, nickel, lead and zinc," *Environ. Sci. Technol.,* **4**(7): 583–86, 1970.

Lotspeich, F. B. "Water pollution in Alaska: present and future," *Science,* **166:** 1239–45, 1969.

Lowman, F. G. "Radionuclides of interest in the specific activity approach," *Bioscience,* **19:** 993–99, 1005, 1969.

Moffet, H. "Troubles brew below a lake's shining surface," *Smithsonian,* 4(2): 68–75, 1973.

Motto, H. L., et al. "Lead in soils and plants: its relationship to traffic volume and proximity to highways," *Environ. Sci. Technol.,* 4(3): 231–37, 1970.

Nielson, E. S., and Wium-Andersen, S. "Copper ions as poison in the sea and in fresh water," *Marine Biol.,* **6:** 93–97, 1970.

Oberle, M. W. "Lead poisoning: a preventable childhood disease of the slums," *Science,* **165:** 991–92, 1969.

Rutzler, K., and Sterrer, W. "Oil pollution," *Bioscience,* **20:** 222–24, 226, 1970.

Task Force on Environmental Health and Related Problems. *A Report to the Secretary of Health, Education, and Welfare,* Washington, D.C.: Government Printing office, 1969.

Tinker, J. "What's happening to Lake Baikal?" *New Scientist,* **58:** 694–95, 1973.

Toms, R. "Monitoring river pollution," *New Scientist,* **43:** 595–96, 1969.

Chapter

12

Pesticides

The various kinds of pollution discussed in the previous two chapters involved the *incidental* introduction of toxic or waste substances into the air sheds or watersheds of particular ecosystems. Except that industry has purposely externalized the elimination of waste products on human societies, the release of such pollutants has not been deliberate. In the case of pesticides, however, highly toxic substances are *intentionally* introduced into given ecosystems in order to reduce the population growth of particular species believed responsible for certain plant and animal diseases, spoilage, or wastage.

Whereas atmospheric and water contamination by various forms of human waste have existed for several millennia—and are probably as old as the urban societies with which they are associated—pollution by pesticides dates essentially from World War II. The human generation now entering middle age is the first in history to have been exposed continuously to the effects of pesticide pollution from birth. If these effects are anything like those observed in other animals, this generation will be extremely resentful of what its predecessors have allowed to occur.

Classes of Pesticide

It is now usual to include under the heading *pesticide* any form of chemical substance used for the regulation of population growth, whether it is technically for the control of herbaceous plants (herbicides), woody plants (arboricides), or insects (insecticides) or has biocidal activity affecting rodents, arachnids, or any other population.

Before World War II, only "natural" pesticides were in common use, apart from specific application of poisons such as sodium arsenite, several forms of sulfur, and various mercurial salts for particular and limited purposes. These latter inorganic substances were not employed on a sufficient scale to become widespread pollutants; the principal organic substances used were derris powder, nicotine, and pyrethrum powder. Interest in these has revived now that the chlorinated hydrocarbons are being phased out by legislative action.

These earliest pesticides are occasionally classified in two groups, stomach poisons and contact poisons. The former included sodium arsenite (paris green), lead arsenate, and nicotine; contact poisons were notably pyrethrum, derris, and certain nicotine preparations.

Contact Poisons

These organic poisons were naturally occurring substances. Derris and other rotenones in particular had a rather widespread distribution in various plant groups and were the active agents in a number of fish poisons traditionally used for stunning fish in confined waters. They were used especially as a dust on agricultural crops. Pyrethrum, obtained from the flowers of a daisy-like genus, *Pyrethrum,* was mostly used in suspension as a knockdown spray for killing household insects such as flies and mosquitoes.

Neither of these organic substances was considered very satisfactory, for their effect was immediate and temporary; their pesticide activity did not persist more than a few hours. Use of them did, however, suggest the idea of more persistent substances, if these could be found or synthesized. During World War II the first of this new class of synthetic insecticides, DDT, was developed.

Chlorinated Hydrocarbons

DDT (1,1,-trichloro-2,2-bis [*p*-chlorophenyl]-ethane or *di*chloro*di*phenyl*tri*chloroethane, hence DDT) was first synthesized in 1874, but its insecticidal properties were not realized until 1939. The first large-scale demonstration of its efficiency as an insecticide came in 1943, when it prevented an epidemic of typhus in Naples, Italy. The Neapolitan population was dusted with DDT to control lice, the vector of typhus, an often fatal human disease.

During the past two decades the use of DDT as a pesticide has become immensely diversified. Its success has prompted the introduction of a number of new compounds in this same class of *chlorinated hydrocarbons.* Although some amounts of these may be carried away in drainage waters, mostly they are distributed by evaporation and subsequent fallout in rain or snow to ecosystems other than the one intended. England, for example, is estimated to receive 40 tons of chlorinated hydrocarbons annually as fallout.

DDT, like all the chlorinated hydrocarbons presently in use, is not readily decomposed by the reducer organisms in natural ecosystems. It persists for many years, either unchanged or modified to degraded substances of similar chemical structure and activity such as DDD and DDE. This is not unexpected, for the reducer organisms of natural ecosystems have had no prior exposure to this synthetic organic chemical, and there has been no selection for populations able to utilize it in their metabolism. By contrast, an extensively used herbicide, 2,4-D, is broken down in soil in a matter of days. 2,4-D is a synthetic compound whose chemical structure closely resembles that of naturally occurring growth-promoting plant hormones, or *auxins.* There has been adequate evolutionary time for some reducer populations to be adapted to utilize this substance, and no fundamental difficulty in overriding the comparatively minor chemical difference.

The first major attempt to describe possible irreversible changes that were resulting from the massive application of DDT and other *broad-spectrum* persistent chlorinated hydrocarbons came with the publication in 1962 of *Silent Spring* by Rachel Carson. This sparked a violent controversy, but was essentially responsible for initiating the public reaction that has culminated in the prohibition of DDT use from 1969 in certain U.S. states, Sweden, and some other countries. Another result was the shift in emphasis from broad-spectrum types of pesticides to *integrated control.*

Since 1962 it has been established repeatedly that DDT residues have become incorporated in all naturally occurring ecosystems. Their presence as pollutants can be demonstrated in soil, water, air, and man. Recently reported effects on various marine phytoplankton suggest that the proportion of oxygen in the atmosphere could be affected.

Selective Concentration

Although DDT may be introduced directly into the soil, water, or air of a particular ecosystem, it is relatively insoluble (1 to 2 parts per billion) and is most commonly distributed through ecosystems by *selective concentration* as it passes unchanged along the successive elements of food chains and food webs, much in the manner previously described for radioactive elements. Thus in Lake Michigan, for example, the concentration of DDT in lake sediments was found to be 0.0085 ppm. Invertebrate primary consumers selectively concentrated this to 0.41 ppm, their fish predators to 3–8 ppm; the fatty tissues of herring gulls preying on the fish were discovered to have no less than 3,177 ppm. Essentially similar findings are reported from marine ecosystems, as illustrated in Table 12-1.

As might be anticipated from theoretical considerations, the effect of selective concentration of such a persistent substance with this broad general toxicity is readily noticeable in top carnivores. The group of this class most susceptible appears to be the carnivorous birds, whose fate is relevant to human ecology in several ways. *Homo sapiens* also often behaves as a top carnivore of this type, as when eating tuna or salmon.

The destruction of top carnivores causes serious imbalance in the ecosystem involved. In a less material sense, the presence of top carnivores, such as the bald eagle, represents part of the indefinable "quality of environment" in terms of human interest and excitement. In regard to other carnivorous birds affected, P. L. Ames has described correlations between the nesting success of ospreys and the presence of DDT residues in the eggs.

Osprey Populations

Ospreys (*Pandion haliaetus*) are large, widely distributed fish-eating birds usually associated with extensive stretches of inland or coastal waters. It is estimated that their numbers have declined during this century by 2 to 3

TABLE 12-1 *Pesticides in a Marine Ecosystem*

Residues in parts per million in order of magnitude concentrations to the nearest multiple of 10. The amounts of residues show a progressive increase at each successive trophic level.

Trophic level	Species involved	DDD (including DDE and TDE)	Dieldrin
Tertiary consumers (top carnivores)	Dolphin	1.0	1.0
	Seal	0.1	0.1
	Duck		
	Gull		
	Shag	0.1–1.0	0.1–1.0
	Cormorant		
	Gannet		
		0.01–0.1	0.01
	Plaice		
Secondary consumers (general carnivores)	Herring		
	Sand eel		
	Cod		
	Whiting		
Primary consumers (herbivores)	Microzooplankton	0.01	0.01
Producers (plants)	*Fucus*	0.001	0.001
	Laminaria		

per cent per year, but recently this rate of decline has greatly increased to about 30 per cent annually.

In 1960 Ames observed in a large breeding colony in Long Island Sound only seven fledglings from 71 active nests. This is a reproductive rate of less than 0.1 young per nest.

By 1965 this colony was down to 12 pairs from an estimated 200 pairs of breeding birds in 1938. Significant amounts of DDT and its derivatives were found in the osprey eggs and also in the fish that formed the osprey diet.

Although no direct evidence on this point was obtained, it can be presumed that pollution of the coastal waters of Long Island Sound with DDT and its derivatives introduced these substances to the marine ecosystem, where they became concentrated first in fish, then in osprey eggs. There, selective concentration has raised the DDT content sufficiently for it to be fatal to chick embryos, although not to adult birds (Figure 12-1).

FIGURE 12-1. *Comparison of DDT residue levels in eggs of osprey populations.* Sample *A* is from a site on the Connecticut River where from 1957 to 1961 the osprey population underwent a very severe decline. Sample *B* is from a population on the lower Potomac River on the western edge of Chesapeake Bay, where ospreys seemed to be maintaining their number. Both samples were examined in 1963. Although the Maryland birds were about 2–2.5 times as successful in hatching their eggs as the Connecticut birds, whose eggs, as seen from the graphs, contained nearly twice as much DDT residue when compared with vigorously reproducing osprey colonies, it would seem that even the lower level of DDT in their eggs is having some effect on their reproductive rate.

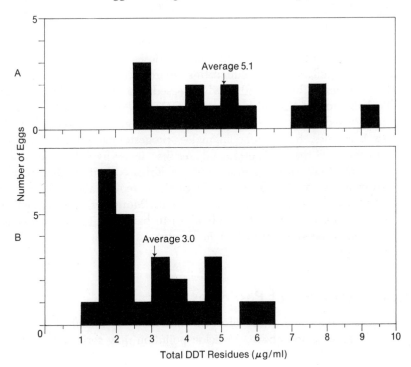

Causes of Population Declines Among Predatory Birds

The first specific evidence as to the causes of population decline in predatory birds is provided in a paper by D. Ratcliffe. British ornithologists had become concerned by falling population numbers of peregrine falcons, sparrow hawks, and golden eagles, beginning in the early 1950s. It was noted that of 109 peregrine falcon nests recorded as examined in Britain in the period 1904–1950, only three contained eggs broken before hatching. During the period 1951–1966, 47 of the 168 nests examined contained eggs broken prematurely.

Attempting to link this significant increase in egg breakage in a causal relationship with the decline in predatory bird populations and the extended use of chlorinated hydrocarbon pesticides, Ratcliffe measured eggshell thickness in peregrines, sparrow hawks, and golden eagles using egg collections dating back sometimes to the beginning of this century. In the peregrine he found a 20 per cent decrease in thickness from 1947, in sparrow hawks 24 per cent, in golden eagles 8 per cent.

In the United States, J. J. Hickey and D. W. Anderson reported a 26 per cent decrease in the eggshell thickness of the East Coast peregrine, 25 per cent in ospreys from New Jersey, and 19 per cent in bald eagles from Florida. They also showed a correlation in herring gull eggs between the decrease in thickness and the amount of chlorinated hydrocarbon residue in the eggs. J. H. Enderson and D. D. Berger established a correlation between chlorinated hydrocarbon residues in prairie falcons, thin eggshells, and poor hatching success.

Chlorinated hydrocarbons dissolve in fatty animal tissues. This was one factor leading to their use as pesticides, for they must be absorbed through the fatty chitinous exoskeleton of insects in order to be effective. R. Kuntzman reported that drug-metabolizing liver enzymes in rats were significantly more active after the rats had nontoxic doses of chlordane. Because of the continuing effect, he concluded this chlorinated hydrocarbon was stored in fatty tissues in the rats and slowly released into the circulatory system. D. B. Peakall showed that in birds given DDT, dieldrin, or both of these chlorinated hydrocarbons, there was an increase in the breakdown rate of the sex hormones, testosterone in males, progesterone in females. As a consequence of these experimental results, it is hypothesized that excess enzyme activity due to ingestion of a chlorinated hydrocarbon breaks down estrogen, producing inadequate calcium mobilization during eggshell formation and therefore thinner eggshells. Moreover, the eggshell effect is only one of a number of metabolic disturbances initiated by the presence of a chlorinated hydrocarbon in the tissues of mammals and birds. The others are still largely obscure.

Secondary Effects of Pesticides

Besides having the direct effect on individual organisms at various trophic levels as indicated above, pesticide pollution may seriously disturb natural ecosystems through selective absorption.

A dramatic example of this has been described by E. G. Hunt and A. I. Bischoff. They investigated the effect of the chlorinated hydrocarbon DDD (dichlorodiphenyldichloroethane) on gnat (*Chaoburus asticipus*) control programs at Clear Lake in northern California. It is rather ironic that these gnats, although somewhat resembling mosquitoes in appearance, are not bloodsuckers, and it appears unlikely that the adults feed at all. They were nevertheless considered a nuisance in this highly popular scenic area, and it was decided to eliminate them.

The first DDD treatment in Clear Lake was carried out in 1949, and the estimated concentration was one part in 70 million parts of lake water. A second treatment was made in 1954, with concentrations reaching one part in 50 million parts of water. In December 1954, 3 months after the second treatment, 100 western grebes were reported dead; in 1955 more dead birds were found.

As gnat populations still increased despite these two pesticide applications, a third DDD treatment was made in 1957, at the same dosage as the second. Three months later 75 dead grebes were reported. Examination of several revealed that DDD was present at the extremely high concentration of 1,600 parts per million in their fatty tissue, a concentration ratio over 100,000 times that of the lake water.

Subsequent examination of fish from Clear Lake revealed DDD concentrations ranging from 40 ppm in carp to 2,500 ppm in brown bullhead, in all cases far over the maximum tolerance level for human consumption of 7 ppm set by the Food and Drug Administration. Generally speaking, smaller fish picked up less DDD than larger, and plankton feeders less than carnivorous fish. The general disturbance to consumer trophic levels in the lake which these pesticides produced caused eutrophication, and Clear Lake became anything but clear. The gnats remained.

Port Clinton Experiment

The effects of treatment of Clear Lake with a chlorinated hydrocarbon were duplicated experimentally on a marsh near Port Clinton, Ohio, by T. J. Peterle. In 1964 this marsh was sprayed with a known amount of DDT into which chlorine-36 had been incorporated to serve as a radioactive tracer to simplify detection of residual pesticide.

The highest level of DDT in surface water was reached about 1 hour after spraying, and at the end of 2 weeks no residues were detectable there. In bottom sediments, however, residue levels remained constant from about 6 weeks after spraying to more than 15 months after. The levels of residual pesticide in marsh populations at various trophic levels during this experiment are summarized in Table 12-2.

Antarctica

That no ecosystem is immune from the insertion of organochloride pollutants is evidenced in a paper by J. O. Tatton and H. A. Ruzicka dealing with DDT and its derivatives in Antarctica, where there is no agriculture, no

TABLE 12-2 *Application of DDT to a Marsh*

Concentrations in parts per million of the pesticide subsequently found at various trophic levels. The temporarily very high pesticide concentrations in water following spraying quite soon disappear and are never reflected in the herbivore populations. Selective absorption occurs in the carnivorous trophic levels and persists there at a high level longest.

Trophic level	Name of species	Time and level of maximum residues	Time and level of residuals at last observation
Tertiary consumers (top carnivores)	Carp	1 month (10)	2nd year (1)
	Snapping turtle		15 months (16)
Secondary consumers (general carnivores)	Red leech (*Erpobdella*)	7 days (13)	2nd year (1.5)
	Crayfish	—	2nd year (<1)
	Leopard frog (tadpole)	4 hours (42)	—
			13 months (42)
	Water snake	—	
Primary consumers (herbivores)	Muskrat	(<1)	
Producers (plants)	Alga (*Cladophora*)	3 days (218)	—
	Rooted plants	—	2nd year (about 1)
	Floating plants	—	2nd year (nil)

insect life, and no use of such pesticides. Penguins and their eggs, skua, shags, and fish from Antarctica were examined. All were found to have significant quantities of various pesticide residues in their bodies, including DDT and its derivatives.

DDT in Fish Populations

The occurrence of DDT in fish populations is not always the result of selective concentration through successive trophic levels after contamination of water. O. B. Cope, who studied the effect of DDT spraying in the Yellowstone River system for spruce budworm control, concluded that the DDT found in fish there was incorporated by the ingestion of dead and dying insects contaminated with the insecticide.

Human Milk

In the case of human populations, although accidental contamination may likewise result in direct intake of pesticides, more commonly and more insidiously it usually results from selective concentration of residues along a food chain. G. Löfroth reports that in areas where use of DDT is widespread, average amounts of DDT in fatty human tissues are 5 to 27 ppm. Such levels are known to result from a daily intake of less than 0.001 milligram of DDT per kilogram body weight, which is ten times less than the highest permissible tolerance level recommended by the United Nations. The average American is estimated currently to have a concentration of 12 ppm DDT in the fatty body tissues.

Work in Sweden cited by Löfroth shows human milk there contains an average 0.117 ppm DDT. This means nursing babies take in 0.017 mg DDT per day per kilogram body weight. That is some 70 per cent over the United Nations' recommended tolerance level. Löfroth considers breast-fed babies in the United Kingdom to have a similar level of intake, and nursed babies in the United States perhaps an even higher one. Comparisons between DDT residues in human and in cows' milk are shown in Table 12-3, which illustrates the concentration factor operating with a carnivore as compared with a herbivore.

As regards dieldrin, the situation is equally disturbing. British and U.S. nursing babies consume about ten times the recommended maximum. In western Australia, Löfroth states, breast-fed babies have an intake of 0.67 ppm dieldrin daily, thirty times the maximum acceptable dosage. The implications of this observation, if it can be confirmed by further and more extensive studies, are quite simply shattering. Indeed, further testing of dieldrin suggested it might be associated with cancer incidence; its manufacture, and also that of the somewhat similar pesticide aldrin, was prohibited in the U.S. from 1974.

Tolerance Levels

One of the acute difficulties in attempting to decide what a *safe* use of DDT would be, if indeed there is such a thing, is determining the *tolerance levels*. Actual lethal effects on mammals appear to require higher dosages than in the case of birds. At the same time very serious effects can occur before such lethal dosages are reached. In mammals these effects appear to be especially associated with the motor controls of the central nervous system.

J. M. Anderson and M. R. Peterson reported a disturbing effect of DDT on the nervous system of brook trout (*Salvelinus fontinalis*). It has been known for some time that fish exhibit behavioral changes following exposure to sublethal levels of pesticides. These two workers have shown that

TABLE 12-3 *Pesticide Residues in Human Milk*

Contrasted with those in cows' milk from the same locality in England.

	DDT	DDE	Total DDT equivalent	Total BHS	Dieldrin
Cows' milk	0.002	0.002	0.004	0.003	0.003
Human milk	0.045	0.073	0.128	0.013	0.006

modifications of two specific responses in fish, the thermal acclimation mechanism and establishment of a visual conditional avoidance response, may result from exposure to sublethal dosages of DDT.

Accumulation of DDT in the Atmosphere

If DDT is being concentrated in the biomass of producer organisms in the ecosphere, it will become ever more concentrated as it passes in succession to the progressively smaller biomasses of primary, secondary, and tertiary consumers (Tables 12-1 and 12-2).

Concentrations of pesticide pollutants might be expected to be greatest in estuares and over the continental shelf, where polluted drainage systems discharge into the ocean. These are precisely the sections of the ecosphere where human marine food is especially harvested.

As if these various concentration factors were not enough, DDT and other chlorinated hydrocarbons are stored in the human body, particularly in the fatty tissues (Table 12-4). A period of prolonged physical stress, which utilizes an extensive amount of these fatty food reserves, could therefore bring the levels of pesticide circulating in the body well above tolerable acceptances. The more serious illnesses, a deliberate fast, or drastic diet could all prove fatal.

Circulation of Chlorinated Hydrocarbons

The detection of chlorinated hydrocarbons in Antarctica, far from any known source of contamination, was explained when it was realized that pesticides are part of the general atmospheric circulation and are carried as air pollutants for thousands of miles. A paper by R. W. Risebrough and colleagues described how dust particles carried 3,000 miles across the North Atlantic by northeasterly trade winds transferred DDT, DDE, and DDD from Africa or Europe to the Caribbean island of Barbados. The pesticides apparently had been adsorbed in vapor form on dust particles,

TABLE 12-4 *Human Residual Pesticide Loads*

In parts per million of DDT/DDE. U.S. citizens eating three meals per day in 1964–66 had an average daily pesticide intake of 0.08 to 0.12 mg. of which some three quarters was DDT. A steady state was reached at this level, the excretion rate balancing the intake rate, with residual accumulations in the body fat of from 5 to 20 ppm, that is, from average to high according to the national figures quoted below.

India	26	Canada	5
Hungary	12	Germany	4
United States	12	England	3
France	9	Alaska	3

which had an annual fallout rate over the Atlantic of approximately two thirds of a ton.

K. B. Tarrant and J. O. Tatton estimated fallout from atmospheric circulation over Britain by determining the amount of DDT, dieldrin, BHC, DDE, and DDD residues in rainwater. Their results are summarized in Table 12-5 together with comparable figures for the United States. The British results confirm that pesticides are in general circulation in the atmosphere and provide an average figure of 1 ton of pesticide in every inch of rainfall (Figure 12-2). This amounts to a pesticide fallout four times greater than the quantity which the Mississippi River annually carries out into the Gulf of Mexico. The magnitude of fallout pattern would appear to be characteristic of the North Atlantic. This massive distribution of pesticides would alone account for their universal appearance in all global ecosystems, but it is probably repeated over other continental masses and related ocean basins. Tables 12-6 and 12-7 show the amounts of pesticides used in the United States and exported, which provides some idea of how the global fallout system is being constantly replenished.

Various experimental determinations suggest that only one half to one third of the pesticide applied actually reaches the ground. The rest volatilizes before the spray droplets fall on their terrestrial target or the soil. From leaf or land surfaces volatilization will in any case usually occur within a few weeks, depending partly on ambient temperatures. Concentrations of pesticides in the atmosphere will tend to be highest over agricultural areas, as shown in Table 12-8. However, once the amount vaporized reaches a steady state with the amount of fallout, updrafts of air from cold fronts, mountain masses, and other mixing mechanisms can carry the pesticides up into the troposphere. There, the temperate westerlies and the tropical easterlies, together with their north-south polar eddies, will carry

TABLE 12-5 *Pesticides in Rainwater*

A comparison of remote, agricultural, urban, and industrial areas in the United Kingdom and United States. Figures express mean monthly concentration in parts per trillion in the rainfall during 1966–67, those for DDT include DDD and DDE.

Location	BHC	Dieldrin	DDT
UNITED KINGDOM			
Lerwick (far north)	145	11	73
Sheffield (industrial)	87	6	87
London (urban)	88	16	93
Maidstone (horticultural)	103	2	97
Camborne (far south)	48	6	115
UNITED STATES			
Ripley, Ohio (agricultural)	50	—	180
Coshocton, Ohio (agricultural)	6	—	75
Cincinnati, Ohio (urban)	20	—	360

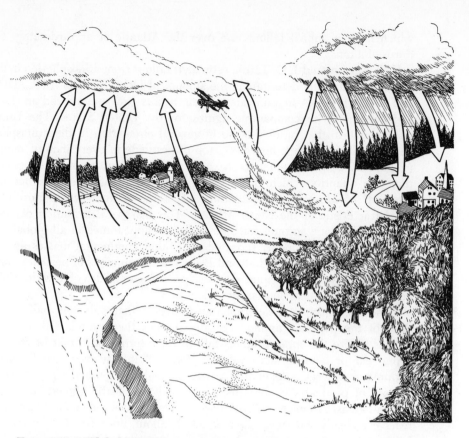

FIGURE 12-2. *Global circulation patterns of pesticides.* Although some local concentration of pesticides occurs near the point of application, and in estuarine and coastal waters, the global circulation is more related to the atmospheric circulation and to the amount of precipitation, that is, rainfall or snow, which will carry particulate matter with vaporized adsorbed pesticides to the ground.

the pesticides in a global pattern. As J. Frost states, the 10 parts per trillion of pesticide residues in the atmosphere over London, England (Table 12-8), can be extrapolated to give a total global figure of 5,700 tons in the atmosphere. This is one tenth the amount of pesticides sold annually in the United States, more than one half of this total amount finding its way into the atmospheric circulation. This latter reservoir might continue to support the atmospheric circulation of pesticides at present levels for a number of years even if the use of pesticides were to be entirely prohibited on a world-wide scale.

Other Chlorinated Hydrocarbons

Other chlorinated hydrocarbons now are marketed under a variety of names; dieldrin, endrin, and heptachlor, together with DDT, are the most commonly used. These and all others of this group listed in Table 12-7 are *persistent pesticides,* not readily broken down in any ecosystem by the

TABLE 12-6 *Annual Amounts of Chlorinated Hydrocarbons Produced in the United States During the Last Decade*

Figures shown are rounded to the nearest 1000 tons: those in parentheses for 1964 indicate the amount believed to have been used by U.S. farmers (see Table 12-7). It is predicted that agricultural needs in underdeveloped areas (which took the major portion of the remaining production) will increase their pesticide applications by six times their use at the end of the last decade.

	1960	1964	1967
DDT	102,000	124,000 (27%)	103,000
Aldrin-toxaphene groups*	85,000	103,000 (52%)	120,000

* Includes aldrin, chlordane, dieldrin, heptachlor, and toxaphene.

decomposer organisms of the reducer trophic level. Eventually microorganisms might be evolved that would cause more rapid breakdown of these substances. Their success as pesticides, however, must be due especially to their novelty in the ecosphere. If organisms at various trophic levels had been long exposed to this group of compounds they would by now have *adapted* to them.

Other Pesticides

The other main group of pesticides is the *organophosphates*, of which parathion and malathion are the best known; others such as phosdrin and chlorthion also are used quite widely. These are highly toxic to human beings, and their application in agriculture requires the wearing of special

TABLE 12-7 *Amounts of Chlorinated Hydrocarbons Used Annually by Farmers in the United States*

Toxaphene	39,000
DDT	37,000
Aldrin	11,000
Strobane	2,700
Endrin	2,200
Methoxychlor	1,400
Lindane	1,000
Dieldrin	900
Chlorane	500
BHC	300
Other	800
Total	96,800

TABLE 12-8 *Pesticides in the Air over Various Cities and Regions*

Expressed in parts per trillion by weight; sampled at different times between 1963 and 1967. Districts like La Jolla, where the prevailing winds are landward, with an unknown admixture of air from neighboring agricultural areas, can have a total pesticide contest in the air a thousand times that of a marine air remote from sites of application as in Barbados.

Locality	Pesticide Concentration			
	DDT	Dieldrin (HEOD)	BHC	Total
United States				
Pittsburgh (heavily industrialized)	0.177	—	—	
La Jolla, Calif. (mean of agricultural areas)	5	—	—	
Caribbean				
Barbados	0.00006	0.000001	—	0.000078
Europe				
London (urban)	10	18	6	

protective clothing. Some of these organophosphates find their way into domestic use and, despite the required printed warnings on the container, are indiscriminately sprayed around the home.

One particular feature of this group is their frequent use as systemic insecticides, as substances that can be taken up into and circulated through plant tissues where they remain toxic for a length of time to any animals consuming these tissues. As a group, organophosphate pesticides are *biodegradable*. They are readily broken down by reducer organisms in natural ecosystems so that they are not long persistent and do not present the same problems as chlorinated hydrocarbons.

Nerve Gases

The main component of many nerve gases is the group of organic phosphorus compounds known as anticholinesterases. They are also the active agents in a number of pesticides, including parathion. These compounds inhibit the activity of cholinesterase, which causes an accumulation of acetylcholine at mammalian nerve endings. Normally the enzymatic breakdown of acetylcholine by cholinesterase occurs almost instantly, so that a nerve impulse for which acetylcholine serves as a chemical messenger has only a momentary effect. When it accumulates instead, the muscle cells are either constantly stimulated or become paralyzed. This is expressed by nausea and vomiting or staggering and twitching in the affected animal.

On March 4, 1968, sheep grazing through the snow near the U.S. Army Chemical and Biological Weapons Testing Center at Dugway near Salt

Lake City were stricken with a "staggering" disease. After a day or so more than 6,000 of them had died, together with rabbits, deer, and some other wildlife. The U.S. Department of Agriculture and the Public Health Service determined that these animals died as a result of exposure to nerve gas released during what the Army described as a routine weekly demonstration test on the previous day, March 3. The Army denied that their test had caused the sheep deaths. The civilian agencies also concluded that sheep rather than other animals had died because sheep have a habit of obtaining their water requirements by eating snow. In the affected area snow was indeed found to be contaminated with nerve gas.

The possible accidental release of this type of colorless and odorless nerve gas during testing, transport, manufacture, or storage has since been regarded with considerable public misgiving, and assurances regarding an overhaul of safety precautions have been sought. That these misgivings may be well founded is illustrated by a number of episodes relating to the use of another type of organophosphate, commonly labelled *Phosvel*. Like all organophosphates used as pesticides, Phosvel is biodegradable in soil and water, but it may persist for several months on various surfaces. It is capable of producing delayed irreversible nerve poisoning and is suspected of causing several episodes of such damage both in domestic stock and in human populations.

Polychlorinated Biphenyls

Although they are not used as pesticides, the environmental effects of various polychlorinated biphenyl (PCB) compounds are so similar to those of broad-spectrum persistent pesticides that they can be discussed here. PCBs have been known and used in industry since before World War II. They were extensively used in the textile and printing industries, for electrical insulation, in heat-exchange and hydraulic systems, and in various other industrial processes. They still are found in polystyrene containers used for food liners and wrappers, drinking receptacles, and frozen-food bags.

Despite the fact that the chemical structure of PCBs closely resembles that of DDT, these substances were not identified as contaminants in any ecosystem until 1966, when they were reported in fish from the Baltic Sea. Further study confirmed that PCBs had not begun to appear in natural ecosystems until after World War II, when their use in industry became widespread. Between 1945 and 1968 U.S. production of DDT totaled nearly 1 billion pounds, the same amount estimated for total PCB production in the last four decades. Whereas DDT is deliberately dispersed into the environment, PCBs one way or another also seem to work into food webs. It seems entirely possible that the amount of PCBs present in the ecosphere may approximate that of DDT residuals.

The effect of PCBs on animal metabolisms and on humans is, as might be anticipated, very similar to that of DDT. They modify sex hormone activities, interfere with liver enzymes, and appear to be potentially carcinogenic. The FDA has set acceptable limits in food for human consumption on the basis of these effects. Barry Commoner reported that by 1973 the Monsanto Company, which produced about half the world's output of PCBs, was withdrawing from sale those PCBs that might enter human ecosystems, either by uncontrollable contamination or by other forms of loss to the environment. Nevertheless, as Commoner reports, totaling all contamination of dumps, air, and fresh and coastal waters, it seems that about 75,000 tons of PCBs annually are still released into our ecosystems.

Resistant Strains

Even as early as 1946 some housefly populations in Sweden had been reported as resistant to DDT. By 1948 there were 12 such species, and by 1957, 76. In 1969 apparently 224 insects pests were resistant to one pesticide or another. Some (Table 12-9) are even resistant to both chlorinated hydrocarbons and organophosphates.

TABLE 12-9 *Some Insect Pests in Which Races Have Evolved Resistant to Both Chlorinated Hydrocarbon and Organophosphate Types of Pesticides*

FLIES

Chrysomyia patoris
Musca domestica (housefly)

TICKS

Boophilus micropus (southern cattle tick)

BEDBUGS

Cimex lectularis (common bedbug)

COCKROACHES

Blattella germanica (German cockroach)

MOSQUITOES

Aedes melanimon
Aedes nigromaculis
Aedes taeniorhynchus (black salt-marsh mosquito)
Culex pipiens (southern house mosquito)
Culex quinquefasciatus
Culex tarsalis

MOTHS

Epiphyas postrittana
Heliothis virescens (tobacco budworm)
Trichoplusia ni (cabbage looper)

HEMIPTERA (bugs and aphids)

Erythroneura lawsoniana
Myzus persicae (green peach aphid)

There appear to be two principal ways in which insect populations adapt to pesticides, one physiological, the other behavioral. Physiological adaptation occurs when races of the insect species population that can destroy or detoxify the pesticide are selected. Selection for genotypes that contain genes producing detoxifying enzymes is quite common. From laboratory experiments with the fruit fly *Drosophila*, it has even been possible to determine the locus of the genes responsible.

The selection of races possessing such enzyme systems is greatly favored by current pesticide application techniques. The spraying of a broad-spectrum insecticide over an extensive area removes not only most pest species but also most predatory and parasitic insects. The few individuals that survive because they possess a particular genotype are then able to increase quickly to form resistant races without either competition or predation. Some economic pests such as the salt-marsh sandfly are reported to have evolved pesticide-resistant races in this way after only three pesticide applications. The more quickly one generation replaces another, the earlier resistant races appear.

Increasing the dosage rate of the pesticide merely delays somewhat the evolution of resistant races. The massive pesticide applications now being used for this reason in the cotton-growing parts of the United States and other western hemisphere countries constitute what must be regarded as a major threat to national as well as global health.

The other form of insect pesticide resistance is developed by behavioral avoidance. Just as selection for individuals with particular enzyme systems occurs, so selection for individuals whose behavioral ecology takes them outside the area of contact with the pesticide will produce resistant races. Thus mosquito populations that originally were primarily house feeders will now bite only outside houses; those that enter houses alight on DDT-sprayed walls and are killed.

This could be called selection for pesticide avoidance, and it has developed in a number of tropical species of *Aedes* and *Culex*. Only careful development of schemes for *integrated control,* as discussed below, will prevent the appearance of many further races or strains of pesticide-resistant insects of a behavioral or any other type.

Biological Control

In any ecosystem there is a considerable amount of production, i.e., producer biomass, removed by primary consumers such as herbivorous insects. This is most obvious when a swarm of locusts visits a corn or millet field and removes the *total* aboveground biomass accumulated by producer organisms. Less conspicuous cropping activities may nevertheless reach formidable dimensions.

Georg Borgstrom estimates that if the insects in tropical Africa which compete in herbivorous activity with human food production could be controlled, it would be possible to feed 2 billion people, eight times as many as now live on that continent.

Figures of similar magnitude are produced by the agricultural industry whenever measures are proposed for prohibiting the use of particular pesticides. A total prohibition of their use, effected immediately, would obviously result inevitably in local if not global famine, as well as in widespread insect-transmitted disease epidemics.

There are several ways in which the use of persistent broad-spectrum pesticides may eventually be superseded with less drastic consequences. One is by development of *selective* pesticides, of which a most promising group is the *ecdysones*. These are the substances that initiate ecdysis or metamorphosis from one stage to another in the larval development of insects. Overstimulation with ecdysones results in repeated metamorphosis without sufficient time for the accumulation of food reserves, so that eventually the larva becomes exhausted and dies. Many plants apparently contain ecdysones or ecdysone analogues in sufficient amount to provide them with immunity from insect attack.

There are also many other insect-plant relationships that provide a balance between herbivore and producer populations. If such relationships had not evolved, the herbivores would literally have consumed the producer populations into extinction. Proper understanding of such steady-state relationships will almost certainly provide new approaches in the search for selective pesticides.

Another aspect of *biological control* is the location and introduction of secondary consumers that prey on herbivorous pests.

Introduction of Predators and Parasites

The earliest application of the principle of biological control by introducing a predator or parasite to regulate the numbers of a herbivore population came long before the development of pesticides. Toward the end of the last century, the citrus industry in California was suffering a massive infestation of a mealybug, cottony cushion-scale (*Icerya purchasi*), introduced from Australia on acacia in 1868. The introduction of two Australian species, the ladybird beetles *Rodolia cardinalis* (vedalia ladybird) and *Cryptochetum iceryae*, provided the necessary predator-prey regulation and reduced the mealybug populations to levels at which they no longer constituted a major pest infestation. Unfortunately, however, as is shown in Figure 12-3, cottony cushion-scale again reached the major pest population levels in California when extensive use of DDT for citrus spraying eliminated the vedalia ladybird locally.

This recurrence emphasized the danger of indiscriminate use of pesticides, which is likely to have a catastrophic effect on population regulatory mechanisms that have held pests below the economic threshold in agricultural microecosystems. Although pesticides momentarily diminish the numbers of a particular pest, they commonly also reduce the numbers of natural enemies in the form of insect predators and parasites. The pest then undergoes a population explosion before population numbers in the natural enemies can recover.

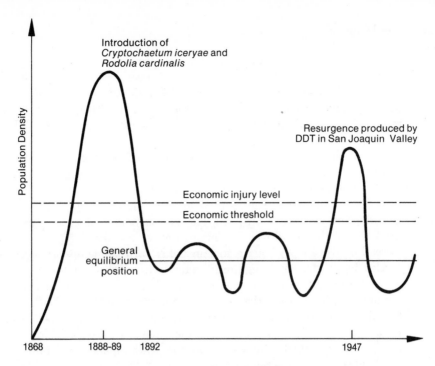

Figure 12-3. *Cottony cushion-scale (Icerya purchasi) incidence on citrus in California.* Introduced in 1868, this pest increased rapidly until two of its ladybird predators were obtained and released in 1889. A resurgence of the pest occurred in 1947 following the use of DDT sprays on citrus, which caused a reduction of the predator populations of ladybirds.

Subsequently to the successful introduction of a biological control measure for cottony cushion-scale, further introductions were made to control other introduced pests, including gypsy moths, Japanese beetles, and European corn borers.

The Sterile Male Technique

A different technique of biological control was introduced in the early 1950s to control the screw-worm fly (*Callitroga hominivorax*), whose larval stage infests sheep. Male flies were rendered sterile by exposure to gamma radiation, which did not, however, reduce their vitality or sexual aggression. The sterilized males were then released in numbers larger than those of males in the natural populations, so that females copulating with them proceeded to lay unfertilized eggs.

By use of this technique, the screw-worm had been eliminated by 1958 from the island of Curaçao and from the southeastern states, the only part of the continental United States in which this pest can overwinter. Further applications of the same technique now are being tried with other pests, combined with the use of chemosterilants and genetically sterile male forms.

The most useful mode of attack for the present, however, would appear to be *integrated control,* which combines several methods in a unified attempt to control a particular pest. The total effect of these combined methods tends to be *synergistic* rather than additive; not only does it reduce the pesticide pollution problem, but also the control obtained is more effective.

One of the earliest applications of integrated control was reported from California. The practice there of mowing alfalfa over a large acreage in a single operation was shown to cause serious disturbance to the predator-prey balance between the spotted alfalfa aphid and its natural enemies. By adopting successive strip cutting over a period of time, parasites and predators of the aphid were maintained at an average of 56 per square foot instead of the 14 per square foot in fields with single-operation cutting.

Also in California, integrated control was applied to the treatment of grape leafhopper after this pest was found to have developed resistance to organophosphate pesticides. It was discovered that the leafhopper was heavily attacked by a wasp (*Anagrus*) that overwinters on blackberry. By interplanting blackberry vines in the vineyards, it is possible to combine the predatory attacks by wasps on the leafhoppers with limited use of new and selective insecticides.

The concept of integrated control has been fully discussed in a paper by V. H. Stern and colleagues. It includes not only appropriate combinations of pesticides, natural enemies, and cultural treatments, but also the use of insect pathogens.

BIBLIOGRAPHY

References

Ames, P. L. "DDT residues in the eggs of the osprey in the northeastern United States and their relation to nesting success," *J. Appl. Ecol.,* 3(suppl.): 87–97, 1966.

Antommaria, P., Corn, M., and Demaio, L. "Airborne particulates in Pittsburgh; association with PIP-DDT," *Science,* 150: 1476–79, 1965.

Baumhover, A. H., Graham, A. J., Bitter, B. A., Hopkins, D. E., New, W. D., Dudley, F. H., and Bushland, R. C. "Screw-worm control through the release of sterilised flies," *J. Econ. Entomol.,* 48: 462–66, 1955.

Biros, F. J., Walker, A. C., and Medbury, A. "Polychlorinated biphenyls in human adipose tissue," *Bull. Environ. Contam. Toxicol.,* 5: 317–23, 1970.

Bitman, J., Cecil, H. C., and Fries, G. F. "DDT-induced inhibition of avian shell gland carbonic anhydrase: a mechanism for thin eggshells," *Science,* 168: 594–96, 1970.

Butler, P. A. "Monitoring pesticides pollution," *Bioscience,* 19: 889–91, 1969.

Commoner, B. "Workplace burden," *Environment,* 15(6): 15–20, 1973.

Cope, O. B. "Effects of DDT spraying for spruce budworm on fish in Yellowstone River system," *Trans. Amer. Fish. Soc.,* 90: 239–51, 1961.

Edwards, C. D. "Insecticide residues in soils," *Residue Reviews,* 13: 83–132, 1966.

Egan, H., Goulding, R., Roburn, J., and Tatton, J. O. "Organo-chlorine pesticide residues in human fat and human milk," *British Med. J.,* 2: 66–69, 1965.

Enderson, J. H., and Berger, D. D. "Pesticides: eggshell thinning and lowered production of young in prairie falcons," *Bioscience,* 20: 355–56, 1970.

Frost, J. "Earth, air, water," *Environment,* 11(6): 15–25, 1969.

Gould, R. F. *Organic Pesticides in the Environment,* Washington, D.C.: Advances in Chemistry Series No. 60, American Chemical Society, 1966.

Hickey, J. J., and Anderson, D. W. "Chlorinated hydrocarbons and eggshell changes in raptorial and fish-eating birds," *Science,* 152: 271–72, 1968.

Hunt, E. G., and Bischoff, A. I. "Inimical effects on wildlife of periodic DDT applications to Clear Lake," *Calif. Fish and Game Bull.,* 46: 91–96, 1960.

Kermode, G. O. "Food additives," *Scientific American,* 226(3): 15–21, 1972.

Kuntzman, R. "Similarities between oxidative drug-metabolizing enzymes and steroid hydroxylases in liver microsomes," *Pharmacol. Exper. Therap.,* 146: 280–85, 1964.

Löfroth, G. "Pesticides and catastrophe," *New Scientist,* 38: 166–67, 1968.

Marwick, C. "Death in Skull Valley," *New Scientist,* 38: 166–67, 1969.

Menzel, D. W., Anderson, J., and Randtke, A. "Marine phytoplankton vary in their response to chlorinated hydrocarbons," *Science,* 167: 1724–26, 1970.

Mrak, G. M. (ed.) *Report of the Secretary's Commission on Pesticides and Their Relationship to Environment and Health,* Washington, D.C.: USDHEW, APO, 1969.

Nisbet, I. C. T., and Miner, D. "DDT substitute," *Environment,* 13: 10–17, 1971.

Peakall, D. B. "Pesticides and the reproduction of birds," *Scientific American,* 222(4): 73–78, 1970.

Peterle, T. J. "Pyramiding damage," *Environment,* 11(6): 34–40, 1969.

Ratcliffe, D. "Decrease in eggshell weight in certain birds of prey," *Nature,* 215: 208–10, 1967.

Risebrough, R. W., Hugget, R. J., Griffin, J. J., and Goldberg, E. D. "Pesticide transatlantic movements in the Northeast Trades," *Science,* 159: 1233–36, 1968.

Schlinger, E. I., and Dietrick, E. J. "Biological control of insects aided by strip-farming alfalfa in experimental program," *Calif. Agriculture,* 14: 8–9, 1960.

Shea, K. P., "PCB," *Environment,* 15(9): 25–28, 1973.

Shea, K. P., "Nerve damage," *Environment,* 16(9): 6–10, 1974.

Steinhaus, E. A. "Concerning the harmlessness of insect pathogens and the standardization of microbial control products," *J. Econ. Entomol.,* **50:** 715–20, 1957.

Tatton, J. O., and Ruzicka, J. H. A. "Organochlorine pesticides in Antarctica," *Nature,* **215:** 346–48, 1967.

Further Readings

Anderson, J. M., and Peterson, M. R. "DDT: sublethal effects on brook trout nervous system," *Science,* **164:** 440–41, 1969.

Bitman, J., et al. "Estrogenic activity of O.P'-DDT in the mammalian uterus and avian oviduct," *Science,* **162:** 371–72, 1968.

Brown, A. W. A. "Insecticide resistance comes of age," *Bull. Entomol. Soc.,* **14:** 3–9, 1968.

Burdick, G. E., et al. "Accumulation of DDT in lake trout," *Trans. Amer. Fish. Soc.,* **93:** 127–29, 1964.

Cottam, C. "The ecologist's role in problems of pesticide pollution," *Bioscience,* **15:** 457–63, 1965.

Curley, A., and Kembrough, R. "Chlorinated hydrocarbon insecticides in plasma and milk of pregnant and lactating women," *Arch. Environ. Health,* **18:** 156–64, 1969.

Diamond, J. B., and Sherbourn, J. A. "Persistence of DDT in wild animals," *Nature,* **221:** 486–87, 1969.

Duggan, R. E., and Weatherwax, J. R. "Dietary intake of pesticide chemical," *Science,* **157:** 1006–10, 1967.

Edwards, C. A. "Soil pollutants and soil animals," *Scientific American,* **220**(4): 88–99, 1969.

Epstein, S. S. "A family likeness," *Environment,* **12**(6): 16–25, 1970.

Falk, H. L., Thompson, S. J., and Koten, P. "Carcinogenic potential of pesticides," *Arch. Environ. Health,* **10:** 848–58, 1965.

Fiserova-Bergerova, V., Radomski, J. L., Davies, J. E., and Davies, J. H. "Levels of chlorinated hydrocarbon pesticides in human tissues," *Industr. Med. Surg.,* **36**(1): 65–70, 1967.

Fonselius, S. H. "Stagnant sea," *Environment,* **12**(6): 2–11, 40–48, 1972.

George, J. L., and Frear, D. E. H. "Pesticides in the Antarctic," *J. Appl. Ecol.,* **3**(suppl.): 155–67, 1966.

Gunther, F. A., Westlake, W. E., and Jaglan, P. S. "Reported solubilities of 738 chemicals in water," *Residue Reviews,* **20:** 1–148, 1968.

Gustafson, C. G. "PCB's—prevalent and persistent," *Environ. Sci. Technol.,* **4**(10): 814–19, 1970.

Hickey, J. H. (ed.) *Peregrine Falcon Populations,* Madison: University of Wisconsin Press, 1969.

Holcomb, R. W. "Insect control: alternatives to the use of conventional pesticides," *Science,* **168:** 456–68, 1970.

Irving, G. W. "Agricultural pest control and the environment," *Science,* **168:** 1419–24, 1970.

Jones, F. J. S., and Summers, D. D. B. "Relation between DDT in diets of laying birds and viability of their eggs," *Nature,* **217:** 1162–63, 1968.

Jukes, T. H. "Antibiotics in animal feeds and animal production," *Bioscience,* **22**: 526–34, 1972.

Laws, E. R., Curley, A., and Biros, E. F. "Men with intensive occupational exposure to DDT," *Arch. Environ. Health,* **15**: 766–75, 1967.

Lichtenstein, E. P., Shulz, K. R., Fuhrmann, T. W., and Liang, T. T. "Biological interaction between plastecizers and insecticides," *J. Econ. Entomol.,* **62**: 761–65, 1969.

Maugh, T. H., II. "DDT: an unrecognized source of polychlorinated biphenyls," *Science,* **180**: 578–79, 1973.

Meeks, R. L. "The accumulation of CI-36 ring-labelled DDT in a freshwater marsh," *J. Wildlife Management,* **32**: 376–98, 1968.

Messenger, P. S. "Utilization of native natural enemies in integrated control," *Ann. Appl. Biol.,* **56**: 328–30, 1965.

Moore, N. W. "A synopsis of the pesticide problem," *Adv. Ecol. Res.,* **4**: 75–129, 1967.

Morgan, K. Z. "Never do harm," *Environment,* 13(1): 28–38, 1971.

Nash, R. G., and Woolson, E. A. "Persistence of chlorinated hydrocarbon insecticides in soils," *Science,* **157**: 924–27, 1967.

Pramer, D. "The soil transforms," *Environment,* 13(4): 43–46, 1971.

Radomski, J. L., et al. "Pesticide concentrations in the liver, brain, and adipose tissues of terminal hospital patients," *Food and Cosmetic Toxicol.,* **6**: 209–20, 1968.

Risebrough, R. W., Menzel, D. B., Martin, D. J., and Olcott, H. S. "DDT residues in Pacific sea birds: a persistent insecticide in marine food chains," *Nature,* **216**: 589–90, 1967.

Robinson, J., Richardson, A., Crabtree, A. N., Couldon, J. C., and Potts, G. R. "Organochloride residues in marine organisms," *Nature,* **214**: 1307–11, 1967.

Ruzicka, J. H. A., Simmons, J. H., and Tatton, J. O. "Pesticide residues in foodstuffs in Great Britain: IV, organochlorine pesticide residues in welfare foods," *J. Sci. Food and Agriculture,* **18**: 579–82, 1967.

Shepard, H. H. *Methods of Testing Chemicals on Insects,* Minneapolis: Burgess, 1968.

Sparr, B. I., Appleby, W. G., DeVries, D. M., Osmun, J. V., McBride, J. M., and Foster, G. L. "Insecticide residues in waterways from agricultural use," *Adv Chem.,* **60**: 146–62, 1966.

Williams, C. M. "Third-generation pesticides," *Scientific American,* 217(1): 13–17, 1967.

Wilson, B. R. (ed.) *Environmental Problems: Pesticides, Thermal Pollution and Environmental Synergisms,* Philadelphia and Toronto: Lippincott, 1968.

Winter, R. *Poisons in Your Food,* New York: Crown, 1969.

Woodwell, G. M. "Toxic substances and ecological cycles," *Scientific American,* 216(3): 24–31, 1967.

Woodwell, G. M., Craig, P. P., and Johnson, H. A. "DDT in the biosphere: where does it go?" *Science,* **168**: 1101–7, 1971.

Woodwell, G. M., Wurster, C. F., and Isaacson, P. A. "DDT residues in an

East Coast estuary: a case of biological concentration of a persistent insecticide," *Science,* **156**: 821–24, 1967.

Wurster, C. F. "DDT reduces photosynthesis in marine phytoplankton," *Science,* **519**: 1474–75, 1968.

Wurster, C. F., and Wingate, D. B. "DDT residues and declining reproduction in the Bermuda petrel," *Science,* **159**: 979–81, 1968.

Chapter

13

Conservation

This survey of the evolution of human populations, their ecological relationships and environmental confrontations, would be incomplete without an attempt to review both what the total effect of our species on our environment has so far been and what our eventual fate will be in the various ecosystems we now occupy. Whatever view is taken of human nature, there is one overriding ecological principle that can be neither denied nor ignored: the pattern of human behavior is relevant only to one time and to one place. Change there always has been, and change there always must be for any species if extinction is to be avoided. The uniqueness of our human population lies in its capacity to adapt to change and in the speed of the response. Within a period of several million years, human populations have provided a striking demonstration of this remarkable adaptability by discovering how to escape from the ecosphere of one planet to invade that of its satellite. We have also learned how to insert artifacts into the biospheres of other planets of our solar system that could irreversibly change any ecosystems that may have evolved there.

General Considerations

There are several implications of this remarkable adaptive capacity that are pertinent to the ideas considered in this text, but first human ecology must be provided with a time dimension by setting it against estimates of a universal chronology. Several calculations now place the origin of the universe at 10 billion years BP. There is increasing evidence, as was considered in the opening chapter, that life on this planet began somewhere between 3½ and 4 billion years ago. Human populations have been recognizable for about 4–5 million years, the present cultural patterns we refer to as *urban civilization* for about 15,000 years.

In the whole universe what is the chance of similar cultural levels having evolved elsewhere, even in this last *one millionth* of universal time? We know from astronomical studies that the evolution from hydrogen and helium of the more than 100 different chemical elements appears to have occurred in every part of the universe that we examine. Could evolutionary response to similar physical environments in other ecospheres prescribe that living systems also had to evolve toward a dominant cultural creature wherever life arose?

We can be fairly confident that if it were so, these other cultural creatures would be totally different from any of the multitude of life forms that have appeared at one time or another on this planet—different in morphology, anatomy, physiology, chemical nature, and genetic processes. All that can be said with any degree of certainty is that if there were dominant cultural forms in other parts of the universe, they would necessarily have evolved some kind of social communication. This would be of a form that our own culture could detect either by our bodily senses or through our mechanical sensors, for sound, light, or some other form of electromagnetic waves would be the most likely transmission medium.

If during our future space travels we were to approach such other cultural forms of life, it is likely that our senses or our sensors would record the encounter. Such a meeting would be a momentous event, for it would again herald competition for the same ecological niche—that of the dominant social heterotroph—but this time on a universal scale. Only cultural displacement could save one or the other of us from extinction.

Extinction Times

In considering the future in such universal ecological terms, we should recall that no dominant form of higher organism has survived more than a short 250 million years on our own planet. We are presently a mere 10,000 to 30,000 years into the latest 250-million-year interval that can be expected to elapse before the next cataclysmic series of ice ages. Yet we have contrived in this brief moment of universal time to exploit and destroy much of this planet's endowments without so far achieving any stability in its human populations or any adequate system for recycling its finite resources. This is poor ecological practice indeed, and it is small wonder that we now are confronted with so many environmental crises.

Resource Exploitation

Through having failed to achieve a proper understanding of territoriality, we are forced to devote huge allotments from our current productivity to defending the regional boundaries of the national subdivisions of our global population. On an even more parochial scale, territoriality is a constant cause of personal friction, extensively violated even to the extreme of mayhem and murder wreaked on our own kind.

In our social evolution we have as yet evolved no better means of developing initiative and encouraging efficiency than the fratricidal struggle called "private enterprise." In the resulting rat-race all resources are up for grabs. No really serious attempt has yet been made to conserve these resources and establish and maintain a steady-state stability in the ecosystems we occupy at a level that will ensure the preservation of a maximum diversity in these systems.

Territorial Dissociation

At the same time the feeling of isolation that human populations have always contrived to adopt in regard to the fate of populations in other territories has intensified. "This is America"—or Vietnam, or France, or Costa Rica; "things like that could not happen here." The ruthless slaughter of Watusi, Ibos, Salvadorans, and Hondurans; Chinese-Russian fighting; Arab-Israeli killings; famine in India and Pakistan; illiteracy in South America— all are images on a television screen that can be turned off or ignored like the commercials and appear to most viewers to be just as fantastic and unreal.

Technological Consequences

Our technological advances have brought some of the populations in our global ecosystem within reach of the utopian dreams of leisure, health, and bounty in the mere two centuries since the industrial revolution began. But these advances have also brought the capacity to destroy all life in the ecosphere. This can be done catastrophically by an atomic holocaust; or the same result can be achieved more slowly and more insidiously, but certainly as effectively, by permitting pollution of one form or another at a level preventing the continuing reproduction of our species.

With all our technological advances we have failed utterly to provide the socioeconomic measures necessary to avoid the threatened onset of a malthusian phase of famine and pestilence that may shortly devastate some of the world's societies if desperate action is not taken. As the great American conservationist Aldo Leopold stated over 40 years ago, *"Human ecology has now become a matter of ethics."*

Carrying Capacity

This chapter has been entitled "Conservation" because the level at which human carrying capacity is placed will determine what can be conserved. There is no doubt that in Advanced Industrial societies the carrying capacity in terms of human populations could be set much higher. The ecosystems of these areas still include many animal and plant populations that are not contributing to human food chains, many in fact that compete with them directly or indirectly. Wilderness areas are still permitted to exist. Populations of inedible birds and insects abound and utilize considerable amounts of energy at one trophic level or another. The case for the conservation in industrial societies of this uneconomic evolutionary diversity was eloquently presented by Evelyn Hutchinson several years ago. For most of us in such societies the destruction of uneconomic portions of our ecosystems for the purpose of increasing human carrying capacity would be totally unacceptable.

By contrast, human population pressures in some of the Advanced Agricultural societies have caused the whole landscape to be reduced to basic ecosystems that relate exclusively to human productivity. Paddy rice fields can cover the whole land, and all surface water is channeled to these fields. No plants but rice grow in the paddies, other than microscopic and planktonic forms. No vertebrate animals but buffalo and chickens are allowed to exist. The food web is expressed simply as

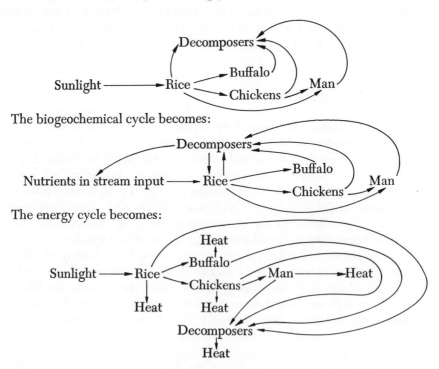

The biogeochemical cycle becomes:

The energy cycle becomes:

Obviously these simple diagrams understate ecosystem reality. A few egrets probably will be included also, a few human parasites such as hookworm, a few rats or mice, and many insects. But the human society in such an ecosystem is unlikely for the present to devote much time to worrying about wilderness conservation or endangered species. If and when it proceeds to later stages of societal evolution, such a society may come to resent the past that forced this extreme exploitation of its territory, and left it with the sole heritage of stripped and degraded agrarian ecosystems.

The Energy Crisis

Many of these issues were recently brought much closer to home for at least two western nations. In the summer of 1972 a 5-week coal strike brought Britain to the verge of collapse. The schemes of societal succession described in Chapter 5 may appear somewhat theoretical, but it should be apparent from them that an Advanced Industrial society differs essentially from previous societal stages only in that it employs an auxiliary energy source. Without this auxiliary energy, the society will not function and must revert to a previous societal stage. When Britain's supply of coal, which furnishes 85 per cent of its energy supplies, is halted by a coal-workers' or a transport-workers' strike, the maintenance of the economy at the level of an Advanced Industrial society becomes impossible.

The fact that we tend to take this auxiliary energy prerequisite for the operation of an industrial society very much for granted was even more forcibly brought home to Americans in the summer of 1973. By June it had become apparent that the nation was faced with a serious oil shortage, and in August the gas stations throughout one state, Colorado, actually closed for a time, stranding summer travelers all over the state. The same situation developed in northern and central California the following September. Elsewhere high prices for gasoline, limits on the amount that could be sold at one time, shorter hours for gas stations, and a number of other inconveniences ultimately forced the U.S. energy crisis to the attention of the general public. Previous warning signs such as local failures in electricity and natural gas supplies, local shortages of fuel oil, and so forth had largely been ignored.

But worse was to come. The next month, October 1973, saw the tragic resumption of fighting in the Middle East. Political consequences precipitated curtailment of Arab oil exports to the United States and several other nations. The energy crisis worsened immediately. To most of the U.S. public it was inconceivable that the richest nation in the world, with the most powerful technology, could run out of anything for very long. What the energy crisis has now forced on the American people is a set of choices that previously have confronted only the populations of poorer and less-developed nations.

The current energy crisis in the United States has two components. First there is the circumstance, now apparent to all, that the world as a whole is running out of fossil fuels; besides this imminent exhaustion, the use of some fuels has been proscribed because of undesirable environmental effects. Second, in the interval before some substitute can be found for these fossil fuels, there will be a transition period when the use of those still unexhausted is continued, but often at some accepted cost to the environment. The choice now before the American people is thus whether the economy is to be permitted to continue more or less at its present level at the certain cost of further environmental degradation or whether economic growth should immediately be curtailed in order to conserve what still remains of the pre-Columbian environment. If U.S. citizens are to have cleaner air, cleaner water, safer foods, better protection from radioactive hazards, and so forth, they will have to opt for a much lower standard of living and a much lower rate of energy utilization than is presently demanded. This lower standard of living is likely to persist for at least a generation, after which it is possible that research and development may once again provide ample and cheap but this time clean sources of energy, and plentiful but this time recycled supplies of nonrenewable resources.

Crude Oil Supplies

It is extremely difficult to obtain any reliable figures for U.S. oil production, import, or use. As a "guesstimate" it can be suggested that the 1973 total consumption of crude oil was at the rate of nearly 5 billion barrels. The proved reserves of oil underground in the United States (*proved reserves* means the reserves the particular industry considers can be economically extracted using current techniques) are estimated at some 100–200 billion barrels. Even without any further increase in the use of crude oil, the U.S. would therefore run out of this fossil fuel somewhere between 1994 and 2014, within our present lifetime or at most that of our children. Unfortunately the total energy use in the United States, actually representing one third of the total energy used in the whole world, doubles every 16 years. During World War II the United States was not only self-supporting in oil but also supplied most of the oil used by its allies. In 1968 it seems to have supplied 3 billion of the 4 billion barrels it used. In 1973 it supplied 3.5 billion barrels of the 5 billion it used, having to import 1.5 billion barrels. Estimated total 1980 requirements will be 8 billion barrels of which one half, 4 billion barrels, can be met from U.S. reserves; the rest will have to be imported. If any of this additional oil has to be met from domestic reserves, the U.S. will definitely exhaust these before the end of the millennium. Offsetting this estimate, however, is the fact that some geologists consider that only about one half of the total oil deposits in the U.S. have as yet been located.

The deficiency in U.S. oil production is presently made good by Canadian and overseas imports. Saudi Arabia, Kuwait, and Iran have by far the most important deposits. Saudi Arabia has 160 billion barrels (mean estimate) of proved deposits. Iran and Kuwait each have about 130 billion barrels (mean estimates); these three nations own about one quarter of the world's crude oil reserves, estimated to be in the region of 2,000 billion barrels. The United States is not the only importing country; other nations such as Japan and France are also actively competing in the world market for oil. It is a seller's market in which the major producers are reacting somewhat differently. Kuwait has decided that as oil is its only source of revenue it should conserve its supply and not augment its exports any further. Iran, by contrast, will sell as much as it can to all who will pay. It converts its oil royalties into assets such as further industrialization of the nation. Saudi Arabia has reached a point where it considers it difficult to spend money at any higher rate on the modernization of the country. It sees little point in continuing to increase production just for the benefit of industrialized nations, especially when some of them are pursuing a foreign policy of which it disapproves. Moreover, if Saudi Arabia stepped up production and could not spend its oil revenues, it would have investment funds capable of demolishing the international monetary system. The funds it could accumulate would, for example, be sufficient to buy up overnight the total assets of a giant multinational corporation like General Motors. The relevant figures are cited in Table 13-1.

A major oil field was discovered on the North Slope of Alaska in 1968, but to exploit this major field it is necessary to build a Trans-Alaskan pipeline to the nearest ice-free port, at Valdez 500 miles away. Because of the protests of environmentalists, the construction of this pipeline was delayed for 5 years. In view of the oil shortage, environmental objections were speedily overruled, and the Senate in 1973 approved the construction of the pipeline. Alaska could supply approximately 1½ million barrels of oil a day. This would continue for 50 years if the top estimate of 25 billion barrels is taken for proved reserves. However, it is pertinent to comment that this amount of oil could be *saved* if the 100 million cars on American roads could be reduced

TABLE 13-1 *Monetary Reserves of Oil-Producing Nations*

The enormous monetary reserves (gold and foreign exchange) accruing to the member nations of the oil producers' cartel (OPEC) is shown in this table. The difference between the 1973 and 1974 position reflects the enormous sums acquired since oil prices were increased more than fourfold in 1973. For comparison, the reserves of the U.S. and the two nonoil-producing nations with the highest reserves are also shown.

	Monetary reserves in billions of U.S. dollars	
	1973	1974
Saudi Arabia	4.1	11.5
Iran	1.0	6.3
Venezuela	1.7	4.9
Nigeria	0.4	4.0
Libya	2.4	3.7
Iraq	1.3	3.0
Algeria	0.6	2.0
Indonesia	0.9	1.6
Kuwait	0.6	1.0
U.S.	14.4	15.7
West Germany	35.0	32.5
Japan	14.0	13.2

from an average weight of 3,500 pounds to 2,500 pounds. That is, there would be no need to use the Alaska oil if everybody in the U.S. would drive a vehicle the size of small foreign imports or of the latest crop of smaller American compacts. Likewise, if all houses were properly insulated against heat and cold, the resulting fuel savings would approximate the estimated total Alaska production.

Coal Reserves

It is unlikely, however, that either of these economy measures could be enforced overnight in the U.S. What is most likely to happen as regards the present energy crisis is a fairly rapid return to more extensive use of coal, not only directly, but also for liquification to oil and gasification to gas.

Some of the difficulties over oil supplies have arisen because of a switch from coal as a fuel for thermal electric power stations. The 1969 Clean Air Act prohibited the use of high sulfur content coal in such installations. Another difficulty with coal that has arisen since 1969 is the National Mines Safety Act which, because of the added safety measures required, doubled the cost of extracting deep-mined coal. But coal is the only fossil fuel of which the U.S. still has a plentiful reserve, and effective technologies may soon be developed to convert some of this into gas and oil. Because of the present-day cost of deep mining, attention has now been focused on the surface coal deposits of the Fort Union Basin, covering an area stretching from Saskatchewan in Canada across Montana and the Dakotas to Wyoming. The coal seams of this basin range up to 40 feet or so thick and contain a quantity of coal estimated as sufficient for 500 years—that is, for the U.S.; world supplies of coal are estimated as sufficient for about 400 years.

In Wyoming, the state with most of the last of the open range, it is estimated that there are some 60 billion tons of coal that could be extracted by strip mining during the remainder of this century. Experience with strip mining in West Virginia, Kentucky, and Illinois has been that the land is left useless. Environmentalists are consequently extremely active in contesting the spread of strip mining to the Rocky Mountain states. Although it is claimed that reclamation of land after strip mining is possible, there appears to be something economically unsound in spending an estimated $6,000 an acre for the reclamation of land that when sold for ranching fetches a meager $20 an acre. Unfortunately even the expenditure of $6,000 an acre does not reconstruct the underground aquifers; serious water problems commonly develop.

As things stand at the moment, 60 per cent of the land in Montana suitable for strip mining is owned by the federal government. In this state some 1 million acres are under lease to 87 private corporations for strip mining operations. In 1973 16 million tons of surface coal were extracted from Montana. Over the next 50 years it is planned to extract a total of some 60 billion tons. The coal concerns claim this operation will destroy only about 100,000 acres of grazing land. The sums of money involved are astronomical. The average cattle ranch in Wyoming covers about 25,000 acres; at $20 per acre it would be worth $500,000. One rancher reported on a television network program in 1973 that he had been offered $15 million for his ranch because it included surface coal deposits. Ranching in the Rocky Mountain states has never been a particularly profitable occupation. Some regions, for example eastern Wyoming, are areas of dying small towns, with unemployment running up to 40 per cent of the labor force, poor schools, and quite inadequate medical services. It seems unlikely that in these circumstances environmental considerations will be allowed to weigh very heavily against the dire needs of the economy for exploitable fossil fuel.

Natural Gas

The situation as regards the supply of natural gas in the U.S. suddenly deteriorated in 1973 much as did the gasoline situation. Because natural gas is clean and inexpensive, something like one half of the homes in the nation have been built or converted so as to be heated by natural gas. One half of the nation's industrial production is fueled by natural gas, one quarter of the electricity is produced by it, and in fact it provides one third of all the energy used in the United States. Over the next 10 years 2 billion liquid feet of liquified natural gas will be obtained from Algeria, $30 billion worth from Russia. However, these imported supplies will not be sufficient to meet estimated added demands.

Natural gas installations are not the sort of thing any community wishes to have in its own backyard. Thus the supply for the states of Ohio, New York, Pennsylvania, and New Jersey comes from Louisiana and Texas, despite the probable existence 300 miles off the south shore of Long Island of petroleum and natural gas deposits sufficient, it is estimated, to carry the nation well into the twenty-first century. Long Island residents do not want the development or even the exploration of this field. In fact, not a single exploratory well ever has been drilled between Florida and the Maine coast.

Experiments are in progress in the U.S. on the gasification of coal to produce gas, as has been done in Europe for very many years. The low-grade coal in the surface deposits of the Fort Hudson Basin would be most economically treated by the erection of gasification plants on the mine site or by the erection there of thermal electric power stations—but nobody wants these, either. The experience of the Southern Edison Power Company with its Four Corners Power Plant in northern New Mexico has illustrated the kind of environmental protests that the erection of such plants in undeveloped regions of the United States can create.

Electricity

Since the famous blackout in the New York area in 1965, the northeastern region of the U.S. has become inured to periodic cuts in electricity supply. It is estimated that such brownouts or whatever they are called will come to the West Coast also by 1975 or 1976. These events underline the now critical situation at peak periods of electricity load. It is not surprising a crisis situation has developed; U.S. demands for electricity have increased at double the rate of the population. From 2,000 kilowatts per person per year in 1950, 6,500 in 1968, an estimated 11,500 kilowatts will be consumed in 1980, some 25,000 kilowatts per person per year by the next millennium. The power companies have been rendered virtually impotent to cope with this situation by various environmental protection measures. The expansion of the capacity of thermal electric power stations has been repeatedly refused on the basis of the loss of air quality that would ensue. Development of nuclear power stations has been resisted on the same grounds, and also because of alleged inadequate safety precautions. Indeed in 1973 legal actions were initiated by Ralph Nader and the Friends of the Earth that if successful could close down the 23 operating nuclear power stations in the U.S. At the same time the Atomic Energy Commission has insisted on still further safety precautions that make the $20-billion-a-year U.S. nuclear industry the safest in the world.

Environmental difficulties associated with nuclear power stations are twofold. First there is the question of the safety factor in the event of internal or external damage to the core elements. The emergency core-cooling system that would automatically come into operation in such an event has never been fully tested; it is impossible to do so safely. The effects of accidents are tested by computer models. Many scientists contend that this emergency system would be ineffective and would not prevent over-heating and the release of a lethal radionuclide cloud. The second problem is that of the disposal of radioactive materials generated in the pile and having a half-life of hundreds of years. In fact, the problems are currently greater in the military than the civilian use of atomic substances. Yet one further but less immediate problem with present fission piles is that the uranium-235 that powers them is estimated to be sufficient at present rates of utilization for only another 40 years. Breeder reactors, which use uranium-238 and thorium-232, are now in the research and development stage but not in commercial operation. They would avoid this last problem, because uranium-235 would be needed only to start them; after that they would be self-fueling. However, the waste problem is even greater with this breeder type of reactor. The only effective nuclear power station, one that will meet many of these objections, is the fusion reactor, but this is not yet even under research and development on a commercial scale.

Fusion reactors are safe because they are powered by atomic fusion, not atomic fission. A number of very common materials could be used to fuel them. The one under consideration at the moment is deuterium, an isotope of hydrogen that occurs in seawater (and fresh water). Two deuterium atoms will fuse to form an isotope of helium, or a deuterium atom will fuse with another istope of hydrogen found in water, tritium, to form another isotope of helium. Both reactions are associated with the release of energy, and are in fact among the reactions that power a hydrogen bomb. It has been calculated that the deuterium in 1 cubic kilometer of seawater would provide the fuel equivalent of about 300 *billion* tons of coal, or nearly 1,500 *billion* barrels of crude oil. Put another way, the removal of 1 per cent of the deuterium in the world's oceans would supply 500,000 times the amount of fuel that all the fossil fuels together can ever provide. If such nuclear fusion stations could be established commercially, the U.S. energy crisis could be resolved for as far into the future as we care presently to look. Many environmental problems would also simultaneously be resolved.

Other Energy Sources

Of the alternatives to the use of fossil fuels as energy sources, some are already established. Hydroelectric dams are cheap, clean, and effective, but mostly have any remaining potential only in the underdeveloped world. Industrialized countries like Norway, Britain, and the United States usually have already tapped all available waters. Only about 4 per cent of total energy output in the U.S. is presently from hydroelectric sources, and this proportion is becoming smaller each year. Windmills and tidal power provide electricity only for small-scale operations. Geothermal power, generally in the form of steam from geysers, can provide often substantial amounts, but there are problems arising from the corrosive action of contained gases and other substances. A few experimental plants produce power from urban trash, but although this method has potential it is not yet beyond the research and development phase.

Under current investigation are several other possible sources. Satellites that collect solar energy appear to have some promise. Anchored in orbit about 20,000 miles from the earth, they would be in perpetual sunlight and could beam power down to any desired location. Magnetohydrodynamics (MHD) is a process that can be applied to combustion reactions, such as the burning of coal, to make them twice as efficient. Oil in a solid form can be extracted from some types of shales, present at the surface in parts of Colorado, Utah, and Wyoming. It is estimated that in producing the equivalent of 100 million barrels of oil from shale, 20 cubic miles of rock waste would accumulate. As with the strip mining of coal, extraction of oil from shale would thus pose a very considerable environmental problem.

The U.S. therefore has many difficult decisions to make before it can resolve the several critical energy crises from which it can only extract itself by taking positive actions. This matter has been described at some length here to emphasize what has already been noted several times, that environmental decisions are not a matter of concern solely for underdeveloped nations. Indeed, as was noted in the case of population growth, what industrialized nations do now to secure their energy needs is most critical for the whole world, because these needs have increased exponentially to extremely high values. The U.S. presently needs an estimated one third of the total amount of energy utilized in the whole world. It may be pondered whether the world can afford such a high carrying capacity for Americans, or, for that matter, for any of the highly industrialized nations. In endeavoring to meet these immense energy needs, industrialized nations can in several decades not only destroy many features of their own environment but also permanently degrade that of any other nation electing to join them in common economic ventures.

PLATE 10. *The economic advantages of monoculture tend to concentrate the cultivation of single crops in increasingly large areas,* as illustrated from this Paarl Valley in South Africa. The vineyards of this valley are said to produce more sherry than any other region in the world. This concentration of grapes, although convenient for many management purposes, provides ideal conditions for the spread of insect pests and diseases.

Water Requirements

Carrying capacity is a complex quantity related to many factors besides energy. Some of these will have additive and others synergistic effects. Regarding factors limiting to any further increase in carrying capacity for human populations other than energy, water is of primary importance outside the polar regions, where temperature becomes a more restrictive factor.

Water requirements of the human population in the United States were considered some years ago by C. C. Bradley. Although, as he observes, a mere 2 quarts of water daily suffice for individual drinking purposes, between 300 and 2,500 gallons a day per person are required to raise the various other necessary consumables. This includes evapotranspiration from the plants consumed, water losses during breadmaking, losses from animal feed, utilization by domestic animals, and so forth.

If all Americans consumed nothing but a minimal vegetarian diet, 300 gallons of water per individual daily would support a population, according to Bradley, of about 17 billion, 85 times the present United States population. However, an Advanced Industrial society such as the United States has additional water needs; industry uses about 1,400 gallons a day per person.

In addition, of the total rainfall over the country, about one quarter is lost in runoff. The rest allows about 13,800 gallons of water per person daily, thus giving a total use of about 15,200 gallons per person per day.

With the present U.S. population and a rainfall of about 5,000 billion gallons per day, the daily individual allowance may be estimated at 28,000 gallons. As we are already using over 60 per cent of this, this country's population obviously could not even be doubled without suffering some reduction in the standard of living or finding some water supply other than rainfall.

On a basis of water requirements, Bradley places the carrying capacity of the United States at 230 million—very close to the population it has already attained. His calculations ignore water supplies from desalination plants and Canadian rivers; nor do they take into account the presently necessary use of water in flushing sewage, cooling industrial plants, and certain other processes from which water could potentially be reclaimed. Nevertheless, it can be readily understood why the policy of the Army Corps of Engineers as regards water projects is now to concentrate on schemes that provide further control of wild rivers or recreational water, preferably both. The days for considering further massive irrigation projects have passed for this reason alone, but also because of the considerations of energy requirements already raised in Chapter 8.

Other Limiting Resources

The relation between carrying capacity and water requirements is singularly clear-cut and factual. So is that of food requirements. The huge food

PLATE 11. *Glacier National Park, Montana.* The effect of human occupation of this ecosystem has been so slight, as yet, that not only mountain sheep and mountain goat populations survive, but also grizzly bears, although these last may be removed.

surpluses that the United States produces and the considerable argicultural subsidies at times paid to farmers so that they do *not* produce certain items illustrate that food is much less a limiting factor on carrying capacity in this country than is water. Distribution of food, nevertheless, still obviously leaves much to be desired, and the situation is even less favorable than it appears because food production must compete for space with other parts of the ecosystem such as cities, recreation lands, roads, and wilderness areas.

The same inconsistencies apply also to space needs, where, for many districts in the great cities, we appear already to have reached or even exceeded the human carrying capacity.

Space Requirements

Space requirements involve many factors. Individual *local* space requirements, for living, are one thing; allocation of *regional* or *national* space—for highways, cities, airports, and so on—is another.

In terms of land use, H. H. Landsberg estimated nearly 10 years ago that cities, for example, presently occupy less than 1.5 per cent of the area of the United States. This modest figure suggests that the number of cities could be increased considerably before, in terms of regional space, they approached any carrying capacity. Likewise, highways, railroads, and airports together still occupy a mere 3 per cent of the area of the country.

Recreational Land

In contrast, as Landsberg remarks, by 1980 it is estimated there will be a need for an estimated 76 million acres of recreational land, rising to 134 million acres by the year 2000. Only 44 million acres presently are available (about 2.5 per cent of the total land area) in the form of national and state parks, monuments, national forests, and other such areas. Because the allocation of 5–6 per cent *more* land suitable for recreational purposes may pose a serious problem, in terms of the availability of such space, the United States may also be approaching its carrying capacity in terms of this resource.

Comparison with Other Countries

On the other hand, population density in the United States still is far below that of most other Advanced Industrial societies. European population densities are 5 to 15 times higher, as shown in Table 13-2. In Japan they are higher still. In certain tropical Advanced Agricultural societies they are highest of all. Carrying capacity in terms of population density is clearly relative, and it is partly a question of what standards a particular population demands. It is very easy to fall into the error of arguing on a basis of what P. R. Ehrlich and J. P. Holdren have aptly named the "Netherlands fallacy" (Table 13-3). Because the Netherlands can take a

TABLE 13-2 *Natural Population Densities*

Expressed in persons per square kilometer estimated from the total population and the total area of the nation for a selection of countries, these can be extremely misleading. No one would have any difficulty appreciating that the top six countries on the list are quite crowded. So also are at least five of the six at the bottom of the list, because extensive mountainous or desert terrain within the national borders permits only the sparsest settlement. However, in terms of conservation there may still be relatively undisturbed ecosystems in these inhospitable areas of such countries that can be selected for preservation. In densely settled countries such as those at the head of the list, significant areas of natural ecosystems have long since disappeared.

Netherlands	333	Uganda	34
Hong Kong	308	United States	29
Belgium	301	Ghana	28
Japan	263	Costa Rica	26
United Kingdom	215	United Arab Republic	26
Italy	172	Mexico	18
India	136	Madagascar	9
Pakistan	100	Venezuela	8
South Vietnam	80	Australia	1
Guatemala	39		

TABLE 13-3 *Population Density: The "Netherlands Fallacy"*

The population density of some of the smaller European countries and of some of the smaller U.S. states is among the highest presently to be found. It cannot be argued from such figures either that the Netherlands is at the top carrying capacity and Belgium and the state of New Jersey are near the maximum or that Denmark and the state of Maryland could accommodate approximately three times their present populations before they even reached Netherlands density. The human carrying capacity of an area depends on particular demographic, social, and geographical features that often vary widely between one region and another. Demographic statistics suggest that in all six of these areas population growth is now tending to level off not far above present density values, although immigration into the individual states tends to elevate the American values a little. There is no significant immigration figure affecting the European countries listed.

Country or state	Land area in thousands of square miles	Population density in persons per square mile in 1970	Percentage increase in density over last decade
Netherlands	13.0	1,002	11
New Jersey	7.8	915	18
Belgium	11.8	822	11
Massachusetts	8.3	689	14
Maryland	10.4	371	26
Denmark	16.6	296	11

carrying capacity leading to a density of 333 persons per square kilometer, does not mean it is possible to extrapolate from this figure to any other geographical area.

In regard to space, most societies lie in some kind of intermediate position. Although schemes have been presented for locating cities underground or even under the sea when we run out of space aboveground, the time when we will have to resort to such edifices still is well ahead. The time for conservation action is, however, already upon us everywhere, as has been admirably expressed by W. H. Whyte, who recently wrote *"The land that is still to be saved will have to be saved in the next few years."* We have no luxury of choice, we must make our commitments now and look to this landscape as the last one. For us it will be.

Levels of Conservation

Clearly, individual concepts of the extent to which the diversity of our ecosystems should be conserved will vary with personal experience. Someone raised in a city might readily understand and support a proposal to institute a world gene pool for the preservation of existing variations in the many breeds of dogs but might fail to appreciate why small isolated herds of mustangs that have managed to survive into this century should not disappear into cans of dog food. Organizations such as the Sierra Club will fight hard to preserve certain wilderness areas but largely ignore the circumstance that the ecosystems included in such areas have frequently reached a steady state in balance with certain human operations such as burning, which are necessary to prevent further change.

It is not sufficient, therefore, to clamor for a reduction of all human family size to a maximum of two children in order to avoid further encroachment on the finite resources and limited diversity of our ecosystems. Initially there must be some determination of acceptable carrying capacities for each ecosystem. This is exceedingly difficult to obtain and probably has to be arrived at by tradeoffs in order to achieve a majority agreement. This dilemma was discussed in Chapter 9 when the problem of deciding on the optimum population was considered. It was brought into the international forum in 1974 both at the Bucharest conference on population and the Rome conference on food. As discussed earlier, at both these conferences many underdeveloped nations gave no outward sign of their recognition of any limited carrying capacity for their own territories. It seemed from the various debates that they were expecting the industrialized countries to bail them out of any temporary predicaments into which they might fall through gross overloading of their national environments.

In achieving a compromise decision about the size of an optimum human population for a particular ecosystem, certain guiding principles can be followed. This problem was considered several years ago by R. J. Kramer, who suggested a logical procedure that can be summarized as follows:

1. Method of population control desired
2. Resource management to provide for population
3. Pollution control

PLATE 12. *Joshua Tree National Monument, California.* Although this appears to be still a natural ecosystem, it is not. The only animals known to eat the joshua tree (*Yucca microflora*), the ground sloths, became extinct about 12,000 BP, along with many other larger animals of the Rancholabrea fauna (see p. 136). Sloth dung, with interwoven yucca fibers, still can be found in fossil form in the Mohave Desert.

Wilderness Reserves

Rephrased and slightly reoriented, these basic requirements can form the principles on which land-use plans may be based for all the ecosystems of a given geographical area. If fortuitously some wilderness is still present, as in many parts of the U.S., such principles will help formulate the criteria by which areas for conservation may be selected. Viable examples of all major naturally occurring ecosystems within a given region should be conserved in order to maintain some portion of the diversity that evolutionary processes have already produced and may be expected to continue to develop. By the same token, all species populations that have evolved should be maintained (wherever feasible) in such a way as to include the whole range of their variation.

E. C. Stone has discussed many of the problems involved in the selection and management of such reserves chosen to illustrate ecosystem diversity. He emphasizes the inevitability of change and the necessity for management to adjust the rate or direction of such change. As to the selection of ecosystems to be preserved and particular sites that are suitable for this, he recommends the use of university groups from campuses with strong ecological and resource management interests. K. S. Norris, who for some years headed one such group, has described the system for site selection and management instituted by the State University in California. This university organized a "Natural Land and Water Reserve System" that has examined some 80 proposed reserve areas typifying the extremely rich ecosystem diversity of that state. When approved, sites are incorporated in the system and maintained as teaching and research reserves by appropriate management. This system also fulfills the conservation need to preserve adequate samples of California's ecosystem diversity. Probably about 40 of the proposed areas eventually will be incorporated in the system; 18 reserves have already been established.

The International Union for the Conservation of Nature and Natural Resources is cooperating with the Conservation of Terrestrial Communities Section of the International Biological Program to initiate measures to preserve examples of ecosystem diversity viewed on a global basis. It may already be too late to rescue large parcels of land from development. In Brazil, for example, the trans-Amazonia highway and its supporting network of roads is fast destroying the largest remaining portion of virgin tropical rain forest.

Synthetic Ecosystems

The value of wilderness reserves as described in the last section will be greatly enhanced if they can be surrounded by *synthetic ecosystems*. This is merely a new name for a concept that foresters and resource management groups have been applying for a number of years with varying degrees of success. The synthetic ecosystem concept envisages an area that has been

PLATE 13. *Cape buffalo* (*Syncerus caffer*) in the Wankie National Park, Rhodesia. The seral communities that this species favors in this area are shrub savannas in the final stages of degradation to grassland through the action of grass fires and grazing. Such seral communities are ephemeral; management practices for the maintenance of a mosaic including this type of community require close control of the buffalo population.

Plate 14. *Greater kudu* (*Strepsiceros strepsiceros*) in the Wankie National Park, Rhodesia. This species, like the Cape buffalo, feeds in shrub savannas. However, being a less aggressive species, the greater kudu prefers a savanna with a much taller woody growth in which it is partially concealed. Unlike the buffalo, its numbers are not sufficient to affect successional processes in this community significantly. Management plans for the preservation of this animal species do not therefore at present involve limitation of its density by culling.

completely planned for maximum and continuing *production* without regard to the maintenance of existing ecosystems, populations, or diversity. Stability and diversity can be ignored in this instance because high productivity can be maintained only if stability and diversity are secured by appropriate environmental controls.

The nearest approach to this concept is what was previously known as "multiple land use." Thus, for example, an imaginative scheme in Wisconsin designates land as "recreational, industrial, or residential" in the hope that it might be used for these designated purposes. In the way multiple land use often has developed, however, it has tended to continue the domination of one preexisting use, e.g., forestry, while adding other activities that do not seriously interfere with this main purpose.

A synthetic ecosystem can be developed after an analysis of all possible activities and their potential productivity, with consideration given to both additive and synergistic effects. This would involve cost-benefit analyses, which might well stress, for example, recreational activities rather than more traditional commercial operations.

Buffer Zones

In more densely crowded areas such as some of the smaller countries of Europe, as already mentioned, there is nothing left but synthetic ecosystems and urban ecosystems. More fortunate areas such as North America still have for the moment some wilderness ecosystems. These could be protected, where they are to be conserved, by the creation around them of buffer zones of synthetic ecosystems, thus separating them from direct contact with agricultural and urban areas. Such buffer zones would prevent marauding invasions of the agricultural areas by cougars, coyotes, bobcats, raccoons, and opossums. It would also discourage drastic change in the behavioral patterns of these animals and a marked tendency for them to become scavengers and scroungers when in direct contact with "civilization." It would also prevent a multitude of weeds spreading directly from agricultural into wilderness areas.

As a result of extensive exploration and discussion, culminating in a national meeting in Yosemite, and an international one in Yellowstone, the U.S. National Parks Service formulated in 1972 a policy approach for its second hundred years. Basically this envisaged conserved areas as starting from open sites like parks and gardens within the cities, extending into peripheral and readily accessible recreational areas, leading finally into the more remote and least disturbed wilderness areas. This policy called for protection of the wilderness area by providing peripheral buffer zones. Accommodations in these zones would be more luxurious, and tourists staying there would be provided with official transport for day-trip visits to the wilderness area itself. Residence provisions within the wilderness areas would be limited to the minimum by accommodating only backpackers. Implementation of this policy was begun in 1973. Zion and Bryce National Parks in Utah are having their cabin accommodations reduced and phased out over the next five years. Similar plans are in preparation for other wilderness preserves, and concessionaires are being persuaded to move out of the parks themselves and to construct tourist facilities on their boundaries thus providing recreational buffer zones.

Range of Activity

It is possible to imagine a reserve area that has fine stands of Douglas fir surrounded by a synthetic ecosystem. This would have a mosaic pattern of rotational clear-felled plantation lumber that would tend to break up man-made wildfires. Open rides would form ski runs in the winter and serve as additional firebreaks in the summer. These grassy areas would be grazed by feral sheep, which could be hunted under license with bow and arrow, and seasonal shotgun hunting of several introduced game birds like the Himalayan partridge could be permitted.

Thinnings in replanted areas would be marketed for Christmas trees, an already existing practice. Dams scattered through the area could be used for recreational fishing, as water reservoirs for firefighting, and as centers for duck and geese shoots. Backpacking trails would be provided with rest-camp stops in which timber trash would be stacked for use in campfires. Limited trailer parks would be available on the perimeter. Berry crops such as loganberry, blackberry, and raspberry would be cultivated at particular stages of regrowth after clear-felling of lumber, and beehives would be moved into the area seasonally. Firing ranges and skeet shooting areas could be located in isolated position, with hill climbs for cars, buggies, and cycles similarly provided.

The planning of such a synthetic ecosystem and its management would need careful control and monitoring, but there is no theoretical reason why such buffer zones should not provide diversified recreational and commercial activities, while maintaining productivity and some semblance of the kind of landscape characterizing the less-disturbed parts of the region. The need for such recreational park areas with ready access from the large U.S. cities has recently and repeatedly been stressed.

Synthetic Ecosystems Elsewhere

In Advanced Agricultural societies that still retain some desirable vestiges of undisturbed ecosystems, buffer zones of such synthetic ecosystems are both feasible and desirable. Africa, for example, contains the richest surviving Pleistocene megafauna of game animals. Some years ago, and with varying success, limited hunting zones were created around some of the game reserves established to preserve examples of this megafauna. Many of these reserves are running into management problems and a review of them should obviously include the practicability and desirability of surrounding them with synthetic ecosystem buffer zones. A French and Belgian colonial concept once applied in tropical Africa was to have each *réserve intégrale* buffered by reserves permitting a limited measure of exploitation. Recent increases in serious poaching activities emphasize the need to establish such buffer zones.

Rare Species

Many biological forms are rare in naturally occurring ecosystems and are especially vulnerable if they are either new or relict populations. These rare populations will often have to be removed and maintained in zoological or botanical gardens if they are to be preserved. An interesting example of individual enterprise in this direction was described in the *New Scientist*. Quoting various publications, this essay provided examples of how living material from various colonies of comparatively rare British plants that are about to be destroyed by one form or another of development were transplanted to safer areas. Similar activities could be cited from many countries in both the Old and New Worlds for animals as well as plants. As will be described, it is now a well-established policy of all the world's larger zoological gardens to include a breeding program for one or more endangered animal species.

Endangered Species

The question of the extinction of species by human interference, either by direct reduction of population size or by disturbance of ecosystem steady state, has been considered earlier in this text in relation to *Pleistocene over-kill*. In recent years attention has again been focused on the possible accelerated disappearance rate of larger animal species as a result of what is basically the same kind of interference, and "red book" and "black book" lists have been prepared. The first names the species—more especially of mammals, such as the giant panda, or larger birds, such as the Californian condor—whose population size is so reduced that they are unlikely to persist unless special measures are taken to preserve them. The "black book" names those species lost since 1600.

These lists have been drawn up by the Survival Service Commission of the International Union for the Conservation of Nature and Natural Resources, which has its headquarters at Morges, Switzerland. The year 1600 was selected as the starting date for these lists because from that time descriptions of animals were usually adequate, especially in regard to color, to permit a diagnosis of the taxonomic entity to which they referred. Moreover, specimen skins began to be retained in regional institutions about this time.

It is estimated that in 1600 there were 4,226 living species of mammals, 36 of which (0.85 per cent) have now become extinct; 120 (2.84 per cent) are in some danger of extinction. The corresponding figures for birds are 94 extinct out of 8,684 species in 1600 (1.09 per cent), 187 (2.16 per cent) now threatened. These figures are further illustrated in Figures 13-1 and 13-2, which are taken from a work by Fisher, Simon, and Vincent, the source for much of this information on endangered species. Other recent figures are cited by L. M. Talbot.

Clearly, before these lists in the red and black books become even longer, some kind of policy must be adopted as to the extent and nature of the measures needed to maintain the present diversity of species in the larger mammalian and bird groups. The analysis of probable causes of extinction included in Table 13-4 suggests that hunting and ecosystem disturbance between them have been responsible for over three quarters of the losses and reductions. Hunting has become much more effective with the introduction of high-velocity rifles and telescopic sights. Ecosystem disturbance takes the form of interference with the predator-prey balance and with the other consumer and producer populations—or abiotic elements—of the ecosystem, like clearing a forest.

FIGURE 13-1. *Threatened species of mammals.* Each number on this map corresponds with a mammalian species or recognized infraspecific mammalian category that is considered so imminently threatened with extinction as to justfy its inclusion in the Red Data Book prepared by the Survival Service Commission of the International Union for the Conservation of Nature and Natural Resources. The key figure on the map is located in the breeding area or part of this area. No area of the globe is immune from the threat of extinction of some mammalian form among the populations of its natural ecosystem. Data on whales are not included in this diagram.

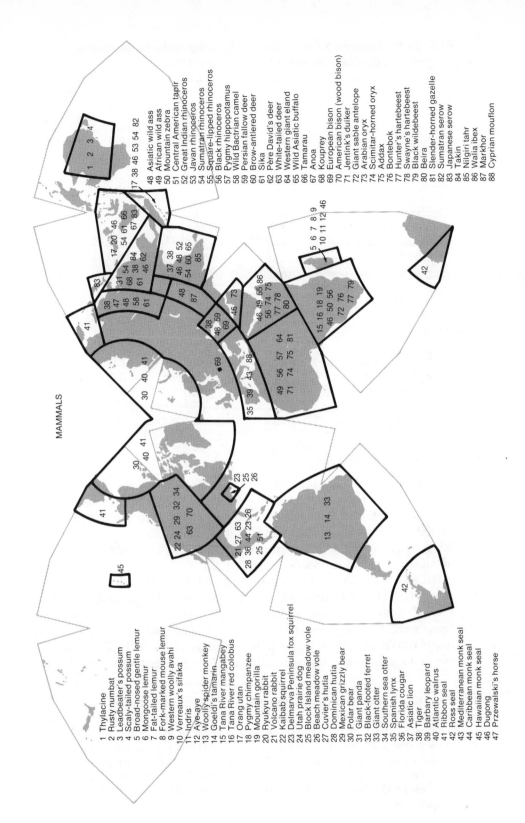

MAMMALS

1 Thylacine
2 Rusty numbat
3 Leadbeater's possum
4 Scaly-tailed possum
5 Broad-nosed gentle lemur
6 Mongoose lemur
7 Fat-tailed lemur
8 Fork-marked mouse lemur
9 Western woolly avahi
10 Verreaux's sifaka
11 Indris
12 Aye-aye
13 Woolly spider monkey
14 Goeldi's tamarin
15 Tana River mangabey
16 Tana River red colobus
17 Orang utan
18 Pygmy chimpanzee
19 Mountain gorilla
20 Volcano rabbit
21 Ryukyu rabbit
22 Kaibab squirrel
23 Delmarva Peninsula fox squirrel
24 Utah prairie dog
25 Block Island meadow vole
26 Beach meadow vole
27 Cuvier's hutia
28 Dominican hutia
29 Mexican grizzly bear
30 Polar bear
31 Giant panda
32 Black-footed ferret
33 Giant otter
34 Southern sea otter
35 Spanish lynx
36 Florida cougar
37 Asiatic lion
38 Tiger
39 Barbary leopard
40 Atlantic walrus
41 Ribbon seal
42 Ross seal
43 Mediterranean monk seal
44 Caribbean monk seal
45 Hawaiian monk seal
46 Dugong
47 Przewalski's horse

48 Asiatic wild ass
49 African wild ass
50 Mountain zebra
51 Central American tapir
52 Great Indian rhinoceros
53 Javan rhinoceros
54 Sumatran rhinoceros
55 Square-lipped rhinoceros
56 Black rhinoceros
57 Pygmy hippopotamus
58 Wild Bactrian camel
59 Persian fallow deer
60 Brow-antlered deer
61 Sika
62 Père David's deer
63 White-tailed deer
64 Western giant eland
65 Wild Asiatic buffalo
66 Tamarau
67 Anoa
68 Kouprey
69 European bison
70 American bison (wood bison)
71 Jentink's duiker
72 Giant sable antelope
73 Arabian oryx
74 Scimitar-horned oryx
75 Addax
76 Bontebok
77 Hunter's hartebeest
78 Swayne's hartebeest
79 Black wildebeest
80 Beira
81 Slender-horned gazelle
82 Sumatran serow
83 Japanese serow
84 Takin
85 Nilgiri tahr
86 Walia ibex
87 Markhor
88 Cyprian mouflon

TABLE 13-4 *Analysis of Causal Factors Leading to the Extinction or Endangering of Bird and Mammal Species Since 1600*

	Birds			Mammals
	Nonpasserine (large) per cent	*Passerine (small) per cent*	*Total per cent*	*Per cent*
CAUSE OF EXTINCTION				
Natural	26	20	24	25
Human				
Hunting	54⎤	13⎤	42⎤	33⎤
Introduced predators	13⎟	21⎟	15⎟	17⎟
Other introductions	—⎬ 74	14⎬ 80	4⎬ 76	6⎬ 75
Habitat disruption	7⎦	32⎦	15⎦	19⎦
	100	100	100	100
CAUSE OF PRESENT RARITY				
Natural	31	32	32	14
Human				
Hunting	32⎤	10⎤	24⎤	43⎤
Introduced predators	9⎟	15⎟	11⎟	8⎟
Other introductions	2⎬ 69	5⎬ 68	3⎬ 68	6⎬ 86
Habitat disruption	26⎦	38⎦	30⎦	29⎦
	100		100	100

Comparison of Extinction Rates

It is interesting to compare the extinction rate of species recorded in the red and black books with postulated rates of Pleistocene overkill. As was discussed earlier in this text, the latter seems to have proceeded between 11,000 and 8,000 BP. Fisher estimates that during this time approximately 50 mammalian and 40 avian species disappeared. This represents a rate of about three species a century.

Because some 36 species of mammals and 94 species of birds have gone since 1600, the extinction rate now is beginning to approach 50 per century. This far higher rate of "kill" suggests that Pleistocene human populations were not very effective hunters or that they evolved totem and taboo rituals against killing of particular species, or both. The comparison serves to emphasize that we shall limit severely the diversity of all our

FIGURE 13-2. *Threatened species of birds.* Each number on this map corresponds with a bird species or recognized infraspecific bird category that is considered so imminently threatened with extinction as to justify its inclusion in the Red Data Book prepared by the Survival Service Commission of the International Union for the Conservation of Nature and Natural Resources. The key figure on the map is located in the breeding area or part of this area. It should be noted that the bird species that breed on islands are especially endangered. Endangered races of species considered safe are not shown.

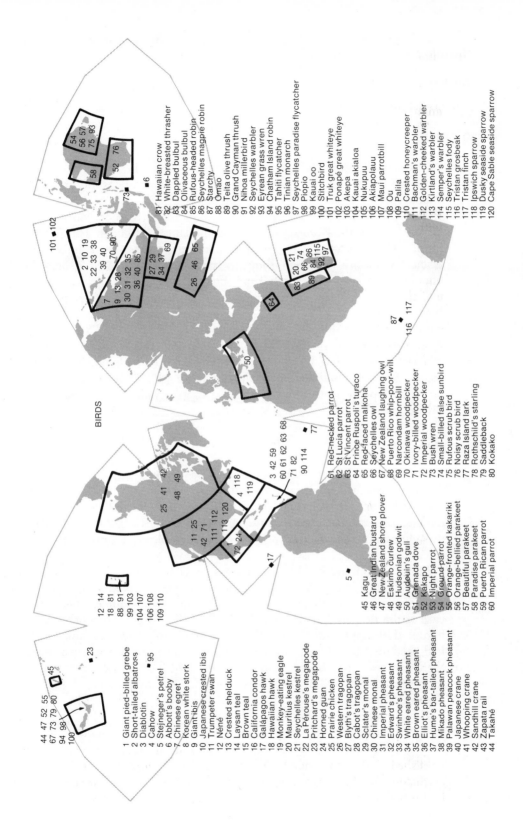

BIRDS

1 Giant pied-billed grebe
2 Short-tailed albatross
3 Diablotin
4 Cahow
5 Steineger's petrel
6 Abbott's booby
7 Chinese egret
8 Korean white stork
9 Giant ibis
10 Japanese crested ibis
11 Trumpeter swan
12 Néné
13 Crested shelduck
14 Laysan teal
15 Brown teal
16 California condor
17 Galápagos hawk
18 Hawaiian hawk
19 Monkey-eating eagle
20 Mauritius kestrel
21 Seychelles kestrel
22 La Pérouse's megapode
23 Pritchard's megapode
24 Horned guan
25 Prairie chicken
26 Western tragopan
27 Blyth's tragopan
28 Cabot's tragopan
29 Sclater's monal
30 Chinese monal
31 Imperial pheasant
32 Edward's pheasant
33 Swinhoe's pheasant
34 White eared pheasant
35 Brown eared pheasant
36 Elliot's pheasant
37 Hume's bar-tailed pheasant
38 Mikado pheasant
39 Palawan peacock pheasant
40 Japanese crane
41 Whooping crane
42 Sandhill crane
43 Zapata rail
44 Takahé

45 Kagu
46 Great Indian bustard
47 Eskimo curlew
48 Hudsonian godwit
49 New Zealand shore plover
50 Audouin's gull
51 Grenada dove
52 Night parrot
53 Kakapo
54 Ground parrot
55 Orange-fronted kakariki
56 Orange-bellied parakeet
57 Beautiful parakeet
58 Paradise parakeet
59 Puerto Rican parrot
60 Imperial parrot
61 Red-necked parrot
62 St Lucia parrot
63 St Vincent parrot
64 Prince Ruspoli's turaco
65 Red-faced malkoha
66 Seychelles owl
67 New Zealand laughing owl
68 Puerto Rico whip-poor-will
69 Narcondam hornbill
70 Okinawa woodpecker
71 Ivory-billed woodpecker
72 Imperial woodpecker
73 Bush wren
74 Small-billed false sunbird
75 Rufous scrub bird
76 Noisy scrub bird
77 Raza Island lark
78 Rothschild's starling
79 Saddleback
80 Kokako

81 Hawaiian crow
82 White-breasted thrasher
83 Dappled bulbul
84 Olivaceous bulbul
85 Rufous-headed robin
86 Seychelles magpie robin
87 Starchy
88 Omao
89 Teita olive thrush
90 Grand Cayman thrush
91 Nihoa millerbird
92 Seychelles warbler
93 Eyrean grass wren
94 Chatham Island robin
95 Tahiti flycatcher
96 Tinian monarch
97 Seychelles paradise flycatcher
98 Piopio
99 Kauai oo
100 Stitchbird
101 Truk great whiteye
102 Ponapé great whiteye
103 Akepa
104 Kauai akialoa
105 Nukupuu
106 Akiapolauu
107 Maui parrotbill
108 Ou
109 Palila
110 Crested honeycreeper
111 Bachman's warbler
112 Golden-cheeked warbler
113 Kirtland's warbler
114 Semper's warbler
115 Seychelles fody
116 Tristan grosbeak
117 Tristan finch
118 Ipswich sparrow
119 Dusky seaside sparrow
120 Cape Sable seaside sparrow

remaining ecosystems if we make no immediate and deliberate further
effort to maintain this still rich species diversity.

Species Conservation

Zoo Banks

As mentioned earlier, all major zoological gardens in the world are now
cooperating in a plan to breed in captivity populations of endangered
animal species. These are known as *zoo banks*.

Two examples will serve to illustrate how effective such zoo banks can
be, the cases of the European bison and the Hawaiian goose. The European
bison (*Bison bonasus*) is regarded as a separate species from the American
bison (*B. bison*), although they are probably both recent geographical
populations descended directly from some Pleistocene ancestor that evolved
into the geographical groups known as *B. antiquus* in North America and
B. priscus in Eurasia (Figure 13-3). *B. antiquus* coexisted in North America
with three other glacial bison populations, *B. latifrons*, *B. alleni*, and
B. crassicornis, and survived them to persist into the Wisconsin glaciation,
when it was hunted by the Early Mongoloid Paleo-Indians. The wood
bison (*B. bison athabascae*) and plains bison (*B. bison bison*) appear to
have developed directly from this species as geographically and ecologically
distinct populations.

European Bison

The European bison, or wissent, likewise evolved apparently into two
separate populations, usually described as subspecies—the Lithuanian or
lowland bison (*B. bonasus bonasus*, Figure 13-3) and the Caucasian or
mountain bison (*B. bonasus caucasius*). Both of these, like the wood bison,
were browsers, thus contrasting with the plains bison (Figure 13-3), which
is a grazer.

At the beginning of this century the lowland bison, whose areas had
originally included western and southern Europe, survived only in the wild
as a single herd in the Bialowieza Forest in Poland, under the personal
care of the Russian imperial family. In 1914 there were 737 animals in this
herd; all disappeared during World War I. However, 45 animals that
had been distributed from this herd survived in various European zoos, and
there was also from this same source a semiwild herd in the Pszcyna Forest
in Upper Silesia, totaling over 70 animals in 1921. By the end of 1921 only
a cow and two bulls were left undestroyed from the Pszcyna herd.

These three animals, together with two cows and another bull purchased
from the remaining zoo-bank animals under semidomesticated conditions,
were used to reestablish the Bialowieza herd. In 1930 a stud book was
published, and this included by 1949 nine purebred cows at Pszcyna, three
at Bialowieza. In 1956 there were enough bison for some to be released
in the Bialowieza Forest; this "wild" herd numbered 57 in 1962. This forest

FIGURE 13-3. *Bison species*. *B*. The lowland form of European bison (*B. bonasus bonasus*) was released into the wild again in 1956 after a number of generations had been maintained in semicaptivity. A wild herd established from these in the Bialowieza Forest in Poland numbered 57 head in 1962. *A*. The wood bison (*B. bison athabascae*) was unfortunately interbred with (*D*) the plains bison (*B. bison bison*) during earlier conservation operations. An isolated wild herd of about 200 head was located in 1957 in Wainwright Buffalo Park in Alberta, but suffered disease losses. This subspecies, which resembles the European bison, is now preserved in three separate wilderness areas in Canada and, like the plains bison, should survive in its various preserve areas. The bison in *C* is *B. antiquus*, which is very commonly represented among the Rancholabrea fossils. It is the probable ancestral form of the plains bison. Although this ancestral form therefore survived until the early Wisconsin, it may have been one of the earliest of the Rancholabrea species to become extinct.

is believed to have a carrying capacity for 100–115 bison under wild steady-state conditions. Other breeding centers have been established in Poland, and there are still animals in the zoo bank, so that the lowland bison now appears safe from extinction.

The Caucasian bison subspecies was less fortunate. This had also been under the personal protection of the Czar of Russia, but in 1924 after the Russian Revolution, only about 15 to 20 animals were estimated to be alive, partly because of destruction by local herdsmen. By 1925 even these animals apparently had gone, but a bull presented as a yearling in 1908 survived to this same year, having sired a number of calves from lowland bison. The Polish Ministry of Forestry is reportedly attempting to reconstitute the Caucasian bison by selecting for its characters from among the fourth- and fifth-generation descendants of these hybrid calves.

The Hawaiian Goose

The Hawaiian goose (*Branta sandvicensis,* Figure 13-4) has a similar history. This goose is estimated to have maintained a population of about 25,000 birds after Polynesian occupation of the Hawaiian Islands and until Europeans arrived following Cook's visit in 1770. By 1850 it occurred only in wilder mountainous areas, and in 1900 it was rare even there. However, from the time of the first recorded breeding in the London Zoo in 1834, a number of zoos had maintained breeding flocks in captivity.

Despite these captive flocks, by 1947 there were only an estimated 50 Hawaiian geese left in the world, and these were in Hawaii, either wild or in captivity. In that year the Hawaiian Board of Agriculture started a breeding program with two captive pairs and a gander, together with a captured wild goose. In 1950 the Board flew two geese from a captive flock on Hawaii back to England and mated them with a gander from the same source. When this gander died in 1963, his progeny had reached more than 230 birds, 50 of which had been released on the Hawaiian island of Maui. By 1964 the total wild and captive birds were estimated as numbering over 500, and the population was doubling every three or four years. Free birds are again occupying the mountainous areas of Hawaii and Maui, where some sanctuaries have been established. The future of the Hawaiian goose, official bird of the fiftieth state, thus seems to have become assured again, partly because of the captive stocks held in aviculture banks.

Other Zoo Banks

Successful attempts have been made to breed in captivity other animals on the red book list of endangered species. These include such mammals as orangutan (*Pongo pygmaeus*), pygmy chimpanzee (*Pan paniscus*), mountain gorilla (*Gorilla gorilla beringei*), giant panda (*Ailuropoda melanoleuca*), Przewalskis horse (*Equus caballus przewalskii*), pygmy hippopotamus (*Choeropsis liberiensis*), Pere David's deer (*Elaphurus davidianus*), and birds like the whooping crane (*Grus americana*) and great Indian bustard (*Choriotis nigriceps*). Increasing attention is being given to the establishment of zoo banks for species of other major groups of animals that have been successfully bred in captivity, including reptiles such as the Galapagos giant tortoise (*Testudo elephantopus*).

FIGURE 13-4. *Three threatened species of birds in the Red Data Book.* A. The New Zealand takahe (*Notornis mantelli*) is a large flightless gallinule, once widespread in both north and south islands as judged from fossil remains. A sanctuary some 200 square miles in extent in a wild and remote area has been established where the main population of 200–300 birds of this relict species has survived. B. The great Indian bustard (*Choriotis nigriceps*), probably the largest flying bird that alights on land, has become rare over all of its previously wide area of distribution in India and is believed to be extinct over much of this former range. Creation of a preserve presents many problems, and perhaps this species will be best maintained, for the present at least, in an aviculture bank. C. Hawaiian goose (*Branta sandvicensis*), male on right, female left; this species is now reestablished in part of its former range on mountainous areas in Hawaii and Maui, by the release of birds from captive flocks.

Unfortunately zoological techniques for the maintenance and breeding of captive animals are not yet by any means routinely established. Taking a list of some 291 species of endangered animals, only 162 are apparently even as yet maintained in captivity. Of these, a mere 75 species have been successfully bred, and only about a half dozen have captive populations of a satisfactory size.

Treatment of Endangered Species

For any threatened species, there is now a fairly well-established procedure. First, its autecology and distributional range are determined. If a preserve area can be set aside for where it appears likely that the species can maintain a viable population, this is the principal action taken, as with the large flightless New Zealand takahe (*Notornis mantelli*) (Figure 13-4).

Sometimes areas from which the animal has disappeared are restocked from surplus breeding areas, as is being done for the white rhino (*Cerato-therium simum*) in Kruger National Park in South Africa from stocks of the Umfolozi and Hlwehlwe game reserves. Unfortunately, where localized geographical forms of a species had evolved, this does not restore them unless adaptation subsequently occurs in the same directions as in the original geographical form. Sometimes the species has become extinct in the wild, as with the European bison or the South African bontebok (*Damaliscus dorcus dorcus*), which also has been reestablished in a national park in a portion of its former distribution area from captive animals. Another alternative is to capture wild animals in areas where eventual extinction is inevitable and translocate them to a new preserve area, as was done in the case of the Persian fallow deer (*Dama mesopotamica*).

Some animal species unfortunately cannot be handled in any of these ways; whales are the best-known example. In such cases the only possible approach is to limit catches, or prohibit them altogether, and hope that the species survives. For several whale species this international action although now taken, may be too late.

The use of primates in medical and other biological research provides a special example of the need to use zoo banks to maintain endangered species. It is considered likely that within the next 5 years all the primates required for such purposes in the U.S. will have to have been bred there. For example, India presently permits the export of 65,000 rhesus monkeys a year but is expected soon to reduce this quota to 30,000. Of the 56,000 primates of all species used for research purposes in the U.S. in 1971, some 26,500 were reported as being rhesus monkeys. The U.S. alone would thus absorb virtually all the animals in a revised quota figure for this species. The situation regarding less prolific and more shy forest species like the gorilla and orangutan is even more critical. Continuing destruction of their forest habitats is inexorably reducing their numbers and even total protection laws do not entirely prevent illegal capture and killing. Although these species are not needed in such great numbers for research, the establishment of zoo banks for them, which is even more urgent from the conservation viewpoint, still presents many problems and difficulties.

Unfortunately the remaining undegraded wilderness habitats frequently border on human ecosystems where the people exist under extremely impoverished conditions. Such is the case, for example, with the Gir Forest in India, a preserve for the last remaining wild population of the Asiatic lion. Conservation then becomes a matter of economics in this as in numerous other instances. Forbidding the sale of wild-captured orangutans, for example, has considerably slowed the terminal destruction of surviving populations. Similar universal action has been taken or will be needed in respect to the sale of skins of the larger cats like the leopard, jaguar, and ocelot if they are to be saved from a similar fate. By 1973 the poaching of African bush elephants in East African reserves for the sake of their ivory tusks reached such a scale it had begun to threaten the survival of this species in the wild. A 75-pound elephant tusk can fetch over \$2,555, which represents a lifetime's cash earnings for an African peasant farmer in many areas. The financial rewards for ivory poaching are so high as to render all conservation laws meaningless and penalties for poaching offenses totally ineffective. Unless economic measures can be devised effectively to prevent the sale of poached ivory on the international market, it is feared the elephant will become extinct in the game preserves of Africa within this decade. Unfortunately it is not only big cats and elephants that are threatened by slaughter for the trade; numbers of alligators and a long list of other animals are rapidly being reduced for the same reason, and only economic action can save them now.

Rare Plants

With plants the problem of conservation is somewhat similar to that for animals, but in certain ways it is very different and fortunately easier. It is similar in that an *ecosystem* rather than a species should if possible be conserved. It is different in that many plants can be propagated by vegetative means, and more or less indefinitely, because plant tissues do not age. This circumstance has permitted an almost world-wide distribution of a famous rare plant once believed to be extinct, the Dawn Redwood, *Metasequoia glyptostroboides*.

Metasequoia

The coniferous tree genus *Metasequoia* was originally described as a fossil plant from Pliocene beds in Japan and was subsequently found in Cretaceous rocks. The genus was therefore at least 60 million years old and was thought not to have any living species. However, about 30 years ago Elmer Merrill, then director of the Arnold Arboretum at Harvard and an authority on Pacific plants, heard rumors of the existence of a rare tree in China. He succeeded in obtaining seed of what proved to be a *Metasequoia* and raised some plants.

In 1948 Ralph Chaney visited the Chinese localities in Szechuan and Hupeh where *Metasequoia* was reported to grow; he found a valley in which the tree was not uncommon and collected seed. From this seed many plants were grown throughout the world. They were further propagated by cuttings, so that the tree is established now in many official

PLATE 15. *Botanical gardens* are commonly stocked with rare or endangered plant species having curious or incomplete understood features. This is the case on both counts with this species of *Leucospermum* (family Proteaceae) growing in the Kirstenbosch Gardens in South Africa.

gardens. Nevertheless it will probably suffer the same fate as the almost equally extensively grown maidenhair tree (*Ginkgo biloba*), which belongs to a genus that geologically is even older. In this instance, the sole surviving species of the genus no longer occurs naturally apart from one locality in China where it may have survived or could equally well have escaped from cultivation and become naturalized.

Anomalous Distributions

In the case of a number of rare plants it is not sufficient merely to insure their survival under cultivation. Their scientific interest essentially lies in accounting for their occurrence in the particular ecosystems of specific areas. F. N. Hepper has listed a number of such rare plants, including *Pitcairnia feliciana* and *Vateria seychellarum*. *Pitcairnia*, which occurs on rocks in Guinea in West Africa, represents the only known occurrence of a member of the family Bromeliaceae outside tropical America. The bromeliads are characteristic epiphytic plants of the tropical forest throughout Central and South America, and it is a phytogeographical enigma why one species alone should be found across the other side of the Atlantic in the Old World. Likewise, *Vateria,* of which only half a dozen specimens survive in the Seychelle Islands off the east coast of Africa, is the only member of a flowering plant subfamily, Dipterocarpoideae, to be encountered outside the Indo-Malaysian area.

PLATE 16. *Botanical gardens* can be developed for other functions than those shown in Plate 15, for example, the maintenance and display of the range of variation in cultivated plants, as in this bed of lettuce varieties in the Montreal Botanical Garden. This imaginatively planned garden facility well illustrates the potential for conservation of genetic diversity, as well as the display, research, and educational aspects of botanical garden functions.

In both these examples, and in many similar ones that could be cited, research needs to be done on the biology of the *whole* relict population in order to be able to offer explanations for these apparent anomalies. Cultivated material propagated from a small sample of the population will not suffice.

Weed Species

The problem of ecosystem conservation when considering individual plant species' survival is greater than that regarding animals. Many of the unique island floras of the world—Juan Fernandez, St. Helena, Madagascar, New Zealand, and Hawaii, to mention only a few—are disappearing before invading species of cosmopolitan *weeds*. Not only are these unique island floras exposed to such invasion, but many continental areas as well.

For example, in Orange County in coastal southern California, out of a total of 778 flowering plant taxa at the species or infraspecific level recorded, no fewer than 191 (or 25 per cent) are introductions, mostly from Eurasia. Such replacement of native by introduced weed floras is accelerated by development. Agricultural clearings not only destroy the plant populations of natural ecosystems but also provide the synthetic ecosystems to whose environmental conditions weeds are specifically adapted. The opening up of communication systems, roads, railroads, waterways, and airports also increases the rate of long-distance dispersal of various invading species. The biology of such colonizing species has been extensively treated in a symposium volume edited by Herbert Baker and Ledyard Stebbins.

The same kind of thing happens in the case of faunas, where certain "weed" animals follow closely behind urban development. The English sparrow (*Passer domesticus*) and the common starling (*Sturnus vulgaris*) had spread through most of the world by the beginning of this century, as had the common or Norway rat (*Rattus norvegicus*), house mouse (*Mus musculus*), housefly (*Musa domestica*), flea (*Pulex irritans*), louse (*Pediculus humanus*), and common bedbug (*Cimex lectularis*). In some instances the introduction of such "weed" animals poses a serious threat to the persistence of natural ecosystems; in all cases they are responsible, directly or indirectly, for ecosystem change.

Migration of the House Mouse

D. S. Fertig and V. W. Edmonds have described the particular features that make one such "weed" species, the feral house mouse, so successful a competitor and colonizer of new ecosystems. These include its unobtrusiveness, owing to its small size and nocturnal habit; omnivorous diet of, if necessary, dry food; ability to exist on metabolic water alone or on seawater; aggressiveness within its own ecological niche; and breeding potential, including its capacity for continuous inbreeding.

Direct consumption of human food by the house mouse, or direct damage to stored materials, results in annual losses running into millions of dollars. Indirectly, its effects are equally serious as a potential vector for plague, tularemia, food poisoning, and other human disorders.

The ancestral form of the house mouse is believed to be a central Asian subspecies, *Mus musculus wagneri*, a wild grass seed-eater still including in its area of distribution the Fertile Crescent where neolithic agricultural settlements first appeared (as described in Chapter 5). Starting from the Fertile Crescent, it has been possible to trace the historical migration of the house mouse through the world (Figure 13-5). It would similarly be possible to account for all the familiar "weed" animal species such as those listed above.

Necessary Limitations on Conservation Activities

Much of this chapter has been devoted to the issue of deciding what our society can attempt to conserve. Having determined what this is, there is then a question of determining the land use of the rest of a particular region under a multiple land-use scheme. Consideration must be given to the extent to which externalization and internalization can be permitted in relation to other such regional units. Although it is essential that development master plans meet all these requirements, inevitably they impose drastic restrictions on both economic growth and private enterprise in the conventional

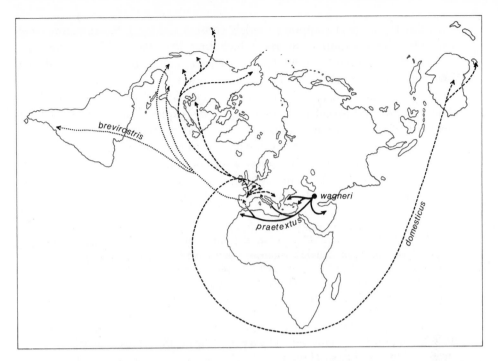

FIGURE 13-5. *Migration of the house mouse* from its adaptation as a commensal of agricultural ecosystems in neolithic settlements of the fertile crescent to its present worldwide distribution. The parent subspecies *Mus musculus wagneri* (bold lines) still occurs wild in west central Asia. Two other subspecies, *M. m. praetextus* and *M. m. domesticus,* out of a further four known, have traveled, supposedly by boat, to the New World, where they have met and interbred in the central United States. (From D. S. Fertig and V. W. Edmunds, "The physiology of the house mouse," *Scientific American,* **221**[4]: 110, 1969. Copyright ©1969 by Scientific American, Inc. All rights reserved.)

speculative meaning of the term. There will have to be other restrictions, such as limiting population migration into an area already exceeding its carrying capacity. Such restrictions are naturally opposed by the individual and sectional interests that stand to lose most from their introduction. One major aspect of conservation is not even considered here, *soil conservation.* This already has been investigated extensively, and restrictive legislation in many countries ensures conformation with basic soil conservation practice, even if it ignores more insidious problems.

The Limits to Growth

Within the last two years the issue of human carrying capacities has been sharpened by the publication of several controversial works. Sponsored by many distinguished scientists and others, *Blueprint for Survival* described what the population limits for the little island of Great Britain could be. It was suggested that the size of the optimum population lay at about one half of the present 55 million persons.

In the U.S., world computer models constructed by J. S. Forrester, later expanded by a group of workers including Donella H. and Dennis L. Meadows, attempted the same kind of exercise on a world scale. The Meadows models sustained numerous attacks, but their principal message remains clear and undisputed, the world population is already too large. The rate of industrialization and resource exploitation that would be achieved if every nation attained the standard of living demanded by the presently industrialized nations, taken in conjunction with the extent of pollution resulting, would be quite unsupportable. Two illustrations (Figures 13-6 and 13-7) show an initial run and one of the final runs of the many variations of the Meadows models. Figure 13-6 illustrates the "stan-

FIGURE 13-6. *The "standard" world model run* assumes no major change in the physical, economic, or social relationships that have historically governed the development of the world system. All variables plotted here follow historical values from 1900 to 1970. Food, industrial output, and population grow exponentially until the rapidly diminishing resource base forces a slowdown in industrial growth. Because of natural delays in the system, both population and pollution continue to increase for some time after the peak of industrialization. Population growth is finally halted by a rise in the death rate resulting from decreased food and medical services.

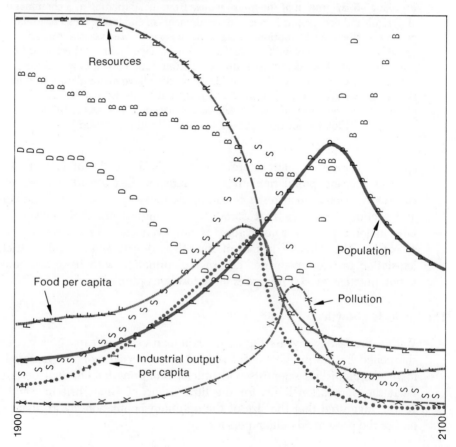

dard" world model in which there is no major change in the physical, economic, or social relationships that in the past have regulated societal evolution. In this base form of the model, population growth is set to proceed at its present rates, pollution to continue to increase to higher levels; resource exploitation is left unchecked. When this model is run, the world system disintegrates several decades before 2100 AD. Resource scarcity catastrophically reduces industrial and agricultural production, which are both in any case inhibited by pollution. Population size eventually plummets under these production shortages and high pollution levels.

In the models shown in Figures 13-8 and 13-9, natural increase is reduced as rapidly as feasible, reaching ZPG by 1975, pollution is maintained at one quarter of its 1970 levels, and the rate of nonrenewable resource exploitation is reduced to one quarter the 1970 value by recycling, reuse, and regeneration. Before 2100 in these models the world system has been stabilized. Industrial and food production per capita are steady, albeit at a lower

FIGURE 13-7. *The problem of resource depletion* in the world model system is eliminated by two assumptions: first, that "unlimited" nuclear power will double the resource reserves that can be exploited and, second, that nuclear energy will make extensive programs of recycling and substitution possible. If these changes are the only ones introduced in the system, growth is stopped by rising pollution.

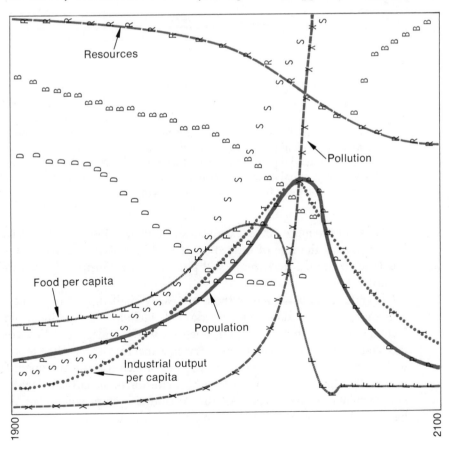

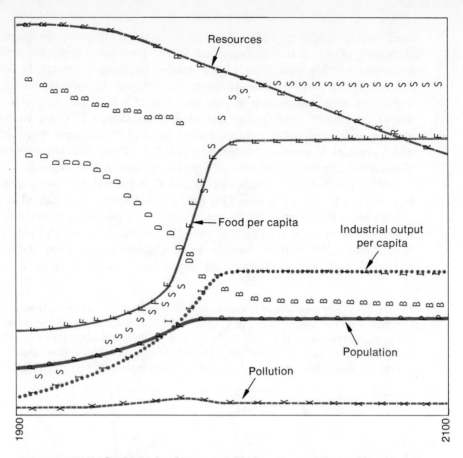

FIGURE 13-8. *Technological policies* are added to the growth-regulating policies of previous runs to produce an equilibrium state sustainable far into the future. Technological policies include resource recycling, pollution control devices, increased lifetime of all forms of capital, and methods to restore eroded and infertile soil. Value changes include increased emphasis on food and services rather than on industrial production. Births are set equal to deaths and industrial capital investment equal to capital depreciation. Equilibrium value of industrial output per capita is three times the 1970 world average.

level than in the more affluent countries of today. Pollution is at a relatively low level, as is the rate of utilization of nonrenewable resources. However, this latter continues and would still pose a vital problem for the stabilized society.

Generalized models such as these are easily challenged on specific items. Moreover, the Meadows models do not include some critical limiting factors discussed in this chapter, such as water and space. Nevertheless, this kind of approach to human carrying capacities is clearly the way we will go in future studies. Only by this kind of exercise can we quantify the information that it is essential to have before making critical decisions on environmental conservation. As the Meadows group points out, society basically has only three alternatives. It can continue unrestricted growth,

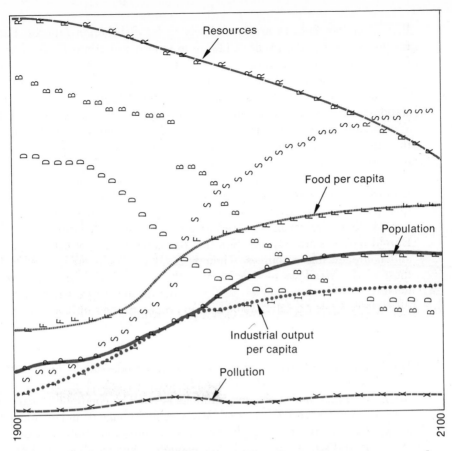

Resources

Food per capita

Population

Industrial output
per capita

Pollution

1900

2100

FIGURE 13-9. *If the strict restrictions on growth of the previous run are removed, and population and capital are regulated within the natural delays of the system,* the equilibrium level of population is higher and the level of industrial output per capita is lower than in Figure 13-8. Here it is assumed that perfectly effective birth control and an average desired family size of two children are achieved by 1975. The birth rate only slowly approaches the death rate because of delays inherent in the age structure of the population.

sustain a naturally imposed limitation, or devise its own limiting controls. The Meadows models plausibly demonstrate the first is impossible for more than one or two further decades. In these pages, the second has been called "malthusian" control, the third "cultural" control. Examination of all available evidence suggests we shall soon have to endure the malthusian controls if we do not right now instigate the cultural ones. One way or the other, the present "r" phase of growth has to be brought into the "K" phase that has characterized all human and nonhuman populations throughout time.

It is possible that much of the criticism of the Meadows models can be attributed to varying interpretations of this time scale. When considering the possibilities of system breakdown, three time periods have to be considered, the present, the immediate future, and the more distant future.

Present system failures are most likely to be *distribution* failures. Famine for the moment is the result of failure in food distribution. There is at least momentarily sufficient food production in the world on a per capita basis to feed everyone at a life-sustaining level if it could be got around to them. By contrast, a society that depends on coal-fired thermal electricity power stations will have a system failure at some time in the more distant future. World supplies of coal are being utilized at an immensely greater rate than this fossil fuel is being replenished. Even allowing for better extractive methods, coal supplies are finite and must eventually be exhausted. Before that happens, in the immediate future, the price of coal might be reduced. Instead of slow reduction of the growth of electricity consumption, it is thus possible that exponential growth in energy use will continue in the immediate future, even as reserves are steadily eroded. Economists like Harold Barnett correctly maintain that this is the history of many renewable and nonrenewable resources. Their findings do not actually conflict with those of the Meadows models because they refer to the immediate future, whereas the Meadows models deal with the distant future when postulated rates of resources depletion will have proceeded to their ultimate conclusion.

Ecological Land Use

Many nations are fiercely independent. This is understandable, considering that they have often paid an extreme price to win their independence. Attempts to plan their behavior, let alone attempt to condition it artificially, are certain to be strongly resisted. Many such feelings of independence are associated with the individual territorial patterns that Robert Ardrey and others have discussed in detail. Not enough is known about territorial behavior in human populations. It may be that it could be just as much a matter of cultural conditioning as is, apparently, social morality. If this is so, some might consider that an element of territoriality could with advantage be added to our training in early infancy, as advocated by H. Eysenck. This would possibly prevent further gross misuse and plundering of the natural ecosystems that still survive today despite all our past exploitation and neglect.

Thus far, we have been concerned with carrying capacity in terms of living standards and with the conservation of portions of ecosystems as extensive and representative samples of existing biological diversity. A complementary but different approach is to view human land use as an integral function of a given ecosystem, which is accommodated within its structure and processes without irreversibly changing their nature.

Landscape Development on an Ecological Basis

The leading protagonist and most eloquent spokesman for this approach is Ian McHarg, who has developed an internationally recognized school of landscape architecture based on this philosophy at the University of Pennsylvania. If such an approach had always been adopted in regional land-use planning, the United States would now be a much more pleasant and interesting place in which to live. It would also of necessity be carrying either a lower population or a more extensively distributed one. The new city of Woodlands, just north of Houston, is the first city to have been developed entirely on the McHarg principles.

As with all conservation issues, it is necessary at some stage to reach a quite arbitrary decision, even with the McHarg philosophy. Human occupation, even the mere human visitation of an area, inevitably imposes some changes in its ecosystems. What can be conserved then, or what degree of diversity and stability can be maintained, must always be a matter of degree.

Future Environments

In 1965 the Conservation Foundation organized an international conference to consider the future environments of North America. The papers presented at this conference and the discussions following their presentation have been published. They represent a valuable compendium of conservation thought and provide interesting background to any consideration of conservation problems. If such an extensive and exhaustive series can be said to express any common views, there are two. First, conservation is basically a matter of economics. Second, as Lewis Mumford emphasized in the closing address at this conference, conservation must start from a study of urban man.

One of the premises developed early in this text was that urban settlements *preceded* the agricultural revolution, and therefore their inception marks the turning point when earlier human populations began for the first time extensively to destroy natural ecosystems. We are committed irrevocably to an urban civilization. As it was the urban revolution that launched the first wholesale attack on our natural ecosystems, so it must now initiate and vigorously pursue the task of conserving that which remains.

BIBLIOGRAPHY

References

Anonymous. "Blasphemy with a trowel," *New Scientist,* **42:** 676–77, 1969.
Ardrey, R. *The Territorial Imperative,* New York: Delta, 1966.
Baker, H. G., and Stebbins, G. L. (eds.) *The Genetics of Colonising Species,* New York: Academic Press, 1965.
Boughey, A. S. *A Checklist of Orange County Flowering Plants,* University of California, Irvine: Museum of Systematic Biology, Res, Ser. No. 1, 1968.
Bradley, C. C. "Human water needs and water use in America," *Science,* **138:** 489–91, 1962.
Cain, S. A. "Ecology: its place in water management," *Water Spectrum,* **1**(1): 10–14, 1969.

Clutterbuck, D. "Industrial come-back for coal," *New Scientist,* **58:** 17–20, 1973.

Conservation Foundation. *National Parks for the Future,* Washington, D.C.: Conservation Foundation, 1972.

Darling, F. F., and Milton, J. P. (eds.) *Future Environments of North America,* New York: Natural History Press, 1966.

Day, A. T., and Day, L. H. "Cross-national comparison of population density," *Science,* **181:** 1016–23, 1973.

Dubos, R. "Replenish the earth, and subdue it," *Smithsonian,* 3(9): 18–29, 1973.

Eaton, R. L. (ed.) *The World's Cats,* Vol. I, *Ecology and Conservation,* Winston, Ore.: World Wide Safari, 1973.

Eisner, T. "The Big Thicket National Park" (editorial) *Science,* **179:** 525, 1973.

Emmelt, J. L., Nuckolls, J., and Wood, L. "Fusion power by laser implosion," *Scientific American,* **230**(6): 24–37, 1974.

Eysenck, H. "The technology of consent," *New Scientist,* **42:** 688–90, 1969.

Fertig, D. S., and Edmonds, V. W. "The physiology of the house mouse," *Scientific American,* **221**(4): 103–10, 1969.

Fisher, J., Simon, N., and Vincent, J. *Wildlife in Danger,* New York: Viking Press, 1969.

Hepper, F. N. "Plants," in J. Fisher et al. (eds.), *Wildlife in Danger,* New York: Viking Press, 1969, pp. 353–60.

Hutchinson, G. E. "Fifty years of man in the zoo," *Yale Review,* **51**(91): 58–68, 1964.

Kenward, M. "The oil's not for burning," *New Scientist,* **58:** 3–4, 1973.

Kramer, R. J. "Strengthening the biological foundations of resource management," *Trans. N. Amer. Wildlife and Nat. Resources Conf.,* **29:** 58–68, 1964.

Landsberg, H. H. "The U.S. resource outlook: quantity and quality," *Daedalus,* **96:** 1034–57, 1967.

Leopold, A. "Conservation ethic," *J. Forestry,* **31:** 634–43, 1933.

McHarg, I. L. "An ecological method for landscape architecture," *Landscape Architecture,* January 1967, pp. 105–7.

McVay, S. "The last of the great whales," *Scientific American,* **215**(2): 13–21, 1966.

Meadows, D. H., et al. *The Limits to Growth,* New York: Universe, 1972.

Metz, W. D. "Fusion: Princeton Tokamak proves a principle," *Science,* **178:** 1274B, 1972.

Nace, R. L. "Arrogance toward the landscape: a problem in water planning," *Bull. Atomic Scientists,* **25**(10): 11–14, 1969.

Norris, K. S. "California's natural land and water reserve system," *Bioscience,* **18:** 415–17, 1968.

Rocks, L., and Runyon, R. P. *The Energy Crisis,* New York: Crown, 1972.

Smith, A. "Breeding endangered species," *New Scientist,* **54**(795): 333–35, 1972.

Stone, E. C. "Preserving vegetation in parks and wilderness," *Science,* **150:** 1261–67, 1965.

Talbot, L. M. "Endangered species," *Bioscience,* **20:** 331, 1970.

Waller, R. "Modern husbandry and soil deterioration," *New Scientist,* **45:** 262–64, 1970.

Whyte, W. H. *The Last Landscape,* New York: Doubleday, 1968.

Wilcox, D. "Fuel from city trash," *Environment,* **15**(7): 36–42, 1973.

Further Readings

A.A.A.S. *Energy,* 184, no. 4134, 1974.

Aubréville, A. *Climats, Forêts et Desertification de l'Afrique Tropicale,* Paris: Muséum d'Histoire Naturelle, 1949.

Averitt, P. *Coal Resources of the United States, January 1, 1967,* Geological Survey Bulletin, Washington, D.C.: U.S. Department of the Interior, 1969.

Boughey, A. S. "Man and the African environment," *Proc. Trans. Rhodesia Sci. Assn.,* **48:** 8–18, 1960.

Cerowsky, J. "Conservation in east Europe," *New Scientist,* **46**(697): 122–26, 1970.

Charter, S. P. R. "Why preserve nature?" *Man on Earth,* **1**(2): 1–8, 1965.

Chedd, G. "Brighter outlook for solar power," *New Scientist,* **58:** 36–37, 1973.

Ciriacy-Wantrup, S. V., and Parsons, J. J. (eds.) *Natural Resources—Quality and Quantity,* Berkeley: University of California Press, 1967.

Committee on Resources and Man. *Resources and Man,* San Francisco: Freeman, 1969.

Conway, W. G. "Zoos: their changing roles," *Science,* **163:** 48–52, 1969.

Egler, F. E. "Wildlife habitat measurement for the citizen," *Atlantic Naturalist,* **22:** 166–69, 1967.

Ehrlich, P. R., and Holdren, J. P. "Impact of population growth," *Science,* **171:** 1212–17, 1971.

Ehrlich, P. R., and Holdren, J. P. "One-dimensional ecology," *Science and Public Affairs: Bull. Atomic Scientists,* **28**(5): 16–27, 1972.

Elton, C. S. "The reasons for conservation," in *The Ecology of Invasions by Animals and Plants,* London: Methuen, 1958, pp. 143–52.

Emmelt, J. L., Nuckolls, J., and Wood, L. "Fusion power by laser implosion," *Scientific American,* **230**(6): 24–37, 1974.

Hediger, H. "Man as a social partner of animals and vice-versa," *Symp. Zool. Soc. London,* **14:** 291–300, 1965.

Henning, D. H. "Comments on an interdisciplinary social science approach for conservation administration," *Bioscience,* **20**(1): 11–16, 1970.

Henning, L. M. "The versatile cockroach," *Smithsonian,* **3**(9): 76–82, 1973.

Hickey, J. J. (ed.) *Peregrine Falcon Populations,* Madison: University of Wisconsin Press, 1969.

Hoffman, L. "Saving Europe's wetlands," *New Scientist,* **46**(697): 120–22, 1970.

Holyoak, D. "Polynesian land birds face H-bombs and malaria," *New Scientist,* **57:** 288–90, 1970.

Leopold, A. S., Cain, S. A., Cottam, C. H., Gabrielson, I. N., and Kimball,

T. L. "Wildlife management in the national parks," *American Forests,* **69**: 32–35, 61–63, 1963.

Love, R. M. "The rangelands of the western U.S.," *Scientific American,* **222**(2): 89–96, 1970.

McHarg, I. L. *Design with Nature,* New York: Natural History Press, 1969.

McKelvey, V. E. "Mineral resource estimates and public policy," *American Scientist,* **60**: 32–40, 1972.

Myers, N. "National parks in savannal Africa," *Science,* **178**: 1255–63, 1973.

Perry, H. "The gasification of coal," *Scientific American,* **230**(3): 19–25, 1974.

Player, I. *The White Rhino Saga,* New York: Stein and Day, 1972.

Quigg, P. W. "New energy sources," *Saturday Review World,* Feb. 8: 47–50, 1974.

Quigg, P. W. "Alternative energy sources" *Saturday Review World,* Feb. 23: 29–32, 1974.

Scorer, R. "In pursuit of ecological wisdom," *New Scientist,* **57**: 652–53, 1973.

Shea, K. P. "The bison's woe," *Environment,* **15**(6): 34–37, 1973.

Tinker, J. "Britain's endangered wildlife," *New Scientist,* **58**: 67, 1973.

Westhoff, V. "New criteria for nature reserves," *New Scientist,* **46**(697): 108–13, 1970.

Whittemore, F. C. "How much in reserve?" *Environment,* **15**(7): 16–20, 31–35, 1973.

Yannacone, V. J. *Energy Crisis,* St. Paul, Minn.: West, 1974.

Chapter

The Future

14

For centuries writers have attempted to foretell the future of human civilization; the credibility gap between these fictional scenarios and reality has diminished with each astonishing new scientific discovery and development. Recently, however, the auguries have tended to become not only more realistic, but more bleak. Ranging from the undogmatic, carefully weighed considerations of *Mankind 2000* to the deliberately provocative doomsday scenarios of Paul Ehrlich, such works are consistently if varyingly pessimistic. A foreboding of environmental disaster now hovers over the bright promise of the future.

There is no shortage of prophets, or of prophecies. In medicine we are promised banks of spare body parts, chemical correction of defective genotypes, larger and more efficient brains, delayed aging, test-tube births, brain transplants, electronically controlled behavior, and even immortality by freezing. Education offers college training for all, sleep learning, instruction by injection, a computer terminal in the "education room" of every home, and computers that think and reproduce themselves.

In the socioeconomic field labor is to be minimized and leisure maximized;

a new boom was promised for the "surging seventies," although these are now well advanced without any visible prospect of such an event. Equality of the sexes in reality as well as theory is projected, with as a consequence a less inhibited, more fancy-free society. Redundancy and unemployment are to disappear because continuing education copes with problems of obsolescence. Everything is to be disposable—clothes, linen, carpets, walls, houses—or automatic—cooking, correspondence, financing, vehicles, travel. Automation is to boost the output of material products until we can luxuriate in a service-oriented economy, which America has been the first nation in history to achieve.

Implemental Failures

In the harsh light of reality some of these possibilities are no nearer achievement for the vast mass of our human population than when the ideas were first presented decades or even centuries ago. The recent history of heart transplants emphasizes how far we still remain from technological perfection. Our fumbling techniques may buy a little time, and for this a few privileged recipients are grateful. Many are doubtless willing to be used as pioneers in areas where we can advance only from actual human experience. But our failures have been on a much more massive scale than this. We have been very poignantly reminded that before we can progress into these new dream worlds there remains the stark question of how our global population can survive even through the remaining three decades of this second Christian millennium. The brief review of environmental crises presented here has suggested some choices we must make, urgently and immediately. As several authors have remarked recently, scientific information is becoming increasingly used as a political shuttlecock, volleyed around in an ever more partisan world. The concerned individual is confronted by adversary factions plausibly arguing the immediate adoption of diametrically opposed courses. The hypotheses presented in these pages, taken together with the experimental evidence on which they are based, should, however, provide inescapable argument for the necessity of immediate positive action to avoid the otherwise certain ecological disasters that we can now already identify. The difficulty of obtaining any consensus as to priority of needs in technological invention and innovation is well exemplified by the report of a National Academy of Sciences panel formed to make recommendations on such requirements. Two of the four objectives considered as soon realizable are concerned with death control, none with population control. The same emphasis will continue into the 1980s according to this report. Yet there seems little point in attempting to preserve the lives of relatively unproductive members of society while resources are inadequate to provide a proper share for those whose efforts will be needed to support them.

Population Ecology: The Problem

As emphasized repeatedly in this text, by far the most critical of our present major environmental crises lies in human population growth. This is one problem that has been with us in its most acute form for a mere 30 years, for reasons discussed in earlier chapters. Every country in the world now needs to take *immediate* steps to insure that *no later than the end of this millennium* its population growth has been reduced to zero. For not a few areas this will still result in too great a population size, and additional measures will have to be taken to bring the population closer to an estimated optimal carrying capacity.

The information presented in Chapters 7 through 9 is intended to show both why this must be done and how it can be done. *All other developments for the future are contingent on immediate population control action in all nations and in all segments of nations.* But this is the beginning, not the end. In our four to five million year history our global population has been in an "r" mode before, but only temporarily and never previously in global numbers so immense. The reason why in the past our own population, as with all natural populations, did not continue long in an "r" phase but tailed off into a "K" phase was because of impending terminal resource depletion. For the same reason, it will be of no avail if we now move our human population growth from an "r" to a "K" one if we do not simultaneously also perform the same operation on our economy. Regrettably we are still permitting our industries to attack all our nonrenewable resources as though there were no tomorrow.

For a number of years now a few economists of repute have abandoned the economic growth ethic of their colleagues, which they have roundly criticized. Several have gone so far as to declare that any current political philosophy whether communist, socialist, or capitalist has built into its system the machinery for self-destruction. The main argument for this supposition appears to reside in the need for all present systems continuously to intake nonrenewable resources, which are soon discarded as waste. One of the most articulate of these critics is John Galbraith, whose economic writings have been concerned more especially with the recent growth of technology and the multinational corporation. Galbraith considers that these giant companies determine for us our needs, values, and social mores, shaping them toward the maximization of already immense profits. Technological growth then becomes an overriding objective, whatever the environmental cost. In recent work Galbraith offers solutions to these economic problems. His solutions are founded on the same basic ideas that have been gradually developed in this text; they are discussed later in this chapter.

Resource Modeling

In outlining these prospects for the future, it might appear that pragmatic political considerations can be allowed to override any biological ones, but this is not the case. Such political decisions are value judgments that have to be taken, but they must be based on ecologically determined criteria. Application of systems analysis techniques to the study of the causal relationships of environmental crises should prove extremely illuminating. As

T. H. Waterman has written, "systems analysis may be defined quite generally as the application of organized analytical and modeling techniques appropriate to explaining complex multivariate systems, many of whose functional components may be initially quite imperfectly measured or even largely unidentified." As a result of such systems research it should be possible to prepare models of given environmental situations. Predictive simulation models are the only logical basis on which all political decisions as to resource utilization can be founded. Despite their freely admitted shortcomings and their initially very mixed reception, the Forrester-Meadows group models pioneered an approach that will prove in the years ahead to be the only way that we can go.

The primary need for the moment is to create accurate predictive models for population growth. Demographic data entered into these will provide reliable estimates, within specified limits, of the manner of growth of the population, its changing age structure, predictable mortalities, sex distribution, mobility, and other required information. In this century demographic projections have provided totally inadequate warnings of both upward and downward population trends. This is because the forecasts were based on simple determinate mathematical models rather than on the more complex stochastic, multivariate forms that can now be constructed with sophisticated computing facilities.

To these predictive population simulation models, once they have been prepared, can be added the various resource limitations and requirements. These further simulations will then provide the first real estimates of national needs in terms of limited resources such as food, fuel, minerals, education, and space.

No simulation model will provide anything but partial information, however, unless it is related to contemporary ecosystems. The ecosystem is indeed the ultimate conceptual reality against which all predictive models have to be measured. If we needed any practical demonstration that our urban ecosystem, any more than a natural ecosystem, cannot be perturbed in one part without precipitating disturbance in the whole, it has been provided by the current energy crisis. As an objective, society has to strive for the maintenance of stability and diversity in all the ecosystems that it occupies or with which it is in contact. Only in this way can we avoid repeated environmental crises following one another with an inexorable domino effect, and the externalization of our problems in time and space. Utilization of resources and discharge of wastes must become internalized procedures which cannot be allowed to extend from one ecosystem into the resources of another. Nor should the producer segment of the population always transfer the cost of internalization to consumer segments. The problem of waste discharge, for example, should be solved by ensuring that cycling processes return all material resources for reuse without creating any spin-off wastes.

René Dubos recently has expressed this need to internalize our economies in the following passage:

all ecological systems, whether manmade or natural, must in the long run achieve a state of equilibrium for several decades. Furthermore, ecological instability is increasing at such an accelerated rate that disasters are inevitable if the trend continues. We cannot afford to delay much longer the development of a nearly "closed" system in which materials will retain their values throughout the system, by being recycled instead of discarded . . . .

This in essence is the message this book is intended to convey, for it would be immoral to present a work on human ecology at this time that left unstated the ecological conclusions to be drawn from our highly critical situation.

International Obligations

Earlier chapters have discussed how societies have evolved—in a way comparable with an ecological succession—through various ecological stages that include a population explosion followed by colonization of other territories.

With the world now totally segmented into a mosaic of national territories, the last frontier has gone, and there is no further possibility of colonization without displacing some previous human occupants of the territory. The first international freedom that has to be relinquished is therefore the right of a nation to expand its boundaries by force. This might seem an elementary assumption, but during the 12 months in 1968–69 when the first edition of this text was being prepared, there were at least four separate and major examples of occupation by one nation of the territory of another. These were an attempted invasion on the Chinese-Russian frontier, another of Honduras by El Salvador, the continuing occupation of the Sinai Peninsula by Israel and of South Vietnam by North Vietnam. During the intervening years two of these situations have persisted. A further one has developed in Cyprus, and India invaded and forced the partition of Pakistan. However arbitrary the methods by which the present international boundaries were achieved, unless they can be readjusted in free negotiations, no change can be acceptable. Time may require some territorial modifications, and negotiation should make these possible.

If we could finally remove the need to defend international frontiers, as we have indeed succeeded in doing locally in such examples as the North American continent and the Scandinavian Union, the tremendous resources previously deflected to frontier defense would be released for other enterprises. There is already a world forum for discussion, negotiation, and law enforcement—the United Nations—and a judiciary body, the International Court of Justice. These could provide all the international law and law enforcement necessary to maintain international accord, if national force could be abandoned as a final arbiter.

A United Nations conference on the human environment was held in 1972 in Stockholm, Sweden. At this conference plans were presented for the International Council of Scientific Unions (ICSU) to create an environmental committee charged with the establishment of an International Center for the Environment. This organization is now established in Nairobi, Kenya, and will explore any scheme appearing likely to cause environmental degradation as well as attack such long-range problems as atmospheric and water pollution. In 1976 a similarly sponsored conference will be held in Vancouver, Canada, on the subject of human settlements. This conference will address itself to urban problems, the area in which our human environments exhibit the most intolerable and totally unacceptable degradation.

Carrying Capacity

Construction of computer simulation models of populations and resources, and prediction of the dimensions of internalized processes, will make possible an estimation of the *carrying capacity* of particular regions in terms of human populations, the optimum population at all levels, national, regional, or urban.

The carrying capacity or optimum population is a somewhat indefinable concept. Unless it is matched against other features of the ecosystem, such as the degree of diversity which it is desirable to maintain, what size the area to which it is applied, and so forth, its estimation will be quite arbitrary. Obviously there must be some tradeoff or compromise between conservation and utilization, and the carrying capacity of a given area must be arrived at after some political decision has been taken as to the extent of diversity which it is agreed should be conserved. This involves economic land-use planning and the acceptance by the whole community of the restrictions which this imposes on freedom of choice. In the U.S. plans for all major developments must be accompanied by an Environmental Impact Report. Although the form of these reports has recently tended to become stylised, and the consequences they predict are sometimes ignored in the face of partisan pressures, their appearance has had a salutary effect on heedless environmental degradation.

Many national populations already have far exceeded the carrying capacity that simultaneously permits the conservation of a wide diversity. Others may well unavoidably do so in the near future, or may wish to do so rather than too rigidly restrict population growth at this stage. Internationally provided incentives must be offered where nations are urged to set a low carrying capacity in order to conserve particular types of diversity, e.g., a surviving megafauna of game animals. By the same token, economic incentives may be offered to individuals or groups within the national territory in order to achieve this same object. Eventually, environmental impact reports will have to become obligatory at all levels of society.

Technological Requirements

The last quantum step in the ecological evolution of human societies was the urban revolution and the concentration of populations in cities. The particular features of urban societies are *specialization, interaction,* and *innovation.* Our urban society must have these ecological characteristics if it is to progress; they are inseparable from urban ecosystems.

These urban ecosystems must, however, be constructed more rationally and not left so much to chance. Cities need to be analyzed in terms of their purposes and their needs, so that their structures can conform more readily to desirable plans for urban living. Land-use planning operations must start from the cities and reach outward. As Ian McHarg has eloquently emphasized, we need to estimate the carrying capacity of a given region in terms of its limited resources and contrast this with what we require in the form of "environmental quality." A city or cities can then be constructed in situations that provide for the optimum attractions for a population at this carrying capacity. To provide the necessary support for this population, industries can be brought in that are based on internalized practice without having to externalize any problems on surrounding ecosystems or communities except by mutual arrangement.

This is the exact reverse of what has been the traditional practice in our industrial societies. Industries have grown up where there have been exploitable resources. People have moved to the industry without reference to their needs in terms of outdoor recreation and other living requirements such as education, shopping, entertainment, and public health. By relating carrying capacity to resources in terms of food and water, and services such as health and education, it is possible to avoid externalizing industrial problems. In this way water and air pollution, poisoning of the ecosystem, and overexploitation of resources are avoided entirely.

Exhaustion of Resources

With the exception of Advanced Industrial societies, where various incidental factors operate to reduce population growth, many existing social groups left to themselves may be expected to ignore all advice and continue unchecked the evolution of their population growth rates until they completely exhaust their environmental resources. The U.S. National Policy Panel of the UN Association, in a report entitled "World Population," stated that high fertility and high population growth rates in developing countries can mean widespread famine, increased illiteracy, unemployment, squalor, and unrest threatening the very foundations of public order.

Although the mounting external pressures and propaganda to which earlier successional societies are now beginning to be exposed might cause some to reduce this fatal rate of population growth, the exhaustion of resources they have already sustained will still inhibit industrialization. No net primary productivity will be available for the continuation of ecological societal succession. The life-support systems of such depleted societies entirely absorb whatever limited gross productivity they can develop, and they have to remain essentially Advanced Agricultural nations.

National Obligations

Within nations themselves, considerable sacrifices will have to be made by various segments of the population. For example, as stressed at the beginning of this chapter, all societies will have to reduce their population growth at least to zero, and for many it will be considered desirable to run at a negative population growth until the population has been reduced to a more realistic size.

To achieve this low level of population growth, it will be necessary to ask the youth of each country to restrict their ambitions as to family size. This cannot be done equitably without requesting some restrictive contribution from the middle-aged and the elderly. The most obvious compensatory contribution from this postreproductive segment of the community lies in taxation, which is the very burden this portion of the population in many Advanced Industrial societies is now unwilling to contribute. This is especially unreasonable when it is precisely the uncontrolled self-propagation indulged in by the older generations that has placed modern youth in the situation where it has to severely limit its own reproduction.

Numerous partisan labels have been bandied about in this century such as socialism, fascism, and communism. Without applying such words, it is obvious that many societies in the world have to go much further than they have at present in resource partitioning within their populations. The distribution of wealth is very spotty, and in many instances hereditary cliques still are privileged and provided with opportunities in great excess compared with the less wealthy members of society.

It will be apparent from even the brief treatment of this subject that has been possible in this text that we are seriously lacking in fundamental knowledge of social behavior in human populations. Indeed, it is only in recent years that we have initiated such experimental studies even in other social primates. Thus we have little understanding of our own behavioral needs and responses in such ecological areas as territoriality, group identity, group defense, pair bonding, social learning, social hierarchies, and aggression. We are quite unable as yet to separate our responses in these areas into phenotypic and genotypic categories, that is, into the cultural reactions resulting from or modified by our early conditioning and the innate behavioral reactions that develop from our genetic inheritance. Perhaps our contemporary advertising media have come closest to obtaining some understanding of certain aspects of these ecological relationships, through an essentially empirical process of trial and error in perfecting marketing techniques. Likewise great orators throughout history have happily (or unhappily, in some instances) hit on the right combination of verbal stimuli to trigger desired ethological responses.

Religious Beliefs

We still are unwilling to face many philosophical conclusions and deductions that may be made from our examination of ecological evolution. The most restrictive aspect of this refusal is undoubtedly the tenacious persistence of *doctrinaire* religious beliefs. Religious organizations, with their well-structured and often tightly disciplined hierarchies, retain a firm hold on many of the resources as well as some of the beliefs of many of the world's societies. The most pernicious of these beliefs, which is still included in some Christian literature, has been the contention that children are a divine punishment visited upon married couples unable to resist the worldly temptation to indulge in sexual activities. However, scientists should not adopt a superior attitude about such matters, for some scientific beliefs too persist long after they have actually been disproved, and sometimes handicap the progressive development of new ideas.

The reason strong efforts are not made to disprove such doctrinaire beliefs may well reside in the fact that few scientists are nihilists. They are hesitant to destroy old hypotheses before they have formulated what they regard as more tenable substitutes. Toward this end Julian Huxley was striving with his "religion without revelation," which he intended to contain enough verisimilitude to be scientifically acceptable but remain sufficiently flexible to permit periodic reconciliation with new discoveries. Even Huxley, however, as is apparent from his work *The Humanist Frame,* cannot avoid the implication of a "human destiny." Few scientists presently seem willing to comment on such issues. L. Goldstein, one of the few who has done so in discussing exemption of churches from taxation in this country, stated: "This kind of discrimination is particularly offensive for nonbelievers (a group to which most biologists belong) . . . ." Such statements have a high emotional content and are likely to provoke strong reactions.

Some individuals doubtless consider that the lingering persistence of doctrinaire religious belief represents an impediment to the introduction of any realistic philosophy of life that would provide an acceptable motivation for future societies. To such individuals, proper motivation can be achieved only when we are sufficiently honest with ourselves to acknowledge that, despite our unique cultural development, there is no more *raison d'être* for our species than there is for any other plant, animal, or microbial population on this planet.

We enjoy living, and the vast majority of us wish to continue living, because over many millennia we have been selected for genotypes that produce phenotypes with such feelings. If they had not, they would have been automatically eliminated from the gene pool of future generations.

There is no scientific justification for the preservation of our human population other than its typification of a unique and extremely interesting development among the diversity of populations that have survived in our contemporary world following selection by ecological evolutionary processes. Even were we to learn how to control the universe, this would not as far as our scientific knowledge goes be fulfilling any divine purpose or realizing any predestined universal plan. We have, as far as we can judge on present evidence, no preordained scientific function, either on this planet or in the universe, and the choice is entirely ours as to whether we continue or discontinue the propagation of our particular species.

Concerning the individual as opposed to the population, there are patently personal feelings both of well-being and of discomfort that are unique to each of us. Partly genetically determined, they seem mostly to be the result of conditioning in the very early months of life. These feelings, like all those expressed as human cultural behavior, are exceedingly complex and still only imperfectly analyzed and understood. For the moment perhaps all that can be said is that in order to ensure continuation of the human species, the individual should feel "good"! If the future of further generations of our species is to be assured, it will have to be associated with better conditions than those that we ourselves have known, so that the good feeling persists. This hedonistic oversimplification presents us with the only rational philosophy of life presently scientifically conceivable: enlarge our territorial concepts until they encompass all mankind and improve conditions for the next generation without simultaneously making ourselves too utterly miserable.

To implement such a philosophy, we are immediately faced with a number of pragmatic decisions based on value judgments. These invariably impose some loss of individual freedom. As long as we have been social animals we have been selected to concede some measure of individual loss of freedom in return for the benefits of social life, the ecological process of "group selection."

There are many modern examples that could be cited of loss of freedom which we take for granted, even though it has not always been so, even when the privileges we have lost pertained until, say, 10 years ago, or even last year. Some of these have been mentioned in the chapter on population control. We cannot in this country run away if ordered to stop by a police officer, whether he is immediately identifiable as such or not, and we may be shot dead if we ignore such a request. Until recently, if we were young, fit, and male, we could not avoid being directed to fight and kill other individuals of our species if we were so instructed, unless we were prepared to endure social censure and/or physical imprisonment. We cannot venture onto a substantial portion of the land of the United States that is in military hands. Nor can we trespass onto much of the rest that is in private hands. We cannot, at least in America, legally practice polygamy, even if it is a tenet of our religious belief as it was for Mormons and is for Moslems.

Such a list of restrictions on freedom, serious or otherwise, could be extended. In view of the many limits we already accept or take for granted, it is interesting to look at the kind of further restrictions that would need to be imposed in this country, for example, to obtain more peaceful social arrangements and a world that had some chance of survival without catastrophic ecological change. The rights that would have to be conceded include the following:

1. The right to have as many children as we wish
2. The right to have any children at all by mates with particular genotypes
3. The right to keep pets that consume food that could directly or indirectly be used for human purposes while populations and individuals elsewhere are stunted from malnutrition or dying from starvation
4. The right to externalize our labor problems by strikes that dislocate societal facilities
5. The right to reside in any locality of our own choice in the United States
6. The right to accumulate wealth entirely disparate with basic needs while many elsewhere still cannot satisfy these, or to transfer inherited wealth

without reduction to an indefinite number of descendant generations
7. The right to monopolize what would otherwise be community facilities
8. The right to exploit national resources to the public detriment
9. The right to possess lethal weapons
10. The right to discharge wastes into the environment

If we fail to recognize that it is ecologically unsound to permit individual members of our population to remain unregulated in regard to the above "rights," we can never achieve the necessary steady-state stability in our human ecosystems. It is hoped that the presentations in this text demonstrate that it is ecologically impossible to avoid population regulation (items 1 and 2), to permit resource depletion (3–8), to allow factional and individual strife (4, 9), and to sanction pollution and contamination (10). We must calmly and rationally prepare a master simulation model of our global ecosystems that will include ecological feedback mechanisms involving at least the periodic or temporary denial of all these ten aspects of our culture that we have traditionally regarded as "inalienable rights." The ecological unsoundness of such an assumption should now be apparent, and the reasons this monumental task must be completed within the next decade ought to be equally clear.

Ecological Evolution

Until we learn to alter the chemistry of genes in an extensive and generally applicable manner, further evolution of our population will occur by the same processes that have determined it over the past 4–5 million years. Even without chemical control of genes we can apply artificial selection to human genotypes; this is still another political decision we must eventually make. Much progress could be made in avoiding "disadvantaged" homozygous genotypes if tests for heterozygosity for these "deleterious" traits could be universally applied to prospective parents, or similar tests applied to embryos. Positive selection could be made, and has indeed been recommended, by many respected authorities for such features as intelligence and creativity. The comparative frequency with which genius occurs, i.e., the number of times we encounter individuals with IQs measured in the region of 200–250 or higher, indicates that we could considerably improve on the general level of intelligence by artificial selection, because to an appreciable extent intelligence is hereditarily controlled. Julian Huxley two decades ago presented proposals relevant to this issue.

There would be some substance to an argument that in order to cope with our increasingly complex technology—particularly as computers evolve from peripheral information retrieval and auxiliary calculating devices to decision-making innovators—the general level of intelligence within our population should be raised. It is more difficult to argue a case for longevity, although by storing the germ plasm of individuals while they were progeny-tested for longevity it would be relatively simple artificially to select for longevity and produce a population with a longer life span. In a technological society that seems increasingly to favor built-in obsolescence, it is possible there would be some resistance to built-in longevity being made the object of artificial selection at this stage of our population's development.

Negative selection is an even more difficult matter to legislate, although simulation models will provide prediction of the rate of spread through the population of genes presently considered detrimental to individuals.

Most people would agree, for example, that mutants which resulted in limbless individuals are not desirable, unless some important compensatory genetic feature is associated with them that could only be maintained in the human gene pool by continuing to propagate genotypes resulting in such a drastic variation. Even this bizarre possibility cannot be entirely ruled out, for as J. B. S. Haldane quite seriously commented, legless astronauts would be more effective for the initial series of space probes.

When we come down the scale to congenital diabetes, it is more difficult to obtain a majority for a political decision against continuing procreation from genotypes known to produce this disorder. Finally we have minor features, such as insensitivity to musical tones or color blindness, which only slightly impair the individual's full appreciation of particular behavioral rituals. Where should negative eugenics stop, or, rather, where could it begin?

It seems doubtful whether the prediction of any computer simulation model will require that immediate decisions be made on such choices; there are more pressing issues. In view of the vital decisions on major restrictions that we must immediately make in terms of population growth and resource exploitation, it would appear advisable to postpone such eugenic decisions for some time, especially as each decade brings us closer to the possibility of chemical adjustment of "undesirable" genetic material.

Moreover, the availability of the power to chemically modify genes, when it is realized, will pose still more political problems. Not only will it offer the possibility of increasing the rate of natural mutations, but it will also provide the opportunity to produce entirely new mutant forms.

Emigration and Immigration

Although national boundaries have been maintained from the beginning of the twentieth century, movement across them always has been restricted. There seems no justification for stopping these limited movements entirely so as completely to prevent emigration and immigration. An exchange of genetic material is biologically desirable if further cultural evolution of the *Homo sapiens* grade and the spread of any further grades that evolve, such as *H. innovatus,* is not to remain restricted. The concentration of particularly innovative, inventive, and creative groups in certain centers may lead, by assortative mating, to the evolution of new forms of our genus. Provided no barriers to gene flow are erected artificially, this kind of concentration and rediffusion is necessary for continuing biological evolution.

The need to permit the concentration of innovative individuals from many countries in areas that facilitate their free expression is illustrated by the national origins of U.S. Nobel laureates. Of those Americans awarded the

Nobel Prize for medicine or for science, 25 per cent have been foreign born, a far greater proportion of this category of citizens than is present in the population at large. Further large-scale immigration into the U.S. may be stopped if Garrett Hardin's recently presented lifeboat metaphor is accepted and acted upon.

Aggression

Although we may hope that territorial aggression in terms of international wars may soon be effectively discouraged, aggression and the associated phenomenon of territoriality on a personal basis will be much more difficult to control. In the so-called Advanced Industrial societies, crime and mayhem flourish alongside a general decline in the adherence to what were once acceptable social mores. This decline is especially dangerous because individuals are externalizing their difficulties on society at large, a process Garrett Hardin has described as the "tragedy of the commons." Individual pilfering from stores and supermarkets is paid for, not by the owner, but by the more conformist customers. By 1973 thefts from U.S. supermarkets had reached an estimated 10 billion dollars annually, increasingly the food bill of every American by $50 per year. Apprehended supermarket thieves rarely pleaded poverty; indeed in California, the most affluent state, losses were four times as high as in midwestern cities like Chicago. Likewise the community at large had to pay for the towels removed from hotels and motels, the insurance to cover theft, the security measures to guard against robbery, the policing of areas to protect life and property. Finally there is the game no one group can ever finally win, the walkout strike, which again is basically almost always an externalization of particular problems on society at large. During 1973 a strike by several thousands of coalminers in Britain brought the life of that nation of 50 million people to the verge of total collapse and was responsible for unrecordable misery, suffering, and loss. Advanced Industrial societies are increasingly exposed to this kind of action. If their political organization is still basically of a democratic type, they have no defense against such vocational extortion.

There have been suggestions that early training methods be programmed to condition individuals to a required level of aggressive behavior. This approach has been associated more especially with the published views of the distinguished Harvard University psychologist B. F. Skinner. The scheme in Figure 14-1 illustrates the fundamentals of this approach. It depicts human behavior, insofar as it is exhibited in terms of individual expressions of aggression, violence, and anger, as either falling within acceptable limits or attaining levels beyond these. The limiting values are set by the particular society, and will vary from one society to another. Unbalanced behavior—that is, behavior that exceeds the tolerance limits set by the society—may be treated by several curative procedures. Adjustments may be achieved by modification of the microenvironment, the approach of the environmental psychologist. The endrocrinologist can attempt to modify glandular secretions, the neurosurgeon to manipulate operationally the level of mental

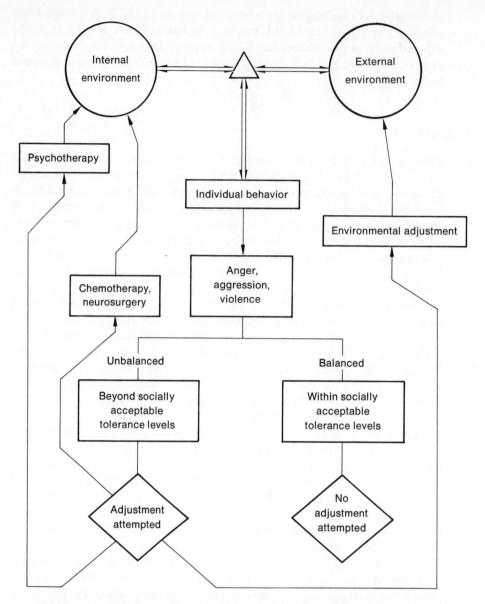

FIGURE 14-1. *A model of Skinner-type methods of controlling aggression.*
Individual human behavior is shown by the model to be determined by
interaction between the internal environment, which originates genetically and
operates through the endrocrinal system, and the external environment, with
which the genetical components continuously interact. Insofar as behavior is
expressed in anger, violence, and aggression, the levels selected over the many
millennia we existed in hunter-gatherer societies may be intolerable in urban
communities. Readjustment to the new levels can be obtained by adjustment of
endrocrinal activity (chemotherapy and neurosurgery), psychological
conditioning (psychotherapy), and modification of the external environment.
The selection of one or a combination of these treatments involves some
surrender of individual freedom.

activity. Finally the psychologist can attempt to produce in the adult the reactions that Skinner-type conditioning could effect in the young. One or a combination of these methods would most likely succeed in bringing the aggressive behavior of an individual within the acceptable societal tolerance levels. In the process the individual has surrendered, voluntarily or involuntarily, some individual freedom, and this circumstance is of deep concern to many people.

Industrialized Societies

The Advanced Industrial nations, contrasted with less evolved societies, may be expected to enter a phase of selection pressure for particular behavioral patterns such as they have never previously experienced and whose nature and intensity have not yet been calculated. These pressures may tend to reproduce specifically those genes that favor "parental" characteristics against the preference for a "swinging" life, or against forms of behavior unlikely to lead to self-propagation as a consequence of sexual activities. There may also be a contrasting selection for "social-service" genes that act against the individual but favor reproduction of the group.

For a number of generations extensive genetic evolution could thus be proceeding simultaneously with cultural evolution. This circumstance, which has not applied to human populations for several millennia, will arise from the almost complete separation of sex and reproduction that population control requires and modern techniques permit. Whereas previously many individuals or couples contributed to the human gene pool whether they really wanted to or not, now only those whose genetic characteristics have disposed them toward parenthood will reproduce. The consequence of such genetic selection will be an increase in "parental" gene frequencies, invoking presumably further resistance to compulsory population control measures.

It would seem advisable to direct some of this heightened social competitive drive into some different form of challenge. Athletic competition as a sublimation technique in maintaining internal order is as old as history; it has served the older nations of Europe well. We need perhaps, especially in international rivalries, to turn to the possibilities of other areas of competitive enterprise. Intellectual problem-solving is the most promising, but with it must come a new social hierarchy. There is already, as Eric Hoffer has observed, a possible conflict between those who work with their hands, producing artifacts, and those who employ mainly their minds, developing mentifacts. In fact, the U.S. has become the first nation in which more of the labor force is engaged in service industries than in industrial production. These two activities of providing services and supplying goods are separated somewhat along the lines of C. P. Snow's *Two Cultures*. There is no need for stress to arise from such social cleavages, provided each type of citizen is content that his individual cultural opportunities are not restricted by covert or overt social hierarchies. Jane Jacobs, for example, envisages the possibility that significant new activities can and will develop even from such "lowly" occupations as janitorial work. Many authorities like Clark

Kerr, one-time President of the University of California, consider that material rewards for work will tend to standardize; the differentials that persist will be in the relative pleasantness of the various forms of this required work, rather than a sliding scale of financial remuneration.

The Creative Society

The political concept of the "creative society" is, perhaps unconsciously, an approximation of this idealized social structure, at least in its aims. The pursuit of new knowledge (mentifacts), the fashioning of new tools (artifacts), and the performance of new activity (sociofacts) are the features that have always characterized our populations and for which we have been subject to selection pressures over perhaps 4 to 5 million years. We cannot simply lose these characteristics unless we deliberately select eugenically for the indolent and the unimaginative. Some means of achieving the essential social structure needed for a peaceful society still possessed of such innate drives must be evolved.

John Platt spelled out very succinctly just what kind of creative technology we now have most urgently to organize. Pointing out that our most critical environmental problems are essentially similar in urgency and complexity to critical wartime problems, Platt postulates that the same problem-solving techniques should be applied to their resolution. He proposes the organization of task forces of fulltime interdisciplinary teams comprised of specialists including natural and social scientists, medical workers, engineers, lawyers, and teachers, bringing together known ideas and by their interaction generating innovative new approaches to social research and development. Both Platt and others such as B. L. Crowe, who reexamined Hardin's "commons" proposition, consider that basic or fundamental research may momentarily at least lose out for one reason or another as we concentrate on the solution of crushing environmental problems. This may well be, but creativity is a self-satisfying objective for any life style, and all forms of inquiry must sooner or later contribute to the sum total of human culture. Whether they are likely to do so immediately or at some time in the more remote future must be determined by priorities dictated by political decisions about the exigencies of any given crisis condition.

It is interesting to compare such suggestions for the future of human society with the recommendations proposed by professional sociologists. John Galbraith, as mentioned earlier, is one of the most liberal of U.S. economists. In his latest book, he has asserted that modern society is a planning system that exploits people counter to their own best interests. He proposes that in order to rectify this situation a "new socialism" should be introduced. The main innovations of the "new socialism" would be as follows:

1. Organization under public ownership of the weaker parts of the present market system, housing, medical care, and transportation
2. Extension of the provision for an adequate minimum wage
3. Encouragement of trade associations for small businessmen; government price regulations

4. Abandonment of the struggle to maintain full employment, correlated with the introduction of universal unemployment assistance
5. Nationalization of major established enterprises, as has been done with the Tennessee Valley Authority, and also of the military-industrial corporations such as Lockheed and General Dynamics
6. Coordination of major utilities under a public planning authority
7. Organization of public support of the arts

There appears to be nothing very original about Galbraith's proposed "new socialism," except that it is new to the U.S. The majority of European nations, for example Great Britain, have adopted most or all of these measures over the last three decades. There may be some substance in Galbraith's statement that the present economic planning system of the U.S. exploits people, but the kind of social state he proposes equally decides what is good for people in terms of its own prosperity rather than theirs. Both systems are classified politically as parliamentary democracies in which majority rule prevails, but in both the mechanics of power tend to become too complex to be perceptible to the general citizen. Future societies would appear to require some more satisfactory organization to ensure the optimization of human involvement and participation in urban societies.

Brain-Drain Effects

We may expect that particular regions of rapid technological advance will continue to exert an irresistible attraction for phenotypes in which enterprise and initiative—and intelligence—are highly developed. These are self-perpetuating characteristics; selection pressures may well continue to operate until these phenotypes have little or no reproductive contact with the parental gene pool from which they were selected originally.

This preferential association of reproducing individuals in relation to the characters their genes determine in them is what geneticists term *assortative mating*. Its chief effect in human populations is to associate genes of similar effect but at different loci. With some measure of reproductive isolation, it is certain that individuals possessing these unique associations will be founders of a new species grade very different from our own in its social hierarchy, reproductive characteristics, intelligence, and harnessing of aggression. Evolution in the several species grades of the genus *Homo* has characteristically taken this form of "cultural speciation"; for this latest species grade of our human populations, the name *Homo innovatus* has been suggested here as a convenient appellation, not to designate a new taxonomic entity.

Homo innovatus

We could expect *Homo innovatus* grade populations to appear first in the megalopoli. In the United States this means the three burgeoning metropolitan areas around the Northeastern Seaboard ("Boswash"), the Great Lakes ("Chipitts"), and the southern Pacific Coast ("Sag"). Because it probably already contains higher gene frequencies for the particular characters involved, the western United States may well prove to be the initial center of origin of this new species grade; perhaps individual clans of *H. innovatus* grade already are established there. Occupied initially by Caucasoid and Mongoloid stocks partially selected in advance for aggressive characteristics, the western United States has continued this selective attraction for aggression at least insofar as it is expressed in the forms of enterprise and initiative.

R. Lynn estimated in 1969 that 5,000 persons with IQs of 130 to 150 annually leave Britain for the United States and that of the 23 per cent of the annual university scientific output who emigrate, a considerable proportion migrate to the American West. Such immigrants are often accompanied by mates and offspring. Similar disproportionate selective immigration occurs from other countries, now especially from Asia, and from other states.

When the relationship between innovation, invention, and basic research is considered it is concluded that this process of change requires an effective coupling of scientific and technological communities. Nowhere is this better exemplified than in the synthetic and essentially immigrant societies of the Western Megalopolis.

Spread of *H. innovatus*

Slightly varying geographical populations of *H. innovatus* grade may appear later in the other two American megalopoli, and also around Moscow, Peking, Tokyo-Osaka-Yokohama, and possibly Sydney, Australia, where similar selection pressures prevail. It is clear that *H. innovatus* everywhere will have multiethnic origins, as did our *H. sapiens* grade. We know from history that all human groups have contained, although not necessarily at the same frequency level, genes that when appropriately combined can produce a Cetawayo, a Genghis Khan, or a Napoleon.

Because no two populations with even partially distinct breeding systems can coexist in the same ecological niche, not only will the populations of *H. innovatus* grade increase in size until they exclusively occupy their centers of origin, but their selected recombinant gene characteristics could introgress slowly into the surviving populations of *H. sapiens* grade until the genotypes of this species have been largely replaced and it has passed to extinction, as did the *H. erectus* grade populations. Or rather, we should say, until the particular combinations of alleles that characterize this particular human grade are not encountered any more: in these later stages of human evolution it is not possible to think of extinction in the same terms as we regard the extinction of, say, the dodo.

Looking again at Figure 14-1, we will perceive that this diagram can be even more generalized by removing the entries "violence" and "anger" and interpreting aggression in its broadest meaning. When we do this we can see that urban *Homo sapiens* may no longer fall within sufficiently competitive levels of aggression. We had many millennia of natural selection for a level of aggression appropriate to our hunter-gatherer societies with their small, tightly knit family bands. There is every reason to suppose that the very different microenvironments of urban ecosystems within our Advanced Industrial societies require an adjustment of our output levels for aggression. Our 15,000 years or so of exposure to these new environments will have been insufficient time for natural selection to make all the necessary changes. We may be able to cover this deficiency by manipulation of our levels of aggression through psychotherapy, chemotherapy, neurosurgery, or environmental design, but these genomes that provide naturally for the appropriate adjustment will still be strongly selected. One of the more frightening prospects in this possibility is that, should Garrett Hardin's recently developed lifeboat metaphor really reflect reality, we may end with a divisive world with two kinds of lifeboats. In one flotilla there will be an ever-increasing preponderance of *Homo innovatus* grade individuals. Roger Revelle labels the possibility "obscene," but determination to survive readily overrides such humane objections. Cannabalism is nothing new to *Homo sapiens* and any succeeding grade cannot be expected to forgo this competitive technique on a societal level, even if there is now a universal revulsion against the act on an individual basis.

The chances of our world becoming deeply divided have worsened because of the formation of national cartels to fix prices for scarce commodities. Table 13-1 lists the monetary reserves suddenly accruing to the first of these cartels, OPEC, the oil-producing nations. Members of OPEC have no economic choice but to invest their immense monetary reserves in the Advanced Industrial nations which provided them with the money in the first place. The industrialized nations, for their part, will have no economic choice but early nationalization of the OPEC purchases, and so the game will go on. Meanwhile the part of the Third World left outside the commodity cartels will try to man its sinking lifeboats while it watches the economic battle between the cartels and the industrial nations.

Economic Considerations

If we look around us, certain features of the nightmare world of George Orwell's *1984* already are beginning to come about. The "haves" and "have nots" are already divided by natural fences through which movement is restricted severely. Unless an entirely new economic approach is established immediately and successfully, the condition of the "have nots" must inevitably deteriorate still further until we reach the Age of Famines.

Although the "haves" increasingly are made aware of many of the distressing circumstances of the "have nots," they still as yet contribute little of their abundant resources to alleviate these dire needs. Desultory and often disorganized efforts occasionally are made by private enterprise and officially through international agencies to bring about some improvement in the condition of these underprivileged world citizens.

Meanwhile, the steady drain of the genotypes that produce a fitter phenotype is allowed to continue, and further progress of the "have-not" population is impeded by this genetic depauperization, coupled with the general lowering of intelligence through early malnutrition. The computer revolution has led to a quantum advance in the industrial progress of the "have" nations, as Garrett Hardin has stressed. The operation of sophisticated computer-based techniques is entirely beyond the present ability of the Advanced Agricultural societies, whose per capita resources providing food, education, medical attention, and communication for their citizens are rapidly becoming less as their population growth continues unabated.

Precisely what kind of effort would be needed in an Advanced Agricultural society to achieve some progress was estimated by S. Enke. He calculated that about one quarter of the population aged 15 through 49 would have to practice contraception to halve the gross reproduction rate in 30 years. In terms of economic welfare this would provide an increase in income per head of 3.0 per cent a year instead of 1.7 per cent and a third more capital per worker after 30 years.

Viewed in isolation, these calculations might be encouraging, but contrasted with estimated material progress in a "have" nation over a comparable period, this still is a depressing picture. Although it may be a source of self-congratulation that some elements of the world population of our present species grade may evolve into a form with greater intelligence, increased aggression, and a greater identity with a more effective social order, this is little consolation for the apparent certainty that for billions among our present species the future holds no more promise than a continuing, desperate struggle to find food and shelter, and perhaps a few forgetful moments of simple happiness. The writing is on the wall for all to see; the species *Homo sapiens* as we know it surely is doomed.

Through cultural impositions or by direct innate and inherited response, we exhibit two kinds of loyalties, one to ourselves, our mates, and our offspring, the other to our society. Our social behavior is determined by an interplay between the often conflicting demands from these two directions. Societal loyalties have resulted from group selection and cultural evolution, but a powerful stimulus is needed to invoke them to the extent that they override personal loyalties. War is the commonest such stimulant, but perhaps our present environmental collapse will promote in the end a similar social reaction.

Finale

The study of human ecology eventually will be able to explain much and to predict even more, but ultimately the decisions that must be taken regarding the future of our species are political and represent value judgments. What we are to be, what we eventually will become, and the effect we are to have on the universe we occupy all lie within our ecological control; but no ecological concept or principle can make the ultimate decisions for us. Nor can we any longer find escape in the excuse that four or five men in the world alone have the absolute power to destroy our world. Whether it takes 10 minutes of atomic warfare or 100 years of overpopulation, overexploitation, and overpollution, our world is just as surely going to be destroyed unless each of us is individually willing to take some immediate positive action to prevent this ultimate catastrophe.

It will be apparent from these pages that our civilized world and our human species throughout almost all of their time have existed in a steady-state, no-growth condition. Occasionally cultural advances have temporarily induced short phases of exponential growth, quickly succeeded again by further long periods of no growth. Over the last few centuries we have again entered a period of exponential growth, the consequence of the various cultural innovations resulting from the urban revolution. We are now at the end of this exponential phase and are about to initiate a new no-growth period. Whether we are precipitated into this by a series of devastating natural disasters of dimensions utterly beyond our control or whether we ease into it painfully but at a governed pace is the only choice which is now available to us. The time for our decision has almost expired.

BIBLIOGRAPHY

References

Brooks, H., and Bowers, R. "The assessment of technology," *Scientific American*, 22: 13–22, 1970.
Clarkson, F. E., Vogel, S. R., Broverman, I. K., Broverman, D. M., and Rosenkrantz, P. S. "Family size and sex-role stereotypes," *Science*, 167: 390–92, 1970.
Crowe, B. L. "The tragedy of the commons revisited," *Science*, 166: 1103–7, 1969.
Daly, H. *Toward a Steady-State Economy*, San Francisco: Freeman, 1973.
Dubos, R. "A social design for science" (editorial), *Science*, 166:(3907), 1969.
Ehrlee, E. B. "California's anti-evolution ruling," *Bioscience*, 20: 291, 1970.
Ehrlich, P. R. *The Population Bomb*, New York: Ballantine Books, 1968.
Eichenwald, H. F., and Fry, P. C. "Nutrition and learning," *Science*, 164: 644–48, 1969.
Enke, S. "Birth control for economic development," *Science*, 164: 798–802, 1969.
Ewald, W. R. (ed.) *Environment for a Man—The Next Fifty Years*, Bloomington: Indiana University Press, 1966.
Galbraith, J. K. *Economics and the Public Purpose*, Boston: Houghton Mifflin, 1973.
Goldsmith, E. (ed.) *Blueprint for Survival*, Boston: Houghton Mifflin, 1972.
Goldstein, L. "Thoughts of tax exemption," *Bioscience*, 19: 871, 1969.
Haldane, J. B. S. "Biological possibilities for the human species in the next thousand years," in G. Wolstenholme (ed.), *Man and His Future*, Boston: Little, Brown, 1963.
Hardin, G. "The tragedy of the commons," *Science*, 162: 1243–46, 1968.
Hardin, G. "Computers and the slave society," *Information Display*, 6: 215–17, 1969.
Hardin, G. "Not peace, but ecology," in G. M. Woodwell and H. H. Smith (eds.), *Diversity and Stability in Ecological Systems*, Brookhaven Symposia in Biology No. 22, 1969, pp. 151–61.

Hardin, G. "Living on a lifeboat," *Bioscience*, **24**: 561–68, 1974.

Heller, A. (ed.) *The California Tomorrow Plan*, Los Altos, Calif.: Kaufman, 1972.

Hoffer, E. *The Temper of Our Time*, New York: Harper & Row, 1967.

Huxley, J. S. *Religion Without Revelation*, New York: Harper & Row, 1957.

Huxley, J. S. (ed.) *The Humanist Frame*, London: Allen and Unwin, 1961.

Jacobs, J. *The Economy of Cities*, New York: Random House, 1969.

Johnson, C. E. (ed.) *Social and Natural Biology: Selections from Contemporary Classics*, Princeton, N.J.: Van Nostrand, 1968.

Jungk, R., and Galtung, J. (eds.) *Mankind 2000*, New York: Allen and Unwin, 1969.

Lynn, R. "Genetic implications of the brain drain," *New Scientist*, **41**: 622–25, 1969.

Michael, D. N. *The Unprepared Society*, New York: Basic Books, 1968.

Mishan, E. *Technology and Growth: The Price We Pay*, New York: Praeger, 1970.

Myrolal, G. *The Challenge of World Poverty*, New York: Pantheon, 1970.

Nussenveig, H. M. "Migration of scientists from Latin America," *Science*, **165**: 1328–32, 1969.

Price, W. J., and Bass, L. W. "Scientific research and the innovative process," *Science*, **164**: 802–6, 1969.

Revelle, R. "The ghost at the feast," *Science*, **186**: 589, 1974.

Rosenfeld, A. *The Second Genesis: The Coming Control of Life*, Englewood Cliffs, N.J.: Prentice-Hall, 1969.

Skinner, B. F. *Beyond Freedom and Dignity*, New York: Knopf, 1971.

Snow, C. P. *The Two Cultures and the Scientific Revolution*, 2nd ed., New York: Cambridge University Press, 1963.

Taylor, G. R. *The Biological Time Bomb*, Cleveland: World Publishing, 1968.

Waterman, T. H. "Systems theory and biology—view of a biologist," in M. D. Mesarovic (ed.), *Systems Theory and Biology*, New York: Springer-Verlag, 1968, p. 4.

Woolridge, D. E. *Mechanical Man: The Physical Basis of Intelligent Life*, New York: McGraw-Hill, 1968.

Further Readings

Allen, G. E. "Science and society in the eugenic thought of H. J. Muller," *Bioscience*, **20**: 346–53, 1970.

Bronowski, J. *The Identity of Man*, New York: Natural History Press, 1965.

Brown, H., and Hutchings, E. (eds.) *Are Our Descendents Doomed?* New York: Viking, 1972.

Brown, L. R. *World Without Borders*, New York: Random House, 1972.

Cloud, P. E. "Realities of mineral distribution," *Texas Quarterly*, **11**: 103–26, 1968.

Dubos, R. *Man Adapting*, New Haven: Yale University Press, 1965.

Dubos, R. *Man, Medicine, and Environment*, New York: New American Library, 1968.

Ehrlich, P. R., and Ehrlich, A. H. *Population, Resources, Environment: Issues in Human Ecology,* San Francisco: Freeman, 1970.

Eiolart, T. "Putting inspiration on tap," *New Scientist,* **58:** 99–102, 1973.

Enke, S., and Zind, R. G. "Effects of fewer births on average income," *J. Biol. Sci.,* **1:** 41–55, 1969.

Falk, R. A. *This Endangered Planet—Prospects and Proposals for Human Survival,* New York: Random House, 1971.

Fermi, L. *Illustrious Immigrants,* Chicago: University of Chicago Press, 1965.

Gabor, D. *The Mature Society,* New York: Praeger, 1972.

Kahn, H., and Wiener, A. J. *The Year 2000: A Framework for Speculation on the Next Thirty-three Years,* New York and London: Macmillan, 1967.

King-Hele, D. *The End of the Twentieth Century,* New York: Macmillan, 1970.

Kyllonen, R. L. "Crime rate vs. population density in United States cities: a model," *Yearbook of the Society for General Systems Research,* **12:** 137–45, 1967.

Lerner, M. I. *Heredity, Evolution, and Society,* San Francisco: Freeman, 1968.

Leshner, A., and Candland, D. "The hormonal basis of aggression," *New Scientist,* **57**(829): 126–28, 1973.

Lincoln, G. A. "Energy conservation," *Science,* **180:** 155–62, 1973.

Parker, E. B. "Social information," *Educom,* **7**(4): 12–14, 1972.

Peterson, M. L. "The space available," *Environment,* **12**(2): 19, 1970.

Platt, J. "What we must do," *Science,* **166:** 1115–21, 1969.

Proshansky, H. M. "Aversive stimuli" review of D. C. Glass, and J. E. Singer, *Urban Stress,* New York: Academic Press, 1972, *Science,* **178:** 1275A–B, 1972.

Shepard, P. "Introduction: ecology and man—a viewpoint," in P. Shepard and D. McKinley (eds.), *The Subversive Science,* Boston: Houghton Mifflin, 1969, pp. 1–10.

Szuprowicz, B. O. "Computers in Mao's China," *New Scientist,* **57:** 598–600, 1973.

Wager, W. W. *Building the City of Man: Outlines of a World Civilization,* New York: Grossman, 1971.

White, L. "The historic roots of our ecologic crisis," *Science,* **155:** 1203–7, 1967.

GLOSSARY

Abductors Muscles that move the upper or lower limbs away from the line of the body; see *Adductors,* below.

Acetabulum Cup-shaped depression on the pelvis into which the head of the femur fits; see Appendix II.

Adductors Muscles that oppose the abductor muscles; see Figure II-2, Appendix II.

Age-specific birth rate The number of *female* live births occurring per 1,000 females in a given 5-year cohort (in the present U.S. population, ranging from a low of about 3 [40–44 cohort] to a high of about 90 [20–24 cohort]).

Agglutination Clumping of cells normally dispersed individually.

Aldehyde Organic compound with a terminal carbonyl group $(C = O)$ on the carbon chain.

Allele One of the two or more forms of a gene.

Allopatric Having a different dispersal area.

Amino acid An organic carbon chain or ring compound with both *amine*

(NH²) and *carboxyl* (COOH) groups, bonding with other amino acids to form proteins.

Aneuploid Karyotypical variant having more or less than the normal number of chromosomes for one or several homologous sets.

Anthropocentric Considering the world only from a human standpoint.

Anthropoid (noun or adjective) Belonging to a suborder of Primates containing monkeys, apes, and men.

Antibodies Proteins in mammalian blood that react with specific antigens and neutralize them.

Antigen An organic substance introduced into an animal body and generally toxic to it.

Arboreal Occupying a tree habitat.

Assortative mating Pair bonding between individuals of largely similar phenotypes.

Autocleansing Self-cleaning by internalized processes.

Autosome Chromosome not directly involved in sex determination.

Autotoxic Toxic to the life processes that produced it.

Autotroph (adjective, autotrophic) An organism that synthesizes its food from inorganic substances.

Autoxidation Self-oxidation, without any external interaction.

Auxin One of a universal group of growth-promoting plant hormones.

Biome A regional category of related ecosystems, e.g., tundra, grassland, rain forest.

Biosphere The totality of the world populations, and the materials and factors with which they interact.

Biota The totality of organisms in one place or time.

Bipedal Locomotion solely by means of the two hind limbs.

BP Before present. For the purpose of dating an artifact or a fossil, *present* is taken as the year 1950 AD.

Brachiation Arboreal locomotion using the upper limbs only, as in gibbons.

Carcinogenic Cancer-inducing.

Cerebellum The portion of the brain controlling muscular coordination.

Cerebrum The portion of the brain in which ideas are coordinated.

Chaparral Vegetation dominated by small-leaved evergreen shrubs, characterizing the Pacific Southwest.

Clavicle The collar bone, controlling articulation between the scapula and the forelimb.

Cline A directional change in gene frequencies in a population along a gradient of continuous variation in some abiotic factor.

Coitus Sexual intercourse resulting in the deposition of sperm in the vagina.

Coprolite Fossilized feces.

Crude birth rate The mean number of live births of both sexes occurring in a year per 1,000 persons in a given population as assessed at midyear (about 17 in the U.S.).

Cultigen A cultivated plant species.

Cultivar An individual variety of a cultivated plant species, often identified by a common name.

Cusp A projection above the upper surface of molar or premolar teeth in primates.

Deme A local, freely interbreeding population.

Dentition The number, form, and arrangement of teeth in an animal.

Dimorphic Having two forms.

Diversification The range of variation in characters exhibited within a species, community, or ecological niche.

Dominance (social) The acceptance of a subordinate social role by all individuals of an animal population except the dominant individual or clique.

Dominant In a bioenergetic sense, the particular species of plant, animal, or microbe at each trophic level through which the major portion of energy transfer in a given ecosystem occurs.

Down's syndrome Trisomy 21; three homologous chromosomes instead of two in set number 21 in a human karyotype, resulting in mental and physical deformity; mongolism.

Ecological niche The totality of biotic and abiotic factors to which a given species is exposed.

Electrophoresis A chemical technique that separates compounds according to their different rates of migration in an electric field.

Epoch A major division of geological time, less than a *period* and greater than an *age*.

Era The first major division of geological time, greater than a *period*.

Estrus The stage of the reproductive cycle in female primates when ovulation, release of one or more eggs from the ovary, occurs.

Eutrophic Adjective applied to a body of water rich in mineral nutrients.

Exogamy Outbreeding; mating that involves no intersibling or parent-offspring crosses.

Extractive efficiency A measurement used in anthropology to quantify the ability of a social group to obtain food from a given ecosystem.

Fallopian tube Paired tube carrying eggs from the ovary to the uterus.

Fecundity As a demographic term, the potential level of child bearing, measured in total live births per female.

Femur The principal thigh bone.

Fertility As a demographic term, the achieved level of child bearing measured in total live births of both sexes per female.

Fertility rate See *General fertility rate, Total fertility rate, Age-specific birth rate, Crude birth rate,* and *Reproduction rate.*

Gene A portion of a DNA molecule in a chromosome that determines one or more characters.

General fertility rate The total of live births of both sexes occurring per 1,000 females between ages 15 and 44 during one year (about 75 in the U.S.).

Genotype The genetic inheritance of an organism as represented by the particular assemblage of genes present in its nuclei.

Geophytic Referring to plants that perennate by means of underground food shortage organs, e.g., tubers, rhizomes, corms.

Gestation The pregnancy period, beginning with implantation and ending with parturition.

Gluten A protein that has the property of increasing cohesiveness in wheat dough.

Hafting The attachment of a previously exclusively hand-held tool or weapon to a wooden shaft.

Hemoglobin The reddish protein pigment in blood cells of higher animals that is involved in oxygen transport.

Heterotroph (adjective, heterotrophic) An organism that obtains its organic food from other organisms.

Heterozygote (adjective, heterozygous) An organism whose genotype has different alleles at the same locus on each homologous chromosome pair. The character for which the organism is heterozygous is usually stated or inferred.

Homeotherm Endothermic, regulating body heat essentially by *physiological* mechanisms.

Hominid The "man-ape" group embracing all forms included in the genera *Ramapithecus, Paranthropus, Homo,* and synonymous taxa.

Homozygote (adjective, homozygous) An organism whose genotype has identical alleles on each homologous chromosome pair. The character for which the organism is homozygous is usually stated or inferred.

Implantation Inserting of the fertilized mammalian ovum in the uterus lining.

Infancy In a mammal commonly taken to mean the period between weaning and the attainment of sexual maturity.

Insectivora A mammalian order often placed with the Primates, containing small insect-eaters like shrews and hedgehogs.

Karyotype The chromosome set characterizing the cells of a particular species.

Keratin A fibrous protein that may be incorporated in epidermal tissues.

Lipid Any kind of fat or wax compound.

Locus The position on a chromosome at which a particular gene is located.

Mandible The lower jaw.

Marsupial Primitive subclass of mammals in which the young complete their development in a pouch provided with mammae.

Melanin A dark pigment present in the skin of many animals.

Mesotrophic Adjective applied to a body of water containing an average amount of mineral nutrients.

Mongolism See Down's syndrome.

Mutualism An association or symbiosis between two or more organisms of different species that appears mutually advantageous.

Niche See Ecological niche.

Nubile Applied to the potentially fertile stage of a woman's life, often taken as age 15–45 years.

Nuchal Referring to the muscles at the back of the neck that hold back the head; see Figure II-1.

Oligotrophic Adjective applied to a body of water poor in mineral nutrients.

Orogeny Mountain chain formation and uplift.

Parturition Childbirth.

Peptide An organic compound formed from two or more amino acids.

Period A major division of geological time, less than an *era* and greater than an *epoch*.

Phenotype Character expression in an organism resulting from the interaction between its genotype and the environment during development.

Photolysis Chemical reaction promoted by light.

Photosynthesis The process of conversion of light energy to chemical energy in green plants by synthesis of various carbon compounds from atmospheric carbon dioxide, releasing oxygen.

Poikilotherm (adjective, poikilothermous) An exothermic animal, regulating body heat essentially by *behavioral* mechanisms.

Polygenic Several genes acting in a complementary manner on the same character.

Polymorphic Having two or more forms of expression of the same character.

Polyploid Organism with more than two sets of chromosomes in its cells.

Pongid (noun or adjective) Fossil or living ape; a tailless anthropoid with canines projecting beyond the general surface of the teeth.

Prosimian Primate at a level of organization below the monkey grade; living forms are usually taken to include tree shrews, tarsiers, lorises, and lemurs.

Reproduction rate *Gross reproduction rate:* the total mean number of *female* live births occurring per female age 15–44 in a given population; *net reproduction rate:* the total mean number of *female* births occurring per female aged 15–44 in a given population, adjusted to allow for any mortality in this age group. A population with an NRR of 1.0 will be *stationary*—that is, will be judged to have achieved zero population growth if there has been sufficient time for it to have acquired a mature structure.

Respiration The aerobic oxidation of food substances releasing utilizable energy and carbon dioxide.

Speciation The evolutionary development of new species.

Steroid A chemical compound having a basic structure of four rings of carbon atoms; vitamin D is an example.

Stratosphere Second region of the atmosphere, above the troposphere and below the mesosphere and the ionosphere.

Sympatric Having an overlapping dispersal area.

Taiga Arctic ecosystem dominated by coniferous trees.

Taxon (pl. taxa) A taxonomic group of unspecified rank.

Total fertility rate (Or simply fertility rate). The mean total number of live births occurring to each female during her lifetime (about 2.1 in the U.S.; or, when left as per 1,000 females, about 2,100).

Trait The particular phenotypic character expression produced by the presence of a given allele in the genotype.

Trisomy Three instead of two homologous chromosomes in one pair of the karyotype.

Troposphere Region of the atmosphere closest to the ground, varying in height according to the season and latitude.

Tundra Arctic ecosystem dominated by low shrubby plants or herbs.

Vas deferens Paired tube carrying sperm from the testes to the penis.

Volcanism Volcanic eruptions and activity.

APPENDIX I

Geological Dating

Until this century the age of rocks and their contained fossils was determined in most instances solely by their position relative to a base series. The study of this base complement, the science of *stratigraphy*, had no absolute reference points, except in the Quaternary Period. Two procedures, *dendrochronology* and observations on *varved clays*, provided these absolute dates, reaching back some 20,000 to 30,000 years BP. The first depends on the fact that many trees, especially temperate coniferous species, produce distinct annual growth rings in their wood. Moreover, the rings vary in size according to the particular growing season. The rings of long-dead trees can thus be matched where they overlap the ring patterns of younger specimens. This permits the backward extension of the dendrological time scale for wood incorporated in domestic articles or buildings to about 10,000 BP.

Varved clays are freshwater sediments, more especially those produced in glacial lakes. During the summer the less turbid water produces fine sediments. In the spring thaw, coarse sediments carried into the lake settle in a distinct layer. Again, the alternating layers of coarse and fine sediments provide a recognizable patterning that can be matched when one alluvial deposit is compared with another.

Such absolute dating methods as these were useful in archeological studies and in Quaternary geology, but the real breakthrough in dating came with the appreciation of the significance of the rate of decay of radioactive isotopes. Calculations based on the rate of change of radium provided the first absolute (but inaccurate) estimation of the age of the earth. It was then discerned by an American chemist, William Libby, how the radioactive isotope of carbon (C^{14}) could be used for absolute dating of the age of organic material.

Carbon-14

Libby (1952) proposed that events in the Holocence and late Pleistocene could be given an absolute date by utilizing C^{14}.

This isotope is created when cosmic rays from the sun strike atoms of nitrogen in the earth's stratosphere. C^{14} has a half-life of over 5,000 years, but its concentration in atmospheric air is only about one part in a trillion. So, although it becomes incorporated in plant tissues by photosynthesis just as do atoms of the commonest carbon isotope C^{12}, its radioactivity is not easy to record on a geiger counter, and an extremely sensitive instrument is required. Because radioactive decay commences the moment the C^{14} is incorporated by photosynthesis into a living plant, a comparison of the C^{12}/C^{14} ratio in dead organic matter with that of the atmosphere will provide an absolute estimate of its age.

After Libby's ideas on C^{14} dating had been confirmed, the 1950s and 1960s saw the technique extensively applied to material dated between 300 and 50,000 BP. Beyond these limits it is experimentally impossible to establish significant differences between the ratios of C^{12} and C^{14}. Any producer and any consumer organism, each of which must directly or indirectly receive a mixture of $C^{12/14}$, therefore leave organic carbon remains that can be dated by this method when their ages fall between these limits.

Potassium/Argon

Other radiometric techniques have been developed to cover periods between the relatively recent interval to which C^{14} dating can be applied and the extended time period of radium. For the purposes of hominid dating the most serviceable of these is potassium/argon dating.

The element potassium occurs naturally as a mixture of the isotopes K^{39}, K^{41}, and K^{40}. The last is radioactive and decays to form calcium-40 and argon-40. K^{40}, which occurs as about 0.0118 per cent of the mixture, has a half-life of 1.30×10^9 years. All organisms contain potassium, so do many rocks. The age of a fossil or rock can therefore be determined by a calculation based on the relative amounts of A^{40} and K^{40} it contains (Carr and Kulp, 1957). Igneous rocks are particularly suitable for dating by the potassium/

argon method, because all the argon gas they might previously have contained has been driven off by the heating during their molten stage. Any argon detected in them therefore has to have been formed subsequent to this; its amount is a measure of their age. All absolute dates for the Pleistocene after 50,000 BP, for the Pliocene, and for the Miocene quoted in this text have been determined by the potassium/argon method. Although this method of dating is suitable for determining the age of igneous rocks, it is not applicable directly to fossil material. To date fossils by this technique, they must lie adjacent to or between igneous deposits whose age can be determined.

Current revised estimates of the geological time scale during the later stages of primate evolution are given in the table on page 541. This differs appreciably from similar tables prepared even 10 years ago, especially as regards the Pliocene-Pleistocene boundary. Potassium/argon dating is gradually providing more established dates for this time, just as radiocarbon dating has permitted a much more accurate determination of events in the late Pleistocene and early Holocene.

The Protein Clock

Radiocarbon dating and the potassium/argon technique were enormously valuable in providing absolute dates for much archeological material; they still, however, left a major gap substantially uncovered, the time span from approximately 40,000 to 500,000 BP. This period is critical particularly for the developing studies of Neanderthals and their relationships with *Homo erectus* and modern *H. sapiens*. Two different techniques quite distinct from radiometric methods have now become available for absolute dating within this time gap, one for biological material, the other for geological substances.

The protein clock for dating once-living material utilizes the phenomenon known to molecular biologists as *racemization*. All amino acids that form the protein molecules of living structures are levulorotary, "left-handed" in popular language. On the death of an organism the structure of its amino acids begins to realign so as to become dextrorotary, or "right-handed." This change is the process to which the term *racemization* is applied. It appears to take place at a rate that gives, for some amino acids at least, a "half-life" of 100,000 years. This "protein clock" thus comfortably spans the 40,000- to 500,000-year gap in present radiometric techniques.

The technique as described by J. L. Bada requires the separation by a complicated series of biochemical steps first of the amino acids from a given archeological material, then of the dextro- and levulorotary portions of each amino acid. By calibration against an amino acid whose rate of racemization is well known, it is possible from the ratio of the two forms to determine the age of the sample.

This analysis can be carried out by an automatic amino-acid analyzer using only a very few grams of the sample material. The commonest amino-acid selected in the technique so far has been aspartic acid. Because the rate of racemization of amino acids is temperature-dependent, it is necessary to calibrate the process for a given locality and make the assumption that the prevailing temperatures remained very similar throughout the time period under consideration. The most convenient way to achieve this temperature calibration is to work on a cluster of sites where variously aged bones that have already been carbon-dated are available for further analysis.

J. L. Bada and R. Protsch have described the dating of various Olduvai Gorge fossil bones by this technique. By far the most dramatic dating by the technique, however, has been the very recent report by Bada, Schroeder, and Carter on the age of several Paleo-Indian skeletons from the American Pacific Southwest. These date to at least 50,000 years BP, making them approximately twice as old as any human remains from the New World previously examined. The implications of these new datings on the theories developed earlier in this text regarding the human occupation of the Western Hemisphere are obviously profound, but it is too early to be specifically defining the extent to which we must revise our ideas on this topic.

Thermoluminescence

The second new dating method, which unlike the protein clock is applicable only to inanimate material, depends on a phenomenon known as *thermoluminescence*. This is a physical phenomenon entailing the production of light by an object because of changes in its atomic structure when it is heated. Certain materials in these inert substances are affected by heating to the extent that the electrons in their constituent atoms unite with the protons of their nuclei, with the release of light energy. As the material cools, the electrons tend to be lost from the nuclei and to resume their original positions in the atom.

To test when a stone was last heated, it is thus only necessary to measure the amount of light released when it is reheated. For any *Homo* population utilizing fire and so building a hearth from stones, it is possible to date the time that has elapsed since the hearth was last used.

Other laboratory techniques are rapidly developing that will inform the archeologist not only of the absolute date of his once-living and inanimate finds, but also as to such features as the extent of activity of animal remains and the prevailing weather conditions.

References

Bada, J. L., and Protsch, R. "Racemization reaction of aspartic acid and its use in dating fossil bones." *Proceedings of the National Academy of Sciences, U.S.A.*, **70**: 1331–34, 1973.

Bada, J. L., Schroeder, R. A., and Carter, G. F. "New evidence for the antiquity of man in North America deduced from aspartic acid racemization," *Science*, **184**: 791–93, 1974.

Carr, D. R., and Kulp, J. L. "Potassium-argon method of geochronometry," *Bull. Geol. Soc. Amer.*, **68**: 763, 1957.

Libby, W. F. *Radiocarbon Dating*, Chicago: University of Chicago Press, 1952.

Oakley, K. P. *Frameworks for Dating Fossil Man*, Chicago: Aldine, 1964.

Scheme Showing the Approximate Currently Accepted Dating and Subdivision of the Portion of Geological Time that Covers Anthropoid Evolution

Epoch	Time of commencement in millions of years BP	Period	Era
Holocene (Recent)	(10,000–30,000 years BP)	Quaternary	
Pleistocene	4		
Pliocene	12		Cenozoic
Miocene	25		
Oligocene	35	Tertiary	
Eocene	55		
Paleocene	60		
	120	Cretaceous	
			Mesozoic

Approximate Time Correlations of the Quaternary Period

The absolute dates on the left margin can now be determined by the several techniques described in this appendix. The other three chronologies in the table are comparative and often extremely variable as regards absolute dating. The glacial periods of the Swiss Alps, the first reference base to be determined, do not completely coincide with episodes elsewhere; the four geological subdivisions of the Pleistocene are arranged each to contain one of the last four European alpine glaciations. Lithic industries have profound cultural significance but are of virtually no account chronologically by themselves.

Thousands of years BP	Alpine glacial periods	Lithic industry	Geological time
10–30	Postglacial	Neolithic Mesolithic	Holocene
		Upper paleolithic	
[35]	Würm		Upper Pleistocene
75		Middle paleolithic	
	Interglaical		
100	Riss		Middle
200	Interglacial		Pleistocene

Approximate Time Correlations of the Quaternary Period (cont.)

Thousands of years BP	Alpine glacial periods	Lithic industry	Geological time
275	Mindel		
		Lower	Lower Pleistocene
	Interglacial		
500	Günz		
		Paleolithic	
	Interglacial		Basal
	Donau and possible earlier glaciations		Pleistocene
First observed glaciation			
	Villafranchian		
4,500			

APPENDIX II

Hominid Models

Throughout the account of human evolution in this text attention has been drawn to the still controversial nature of many of the theories presented. By modern scientific standards, supporting material evidence is too often fragmentary, statistically significant measurement minimal, and speculation free.

In recent years, however, electronic data-processing techniques have made possible comparisons between extensive series of detailed measurements, which has greatly stimulated quantitative exercises in anthropology and paleontology. Resolution of differences of opinion has been raised from the level of authoritarian pronouncements to verifiable statements of statistical validities. These techniques made possible, for example, positive allocation of the Steinheim skull to the *Homo sapiens* grade. Armed with a radio-metrically determined date of approximately 250,000 BP, an anthropologist can state with confidence that at least one individual with cranial characteristics falling within the range of modern man was present in Europe a quarter of a million years ago. Chances are there were many more, as further finds may confirm. We need to have similar statistics about these others to provide *population* data. Only when we have physical information as to population ranges and means, and cultural information as to associated remains and artifacts, can we begin to be sure that further chance finds will not completely upset our first conjectures.

Among the most speculative of these conjectures is the kind of flesh that once clothed the bare bones of fossil finds. This is still largely decided from experience, skill, and intuition. Figures II-1 and II-2 illustrate the major musculature of modern man, which medical anatomists know in very great detail. From the size and position of the muscle attachment areas on the bones, they can estimate the size and position of the muscles and the relative movement capacities of the anatomical parts they controlled.

It is therefore possible for an anatomist to clothe the cast of a skull or skeleton with clay muscles and to surround them with a simulated skin. The various body parts can be related to one another on a similar basis to provide a postured form. Intelligent guesses can be made, for example, that the skin cover of all tropical hominids was heavily pigmented. Because only short kinky head hair is found in modern tropical African aborigines, African hominid models can be given kinky hair, which, like the skin, was just as surely black as the eyes were dark. Other traits at least partially environmentally determined can also be inserted, such as wide flared nostrils and a squat nose on a tropical rain forest hominid.

FIGURE II-I. *The arrangement of muscles in the contemporary human head and neck.* An intimate knowledge of this musculature system enables medical anatomists to build up a reconstruction of the system associated with particular fossil skulls. It is from such skillfully prepared reconstructions that authenticated hominid models and illustrations can be prepared.

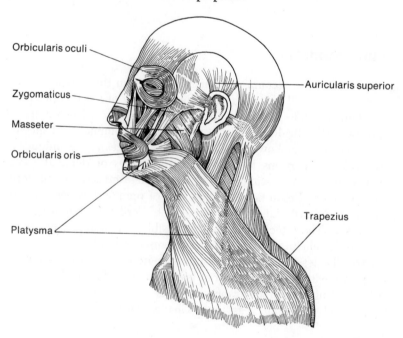

Thus although hominid models prepared even three decades ago now appear to us incredibly naive, contemporary models, many of which are featured in this text, have a firmer basis for both their scientific and their more conjectural elements. Some conjectural elements, however, we shall never be able to eliminate completely, whatever our technical advances.

FIGURE II-2. *The arrangement of muscles in the contemporary human pelvic area.* Drawn from behind. The size and position of these muscles can be assessed by medical anatomists working with fossil hominid material, as in the case of the head and neck muscles illustrated in Figure II-1. This provides information not only about morphology but also about posture and gait.

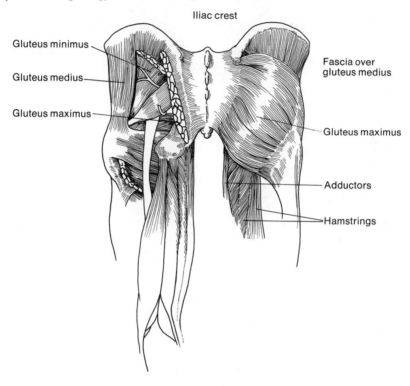

Iliac crest

Gluteus minimus

Gluteus medius

Gluteus maximus

Fascia over gluteus medius

Gluteus maximus

Adductors

Hamstrings

ACKNOWLEDGMENTS

Chapter 1

Figures

FIGURE 1-1. After G. M. Van Dyne, Oak Ridge National Laboratory Report 3957, p. 17, 1968, reproduced with permission.

Chapter 2

Figures

FIGURE 2-1. Reproduced by permission of the published from C. E. P. Brooks in *The Compendium of Meteorology*, Boston, Mass.: American Meteorology Society, p. 1017.

FIGURE 2-5. Based on a drawing in Eimerl and DeVore, 1965.

FIGURE 2-10. From "The Common Langur of North India" by Phyllis Jay, from *Primate Behavior: Field Studies of Monkeys and Apes,* edited by Irven DeVore. Copyright © 1965 by Holt, Rinehart and Winston, Inc. Reprinted by permission of Holt, Rinehart and Winston, Inc.

FIGURE 2-11. Based partly on a drawing by I. DeVore, 1965.

FIGURE 2-12. Based partly on a drawing in Eimerl and DeVore, 1965.

FIGURE 2-13. Based partly on a drawing in Eimerl and DeVore, 1965.

FIGURE 2-14. Based partly on a drawing in Eimerl and DeVore, 1965.

FIGURE 2-16. Reproduced with the permission of the authors and publisher from V. M. Sarich and A. C. Wilson, *Science,* **158:** 1201, 1967. Copyright 1967 by the American Association for the Advancement of Science.

Chapter 3

Figures

FIGURE 3-5. Based on Matternes, 1965.

FIGURE 3-6. Based on Matternes, 1965.

FIGURE 3-7. Reproduced with modifications by permission of the publisher from W. M. Schaffer, *American Naturalist,* **102:** 928, 1968.

FIGURE 3-8. Based on Kohts, 1935.

FIGURE 3-9. Modified from E. J. Clegg, *The Study of Man,* New York: American Elsevier, 1968, p. 76; reproduced with permission of the original publisher, The English Universities Press, London.

FIGURE 3-10. Based partly on Clark Howell, 1965.

FIGURE 3-14. Reproduced in part by permission of the publisher from D. J. Morton, *The Human Foot,* New York: Columbia University Press, 1964, p. 41.

FIGURE 3-15. Reprinted by permission of Quadrangle Books, Inc., from *The Antecedents of Man* by W. E. Le Gros Clark, copyright © 1959 by Edinburgh University Press, p. 216.

Chapter 4

Figures

FIGURE 4-1. After painting by Jay H. Matternes, © National Geographic Society, 1965. Reproduced with permission of the publisher.

FIGURE 4-3. Adapted with the permission of the author from J. D. Clark, *The Prehistory of Southern Africa,* London: Penguin, 1959, p. 138.

FIGURE 4-4. Adapted with the permission of the author from J. D. Clark, *The Prehistory of Southern Africa,* London: Penguin, 1959, p. 134.

FIGURE 4-6. Published by permission of the Trustees of the British Museum (Natural History).

FIGURE 4-7. Published by permission of the Trustees of the British Museum (Natural History).

FIGURE 4-16. Based on a painting by Z. Burian. Reproduced with permission of the artist and *Artia*, Foreign Trade Corporation for Import and Export of Cultural Commodities, Prague, Czechoslovakia.

FIGURE 4-17. Based on a painting by Z. Burian.

FIGURE 4-20. Reproduced with the permission of the author and publisher from C. V. Haynes, Jr., *Science*, **166:** 710, 711, 1969. Copyright 1969 by the American Association for the Advancement of Science.

FIGURE 4-23. Reproduced by permission of the publisher from P. S. Martin and H. E. Wright (eds.), *Pleistocene Extinctions*, New Haven: Yale University Press, 1967.

FIGURE 4-24. From a painting by Jay H. Matternes, reproduced by permission of the National Geographic Society.

FIGURE 4-25. Reproduced by permission of the publisher from P. S. Martin and H. E. Wright (eds), *Pleistocene Extinctions*, New Haven: Yale University Press, 1967, p. 95.

FIGURE 4-26. Reproduced with the permission of the author from J. D. Clark, *The Prehistory of Southern Africa*, London: Penguin, 1959, p. 108.

FIGURE 4-27. Reproduced by permission of the author and publisher from C. A. Fleming, *Notornis*, **10:** 114, 115, 1962.

Tables

TABLE 4-1. Reprinted by the permission of the publisher from P. S. Martin, *Nature*, **212:** 340, 1966.

TABLE 4-2. Reprinted by permission of the publisher from P. S. Martin *Nature*, **212:** 339, 1966.

Chapter 5

Figures

FIGURE 5-1. Reproduced with the permission of the author and publisher from J. B. Birdsell, *American Naturalist*, 87: 179, 1953.

FIGURE 5-2. Reproduced with the permission of the authors and publisher from M. D. Coe and K. V. Flannery, *Science*, 143: 652, 1964. Copyright 1964 by the American Association for the Advancement of Science.

FIGURE 5-3. Reproduced with the permission of the authors and publisher from M. D. Coe and K. V. Flannery, *Science*, 143: 652, 1964. Copyright 1964 by the American Association for the Advancement of Science.

FIGURE 5-6. From James Mellaart, "A Neolithic City in Turkey," *Scientific American*, 210(4): 95, 1964. Copyright © 1964 by Scientific American, Inc. All rights reserved.

FIGURE 5-10. From James Mellaart, "A Neolithic City in Turkey," *Scientific American*, 210(4): 100, 1964. Copyright © 1964 by Scientific American, Inc. All rights reserved.

Chapter 6

Figures

FIGURE 6-8. After N. T. Newman, *American Anthropologist,* **55:** 314, 1953; reproduced with the permission of the publisher.

FIGURE 6-11. Reproduced with modification by permission of the publisher from C. L. Brace and M. F. Ashley Montagu, *Man's Evolution,* New York: Macmillan, 1965, p. 272.

FIGURE 6-13. Reproduced by permission of the publisher from C. L. Brace and M. F. Ashley Montagu, *Man's Evolution,* New York: Macmillan, 1965, p. 300.

FIGURE 6-14. Reproduced by permission of the publisher from C. L. Brace and M. F. Ashley Montagu, *Man's Evolution,* New York: Macmillan, 1965, p. 396.

FIGURE 6-17. Reproduced by permission of the publisher from J. Buettner-Janusch, *Origins of Man,* New York: Wiley, 1966, p. 542.

FIGURE 6-18. Reproduced by permission of the publisher from J. Buttner-Janusch, *Origins of Man,* New York: Wiley, 1966, p. 546.

FIGURE 6-19. Reproduced with the permission of the author and publisher from M. G. Bulmer, *Annals of Human Genetics,* **23:** 455, 1958–59. Copyright 1958, Cambridge University Press, New York.

Tables

TABLE 6-4. Reprinted by permission of the publisher from S. Schreider, *Evolution,* **18:** 4, 1964.

TABLE 6-6. Reprinted with permission of the publisher after G. A. Harrison, J. S. Weiner, J. M. Tanner, and N. A. Barnicot, *Human Biology,* New York and Oxford: The Clarendon Press, 1964, p. 459.

TABLE 6-7. Reprinted by permission of the publisher from A. E. Mourant, *The Distribution of the Human Blood Groups,* Oxford: Blackwell Scientific Publications, 1954, pp. 383, 393, 394.

TABLE 6-8. Reprinted with permission of the publisher from G. A. Harrison, J. S. Weiner, J. M. Tanner, and N. A. Barnicot, *Human Biology,* New York and Oxford: The Clarendon Press, 1964, p. 275.

TABLE 6-9. Reprinted by permission of the publisher from M. G. Bulmer, *Annals of Human Genetics,* **24:** 121–25, 1960. Copyright 1960, Cambridge University Press, New York.

Chapter 7

Figures

FIGURE 7-7. From Kingsley Davis, "Population," *Scientific American,* **209**(3): 68, 70, 1963. Copyright © 1963 by Scientific American, Inc. All rights reserved.

FIGURE 7-9. Reproduced with the permission of the publisher from H. F. Dorn, *Science,* **135:** 288, 1962. Copyright 1962 by the American Association for the Advancement of Science.

FIGURE 7-10. After H. Brown, 1954. From *The Challenge of Man's Future* by Harrison Brown, copyright 1954 by Harrison Brown. Reprinted by permission of The Viking Press, Inc.

Tables

TABLE 7-1. Partially from H. F. Dorn, *Science,* **135:** 283–90, 1962, by permission of the American Association for the Advancement of Science.

TABLE 7-5. From 1973 World Population Data Sheet, Population Reference Bureau, Washington, D.C. Reprinted by permission.

TABLE 7-6. From 1973 World Population Data Sheet, Population Reference Bureau, Washington, D.C. Reprinted by permission.

TABLE 7-7. From 1973 World Population Data Sheet, Population Reference Bureau, Washington, D.C. Reprinted by permission.

TABLE 7-9. Partially from Makeshwari, *Science and Culture,* **32:** 104–14, 1966.

TABLE 7-11. From 1973 World Population Data Sheet, Population Reference Bureau, Washington, D.C. Reprinted by permission.

TABLE 7-12. Reprinted by permission of the publisher from H. F. Dorn, *Science,* **135:** 283–90, 1962, and the Population Reference Bureau. Copyright 1962 by the American Association for the Advancement of Science.

Chapter 8

Figures

FIGURE 8-1. Reproduced by permission of the publisher from L. R. Brown, *Science,* **158:** 606, 1967. Copyright 1967 by the American Association for the Advancement of Science.

FIGURE 8-2. From Theodosius Dobzhansky, "The Present Evolution of Man," *Scientific American,* **203:** 208, 1960. Copyright © 1960 by Scientific American, Inc. All rights reserved.

FIGURE 8-5. Courtesy J. B. Pick from unpublished data.

FIGURE 8-6. Courtesy J. B. Pick from unpublished data.

FIGURE 8-7. Courtesy J. B. Pick from unpublished data.

FIGURE 8-8. Courtesy R. Wong from unpublished data.

FIGURE 8-9. Reproduced with the permission of the author and publisher from H. Frederiksen, *Science,* **166:** 839, 1969. Copyright 1969 by the American Association for the Advancement of Science.

FIGURE 8-11. Courtesy J. B. Pick from unpublished data.

FIGURE 8-12. Courtesy J. B. Pick from unpublished data.

Tables

TABLE 8-1. From Maheshwari, *Science and Culture,* **32:** 104–14, 1966.

TABLE 8-2. Figures extracted from *Famine 1975! America's Decision: Who*

TABLE 8-3. Reprinted by permission of the publisher from T. Kristensen, *International Journal of Agrarian Affairs,* **5:** 139, 1967.

TABLE 8-6. From Don Paarlberg, "Food for More People and Better Nutrition," in Clifford M. Hardin (ed.), *Overcoming World Hunger.* © 1969 by The American Assembly, Columbia University. By permission of Prentice-Hall, Inc.

TABLE 8-7. From *Famine 1975! America's Decision: Who Will Survive?* Copyright © 1967 by William and Paul Paddock, p. 16. Reprinted by permission of Little, Brown and Co., Boston, Mass. Figures from the 1970 World Population Data Sheet by permission of the Population Reference Bureau.

TABLE 8-8. From Maheshwari, *Science and Culture,* **32:** 104–14, 1966.

Chapter 9

Figures

FIGURE 9-1. Reprinted by permission of the publisher from E. Van de Walle, *Daedalus,* **97**(2): 492, Journal of the American Academy of Arts and Sciences, Cambridge, Mass.

FIGURE 9-6. Courtesy J. B. Pick from unpublished data.

FIGURE 9-7. Courtesy J. B. Pick from unpublished data.

FIGURE 9-8. Courtesy J. B. Pick from unpublished data.

FIGURE 9-9. Courtesy J. B. Pick from unpublished data.

FIGURE 9-10. Courtesy J. B. Pick from unpublished data.

FIGURE 9-11. Courtesy J. B. Pick from unpublished data.

Tables

TABLE 9-3. Figures summarize data extracted from Kourides, 1967, and Pincus, 1966.

TABLE 9-4. Reprinted by permission of the publishers from Theodosius Dobzhansky, *Mankind Evolving,* New Haven: Yale University Press, 1967, p. 33.

TABLE 9-6. Reprinted by permission of the publisher and author from J. Blake Davis, *Science,* **164:** 524, 1969. Copyright 1969 by the American Association for the Advancement of Science.

TABLE 9-7. Reprinted by permission of the publisher and author from J. Blake Davis, *Science,* **164:** 526, 1969. Copyright 1969 by the American Association for the Advancement of Science.

TABLE 9-8. Reprinted by permission of the publisher and author from O. Harkavy, F. S. Jaffe, and S. M. Wishik, *Science,* **165:** 372, 1969. Copyright 1969 by the American Association for the Advancement of Science.

Chapter 10

Figures

FIGURE 10-2. Reprinted by permission of the publisher from J. Gentilli, *A Geography of Climate*, Perth: University of Western Australia Textbooks Board, 1952, p. 38.

FIGURE 10-3. After A. J. Haagen-Smit, *Scientia*, 103(673, 674): 21, 35, 1968; reproduced with the permission of the publisher.

FIGURE 10-4. Reproduced by permission of the publisher from F. N. Frenkiel, *Scientific Monthly*, 82: 198, 1956.

FIGURE 10-5. Reproduced by permission of the publisher from J. Spar, *Weatherwise*, 7: 149, 1954.

FIGURE 10-8. Reproduced by permission of the publisher from L. D. Stamp, *The Geography of Life and Death*, London: Fontana Paperbacks, 1964, pp. 118 and 123, and reprinted from L. Dudley Stamp, *The Geography of Life and Death*. © L. Dudley Stamp, 1964. Used by permission of Cornell University Press.

FIGURE 10-10. Modified from Annual Report 1968, California Air Resources Board, reproduced with permission.

FIGURE 10-11. Prepared from 1968 Annual Report on *Air Pollution Control in California*, Air Resources Board, 1969, p. 19, reproduced with permission.

Tables

TABLE 10-1. Reprinted by permission of the publisher from J. L. Bregman and S. Lenormand, *The Pollution Paradox*, New York: Spartan Books, 1966, p. 54.

TABLE 10-3. After Haagen-Smit, 1956; reproduced with the permission of the publisher.

TABLE 10-4. After Haagen-Smit, 1968; reproduced with the permission of the publisher.

TABLE 10-5. Reprinted by permission of the publisher from E. Sawicki, W. C. Elbert, T. R. Hauser, F. T. Fox, and T. W. Stanley, *American Industrial Hygiene Association Journal*, 21: 447, 1969.

TABLE 10-6. From R. Scott Russell (ed.), *Radioactivity and Human Diet*, Oxford: Pergamon Press, 1966, p. 15.

TABLE 10-7. From R. Scott Russell (ed.), *Radioactivity and Human Diet*, Oxford: Pergamon Press, 1966, p. 20.

Chapter 11

Figures

FIGURE 11-1. Based on Sawyer, 1966.

FIGURE 11-2. Reproduced with the permission of the author and publisher from A. D. Hasler, *Bioscience*, 19(5): 427, 1969.

Figure 11-3. Based on Hasler, 1947.

Figure 11-4. From C. C. Anderson, Trans. 1960 Seminar, R. A. Taft San. Eng. Center, U.S.P.H.S., Cincinnati, Ohio, p. 63, 1960; reproduced with permission of the publisher.

Figure 11-5. Based on Sawyer, 1966.

Figure 11-6. Reproduced by permission of the publisher from A. M. Beeton, *Limnology and Oceanography*, **10:** 247, 1965.

Figure 11-7. From N. S. Baldwin and R. W. Saalfield, Great Lakes Fisheries Commission Technical Report No. 3 plus supplement, 1962, p. 37; reproduced with the permission of the publisher.

Figure 11-8. Reproduced with the permission of the publisher from J. J. Davis and R. F. Foster, *Ecology*, **39:** 531, 1958.

Tables

Table 11-2. Reprinted by permission of the publisher from A. M. Beeton, *Limnology and Oceanography*, **10:** 241, 1965.

Table 11-3. Reprinted by permission of the publisher from A. M. Beeton, *Limnology and Oceanography*, **10:** 242, 1965.

Table 11-4. Reproduced from H. J. Dunster, R. J. Garner, H. Howells, and L. F. U. Wix, *Health Physics*, **10:** 361, 1964 by permission of the Health Physics Society.

Table 11-6. From R. Scott Russell (ed.), *Radioactivity and Human Diet*, Oxford: Pergamon Press, 1966, p. 436.

Table 11-7. From G. Westoo, *Nord Hyg. Tidskr.*, 1969; reproduced with the permission of the publisher.

Chapter 12

Figures

Figure 12-1. Reproduced with the permission of the publisher from P. L. Ames, *Journal of Applied Ecology*, 3(Suppl.): 92, 1966.

Figure 12-3. From V. M. Stern et al., *Hilgardia*, **29:** 93, 1959, reproduced by permission of the publisher.

Tables

Table 12-1. Reprinted by permission of the publisher from J. Robinson, A. Richardson, A. N. Crabtree, J. C. Couldon, and R. G. Potts, *Nature*, **214:** 1308, 1967.

Table 12-2. Table prepared from figures extracted from Peterle, 1969.

Table 12-3. Reprinted by permission of the publisher from H. Egan et al., *British Medical Journal*, **2:** 68, 1965.

Table 12-4. Reprinted by permission of the publisher from H. Egan, R. Goulding, J. Roburn, and J. O'G. Tatton, *British Medical Journal*, **2:** 68, 1965.

TABLE 12-5. Reproduced with permission of the publisher from K. R. Tarrant and J. O'G. Tatton, *Nature,* **219:** 726, 1968, and J. M. Cohen and C. Pinkerton, *Advances in Chemistry,* **60:** 163–67, 1966.

TABLE 12-6. Figures compiled from U.S. Tariff Commission Pub. 295, 1969, and U.S. Department of Interior Agric. Econ. Rpt. No. 131, 1968.

TABLE 12-7. Figures from Agric. Econ. Rpt. No. 131. Reproduced with permission of the Economic Research Service, U.S. Department of Agriculture.

TABLE 12-8. Sources: P. Antommaria, M. Corn, and L. Demaio, *Science,* **150:** 1476, 1965. R. W. Risebrough, R. J. Hugget, J. J. Griffin, and E. D. Goldberg, *Science,* **159:** 1233, 1968. D. C. Abbott, R. B. Harrison, J. O'G. Tatton, and J. Thomson, *Nature,* **211:** 259, 1966.

Chapter 13

Figures

FIGURE 13-1. After J. Fisher et al., *The Red Book: Wildlife in Danger,* New York: Viking Press. © IUCN 1969.

FIGURE 13-2. After J. Fisher et al. *The Red Book: Wildlife in Danger,* New York: Viking Press. © IUCN 1969.

FIGURE 13-4. Adapted from J. Fisher et al. *The Red Book: Wildlife in Danger,* New York: Viking Press. © IUCN 1969.

FIGURE 13-5. From D. S. Fertig and V. W. Edmonds, "The Physiology of the House Mouse," *Scientific American,* **221**(4): 110, 1969. Copyright © 1969 by Scientific American, Inc. All rights reserved.

FIGURES 13-6 to 13.9. From THE LIMITS TO GROWTH by Donella H. Meadows, Dennis L. Meadows, Jørgen Randers, William Behrens, III. A Potomac Associates Book published by Universe Books, New York, 1972. Graphics by Potomac Associates.

Tables

TABLE 13-3. From J. Fisher et al., *The Red Book: Wildlife in Danger,* New York: Viking Press. © IUCN 1969.

Appendix II

Figures

FIGURE II-1. Based on J. Buettner-Janusch, 1966.

FIGURE II-2. Based on J. Buettner-Janusch, 1966.

INDEX OF NAMES

Bates, Marston, 293, 294, 327
Baumhover, A. H., *458*
Beeton, A. M., *435*
Bega, R. V., *405*
Behrman, S. J., *368*
Belden, K. H., *247*
Bell, M. A. M., *436*
Bennett, I. L., 297, 302, 303, 327
Benson, D., *405*
Benson, F. B., *405*
Berelson, B., 342, 353, 367, 368
Berger, D. D., 444, *459*
Berger, R., *149, 187, 188*
Bernard, B., *290*
Bertine, K. K., *406*
Binford, L. R., *150*
Binford, S. R., 149, *150*
Birdsell, J. B., 73, 98, *149,* 154, *187,* 250, 253, 255, 289
Birket-Smith, K., *246*
Biros, E. F., *461*
Biros, F. J., *458*
Bischoff, A. I., 445, *459*
Bishop, A., *64*
Bishop, W. W., *99*
Bitman, J., *458, 460*
Bitter, B. A., *458*
Black, Davidson, 77
Blackburn, R. D., *436*
Blake, Judith, 3, 361, *367*
Bliss, E. L., *99*
Bloom, H. F., *248*
Blum, H. F., 91, *98*
Boffey, P. M., *404*
Boggs, D. H., *404*
Bond, G., *64*
Bordaz, J., *149*
Borgstrom, George, 285, 289, 361, 363, 455
Bormann, F. H., *23, 405*
Boucher de Perthes, M., 121
Boughey, A. S., *23, 64,* 314, *435, 503, 505*
Bowen, D. H. M., *435*
Bowers, R., *527*
Boyd, W. C., 114, *149, 246, 248*

Boyko, H., *327*
Brace, C. L., *149, 150,* 248
Bradley, C. C., 474, 475, 503
Braidwood, R. J., *150,* 160, *187,* 258, 289
Brain, C. R., *98*
Branham, J. M., *435*
Bregman, J. L., *404*
Breidenbach, A. W., *404*
Bresler, J. B., 104, *149*
Breuil, H., *150*
Bridges, K. W., *436*
Brodine, V., *404, 406*
Broecker, W. S., *404*
Bronowski, J., *528*
Brooks, C. E. P., 26, *64*
Brooks, H., *527*
Brose, D. S., *149,* 200, 246
Broverman, D. M., *369,* 527
Broverman, I. K., *369,* 527
Brown, A. W. A., *460*
Brown, H., *528*
Brown, L. R., 297, 327, 328, *528*
Brown, R., *404*
Buettner-Janusch, J., *64, 246*
Buffington, J. D., *150*
Bulmer, M. G., *246*
Bumpass, L., *369*
Burdick, G. E., *460*
Burgess, E. W., 182, 183, 187
Burkhill, I. H., *188*
Bush, A., *405*
Bushland, R. C., *458*
Butler, P. A., *458*
Butzer, K. W., *149*
Byers, D. S., *188*

Cadien, J. D., *247*
Cadle, R. D., *404*
Cain, S. A., *188, 437, 503, 505*
Cairns, J., *435*
Calderone, M. S., *367*
Cambel, H., 160, *187*

Campbell, A. A., *290*
Campbell, B. G., 66, 98, 99, 150, 248
Campbell, T. H., *435*
Candland, D., *529*
Cann, J. R., *151*
Cardiff, E. A., *406*
Carpenter, C. R., 65, 99
Carr, D. R., *540*
Carr-Saunders, A. M., 332, 367
Carson, Rachel, 441
Carter, G. F., 540
Carter, L. J., *404,* 430, 435
Cartmill, Matt, 40, 65
Caspari, E., *100*
Caswell, C. A., *435*
Cavalli-Sforza, L. L., *247*
Cecil, H. C., *458*
Cerowsky, J., *505*
Chance, M., *65*
Chandler, R. F., 301, 302, 327
Chaney, Ralph, 493
Chang, *149, 151*
Changnon, S. A., Jr., *404*
Charlson, R. J., *404*
Charnell, R. L., *437*
Charter, S. P. R., *505*
Chase, R., 2, 3, 23
Chedd, G., 327, *505*
Chesher, R. H., *435*
Chevalier, Auguste, 260, 261, 289
Chiscon, J. A., *65*
Ciriacy-Wantrup, S. V., 505
Clark, Desmond, 104, 105
Clark, J. D., 99, *149, 151*
Clark, J. G. D., *151*
Clark, J. R., *437*
Clarke, R. J., *98*
Clarkson, F. E., *369, 527*
Clawson, M., *327*
Cloud, P. E., *528*
Clutterbuck, D., *504*
Coale, A. J., 328, *367*
Coan, G., *435*
Cockrill, W. R., *188*
Coe, M. D., *187*
Cohen, A., *404*

May, J. M., 310, *328*
May, R. M., 21, *23*
Mayer, D. L., *424*
Mead, M., 3, *23*
Meadows, D. H., *498–502, 504*
Meadows, D. L., *498–502*
Medawar, P., *100*
Medbury, A., *458*
Meeks, R. L., *461*
Meier, R. L., *369*
Mellaart, J. A., 161–63, 170, *188*
Menzel, D. B., *461*
Menzel, D. W., *459*
Merrill, E., *493*
Mesarovic, M. D., *528*
Messenger, P. S., *461*
Metz, W. D., *504*
Michael, D. N., *528*
Middleton, J. T., *405*
Midlo, C., 227, *247*
Mihursky, J. A., *437*
Miller, J. A., *99*
Miller, M. E., *406*
Miller, P. R., *406*
Miller, R. W., *406*
Milton, J. P., *504*
Miner, D., *459*
Mishan, E., *528*
Moffet, H., *438*
Molte, H. L., *436*
Montagu, A. M. F., *247, 248*
Monti, D. R., *406*
Moore, N. W., *461*
Moore, W. J., *65*
Mora, P. M., *290*
Morgan, K. Z., *461*
Morgan, T. R., *404*
Morris, D., 105, *150*
Morton, D. J., *100*
Morton, S., *150*
Motto, H. L., *438*
Mourant, A. E., *150, 248*
Mrak, G. M., *459*
Muckenhirn, N. A., *65*
Mulvaney, D. J., 148, *150*
Mumford, L., *189*
Munn, R. E., *406*
Myers, N., *506*
Myrolal, G., *528*

Nace, R. L., *504*
Nader, R., *472*
Napier, J., *66, 67, 100*
Napier, P. H., *67*
Nash, R. G., *461*
Neff, T., *405*
Nelson, A., *405*
Nelson, H., *150*
New, W. D., *458*
Newell, N. D., *67*
Newell, R. E., *407*
Newman, M. T., *247*
Nielson, E. S., *438*
Nisbet, A., *436*
Nisbet, K. T., *459*
Norris, K. S., 479, *504*
North, W. S., *423*
Novick, S., 429, *436*
Nuckolls, J., *504, 505*
Nussenveig, H. M., *528*

Oakley, K. P., *100, 540*
Oberle, M. W., *438*
Odum, E. P., *23*
Odum, W. E., *436*
O'Flaherty, C. A., *189*
Olcott, H. S., *461*
Organski, H. F. K., 320
Organski, K., 320
Orians, G. H., 257, *289*
Orwell, G., 313, *525*
Osmun, J. V., *461*

Paddock, P., 297, 301, 304, *328, 344, 368*
Paddock, W., 297, 301, 304, 305, *328, 344, 368*
Panofsky, H. A., *407*
Park, R. E., *187*
Parker, E. B., *529*
Parmeter, J. B., *405*
Parmeter, J. R., *406*
Parr, E. L., *368*
Parsons, J. J., *505*
Pasteur, L., *309*
Peakall, D. B., *436, 444, 459*
Pearlman, M. E., *405*
Peel, J., *290*
Perkins, D., Jr., *188*
Perry, H., *506*

Peterle, T. J., 445, *459*
Peterson, M. C., *529*
Peterson, M. R., *447–48, 460*
Petter, J. J., *65, 66*
Pichat-Bourgeois, J., *329*
Pick, James, 279, 314
Pieper, W. A., *247*
Pierce, P. E., *437*
Pilat, M. J., *404*
Pilbeam, D., 60, *66, 67, 73, 99*
Pincus, G., *368*
Pinson, E. N., *100*
Pirie, N. W., 302, 307, *328, 329*
Platt, J., 522, *529*
Player, I., *506*
Poliakoff, S. R., *369*
Postgate, J., *329*
Potts, G. R., *461*
Potts, M., *329*
Pradervand, P., *369*
Pramer, D., *461*
Premack, A. J., 87, *99*
Premack, D., 87, *99*
Price, D. O., *290*
Price, W. J., *528*
Proshansky, H. M., *529*
Protsch, R., *188, 540*

Quigg, P. W., *506*

Race, R. R., *151, 248*
Radonski, J. L., *460, 461*
Rainwater, L., *290*
Randtke, A., *459*
Rankin, R. E., *405*
Ransil, B. J., *369*
Rappaport, R. A., 181, *188, 257, 289, 294, 328*
Ratcliffe, D., 444, *459*
Rawson, D. S., *436*
Reed, C. A., *150, 189, 258, 289*
Reed, T. E., *247*
Regier, H. A., *436*
Renfrew, C., *151*
Revelle, R., *290, 328, 329, 333, 368, 525, 528*
Reynolds, V., *100*
Rhyther, J. H., *329, 436*

Ucko, P. J., *152, 189*
Uzzell, T., 60, *66*

Valentine, D. W., *436*
van de Walle, E., *368*
Van Dyne, G. M., *23*
van Foerster, H., *290*
Van Koenigswald, G. H.
R., *149*
Van Lawick-Goodall, J.,
99
van Tienhoren, A., *367*
Van Valen, L., *67*
Vavilov, N. I., *167, 172,
188*
Vayda, A. P., *150, 189*
Vincent, J., *504*
Vogel, S. R., *369, 527*

Waddington, C. H., *238,
248*
Wager, W. W., *529*
Walker, A. C., *458*
Waller, R., *505*
Walpoff, M. H., 200, *246*
Ward, R. G., *247*
Warner, W. L., *152*
Washburn, S. L., 65–67,
*99, 100, 150, 152, 247,
248*
Waterbolk, H. T., *330*
Waterman, T. H., 510,

528
Watson, P. J., *187*
Watt, K. E. F., *23*
Wayne, W. S., *407*
Weatherwax, J. R., *460*
Webb, J. W., *247*
Weidenreich, F., 77
Weiner, J. S., *149, 150,
247*
Weinstock, B., *407*
Weldon, L. W., *436*
Wells, H. G., 309
West, R. G., *100*
Westberg, K., *406*
Westhoff, C. F., *290, 369*
Westhoff, V., *506*
Westlake, W. E., *460*
Weyermuller, G. H., *407*
Wheeler, A., *436*
White, L., 529
Whittaker, R. H., *23*
Whittemore, F. C., *506*
Whyte, W. H., 478, *505*
Wiener, A. J., *529*
Wilcox, D., *505*
Williams, C. M., *461*
Williams, R., *330*
Willie, A., *407*
Willing, M. K., *369*
Wilmsen, E. N., *152*
Wilson, A. C., 57–59, *66*
Wilson, B. R., *436, 461*

Wilson, E. O., 63, *66*
Wilson, K. W., *406*
Wingate, D. B., *462*
Winter, R., *461*
Wise, W., *406*
Wishik, S. M., *368*
Wium-Andersen, S., *438*
Wolpoff, M. H., *100, 149*
Wood, A. E., *67*
Wood, L., *504, 505*
Woodwell, B. M., *23, 407,
437, 461*
Woolridge, D. E., *528*
Woolson, E. A., *461*
Wortman, S., *327*
Wright, H. E., *150*
Wright, H. E., Jr., *189*
Wright, H. F., *188*
Wrigley, E. A., *330*
Wurster, D. F., *461, 462*
Wynne-Edwards, V. C.,
35, *66*, 332, *368*

Yannacone, V. J., *506*
Young, G., *328*
Young, J. Z., *23, 152*

Zachary, D., *189*
Zeuner, F. E., *152*
Zind, R. G., *529*
Zohari, D., *188*
Zorich, T. M., *437*

INDEX OF SUBJECTS

Diet, 41–42, 53–55, 157, 253, 308–309;
  See also Cooking; Nutrition
Dingo, 147, 168, 255
Disease, 294–97, 309–11, 388–93
Displacement, speech, 87
Distribution. *See* Population
Diversification, niche, 15, 16
Diversity, 15, 16, 20, 179–80, 239
Division of labor, 108–**11**, 157–59, 180
DNA, 59, 60, 63, 64
Dodo, 144
Dogs, 147, 157, 171, 255
Domestic animals, 157, 164, 167–71, 175
Dominance, 46–51, 180
Dose–response relation, nonlinear, 18, 19
Downs syndrome, **194**
Dravidians, **120**, 204, 207
Dromedary, 171
Dryopithecines, 56–57, 60
Dryopithecus genus, 42, 43
Duck, 171, 175
Dynastic Period, Early, 177

Ears, 227
Earth
  cyclic cataclysms of, **27**
  origin of life on, 26
  See also Atmosphere; World
Ecdysones, 456
Ecological equivalents, 20
Ecological rules on temperature responses,
    209–15
Ecological succession, 14–16
Ecology, 3–**4**
  of city, 182–86
  human diversity and, 238–41
  development of human societies and,
    179–82
  in future, 517–18
Economic incentives to discourage child-
    bearing, 346–48
Economy, future, 525–26
Ecosphere, 8
Ecosystems, 2–22
  agricultural, simple, model of, **17**
  autotrophic, 9
  biotic and abiotic elements in, 8, **13**,
    14
  concept of, 3–5, 8–12
  disruption of, sudden, 18–19
  of lakes, and pollution, 410–23
  organization of, 9–11

structure of, 16–22
synthetic, 480–83
urban, **359**
  See also Urban civilization
Education, impact of, on family size and
    fertility, 344–**45**, 346
Electricity, 471–73
Elephants, 34, 70, 129, 137, 171, 175
Electrophoretic examination, 196
Elmenteita, **203**
Eltonian pyramid, **11**, 12
Emigration, 518–19; *See also* Immigra-
    tion; Migrations
Emissions, pollutant, 379–85, 388, 400–
    402
Emphysema, 390
Energy, 9, **10**, **11**, 12, 467–73
Energy differential, 180
Energy flow between various trophic
    levels, 12
Environment, patchy, 62
Environmental confrontations, 186–87
Environmental control agencies, 401–402
Enzymes, 196
Eocene, 31, 37, 541
EPA (Environmental Protection Agency),
    402
Erectus-sapiens. See Homo erectus-
    sapiens
Eskimos, 91, 145–47, **213**, **217**, **220**, **222**,
    227, **252**
Ethnic components of populations, 243,
    244, **359**, **361**, 364
Eutrophication, **410**–16
Evolution, cultural, 15–16
  ecological, future, 517–18
  ecological succession and, 15–16
  of ecosystems, 11–12
  of hominids, 25–67
  human, 61, 177–79
  of language, 81–85
  morphological, 91–94; *See also* Mor-
    phology
  of primates, 30–40
  sequence and absolute timing of, 59
  of tropical rain forest, 28–30
Exogamous mating, 107–**109**
Experimental City, 185–86
Extinction, 12, 137–45, 486; *See also*
    Overkill
Extinction times, 464

Eye color, 218
Eye fold, Mongoloid, 224–27

Facial expressions of chimpanzees, **88**
Facial form, 220
Family, 43, 250
    size of, **345**, 361–66
    See also Division of labor
Famine, 186, 293–94, 297–305
Farming villages, 160–61, 176; *See also*
        Agriculture
Fertile Crescent, 160–**61**, 176, 496
Fertility, 276, 280, 285–87, 333–35, 342–
        44, 346, **355, 356,** 361–66
Fertilizer, 301, 302, 413
Filipino, **206**
Fire(s), 15, 109–12, 254
Fish, 430, 446
Fishing, 418, **419**, 420
"Fitter" types, 20, 199, 241, 245
Flat-faced, 220
Florisbad, **203**
Flu, 295–96
Fluctuation. *See* Population, fluctuations
        in
Folsom man, **135, 138**
Food, 155–57, **165**
    world shortage of, 297–309
    See also Diet
Food additives, as pollutants, 434
Food requirements, 475–76
Food sharing, primate, 52
Food webs, 16
Forests, 29–30, **258, 259,** 260, 262
Founder principle, 132
Freon, 400
Fuel, 367–74, 382–85, 400
Fungicides, 428
Future environments, 503, 507–27
Future shock, 246

Gamblian pluvial phase, Lower, **107**
Garbage, 372; *See also* Waste
Gas, natural, 471
Gasoline, pollution from, 373, 374, 379–
        83, 385, 388, 400–402
Geese, 171, 175, 490
Genes, 196, 229, 230
Genetic continuity, 113
Genetic drift, 239
Genetics, 58, 60, 64, **109,** 113–15, 193–
        96, 219, 229–32, 238, 310

Genotypes, 208
Geographical forms of *Homo erectus,*
        112–16
Geological cataclysms, 26–28
Geological dating, 70–71, 537–42
Geological eras and evolution, 15–16, 26–
        31, 58, 70; *See also names of
        eras*
Gestation period, 33, 34
Gibbons, **31,** 34, 38, **47,** 51–52, 58, 64,
        193
Giraffe, 34
Glacial episodes, **116, 120;** *See also* Ice
        Ages
Glaciations, 71, 72, 78, 122, **128, 135,**
        385, 541, 542
Glacier National Park, Montana, **475**
Goats, 171, 175
Gorillas, **31,** 34, **36,** 38, 39, 58, **93, 94,
        96,** 193
Grains, 297–99, 426, 428; *See also* Seeds;
        Wheat
Grasping abilities, **32**
Grasslands, 40, **41,** 42, 53–55
Great Lakes, 415–23
Greenhouse effect, 383–85
Gross national product (GNP), 181
Groundnut, **165**
Groups, as term, 178
Group selection for defense, 47
Guinea fowl, 171, 175
Guinea pig, 34
Guinea savanna zone, 262

Habitat, virgin, 15
Hadendoa tribes, 201, **203**
Hafting, 254–55
Hair, 102–104, 217–19
Hamites, 201, **203**
Hamster, 34
Handedness, 124
Hawaii, 242, 243
Head and neck, 93, **94, 544;** *See also*
        Brain; Skull
Health, effects of smog on, 388–93; *See
        also* Disease
Hearing, 82
Heidelberg man, 78, 113
Hemoglobins, abnormal, 229–33
    in evolution, 59
Herbivores, 10, 12
Herbs, annual, 40, 41

Heredity, 49–51, 236; *See also* Genetics
Heterotrophic elements in ecosystem, 9
Hippopotamus, 34
Holistic approach to ecology, 3–5
Holocene period, 71, **169**, 201, **203**, 538, 539, 541
Homeostasis in ecosystem, 8
Homestead, in undisturbed ecosystem, **7**
Hominids, 25–67, 543–45
   definition of, 43
   Pliocene and Pleistocene, 69–100
   separation of pongid and, 57–64
Hominoids, 37, 39, 43, 57
Homo, 43, 58, 59, 73–75, 78, 79, 87, 91, 93, 95, 101, 128, 201, 252
   africanus, 58, 71–76, 79, **81**, **92–95**
   erectus, 58, 71, 73, 77–80, 82, 87, 88, **92–95**, **97**, 98, 101–52, 159, 197–201, 203–205, 207, 215, **221**, **228**
   erectus-sapiens, **107–109**, 110, 115, **116**, 117, **118**, **123**, 124, 127–29, 138, 159, 197, 198, 200, 201, **203**, **206**, 215, 224, 227, 230, 238–40, 251, 253–55, 281
   habilis, 58, 75–76, 78, 79, 81, **92**, 94, 95, 104, 105, 250
   heidelbergensis, 78, 113
   sapiens, 25, 34, **36–39**, **56**, 58, 64, 71, 73, **74**, 88, **93–97**, 98, 110, **112**, **118–21**, 122, 129, 137, 147, 148, 156, 159, 193, **194**, 197–204, 207, 209, 246, 524, 526, 543–45; *See also* Humans
   sapiens innovatus, 245, 246, 523–25
   sapiens neanderthalensis, 117–27; *See also* Neanderthals
   sapiens neanderthalis, 58, 192
   sapiens sapiens, 58, 87, 110, 115, 124, 127, 148, 153, 192, **221**, 246
   soloensis, 201
Honey, gathering, **108**
Hopefield, **203**
Hottentots, 113, 201, **203**, 216, 227, **228**
Horde, 250
Horses, 70, 129, 137, 171, 175
Human ecology, purpose of, 1–8
Humans, 191–248; *See also* Homo sapiens
Human evolution, scheme of, 61
Hunter-gatherers, 105, **108**, **109**, 124–25, 132, 137, 153, 154, **158**, 166,

178, 198, 251–52
Hunting, 104–105, **107–10**, 122–23, 138–39, **141**, 180, 418, 420, 484, 486
Hybrids, 245
Hydrocarbons, 382, 383, 388, 400
   chlorinated, 440–41, 444, 448–51, 454
Hypervitaminosis D, 90, 104
Hypolimnions, 416

Ice Ages, 26, 28, 40, 63, 116, 387
Immigration, 311–17, 518–19; *See also* Emigration; Migrations
Immunity to disease, 309
Immunological cross-reactions, 57–59
Immunological work, 229
Incest taboos, 108
Income and family size, 361–66
India, 275, 299–303
Indians. *See* Amerindians
Indo-Malayans, **206**
Indris, 39
Industrial Revolution, 182
Industrial societies, 179, 181, **274**, 359, 466, 467, 521–26
Infanticide, 332
Infants, 33, 36, 45, 48, 50–51, 194, 195, 272–73, 392–94, 397–98; *See also* Birth rates; Child rearing
Information accumulation in ecosystems, 11–12
Information content of ecosystem, 15
Information exchange, 177
Influenza, 295–96
Insect pests, 454–58
Intelligence, 236, 237, 523, 524
Intermarriage, village, **242**
International obligations, future, 511–12
International World Plan (IWP) for Agricultural Development, 305–306
Intrauterine devices (IUD), 337–**38**, 339
Inversions, 376–78, 383
   radiation, 387
IQ, 236
Island migrations, 148

Jarmo, 160, 174
Java man, 77–78
Joshua Tree National Monument, California, **479**
Jurassic, 29

Neanderthals, 115, 117–27, 197, 200, 204, **220**, 539; *See also Homo sapiens neanderthalensis*

Neck muscles, 93, **94, 544**

Negritos, 205, **206**

Negroes, 90–91, **119**, 201, **203**, 216

Negroids, **89**, 115, 127, **158**, 197–99, 201–204, 207, 211, 216, 218, 219, 221, 223, 227, 230–32, 234, 235, 237–39, 241, 242, 244

Neolithic, 163, 541

Nerve gases, 452–53

"Netherlands fallacy," 476–78

Net production rate, 280

Neurosis in primates, 45

New World, 129–47, 165–66

Niche, ecological, 17–18

Niche diversification, 15, 16, 180

Nilotics, 213–15

Nitrates, 21–22, 410–12, 421–23

Nitrogen, 21

Nitrogen oxides, 378, 379, **381**–83, 392, 400

Nitrophilous plants, 160

Noise pollution, 403

Nomadism, 154, 257

Nonwhites, 364; *See also* Blacks

Nose form, 223–24

Nuclear testing, 393–98

Nuclear power, 425, 472

Nunatak, 134

Nutrients, 9, **10, 17**, 22, 413

Nutrition, 308–309; *See also* Diet

Obsidian, 161, 163, 164

Ocean, pollution in, 423–25, 427, 429, 431

Oceania, 207, **208**

Oceanic groups, 205–209, 243

Oil, 468–70, 525

Oil palm, 166

Olduvai, 73, 75–76, 94, 95, 101, **103**, 104, 540

Oligocene epoch, 52, 58, 541

Oligotrophic lakes, **410**, 411

Omo-Rudolf area, 73, 74

Operation Cat Drop, 18–19

Opossum, 34

Orangutan, **31**, 34, 38, **56**, 58

Order, 43

Organophosphates, 451, 454

Osprey, 442–43

Overkill
 African, 137–38
 Neanderthal, 125, 126
 Pleistocene, 137, 139–44, 255–56, 484

Ox, **136**

Oxygen in lakes, 415–17

Ozone, 380–81, 399, 400

Pair bonding, 105, 107, 110

Paleo-Indians, **121**, 129, **130, 131, 134, 138, 158**, 540

Paleocene, 31, 37, 541

Paleolithic times, 200, 541, 542

Paleozoic time, 27, 29, **41**

Palm-wine preparation, **259**

Pan, species, 58

Papuans, 201, 205, 207, **208**

Paranthropus, 71, 73–**80, 92–95, 103**, 128, 201
 boisei, 58
 robustus, 58, 79

Parasites, 456–57

Parental care. *See* Child rearing; Infants; Mothers

Particulate matter, 373, 378, 379, 388, 389

Pastoral group, nomadic, 257

Patterning, duality of, 86–87

PCB (polychlorinated biphenyls), 453–54

Pebble tools, **75**–76

Peking man, **77**, 110, **111**, 113, **121**

Pelvic area, human, **545**

Peracyl nitrate (PAN), **381**, 399

Perennials, geophytic, 40, 41

Permanent occupation sites, **155–56, 158**

Permian, 29

Permocarboniferous period, 28

Peru, 356

Pesticides, 18–19, 186, 439–62

Phenotypic acclimation, 238–39

Phenotypes, 208

Phosphates, 410–12, 421–23

Photochemical reactions, 380–81

Physical characteristics, 91, **92–95**, 102–10, 208–28, **544, 545**

Physiochemical conditions, 416

Phytoplankton, 423, 429

Pigeons, 171

Pigmentation, 54–55, 89–91, 104, 112, 113, 215–19

Pigs, 171, 257
Pill, the, 339–40
Pioneer community, 14
Pithecanthropus, 77
   erectus, 77, 78
Plague, 294
Plants, 28, 40, 159–60, **165, 167,** 172
Pleistocene, 28, 43, 57, 58, 63, 105, **116,**
      117, 129, **133, 134,** 147, 155,
      200, 201, **203,** 204, **205, 228,**
      251, 255, 256, 260, 483, 538,
      539, 541, 542
  hominids of Pliocene and, 69–100
Pleistocene overkill, 137, 139–44, 255–56,
    484
Pliocene, 43, 55–60, 539, 541
  hominids of Pleistocene and, 69–100
Poisons in ecosystem, 18–19, 21, 425–26,
    428–29, 440; *See also* Pesti-
    cides; Pollution
Politics, in population control, 348–51
Pollution
  air, 371–407
  biological, 430–33
  episodes of, 387–88
  noise, 403
  primary, secondary, and major, 378–80
  synergistic effects of, 391–92
  thermal, 429–30
  water, 409–38
Polymorphic genes, 196
Polynesians, 148, **206,** 207, **208,** 242
Pongids, 37, 39, 43, 57–64, **194**
Population
  of cities, 183, **184,** 313, **316–18, 358**
  control of, 331–69, 498–502
  density of, 250, 251, 477
  distribution of human, 198–209
  in ecosystem, 10–11, 14, 17, 18
  fluctuations in, 291–330
  future, 508–509
  growth of, 186, 187, 249–90
    rate of, 270–77
    theoretical, arithmetic, and geo-
      metric, 264, **265, 266**
      zero (ZPG), 280, 313, **318,** 348–
      49
  movements of. *See* Migrations
  stabilization of, 265, 266, 279–81, 356–
    57, 365–66
  stationary, 20–21, 265, 266, 279, **362**
  structure of, 27–80

Population pyramids, 278–**79**
Porpoise, 34
Port Clinton, 445
Potassium/argon, 538–39
Potassium/argon dating, 58, 70
Potato, **165**
Poverty and family size, 361–66
Preclassic Period, 166
Predation, 18, 55, 104
Predators, 456–57
Pregnancy, 396
  avoidance of, 335–41
Prenatal effects of radioactive fallout,
    396–98
Prey switching, 18
Primates, 2, 29–40, 43, 58
Proconsul, 42–44, 46, 51–60, 62, 71, 81,
    92, 93, **96, 97**
Producers, in ecosystem, 9
Productivity, in ecosystem, 12–14
  of society, 181
  speech, 85, 87
Progestin, 341
Prognathy, **93, 220**
Projectile tip tools, **135,** 137
Prosimians, 37, 39
Protein, 58, 59, 302, 307–309
Protein clock, 539–40
Protogiraffe, **136**
Protohominoid, 43
Protoliterate period, 177
Pygmies, **119,** 201, **205, 214,** 215, 227

Quaternary Period, 537, 538, 541, 542

Rabbits, 34, 171
Race, as term, 215
Racemization, 539
Radiation, solar, 9
Radioactive fallout, 393–99, 427
Radioactive isotopes, 424–27
Radioactivity, 22, 394, **395,** 396, 398,
    472
Rain forest, tropical, 28–30
Rainfall, 41, **154,** 475
Rainwater, pesticides in, 449
Ramapithecus, **56–57,** 69, 71, **92,** 93, 95
  brevirostris, 57
  punjabicus, 58
  wickeri, 58
Rancholabrea, 135, **136,** 137, 143, 144
Rats, 34

Recycling, 9, 410
Reducers, in ecosystem, 9
Redwoods, 493–94
Religion, 168–69, 349–50, 365, 514–17
Research, primates used in, 492
Resistant strains, 454–55
Resource depletion, **499**, 513
Resource exploitation, 465
Resource modeling, 509–11
Resource partitioning, 17
Respiratory diseases, 388–93
Rhesus system, 229, 230
Rhodesian man, 119, 126, 127, **203**, 205
Rhythm method, 336
Rice, 166, 301–302
Rickets, 90, 91
Ritual, 25
Root crops, 166, **167**

Sahel savanna, 262–63
Saprobes, 9
Savannas, 62–63, 262–63, **480, 481**
Screw worm fly, 457
Sea-exploration grade, 204
Seawater, pollution of, 423–25, 427, 429, 431
Seeds, 42, 53–55, 159, 163, 164, 426, 428; *See also* Grain
Selection pressures, 86, 219–23, 227, 243, 521
Selective concentration, 21, 441–44
Settlements, 154–59, 161–64; *See also* Communities
Sewage treatment and disposal, 411–13, 420–23
Sexual behavior, 52, 105, 107–108, 227, 352
Sheep, 34, 171, 175
Siamangs, 38, 58
Sickle-cell anemia, 196, 231, 232
Sinanthropus pekinensis, 77
Sivapithecus, 57
Skin color, 54–55, 89–91, 104, 112, 113, 215–19
Skin patterns, 227
Skin reflectance, **89**
Skulls, **93, 94, 118, 123,** 127, **207**
Sloth, **136**
Smell, sense of, 82
Smog, 373–74, 378–81, 388–93, 399
Smoking. *See* Cigarette smoking
Social life, 36, 45–46, 50, 51; *See also* Communication

Sociality, of primates, 34
Society(ies), ecological development of, 179–82
    evolution of human, further, 177–79
    synthetic, 241–46
    as term, 178
    See also Agricultural societies; Industrial societies
Socio-economic institutions, and fertility, 347–48
Sociofacts, 239, 240
Soil, fallout in, 397
Solar energy, 473
Solar radiation, 9
Solo man, **120,** 127, 205, 207
Sorghum, **167**
Space requirements, 476–78
Specialization, 180
Species, in ecological niche, 17
    rare and endangered, 483–95
Speech, origin of, 84–85; *See also* Language
Spermatocides, 337
Squirrels, 34
Starfish, 432–33
Starvation. *See* Famine
Steatopygia, 227, **228**
Steinheim skulls, **118**
Sterile male technique, 457
Sterilization, 344
Stone-axe culture, 122, **123**
Stratosphere, 376, 377, 399, 400
Strontium-90, 394–**96,** 397, 398
Subhominid, 43
Suburbs, 185
Succession, 14–16, 180–82
Sudan savanna, 262
Sulfates, 383
Sulfur oxides, 378–80, 382, 383, 399
Survival problems, 62
Swanscombe man, **112, 118, 120**
Sweat glands, 227–28, 238–39
Sweet potato, **165**
Swine, 257
Synthetic societies, 241–46

Tail, prehensile, 60
Tainus, **165,** 166
Tar pits, 135
Taro, 166
Tarsiers, **30, 37,** 39, **432**

Tasmanians, **120**, 147, 148, 201, **206**, 254
Tax benefit to discourage childbearing, 346–47
TDE, 442
Technology, 465, **500**, 512–13
Teeth, 35, 38–40, 53–54, **56**, 57, **93**, 191, 220–**22**, 223
Temperature responses, 209–15
Terrestrial primates, 46–52
Territorial behavior, 80
Territorial dissociation, 465
Territory size, **154**
Tertiary Period, 40, 42, 43, 53, 54, 56, 57, 59, 60, 137
Thalassemia, **232, 233**
Thermal pollution, 429–30
Thermoluminescence, 540
"Threat thermometer," in macaques, 47, **48**
Time sequences, tropical, 71–72
Tolerance levels, 447–48
Tomato, **165**
Tool-using, 43, 53, 74, 75, 79, **81**
Tools, **75**–76, 104–105, **107**, 109, 122, **123**, **135**, 163, 254–55
Topography, and smog, 378
Transmission, traditional, in speech, 86, 87
Transportation, urban, 184–85
Tree shrews, **30, 32**, 34, 37, 39
Triage system, 304
Tribe, 250
Troop size, 50
Trophic displacement, 78–**82**
Trophic levels, 10, **11**, 12, 16
Trophic position, 125–27
Trophic relationships, 253–54
Tropical areas, **41**, 166
Tropical forests, 28–30, 110–11
Troposphere, 376, 377
Tubal ligation, 340–41
Tuberculosis, **311**
Tuinplaats, **203**
Turkey, 171
Twinning rates, 233–**35**

Ubaid period, 177
Ultraviolet reflectance, 90–91
Urban civilization, 153–89
Urban crisis, 313–18
Urban ecosystem, **7**
    demographic features in, **359**

Urban growth, 288
Urban mobility, 314–17
Urban revolution, 175–77; *See also* Cities
Uruk, 177
Urus, 168, **169**
United States, air pollution in, 373–403
    energy in, 467–73
    immigration, mobility, and demographic transition in, 312–18, **322**
    pesticides in, 451–53, 456–58
    population of, 285–88, 361–66
    as synthetic society, 243–45
    water pollution in, 412, 414–26
United Nations, 303, 305–309, 352–60, 511, 513

Variations, continuous versus discontinuous, 192–93
Vasectomy, 340–41
Vedda, 204, 207
Villafranchian, 70
Villages, 154–61, 176, **242**
Virgin habitat, 15
Viruses, 295–96
Vision, 33, 102
Vitamin D, 87–91, 215, 216, 219

Walda period, 177
War, 294–95, 324–25
Wasatch Mountains of Utah, **6, 7**
Waste, 21–22, 372, 411–12, 420–25, 429, 433–34
Waste impact index, 424
Water, drinking, 155
    pollution of, 409–38
    as resource, 474–75
Water buffalo, 171
Watusi, **214**, 215
Weapons testing, 393–98, 427, 452–53
Weasels, 34
Weather modification, 385–86, 387; *See also* Climate
Weed species, 495
Welfare benefits to discourage childbearing, 346–47
Whales, 34
Wheat, 160, 172–75, 300, 304
Whites, **359, 361, 364**
Wilderness reserves, 479–81, 492–93
Wildfires, 15
Wisconsin, 412

Women, in labor market, 347
  See also Mothers
World Health Organization (WHO), 19
World models
  Meadows, 500–502
  "standard," 498–99
World population, 281–85, 354

X-rays, 396

Yaks, 171, 175
Yam (tainus), **165**, 166
Yautia, **165**
Youth, 3

Zinjanthropus boisei, 75
Zoo banks, 488, 490–92